Logic Colloquium '02

LECTURE NOTES IN LOGIC

A Publication of
THE ASSOCIATION FOR SYMBOLIC LOGIC

LECTURE NOTES IN LOGIC 27

Logic Colloquium '02

*Proceedings of the Annual European Summer Meeting of
the Association for Symbolic Logic and the Colloquium Logicum,
held in Münster, Germany, August 3-11, 2002.*

Edited by

Zoé Chatzidakis
Department of Mathematics
University of Paris 7

Peter Koepke
Mathematical Institute
University of Bonn

Wolfram Pohlers
Institute for Mathematical Logic and Foundational Research
University of Münster

ASSOCIATION FOR SYMBOLIC LOGIC

CRC Press
Taylor & Francis Group
Boca Raton London New York

CRC Press is an imprint of the
Taylor & Francis Group, an **informa** business

AN A K PETERS BOOK

CRC Press
Taylor & Francis Group
6000 Broken Sound Parkway NW, Suite 300
Boca Raton, FL 33487-2742

First issued in paperback 2020

ISBN-13: 978-1-56881-300-4 (hbk)
ISBN-13: 978-1-56881-301-1 (pbk)

Visit the Taylor & Francis Web site at
http://www.taylorandfrancis.com

and the CRC Press Web site at
http://www.crcpress.com

Library of Congress Cataloging-in-Publication Data

Logic Colloquium '02 (2002 : Münster, Germany)
    Logic Colloquium '02 : proceedings of the Annual European Summer Meeting of the Association for Symbolic Logic and the Colloquium Logicum, held in Münster, Germany, August 3-11, 2002 / edited by Zoé Chatzidakis, Peter Koepke, Wolfram Pohlers.
        p. cm. – (Lecture notes in logic ; 27)
    Includes bibliographical references.
    ISBN-13: 978-1-56881-300-4 (hardcover : alk. paper)
    ISBN-10: 1-56881-300-7 (hardcover : alk. paper)
    ISBN-13: 978-1-56881-301-1 (pbk. : alk. paper)
    ISBN-10: 1-56881-301-5 (pbk. : alk. paper)
    1. Logic, Symbolic and mathematical–Congresses. I. Chatzidakis, Zoé Maria. II. Koepke, Peter. III. Pohlers, Wolfram. IV. Colloquium Logicum 2002 (2002 : Münster, Germany) V. Title. VI. Series.

QA9.A1L64 2002
511.3–dc22

2006003625

Publisher's note: This book was typeset in LaTeX. by the ASL Typesetting Office, from electronic files produced by the authors. using the ASL documentclass `asl.cls`. The fonts are Monotype Times Roman. The cover design is by Richard Hannus, Hannus Design Associates, Boston, Massachusetts.

Addresses of the Editors of Lecture Notes in Logic and a Statement of Editorial Policy may be found at the back of this book.

Association for Symbolic Logic: Sam Buss, Publisher, Department of Mathematics, University of California, San Diego La Jolla, California 92093-0112, USA

TABLE OF CONTENTS

PREFACE

This volume contains the joint proceedings of two major international logic meetings which took place at the University of Münster, Germany in August 2002: Logic Colloquium '02, the 2002 European Summer Meeting of the Association for Symbolic Logic and Colloquium Logicum 2002, the bi-annual meeting of the German Association for Mathematical Logic and the Foundations of Exact Sciences. The conferences were attended by international audiences of together more than 300 scientists from five continents, who turned Münster into a logic metropolis for 10 days from August 3 to August 12. The conferences covered all areas of contemporary mathematical logic; philosophical logic and computer science logic were presented as well. The conference programs were densely filled with invited tutorials and plenary talks, several special sessions and a large number of contributed talks. These proceedings contain elaborated and extended versions of a number of invited plenary talks and tutorials. All contributions were subjected to the refereeing standards of the Association of Symbolic Logic.

The town of Münster, located in north-western Germany, is a pleasant middle-sized administrative and university town of 280,000 residents. The city center is characterised by historic buildings, churches, squares and parks, and by an extraordinary number of bicycles on designated cycle paths. In the 1200 year history of Münster, the single most important date was the signing of the Peace of Westphalia in 1648 which marked the end of the Thirty Years War in Europe and can be seen as a very early step towards a united Europe. The University of Münster was first founded in the 18th century. It is now spread out throughout the city and has nearly 40,000 students.

Within the University of Münster formal and mathematical logic are taught and researched at the Institute for Mathematical Logic and Foundational Research which belongs to the department of mathematics and computer science. The institute was founded and developed by Georg Heinrich Scholz who was appointed to a professorship in philosophy in 1928. The character of formal logic at Münster was shaped by Herrmann Ackermann, Justus Diller, Gisbert Hasenjäger, Hans Hermes and Dieter Rödding, with an emphasis on proof theory and recursion theory. Currently the professorships for logic are held by Wolfram Pohlers and Ralf-Dieter Schindler.

Logic Colloquium '02 and Colloquium Logicum 2002 were organized as a collaborative project of logicians in Münster and Bonn. The program committee of Logic Colloquium '02 consisted of K. Ambos-Spies, S. Buss, Z. Chatzidakis, A. Kechris (Chair), P. Koepke, P. Komjath, M. Lerman, V. McGee, W. Pohlers, M. Rathjen, K. Segerberg, and B. Zilber. Abstracts of all talks at the Logic Colloquium can be found in The Bulletin of Symbolic Logic 9(1), 2003. The program committee of Colloquium Logicum consisted of: J. Diller, P. Koepke, B. Löwe, W. Pohlers (Chair), Chr. Thiel, W. Thomas, and A. Weiermann. The conferences were mainly funded by the Association for Symbolic Logic (ASL), the Deutsche Forschungsgemeinschaft (DFG), the National Science Foundation (NSF), the Westfälische Wilhelms-Universität, and the Deutsche Vereinigung für Mathematische Logik und für Grundlagen der Exakten Wissenschaften (DVMLG).

As editors of the proceedings we would like to thank the many institutions and individuals who made both conferences possible. We hope that the scientific level and spirit of the Münster meetings is reflected in this volume.

The Editors
Zoé Chatzidakis
Paris
Peter Koepke
Bonn
Wolfram Pohlers
Münster

PARTICIPANTS PHOTOGRAPH

Photograph by Jürgen Kottsieper, 2002

GENERIC ABSOLUTENESS FOR Σ_1 FORMULAS AND THE CONTINUUM PROBLEM

DAVID ASPERÓ

Abstract. This is a survey paper on the problem whether bounded forcing axioms decide the size of the continuum.

§1. Introduction. Bounded forms of forcing axioms are natural principles extending ZFC. Their original formulation (see [17], [7], [1], [5]) is the following.

DEFINITION 1.1. Given a class Γ of partially ordered sets (henceforth posets for short), the *bounded forcing axiom for* Γ, which I will denote by $BFA(\Gamma)$, is the assertion that for every $\mathbb{P} \in \Gamma$ and every collection $(A_i : i < \omega_1)$ of maximal antichains of $\mathbb{P}$ all of which have size at most $\aleph_1$ there is a filter $G \subseteq \mathbb{P}$ intersecting each A_i. G is said to be *generic for* $\{A_i : i < \omega_1\}$.

$BPFA$, $BSPFA$ and BMM denote $BPFA(\Gamma)$, where Γ is, respectively, the class of proper posets, the class of semiproper posets, and the class of stationary–set–preserving posets.[1] Notice that BMM implies $BSPFA$, which in turn implies $BPFA$. $BPFA$ and $BSPFA$ stand for *Bounded Proper Forcing axiom* and *Bounded Semiproper Forcing Axiom*, respectively. BMM stands for *Bounded Martin's Maximum*, since it is the bounded version of the maximal forcing axiom *Martin's Maximum* (MM) from [13].

It seems that bounded forcing axioms first entered the literature in an explicit way in 1995 in [17]. In the 1992 paper [14], S. Fuchino had proved that the unbounded forcing axiom for a regular[2] class Γ of posets and for collections of κ–many maximal antichains (κ being a cardinal) is equivalent to the assertion that a first order structure $\mathcal{A}$ of size κ can be embedded into another fixed structure $\mathcal{B}$ if and only if such an embedding is forced by some poset in Γ.

This work was supported by the Austrian Fonds zur Förderung der Wissenschaftlichen Forschung, Projekt 13983.

[1] See the end of this section for general definitions.

[2] This means that whenever $\mathbb{P}$ and $\mathbb{Q}$ are posets such that $\mathbb{Q}$ belongs to Γ and $\mathbb{Q}$ is a dense suborder of $\mathbb{P}$, then $\mathbb{P}$ also belongs to Γ.

Logic Colloquium '02
Edited by Z. Chatzidakis, P. Koepke, and W. Pohlers
Lecture Notes in Logic, 27

There he was asking whether MM is also equivalent to the statement that two first order structures of size $\aleph_1$ are isomorphic exactly when they can be made isomorphic by stationary–set–preserving forcing. Answering the corresponding question about PFA, M. Goldstern and S. Shelah introduced in [17] $BPFA$, as well as the general notion of bounded forcing axiom. In this paper, they showed that this statement is rather weak in terms of consistency. They proved, on the one hand, that if κ is a *reflecting cardinal* — meaning that κ is a regular cardinal such that $V_\kappa \preccurlyeq_{\Sigma_2} V$ (i.e., V_κ is correct about Σ_2 facts) — then there is a proper forcing iteration $\mathbb{P} \subseteq V_\kappa$ of length κ forcing $BPFA$. On the other hand, $BPFA$ implies that the real ω_2 is reflecting in L. Reflecting cardinals lie quite low in the large cardinal hierarchy. In fact, it is easily seen that if λ is Mahlo, then in V_λ there are stationarily many reflecting cardinals. Hence, $BPFA$ is much weaker than the full PFA. Goldstern and Shelah also showed that from $BPFA$ it follows that the existence of a proper poset forcing an isomorphism between two given structures $\mathcal{A}$ and $\mathcal{B}$ of size $\aleph_1$ implies the existence of such an isomorphism. Hence, the answer to the above version of Fuchino's question is negative.

Soon it was clear that bounded forcing axioms fail to have all the consequences for the structure of the continuum that their unbounded counterparts have. Before seeing an example I will introduce some notation, part of which I will need in Sections 3 and 5. Given any two functions f and g from ω into some set X, $\Delta(f, g) = \min\{n : f(n) \neq g(n)\}$. Let $<^*$ denote the relation of eventual dominance on the Baire space ω^ω (i.e., given two elements f and g in ω^ω, $f <^* g$ iff there is some $n < \omega$ such that $f(m) < g(m)$ for all $m > n$). Recall that, given two ordinals α and β, an (α, β)–*gap* on ω^ω is a pair $(\langle f_\xi : \xi < \alpha \rangle, \langle g_\xi : \xi < \beta \rangle)$ of sequences of members of ω^ω such that for all $\xi_0 < \xi_1 < \alpha$ and all $\xi_0' < \xi_1' < \beta$, $f_{\xi_0} <^* f_{\xi_1} <^* g_{\xi_1'} <^* g_{\xi_0'}$ and such that there is no $h \in \omega^\omega$ such that $f_\xi <^* h <^* g_{\xi'}$ for all $\xi < \alpha$ and $\xi' < \beta$. The *bounding number* ($\mathfrak{b}$) and the *dominating number* ($\mathfrak{d}$) are, respectively, the least size of an $<^*$–unbounded subset of ω^ω and the least size of a *dominating family*, i.e., of a $<^*$–cofinal subset of ω^ω. Clearly, $\mathfrak{b} \leq \mathfrak{d}$ and, under $MA_{\aleph_1}$, $\aleph_1 < \mathfrak{b}$. A scale is a sequence of members of ω^ω, well–ordered by $<^*$, whose range is a dominating family. A consequence of Todorčević's Open Coloring Axiom, which in turn follows from PFA, is that there are no (ω_2, ω_2)–gaps on the Baire space. However, in [5] J. Bagaria and I proved that bounded forcing axioms are compatible with the existence of such gaps by building, starting from the optimal hypothesis in the case of $BPFA$ and $BSPFA$, a forcing iteration producing a model of the desired bounded forcing axiom in which there is an (ω_2, ω_2)–gap. Subsequently, it was realized by Todorčević [33] that given any model of any bounded forcing axiom $BFA(\Gamma)$ extending $MA_{\aleph_1}$, one can force over it with an ω_2–closed poset — which therefore does not add new subsets of ω_1 and thus, by Characterization 1.1 below, preserves $BFA(\Gamma)$ if Γ

is closed under composition and extends the class (say) of proper posets —, while at the same time adding an (ω_2, ω_2)-gap.

Consider the following principle of generic absoluteness for Σ_1 formulas.

DEFINITION 1.2. Given a class Γ of posets, $Abs(\Gamma)$ denotes the statement that for every $a \in H_{\omega_2}$ and every Σ_1 formula $\varphi(x)$, $H_{\omega_2} \models \varphi(a)$ iff there is some poset $\mathbb{P}$ in Γ such that $\Vdash_{\mathbb{P}} H_{\omega_2} \models \varphi(\check{a})$.

$Abs(\Gamma)$ is a principle which assert that H_{ω_2} is complete with respect to the class Γ in a certain precise way. It says that a Σ_1 formula with parameters in H_{ω_2} is true in H_{ω_2} exactly when there is some partial order in Γ which forces this to happen.

The thesis that bounded forcing axioms are natural principles extending ZFC can be argued for on the ground of the following characterization of them in terms of generic absoluteness, which was first proved by Bagaria in the mid-90s.

CHARACTERIZATION 1.1. [7] Let Γ be a class of complete Boolean algebras. Then, $BFA(\Gamma)$ holds if and only if $Abs(\Gamma)$ does.

Notice that the application of $BPFA$ to the establishment of the existence of an isomorphism between two given structures of size $\aleph_1$ mentioned above follows trivially from Characterization 1.1. Further extensions of this result to stronger fragments of forcing axioms were subsequently found by T. Miyamoto [25]. Also, [1] contains results generalizing those of Miyamoto.

In view of the naturalness of bounded forcing axioms mentioned above, the problem of finding out whether any of them decides the size of the continuum was soon regarded by several people as of greatest importance for set theory. One of the earliest results concerning MM was that it implies $2^{\aleph_0} = \aleph_2$ [13].[3] The proof of this result used a relatively simple coding of reals by ordinals less than ω_2, which involved a certain object of size $\aleph_2$. Dozens of different arguments for showing $2^{\aleph_0} = \aleph_2$ from MM have been found in the meantime, all of them involving some parameter of size $\aleph_2$. This was in contrast with the difficulty to find a coding of reals involving just objects in H_{ω_2}, which was found only very recently, in the general case, by Todorčević (see Theorem 5.11).[4] This difficulty added considerable interest in the above question, and in particular in the question whether BMM decides the size of the continuum.

---

[3]Martin's Maximum was the first forcing axiom known to decide the size of the continuum. Later this was proved by Todorčević and by Veličković, using a much harder argument, for PFA.

[4]Incidentally, the proof from [13] that MM decides $2^{\aleph_0}$ actually shows that the "next" fragment of MM past BMM, namely the forcing axiom guaranteeing the existence of generic filters for all collections of $\aleph_1$-many maximal antichains of any given stationary–set–preserving poset, all of size at most $\aleph_2$ (call this statement $BMM_{<\aleph_3}$) implies $2^{\aleph_0} = \aleph_2$. It is not known whether the same is true for the corresponding fragment of PFA, although one can prove that the analogous statement for collections of maximal antichains of size $\mathfrak{b}$ does imply that both $\mathfrak{b}$ and $2^{\aleph_0}$ are equal to $\aleph_2$.

It is clear that if $\mathbb{P}$ is any poset adding a new real, then $Abs(\{\mathbb{P}\})$ implies $\neg CH$. Also, notice that if Γ is a class of posets which is closed under composition, some poset $\mathbb{P}_0$ in Γ forces the ground model continuum to be of size $\aleph_2$ without adding new subsets of ω_1, and $Abs(\Gamma)$ decides the value of $2^{\aleph_0}$, then this value is at most $\aleph_2$.[5] Notice that most natural classes of posets (like the class of stationary–set–preserving, of semiproper posets or the class of proper posets) satisfy the first two conditions above. Finally, any bounded forcing axiom extending $MA_{\aleph_1}$, which itself is a bounded forcing axiom, implies $2^{\aleph_0} = 2^{\aleph_1}$ [24].

It is simple to see that if $\mathbb{P}$ is a poset forcing that some stationary subset of ω_1 from the ground model fails to be stationary (even if there is no fixed stationary $S \subseteq \omega_1$ such that $\Vdash_{\mathbb{P}} \check{S}$ is not stationary), then there is some collection of size $\aleph_1$ of maximal antichains of $\mathbb{P}$ for which there cannot be any generic filter (see [13]). This shows that MM, which is consistent relative to the existence of a supercompact cardinal [13], is a maximal forcing axiom.[6] It is not clear whether BMM is itself a *maximal* bounded forcing axiom in the same sense that MM is maximal among forcing axioms. However, I proved that, letting Γ_0 and Γ_1 be, respectively, the class of stationary–set–preserving posets and the class of all those posets $\mathbb{P}$ with the property that for every set $\mathcal{A}$ of size $\aleph_1$ consisting of stationary subsets of ω_1 there is some condition $p \in \mathbb{P}$ such that $p \Vdash_{\mathbb{P}}$ "Every $S \in \mathcal{A}$ is stationary", it turns out that $Abs(\{\mathbb{P}\})$ fails for every $\mathbb{P}$ not belonging to Γ_0, that $\Gamma_0 = \Gamma_1$ if and only if the nonstationary ideal over ω_1 is $\aleph_1$–dense and that, given the existence of a reflecting cardinal κ which is a limit of strongly compact cardinals, there is a semiproper iteration $\mathbb{P} \subseteq V_\kappa$ forcing $Abs(\Gamma_1)$ [2]. Hence, $Abs(\Gamma_1)$ is a (consistent) maximal form of $Abs(\Gamma)$. It is not known whether BMM — i.e., $Abs(\Gamma_0)$ — implies $Abs(\Gamma_1)$. BMM certainly implies that Γ_0 is strictly contained in Γ_1.

Being as it is one of the strongest bounded forcing axioms one can think of, BMM became a natural candidate to yield a positive answer to the question wether any bounded forcing axiom decides $2^{\aleph_0}$. In view of the remarks in the previous paragraph, if this were the case, then the value of $2^{\aleph_0}$ given by BMM would necessarily be $\aleph_2$.

In this paper I will review my version of the history of this problem. Many of these results are of the form 'BMM + (some condition) implies (some statement implying) $2^{\aleph_0} = \aleph_2$'.[7] Of course, the most outstanding event in this history was the discovery in April 2002, due to Todorčević, that

---

[5]This is because $V^{\mathbb{P}_0} \models Abs(\Gamma)$.

[6]A recent result of Shelah [28] shows, for example, that the natural candidate for a maximal forcing axiom for collections of $\aleph_2$–many maximal antichains, namely the axiom for the class of those posets that preserve stationary subsets of ω_1 and stationary subsets of ω_2, is false.

[7]There are also a couple of results in this paper which take as hypothesis the assumption (in one case supplemented with some additional assumption) that $(H_{\omega_2}, \in, NS_{\omega_1})$ is a Σ_1–elementary substructure of the corresponding structure obtained after forcing with any $\mathbb{P} \in \Gamma$ (for a relevant

BMM — without any additional hypothesis — decides the size of the continuum. However, I will give proofs of older partial results for several reasons. One of them is to illustrate how the developments leading towards the final result took place. Another reason is that the conclusion of several of these partial results is a certain quotable principle ψ_{AC} to be defined in Section 3, which is equiconsistent with the existence of an inaccessible limit of measurable cardinals and which is not known to follow from *BMM* in general. It is also worth mentioning at this point that Todorčević's argument for showing $2^{\aleph_0} = \aleph_2$ from *BMM* goes through by proving that *BMM* implies a certain principle which he calls θ_{AC} — which is easily seen to imply $2^{\aleph_1} = \aleph_2$ —, and from which no large cardinal consequences are known to follow. Finally, I find some of the arguments used in the proofs of the partial results beautiful and interesting in themselves, and they may even happen to have further applications. I have chosen not to give full proofs of most results which appear elsewhere.

I will finish this introductory section giving some definitions and pieces of notation I have not mentioned yet. The notation and terminology I will be using is intended to be standard. For example, in the context of forcing I will mainly use the notation in [22]. In particular, if $\mathbb{P}$ is a poset and $p, q \in \mathbb{P}$, $q \leq p$ means that q gives at least as much information about the generic object as p (q is stronger than p or q extends p).

Given a cardinal κ, H_κ denotes the set of all sets hereditarily of size less than κ, i.e., such that $|TC(x)| < \kappa$ ($TC(x)$ is the transitive closure of x). Given a set $\mathcal{X}$, $[\mathcal{X}]^{\aleph_0}$ and $[\mathcal{X}]^{<\omega}$ are the sets of countable subsets of $\mathcal{X}$ and of finite subsets of $\mathcal{X}$, respectively. $D \subseteq [\mathcal{X}]^{\aleph_0}$ is a club if and only if the union of any countable $\subseteq$–increasing sequence of members from D is in D and if D is $\subseteq$–unbounded in $[\mathcal{X}]^{\aleph_0}$. Recall (see [21]) that, given any set $\mathcal{X}$ and any club $D \subseteq [\mathcal{X}]^{\aleph_0}$ there is a function $F : [\mathcal{X}]^{<\omega} \longrightarrow \mathcal{X}$ such that every countable subset x of $\mathcal{X}$ closed under F — i.e., such that $F``[x]^{<\omega} \subseteq x$ — belongs to D. This function F is said to generate D. $A \subseteq [\mathcal{X}]^{\aleph_0}$ is a stationary subset of $[\mathcal{X}]^{\aleph_0}$ if A intersects every club of $[\mathcal{X}]^{\aleph_0}$.

Recall that a poset $\mathbb{P}$ is proper if and only if for every regular cardinal $\theta > 2^{|TC(\mathbb{P})|}$ there is a club $D \subseteq [H_\theta]^{\aleph_0}$ such that every $N \in D$ is an elementary substructure of H_θ containing $\mathbb{P}$ and such that for every $p \in \mathbb{P} \cap N$ there is some condition q extending p such that q is an $(N, \mathbb{P})$–generic condition, i.e., $q \Vdash_{\mathbb{P}} \tau \in \check{N}$ for every $\mathbb{P}$–name τ for an ordinal, $\tau \in N$. Being semiproper is defined similarly, except that only $(N, \mathbb{P})$–semigeneric q's are guaranteed to exist, meaning that $q \Vdash_{\mathbb{P}} \tau \in \check{N}$ for every $\mathbb{P}$–name τ in N for a countable ordinal. Finally, a poset $\mathbb{P}$ is said to be stationary–set–preserving if and only if every stationary subset of ω_1 remains stationary after forcing with $\mathbb{P}$.

---

class Γ of posets). There is no class Γ for which I known $BFA(\Gamma)$ to be consistent and equivalent to its corresponding strengthening as above.

Note that every proper poset is semiproper and that every semiproper poset is stationary–set–preserving. Given a set X, $\mathrm{Coll}(\omega_1, X)$ is the σ–closed forcing collapsing X to ω_1 with countable conditions.

For a set of ordinals X, $ot(X)$ denotes the order type of X and, if α is an ordinal, $cf(\alpha)$ is the cofinality of α.

If X is a set and $\mathcal{I} \subseteq \mathcal{P}(X)$ is an ideal, $\mathcal{I}^+$ is the set of $\mathcal{I}$–positive subsets of X, i.e., $\mathcal{I}^+ = \mathcal{P}(X)\backslash\mathcal{I}$. For a regular cardinal κ, NS_κ denotes the nonstationary ideal over κ. If G is generic for $\mathcal{P}(\kappa)/NS_\kappa$ (where $[T]$ is stronger than $[S]$ iff $T\backslash S$ is nonstationary), one can build in $V[G]$ the generic ultrapower derived from G. This is $(V^\kappa \cap V)/G$ embedded with the relation $\in_G$, where $[f]_G \in_G [g]_G$ if and only if $\{v < \kappa : f(v) \in g(v)\}$ is a set in G. In a context involving forcing with $\mathcal{P}(\kappa)/NS_\kappa$, M and j will denote, respectively, the generic ultrapower derived from the generic filter G and the corresponding elementary embedding $j : V \longrightarrow M$ given by $j(x) = [c_x]_G$, where c_x is the constant function on κ with value x. $\mathcal{P}(\kappa)/NS_\kappa$ is equivalent, as a forcing notion, to the preorder NS_κ^+ (i.e., the set of stationary subsets of κ) with the ordering given by $T \leq S$ iff $T\backslash S \in NS_\kappa$.

§2. **A general setting for** BMM **vs** $2^{\aleph_0}$. It turns out that the proofs of almost all results in this paper involve the following steps:

(1) Given a real r, find a suitable "large" (this is a certain combinatorial notion to be defined below) subset A of $[\mathcal{X}]^{\aleph_0}$ for some set $\mathcal{X}$ which is Σ_1 definable with some set in H_{ω_2} and maybe some tuple of ordinals in $\mathcal{X}$ as parameters.

(2) The largeness of the set in (1) ensures that the standard poset for shooting an ω_1–club of elements of A with countable conditions covers the whole $\mathcal{X}$ and is stationary–set–preserving.

(3) An application of BMM to the poset in (2) yields an ordinal $\delta_r < \omega_2$, an ω_1–club of $[\delta_r]^{\aleph_0}$ and possibly some tuple $(\alpha_0^r, \ldots, \alpha_n^r)$ of ordinals below δ_r which, together with the defining parameter of A in H_{ω_2} and if the Σ_1 definition of A was chosen carefully, code our real r, meaning that for another real r', if $\delta_{r'}$ and $(\alpha_0^{r'}, \ldots, \alpha_n^{r'})$ are the corresponding objects obtained from performing (1) and (2) with r', then $\delta_r \neq \delta_{r'}$ or $(\alpha_0^r, \ldots, \alpha_n^r) \neq (\alpha_0^{r'}, \ldots, \alpha_n^{r'})$.

Hence, given a real r, some large subset of $[\mathcal{X}]^{\aleph_0}$ for an arbitrarily large set $\mathcal{X}$ "reflects" into an ω_1–club of some set in H_{ω_2}.

The precise largeness condition I am referring to in (1) is the following strengthening of the notion of stationary subset of $[\mathcal{X}]^{\aleph_0}$. It was defined in [12].

DEFINITION 2.1. Given a set $\mathcal{X}$ such that $\omega_1 \subseteq \mathcal{X}$, a set $A \subseteq [\mathcal{X}]^{\aleph_0}$ is a *projective stationary subset of* $[\mathcal{X}]^{\aleph_0}$ if and only if $\{X \in A : X \cap \omega_1 \in S\}$ is a stationary subset of $[\mathcal{X}]^{\aleph_0}$ for every stationary $S \subseteq \omega_1$.

Given a set $\mathcal{X}$ and $A \subseteq [\mathcal{X}]^{\aleph_0}$, the standard poset for shooting an ω_1–club through A with countable conditions referred to in (2) is the following poset $\mathbb{P}_A$: $p \in \mathbb{P}_A$ if and only if p is a strictly $\subseteq$–increasing and $\subseteq$–continuous (i.e., if $v \in dom(p)$ is a limit ordinal, then $p(v) = \bigcup_{v' < v} p(v')$) $\xi + 1$–sequence of elements of A for some countable ordinal ξ. $q \leq p$ if and only if $p \subseteq q$.

The following fact is proved in [12].

LEMMA 2.1. *Let $\mathcal{X}$ be a set and let A be a stationary subset of $[\mathcal{X}]^{\aleph_0}$. Then, $\mathbb{P}_A$ does not add any new ω–sequences and forces the existence of a strictly $\subseteq$–increasing and $\subseteq$–continuous sequence $\langle X_v : v < \omega_1 \rangle$ of elements of A such that $\mathcal{X} = \bigcup_{v < \omega_1} X_v$. Suppose that $\omega_1 \subseteq \mathcal{X}$. Then $\mathbb{P}_A$ is stationary–set–preserving if and only if A is a projective stationary subset of $[\mathcal{X}]^{\aleph_0}$.*

In [12], a reflection principle for projective stationary sets is introduced. The authors call it *Projective Stationary Reflection (PSR)* and asserts that for every projective stationary set $A \subseteq [\mathcal{X}]^{\aleph_0}$ there is a strictly $\subseteq$–increasing and $\subseteq$–continuous ω_1–sequence of elements of A. *PSR* is equivalent to Todorčević's Strong Reflection Principle and follows from Martin's Maximum. Also, most consequences of *MM* in [13] already follow from *PSR*. From Lemma 2.1 we get that the following bounded form of *PSR* is a consequence of *BMM*.

DEFINITION 2.2. *BPSR* is the following statement:

Suppose γ is an ordinal, $a \in H_{\omega_2}$, $\alpha < \gamma$ and $A \subseteq [\gamma]^{\aleph_0}$ is a projective stationary subset of $[\gamma]^{\aleph_0}$ which is Σ_1 definable with a, α and γ as parameters (i.e., there is some Σ_1 formula $\varphi(x, y, z)$ such that, for every set X, $X \in A$ iff $\models_1 \varphi(X, a, \alpha, \gamma)$, where $\models_1$ denotes the definable satisfaction relation for Σ_1 formulas). Then there are some $\bar{\alpha} < \delta < \omega_2$ and some strictly $\subseteq$–increasing and $\subseteq$–continuous sequence $\langle X_v : v < \omega_1 \rangle$ such that $\delta = \bigcup_{v < \omega_1} X_v$ and, for every $v < \omega_1$, $H_{\omega_2} \models \varphi(X_v, a, \bar{\alpha}, \delta)$.

If the condition of A in (1) at the beginning of this section of being a projective stationary subset of $[\mathcal{X}]^{\aleph_0}$ is strengthened, then $\mathbb{P}_A$ is actually semiproper. If this is the case, an application of *BSPFA* in (3) suffices to yield the desired coding. The precise strengthening is the following.

DEFINITION 2.3. Given a set $\mathcal{X}$ such that $\omega_1 \subseteq \mathcal{X}$, a set $A \subseteq [\mathcal{X}]^{\aleph_0}$ is a *semiproper subset of $[\mathcal{X}]^{\aleph_0}$* if and only if for every club D of $[\mathcal{X} \cup 2^{|\mathcal{X}|}]^{\aleph_0}$ there is s a club $D' \subseteq D$ such that for every $X \in D'$ there is some Y such that

(a) $Y \in D$,
(b) $X \subseteq Y$,
(c) $X \cap \omega_1 = Y \cap \omega_1$, and
(d) $Y \cap \mathcal{X} \in A$.

The following Lemma is quite an easy exercise.

LEMMA 2.2. *Given a set $\mathcal{X}$ such that $\omega_1 \subseteq \mathcal{X}$ and $A \subseteq [\mathcal{X}]^{\aleph_0}$, $\mathcal{P}_A$ is a semiproper poset iff A is a semiproper subset of $[\mathcal{X}]^{\aleph_0}$.*

For the left to right implication one can use at one point the fact that if $\mathbb{P}_A$ is semiproper, then in particular it does not add new ω–sequences, and so if $N[G]$ is in the extension a set of the form $\{\tau[G] : \tau \text{ is a } \mathbb{P}_A\text{–name in } N\}$, $N[G]$ is already in the ground model.

§3. **Coding reals into ordinals inside H_{ω_2}.** Many of the results I will discuss in this paper go through by showing — provided $BSPFA$, BMM, or some strenghtening of BMM holds in some context — that some combinatorial statement holds, which not only implies $2^{\aleph_1} = \aleph_2$, but in fact implies that in H_{ω_2} there is a definable well–order of $\mathcal{P}(\omega_1)$ in order type ω_2. This implies in particular, of course, that $L(\mathcal{P}(\omega_1))$ satisfies the Axiom of Choice, and that is the justification for the AC subscript which these principles tend to carry. In this section I will give some information about them.

The first two statements I will mention were introduced by Woodin in [36] in the context of his $\mathbb{P}_{\max}$ theory. It should be mentioned that in [36] Woodin is interested in a general possibly choiceless context, and this is apparent in his formulation of ϕ_{AC} below. This is not the case in the current paper, where only situations in which ZFC holds are considered.

DEFINITION 3.1. [36, Definition 5.3] ϕ_{AC} is the statement that there is an ω_1–sequence of distinct reals and that whenever $(S_i)_{i<\omega}$ and $(T_i)_{i<\omega}$ are sequences of pairwise disjoint subsets of ω_1 such that the S_i are stationary and $\omega_1 = \bigcup_i T_i$, there exists an ordinal $\gamma < \omega_2$ and a continuous increasing function $F : \omega_1 \longrightarrow \gamma$ with cofinal range such that, given any $i < \omega$, any $\xi \in T_i$ and any surjection $e : \omega_1 \longrightarrow F(\xi)$, $\{\nu < \omega_1 : ot(e``\nu) \in S_i\}$ contains a club of ω_1.

Woodin devised ϕ_{AC} in order to prove that $L(\mathcal{P}(\omega_1)) \models ZFC$ follows from his axiom (∗), namely that AD holds in $L(\mathbb{R})$ and that $L(\mathcal{P}(\omega_1))$ is a $\mathbb{P}_{\max}$ extension of $L(\mathbb{R})$. He proved that (∗) implies ϕ_{AC} [36, Lemma 5.5 and Corollary 5.7].

DEFINITION 3.2. [36, Definition 5.12] ψ_{AC} is the statement that if S and T are stationary and co-stationary subsets of ω_1, then there is an ordinal δ, a surjection $e : \omega_1 \longrightarrow \delta$ and a club $C \subseteq \omega_1$ such that $S \cap C = \{\nu \in C : ot(e``\nu) \in T\}$.

ψ_{AC} is a (simpler) variation of ϕ_{AC}, also implied by (∗) [36, Lemma 5.18]. It was used in some of the variations of $\mathbb{P}_{\max}$.

Before proceeding, I will give two more pieces of notation. Given a sequence $(r_\nu)_{\nu<\omega_1}$ of length ω_1 and a countable set of ordinals X, r_X will denote $r_{ot(X)}$. Note that if x, y and z are pairwise distinct elements of the Cantor space 2^ω, then $\{\Delta(x, y), \Delta(x, z), \Delta(y, z)\}$ contains exactly two elements. Let us denote this set by $\Delta(x, y, z)$. The following statement was isolated by Todorčević from his proof that BMM implies $2^{\aleph_0} = \aleph_2$ [32].

DEFINITION 3.3. θ_{AC} is the statement that for every sequence $(r_\xi)_{\xi<\omega_1}$ of pairwise distinct elements of 2^ω and every set $Y \subseteq \omega_1$ there are $\alpha < \beta < \delta < \omega_2$ and canonical functions g_α, g_β and g_δ for α, β and δ, respectively, such that for all v in a club of ω_1,

$$v \in Y \text{ iff } \Delta\big(r_{g_\alpha(v)}, r_{g_\beta(v)}\big) = \max \Delta\big(r_{g_\alpha(v)}, r_{g_\beta(v)}, r_{g_\delta(v)}\big).$$

The conclusion of Theorem 5.1 is a further example of a combinatorial statement implying the existence of a well–order of $\mathcal{P}(\omega_1)$ definable in H_{ω_2}, and yet another such statement can be extracted from the proof of Theorem 5.8. Fragments or versions of ψ_{AC} or ϕ_{AC} implying the existence of a definable well–order of $\mathcal{P}(\omega_1)$ are $\psi_{AC,local}$, defined in Section 4, and a statement which can be isolated from the proof of Lemma 5.5, specifically from the part of that proof in which the case that clause (B) there fails is considered.

It is well–known (see for example [19, p. 445]) that given any ordinal $\delta < \omega_2$ and any surjective function $e : \omega_1 \longrightarrow \delta$, the function $g : \omega_1 \longrightarrow \omega_1$ given by $g(v) = ot(e``v)$ represents δ in the generic ultrapower derived from forcing with $\mathcal{P}(\omega_1)/NS_{\omega_1}$, i.e., $\mathcal{P}(\omega_1)/NS_{\omega_1}$ forces that the set of M–ordinals below the class of g in M is well–ordered in order type δ. Such a function g is called *the canonical function for δ (derived from e)*. This terminology is justified by the easily checked fact that if $e' : \omega_1 \longrightarrow \delta$ is another surjection and g' is defined as g with e' instead of e, then e' and e agree on a club. Throughout this paper, by a canonical function I will mean a canonical function for some ordinal less than ω_2.

It is easy to see that ψ_{AC} is equivalent to the statement that for all stationary and co-stationary $S, T \subseteq \omega_1$ there is some $\delta < \omega_2$ such that $S \Vdash_{NS_{\omega_1}^+} \delta \in j(T)$ and $\omega_1 \backslash S \Vdash_{NS_{\omega_1}^+} \delta \notin j(T)$, i.e., letting $\mathbb{B}$ be the regular open completion of $NS_{\omega_1}^+$, the class $[S]_\mathbb{B}$ — under the natural embedding $NS_{\omega_1}^+ \longmapsto \mathbb{B}$ — of S is equal to $[[\check{\delta} \in j(\check{T})]]_\mathbb{B}$, the Boolean value of the formula $\delta \in j(T)$. Similarly, the second part of the statement of ϕ_{AC} is equivalent to the assertion that whenever $(S_i)_{i<\omega}$ and $(T_i)_{i<\omega}$ are sequences of pairwise disjoint subsets of ω_1 such that the S_i are stationary and $\omega_1 = \bigcup_i T_i$, there exists an ordinal $\gamma < \omega_2$ and a continuous increasing function $F : \omega_1 \longrightarrow \gamma$ with cofinal range such that, given any $i < \omega$ and any $\xi \in T_i$, $F(\xi) \in \tilde{S}_i$, where $\tilde{S}_i = \{\alpha < \omega_2 : \Vdash_{\mathcal{P}(\omega_1)/NS_{\omega_1}} \alpha \in j(S_i)\}$.

From the above paragraph one can easily conclude that ψ_{AC} implies, given a stationary and co-stationary $T \subseteq \omega_1$ and a partition $(S_\xi)_{\xi<\omega_1}$ of ω_1 into stationary sets, that there is a well–order of $\mathcal{P}(\omega_1)$ which is Δ_1 definable over $(H_{\omega_2}, \in, NS_{\omega_1})$ with T and $(S_\xi)_{\xi<\omega_1}$ as parameters [36, Lemma 5.13]. To see this, let S_Y be, for every nonempty $Y \subseteq \omega_1$, $Y \neq \omega_1$, $\bigcup_{\xi\in Y} S_\xi$. Then S_Y is stationary and co-stationary and by ψ_{AC} there is some δ_Y such that $S_Y \Vdash_{NS_{\omega_1}^+} \delta_Y \in j(T)$ and $\omega_1 \backslash S_Y \Vdash_{NS_{\omega_1}^+} \delta_Y \notin j(T)$. Suppose $Y' \neq Y$ and $\delta := \delta_Y = \delta_{Y'}$. Take ξ in the symmetric difference of Y and Y'. Then, on the one hand, $S_\xi \Vdash \delta \in j(T)$, and on the other hand $S_\xi \Vdash \delta \notin j(T)$. This contradiction shows that the ordinal δ_Y codes the subset Y with respect to T and $(S_\xi)_\xi$. Moreover, it is easy to see that this coding is Δ_1 definable over H_{ω_2} with the above parameters, and therefore a well–order of $\mathcal{P}(\omega_1)$ in order type ω_2 is Δ_1 definable over $(H_{\omega_2}, \in, NS_{\omega_1})$ with the same parameters. Similarly one can prove (see [36, Lemma 5.4]) that ϕ_{AC} implies the existence, for every nonempty $a \subseteq \omega$, of an ordinal $\gamma < \omega_2$ which codes a in the sense that $\tilde{S}_k \cap \gamma$ is a stationary subset of γ exactly when $k \in a$ (where the $\tilde{S}_k$ are defined as in the above paragraph in terms of a fixed sequence $(S_i)_{i<\omega}$ of pairwise disjoint stationary subsets of ω_1). From this, the existence of a Δ_1 definable over $(H_{\omega_2}, \in, NS_{\omega_1})$ well–order of $\mathbb{R}$ easily follows.

According to [36], the idea for using certain subsets of ω_2 (the $\tilde{S}_i$ from the above paragraph) to define a well–order of the reals originated in the original proof of [13, Theorem 10] — using stationary subsets of ω_1 and of $\{\alpha < \omega_2 : cf(\alpha) = \omega\}$ as parameters — that MM implies $2^{\aleph_0} = \aleph_2$. ϕ_{AC} and ψ_{AC} both follow from MM (see [36, Theorem 5.9 and 5.14], respectively). The first result follows from the proof in [13], Theorem 10 plus the fact that under MM, $\{\alpha < \omega_2 : \Vdash_{\mathcal{P}(\omega_1)/NS_{\omega_1}} \alpha \in j(S)\}$ is a stationary subset of ω_2 for every stationary $S \subseteq \omega_1$ [36, Lemma 5.8]. That MM implies ψ_{AC} follows for example from Theorem 4.1 together with the fact that MM implies the saturation of NS_{ω_1}.

ψ_{AC} also implies $2^{\aleph_0} = 2^{\aleph_1}$ [36, Theorem 3.51]. Another consequence of ψ_{AC} is that every function from ω_1 into ω_1 is dominated on a club by a canonical function [6]. This was proved by me building on a previous result of P. Koepke showing that ψ_{AC} implies the so-called weak Chang's Conjecture, a weakening of the above bounding principle equiconsistent with a large cardinal notion of strength below that of an ω_1–Erdős cardinal. By [10], the bounding of every function from ω_1 into ω_1 by a canonical function implies the existence of an inner model with an inaccessible limit of measurable cardinals. This fact can be turned into an actual equiconsistency result for ψ_{AC} since, on the other hand, if κ is an inaccessible limit of measurable cardinals, then there is a semiproper forcing iteration of length κ forcing ψ_{AC} [4].

The fact that ψ_{AC} implies the existence of an inner model with an inaccessible limit of measurable cardinals was exploited in [6] to produce a model separating $BSPFA$ and BMM. It had been proved by Shelah [27] that MM and $SPFA$, the forcing axiom for the class of semiproper posets, are equivalent statements. However, it was not known whether the corresponding bounded forms of these forcing axiom formulations could be proved to be equivalent.[8]

---

[8]Arguing in terms of consistency strength, it is not difficult to see that the fragments of the formulations $SPFA$ and MM asserting existence of generic filters for $\aleph_1$–sized families of maximal antichains of size at most $\aleph_2$ are not equivalent; the corresponding statement referring to the class of semiproper posets can be forced over L [25], whereas it is easy to see that $BMM_{<\aleph_3}$

In [6] it is shown that BMM, together with the presence of some relatively small large cardinal of consistency strength below that of an ω_1–Erdős cardinal[9], gives ψ_{AC}. On the other hand, there is a model of ZFC, $BSPFA$ and this mild large cardinal hypothesis in which there is no inner model with an inaccessible limit of measurable cardinals. Hence, by the Deiser–Donder result, in this model BMM fails.[10]

Woodin also considers the variants of ψ_{AC} which he calls ψ^*_{AC} and $\psi_{AC}(I)$ — for a given normal and uniform ideal I on ω_1 — in [36, Definition 5.41 and Definition 7.16], respectively. ψ^*_{AC} is equivalent to ψ_{AC} modulo ZFC. The reason for introducing ψ^*_{AC} is the fact, which need not be true for ψ_{AC}, that if $(M_i)_{i<\omega}$ is an iterable sequence (in the sense of [36, Definition 4.15]) such that $\bigcup_i M_i \models \psi^*_{AC}$ and $(M^*_i)_{i<\omega}$ is an iterate of $(M_i)_{i<\omega}$, then $\bigcup_i M^*_i \models \psi^*_{AC}$. $\psi_{AC}(I)$ is the natural relativization of ψ_{AC} to I and, unlike ψ_{AC}, it may be consistent with CH [36, Remark 7.17]. Finally, it should be remarked that there is no obvious dependence between ϕ_{AC} and ψ_{AC}. For more information about ϕ_{AC} and ψ_{AC} the reader is referred to [36].

The proof that θ_{AC} implies the existence of well–order of $\mathcal{P}(\omega_1)$ which is Δ_1 definable over $(H_{\omega_2}, \in, NS_{\omega_1})$ with parameters is also simple.[11] Fix a sequence $(r_\xi)_{\xi<\omega_1}$ of distinct elements of 2^ω and a partition $(S_\zeta)_{\zeta<\omega_1}$ of ω_1 into stationary sets. Fix $Y \subseteq \omega_1$ and let S_Y be $\bigcup_{\zeta\in Y} S_\zeta$. Let α_Y, β_Y, δ_Y be, respectively, the α, β and δ given by an application of θ_{AC} to $(r_\xi)_{\xi<\omega_1}$ and S_Y. Now suppose $Y \neq Y'$ but $(\alpha_Y, \beta_Y, \delta_Y) = (\alpha_{Y'}, \beta_{Y'}, \delta_{Y'})$. Pick $\zeta \in Y \triangle Y'$. Then there are canonical functions g_{α_Y}, g_{β_Y} and g_{δ_Y} for α_Y, β_Y and δ_Y, respectively, and club–many v such that, if $v \in S_\zeta$, then $g_{\alpha_Y}(v) <$ $g_{\beta_Y}(v) < g_{\delta_Y}(v)$ and both $\Delta(r_{g_{\alpha_Y}(v)}, r_{g_{\beta_Y}(v)}) = \max\Delta(r_{g_{\alpha_Y}(v)}, r_{g_{\beta_Y}(v)}, r_{g_{\delta_Y}(v)})$ and $\Delta(r_{g_{\alpha_Y}(v)}, r_{g_{\beta_Y}(v)}) = \min\Delta(r_{g_{\alpha_Y}(v)}, r_{g_{\beta_Y}(v)}, r_{g_{\delta_Y}(v)})$. But this is a contradiction, since $r_{g_{\alpha_Y}(v)}$, $r_{g_{\beta_Y}(v)}$, $r_{g_{\delta_Y}(v)}$ are pairwise distinct reals. I do not know of any large cardinal consequence of θ_{AC}.

§4. Results involving ψ_{AC}.
Given an ordinal γ and $Y, Z \subseteq \omega_1$, set

$$S^\gamma_{Y,Z} = \left\{ X \in [\gamma]^{\aleph_0} : X \cap \omega_1 \in Y \text{ and } ot(X) \in Z \right\}.$$

By the general remarks in Section 2 it is clear that the existence of some

---

(see footnote 4 on page 3) implies the $\aleph_2$–saturation of NS_{ω_1}. Incidentally, neither BMM alone nor ψ_{AC} imply the saturation of NS_{ω_1} [36, Theorem 10.99, Lemma 10.102].

[9]This is the kind of cardinal considered in Theorem 4.3.

[10]More recently (see [4]), in late 2002, I proved that there can be no semiproper forcing extension of L satisfying BMM. Thus, using the Goldstern–Shelah results, a model separating $BSPFA$ from BMM can be produced starting from the optimal hypothesis (the consistency of $BSPFA$).

[11]Todorčević also proved in [32] that θ_{AC} implies $2^{\aleph_0} = 2^{\aleph_1}$ by showing that it implies the failure of Devlin-Shelah's weak diamond principle from [11].

ordinal $\gamma \geq \omega_2$ such that

$$\{X \in [\gamma]^{\aleph_0} : X \cap \omega_1 \in S \text{ iff } ot(X) \in T\}$$

is a projective stationary subset of $[\gamma]^{\aleph_0}$ is sufficient to give, in the presence of *BPSR*, ψ_{AC}. Hence, if both the following statement $\mathcal{S}$ and *BPSR* hold, then ψ_{AC} also does.

DEFINITION 4.1. [4] Let $\gamma \geq \omega_2$ be an ordinal. Then $\mathcal{S}^\gamma$ is the statement that for all stationary S, $T \subseteq \omega_1$, $\mathcal{S}^\gamma_{S,T}$ is a stationary subset of $[\gamma]^{\aleph_0}$. Also, $\mathcal{S}$ is the statement that there is some $\gamma \geq \omega_2$ such that $\mathcal{S}^\gamma$.

Notice that, by Lemma 2.2, a sufficient condition to prove ψ_{AC} in a context in which *BSPFA* holds is to find some cardinal $\gamma \geq \omega_2$ such that for every stationary subset T of ω_1 and every club $D \subseteq [2^\gamma]^{\aleph_0}$ there is a club $D' \subseteq D$ such that for every $X \in D'$ there is an end–extension Y of X in D such that $ot(Y \cap \gamma) \in T$.

The first result on the effect of bounded forcing axioms on the continuum problem that I am aware of is the following, due to Woodin [36, Lemma 10.95]. Recall that NS_{ω_1} is said to be *precipitous* if and only if the generic ultrapower obtained from forcing with $\mathcal{P}(\omega_1)/NS_{\omega_1}$ is always well–founded.

THEOREM 4.1 (Woodin). *Suppose that BMM holds and that either there is a measurable cardinal or NS_{ω_1} is precipitous. Then, ψ_{AC} holds and every function from ω_1 into ω_1 is dominated on a club by a canonical function.*

A full proof of the theorem for the case that NS_{ω_1} is precipitous is given in [36]. There it is actually shown that $\mathcal{S}^\kappa$ holds for $\kappa = (2^{2^{\aleph_1}})^+$ in case NS_{ω_1} is precipitous. Note that an immediate corollary of this result is that $BMM_{<\aleph_3}$ implies ψ_{AC}.

A related result due, independently, to P. Larson and Woodin (see [23]), is that $\mathcal{S}^{\omega_2}$ holds in case NS_{ω_1} is precipitous and $\Vdash_{NS^+_{\omega_1}} j(\omega_1^V) = \omega_2^V$.

If γ is measurable, then it has the property stated at the beginning of this section. To see this, fix a stationary set $T \subseteq \omega_1$, a club $D \subseteq [2^\gamma]^{\aleph_0}$, let θ be some large enough cardinal, let $\leq_\theta$ be a well–order of H_θ and let D' be a club consisting of intersections with 2^γ of countable elementary substructures of $(H_\theta, \in, \leq_\theta)$ containing D and some normal γ–complete ultrafilter $\mathcal{U}$ on γ. Pick $X \in D'$ and $N \prec H_\theta$ witnessing this. We will build a strictly $\subseteq$–increasing and $\subseteq$–continuous sequence $(N_\xi)_{\xi < \omega_1}$ such that

(a) $N_0 = N$,
(b) $N_\xi \prec (H_\theta, \in, \leq_\theta)$ for each ξ, and
(c) $N_\xi \cap \gamma$ is a proper initial segment of $N_{\xi+1}$ for each ξ.

Since then $(ot(N_\xi \cap \gamma) : \xi < \omega_1)$ is a strictly increasing and continuous enumeration of countable ordinals, it follows that there is some ξ_0 such that $ot(N_{\xi_0} \cap \gamma) \in T$. Since N_{ξ_0} is an end–extension of N whose intersection with 2^γ is in D (as $D \in N_{\xi_0} \prec H_\theta$), this shows that D' is a witness for the desired property of γ applied to D.

There remains to see how to obtain $N_{\xi+1}$ from N_ξ for any given ξ. Let $\eta = \min \bigcap \mathcal{U} \cap N_\xi$ and let $N_{\xi+1}$ be the collection of all $f(\eta)$, where $f \in N_\xi$ is a function with domain γ. Then, since H_θ has definable Skolem functions, it is easy to see that $N_{\xi+1}$ is a countable elementary substructure of $(H_\theta, \in, \leq_\theta)$. To see that $N_\xi \cap \gamma$ is a proper initial segment of $N_{\xi+1}$, note first that $\eta = id_\gamma(\eta) \in N_{\xi+1}$ lies above all members of $N_\xi \cap \gamma$. Also, given any $\alpha \in \eta \cap N_{\xi+1}$, there is some function f in N with domain γ such that $f(\eta) = \alpha$. Then, f is regressive on some set A in $\mathcal{U} \cap N_\xi$. Finally, since $A, f \in N_\xi$ and $\mathcal{U}$ is a normal ultrafilter, there is a unique $B \subseteq A$ and α_0, $B \in \mathcal{U} \cap N_\xi$, such that $f(\gamma) = \alpha_0$ for all $\gamma \in B$. But then, $\alpha = f(\eta) = \alpha_0 \in N_\xi$.

The above proof shows that in fact ψ_{AC} follows from $BSPFA$ plus the existence of a measurable cardinal.

The next result I will mention was proved in 2000 [3].

THEOREM 4.2 (Asperó). *Suppose $r^\sharp$ exists for every real r, $u_2 = \omega_2$ and* $\Vdash_{NS^+_{\omega_1}}$ *"Ord^M has an initial well–founded segment of length $\breve{\omega}_2 + 1$". Assume BPSR. Then ψ_{AC} holds.*

Recall that a *uniform indiscernible* is defined — in the situation that $r^\sharp$ exists for every real r — as an ordinal which is a Silver indiscernible for $L[r]$ for every real r, that the first uniform indiscernible (u_1) is ω_1 and that the second uniform indiscernible u_2 is at most ω_2. Also, notice that, by the remarks on canonical functions made in Section 3, the generic ultrapower derived from forcing with $NS^+_{\omega_1}$ always has an initial well–founded segment of length ω_2^V. It is an unpublished result of Hajnal that the extra well–foundedness hypothesis in the statement of Theorem 4.2 does not hold in L (i.e., in L, $NS^+_{\omega_1}$ forces that the generic ultrapower is only well–founded up to ω_2^V). On the other hand, it can be forced over any ZFC model that $NS^+_{\omega_1}$ forces it [20]. The proof of Theorem 4.2 uses the observation, due jointly to Todorčević and myself, that if $BPSR$ holds and $NS^+_{\omega_1}$ forces the well–foundedness of the generic ultrapower up to $\omega_2^V + 1$, then every function from ω_1 into ω_1 is dominated on a club by a canonical function. From this boundedness — actually $\Vdash_{NS^+_{\omega_1}} j(\breve{\omega}_1) = \breve{\omega}_2$ suffices — and from $u_2 = \omega_2$, a variation, using the proof of Lemma 3.16 of [36], of Woodin's argument for proving S^κ for $\kappa = (2^{2^{\aleph_1}})^+$ from the precipitousness of NS_{ω_1}, shows S^κ for κ as above.

Also in 2000 I improved the first half of Theorem 4.1 by proving that $BPSR$, together with the existence of an ω_1–Erdős cardinal, implies ψ_{AC}. Then, P. Welch was able to weaken the large cardinal assumption in this result to that of the existence of a so-called ω–*closed cardinal* [35]. Finally, in 2001 he proved this conclusion for the yet weaker assumption that $BPSR$ holds and there exists a cardinal κ satisfying a certain weak Erdős property [6].

THEOREM 4.3 (Welch). *Suppose there is a cardinal with the partition property* $\kappa \longrightarrow (< \omega_1)_{2^{\omega_1}}^{<\omega}$, *meaning that for every first order structure* $\mathcal{A} = \langle L_\kappa[A], \in, A\rangle$ *there is a sequence* $(I_\alpha : \alpha < \omega_1)$ *such that*

(i) *for all* $\alpha < \omega_1$, I_α *is a set of indiscernibles for* $\mathcal{A}^+$ *of order type* $\omega\alpha$, *where* $\mathcal{A}^+ = (\mathcal{A}, \dot{\xi})_{\xi < \omega_1}$,

(ii) *for all* α, $\beta < \omega_1$, *all formulas* $\varphi(v_0, \ldots, v_{n+m-1})$, $\vec{\xi} \in \omega_1^n$, $\vec{\gamma} \in I_\alpha^m$ *and* $\vec{\gamma}' \in I_\beta^m$, *if* $\vec{\gamma}$ *and* $\vec{\gamma}'$ *are strictly increasing,*

$$\mathcal{A}^+ \models \varphi(\vec{\xi}, \vec{\gamma}) \longleftrightarrow \varphi(\vec{\xi}, \vec{\gamma}').$$

If BPSR holds, then so does ψ_{AC}.

The original proof of this result goes through by showing that if κ is such that $\kappa \longrightarrow (< \omega_1)_{2^{\omega_1}}^{<\omega}$, then there is an inaccessible cardinal $\gamma < \kappa$ such that S^γ.[12] As to the consistency strength of this partition relation, note that if κ is ω_1–Erdős, then in V_κ there are cardinals γ such that $\gamma \longrightarrow (< \omega_1)_{2^{\omega_1}}^{<\omega}$.

The following result is a further generalization of the first half of Woodin's Theorem 4.1, which I obtained in late 2001 [4]. It involves the notion of a certain game on $\mathcal{I}^+$ associated to a given ideal $\mathcal{I}$ over some cardinal, as defined in [15].

DEFINITION 4.2. Let κ be an infinite cardinal and let $\mathcal{I}$ be an ideal over κ. $G_\mathcal{I}$ is the following ω–length game with two players I and II, I moving first: I and II alternately choose $\mathcal{I}$–positive subsets S_i of κ such that $S_{i+1} \subseteq S_i$ for all i. I wins if and only if $\bigcap_i S_i$ is empty.

In [15] it is shown, for example, that player I fails to have a winning strategy in $G_\mathcal{I}$ if and only if $\mathcal{I}$ is precipitous, that player II does not have any winning strategy for $G_\mathcal{I}$ if $\kappa \leq 2^{\aleph_0}$, and that player II never has a winning strategy for G_{NS_κ} for any regular cardinal $\kappa \geq \omega_1$.

Also, given κ and $\mathcal{I}$ as in Definition 5.3, let $G_\mathcal{I}'$ be a game exactly as $G_\mathcal{I}$, except that player II moves first. Obviously, if player II has a winning strategy in $G_\mathcal{I}$, then she has one for $G_\mathcal{I}'$.

Simple variants of the proofs in [15] show that the negative results for player II in the games $G_\mathcal{I}$ mentioned above also hold for her in the corresponding games $G_\mathcal{I}'$. Also, player I has a winning strategy in $G_\mathcal{I}'$ if and only if $\Vdash_{\mathcal{I}^+}$ "The generic ultrapower is ill–founded".

For any class N and any partial order $\mathbb{P}$ in N, $G \subseteq N$ is $(N, \mathbb{P})$–generic if G is a filter of $N \cap \mathbb{P}$ meeting every dense subset of $\mathbb{P}$ which is in N. A sequence is $(N, \mathbb{P})$–generic if its range is so. The proof of the following lemma is straightforward.

LEMMA 4.4. *Let* κ *be a cardinal and let* $\mathcal{I}$ *be an ideal over* κ. *If player* II *has a winning strategy* σ *for* $G_\mathcal{I}'$, *then for every large enough cardinal* θ

---

[12]The proof shows in fact also S^κ.

$(i.e., \theta \geq (2^{2^{\kappa}})^+)$ *and every countable* $N \preceq H_\theta$ *containing* $\mathcal{I}$ *and* σ *there is some* $(N, \mathcal{I}^+)$*–generic* G *such that* $\bigcap G$ *is nonempty.*

It can be shown as well that player II has a winning strategy in $G_{\mathcal{I}}$ in case every $(N, \mathcal{I}^+)$–generic sequence has nonempty intersection for all N in some club of $[H_\theta]^{\aleph_0}$. Also, if there are stationarily many countable $N \preceq H_\theta$ for which every $(N, \mathcal{I}^+)$–generic sequence is nonempty, then $\mathcal{I}$ is precipitous. Finally, if $\mathcal{I}$ is precipitous, then the set of countable $N \preceq H_\theta$ such that there is some $(N, \mathcal{I}^+)$–generic sequence with nonempty intersection, is a stationary subset of $[H_\theta]^{\aleph_0}$.

LEMMA 4.5. *Suppose* κ *is a regular cardinal carrying a* κ*–complete ideal* $\mathcal{I}$ *such that player* II *has a winning strategy in* $G'_{\mathcal{I}}$. *Then there is a normal* κ*–complete ideal* $\mathcal{J}$ *over* κ *such that player* II *has a winning strategy in* $G'_{\mathcal{J}}$.

PROOF. $\mathcal{I}^+$ does not force that the generic ultrapower is ill–founded. Therefore there is some function $f : \kappa \longrightarrow \kappa$ and some $\mathcal{I}$–positive set S forcing that f represents κ in the generic ultrapower. Let $\mathcal{J} = f_*(\mathcal{I}) \cap S$, i.e., given $A \subseteq \kappa$, $A \in \mathcal{J}$ iff $f^{-1}(A) \cap S \in \mathcal{I}$. It is a standard easily checked fact that such a $\mathcal{J}$ is always a κ–complete ideal over κ and that, since f represents — modulo S — κ in the generic ultrapower, it is also normal. Let σ be a winning strategy for player II in $G_{\mathcal{I}}$. There remains to describe a winning strategy for player II in $G'_{\mathcal{J}}$. She starts playing $f``S$. $I$ replies with some $\mathcal{J}^+$–positive $T_1 \subseteq f``S$. Then, player II moves $f``S_1$, where $S_1$ is $II$'s response, according to $\sigma$, to $f^{-1}(T_1) \cap S$ (by the definition of $\mathcal{J}$ as $f_*(\mathcal{I}) \cap S$, it readily follows that $f^{-1}(T_1) \cap S$ is $\mathcal{I}$–positive). As the game progresses, players $II$ and $I$ will have moved $\langle f``S, T_1, \ldots, T_n \rangle$, where T_n is a move of player I. This will correspond to a sequence $\langle f^{-1}(T_1) \cap S, S_1, f^{-1}(T_3) \cap S_1, \ldots, S_{n-1} \rangle$ of $G_{\mathcal{I}}$ in which player II follows σ. S_{n-1} is a move of player II. Now, player II moves $f``S_{n+1}$, where $S_{n+1}$ is the move, according to $\sigma$, of player $II$ after player $I$ moves now $S_n := f^{-1}(T_n) \cap S_{n-1}$ in the play of $G_{\mathcal{I}}$. Again, since $T_{n-1} = f``S_{n-1}$, S_n is an $\mathcal{I}^+$–positive subset of κ. After ω–many moves, the game $\langle f^{-1}(T_1) \cap S, S_1, f^{-1}(T_3) \cap S_1, S_2, \ldots, \rangle$ is won by player II, and so there is some $\alpha \in \bigcap_n S_n$. But then, $f(\alpha) \in \bigcap_i T_{2i} = \bigcap_i T_i$. This shows that the above described procedure is a winning strategy for II in $G'_{\mathcal{J}}$.   ⊣

Consider the following generalization of the Strong Chang's Conjecture.

DEFINITION 4.3. Given an ordinal γ of uncountable cofinality, $CC^*_{\omega_1}(\gamma)$ is the statement that given any club $E \subseteq [\gamma]^{\aleph_0}$ there is a club E' of $[\gamma]^{\aleph_0}$, $E' \subseteq E$ such that for every $X \in E'$ there is some $Y \in E'$ such that $X \cap \omega_1 = Y \cap \omega_1$ and X is a proper initial segment of Y.

Obviously, $CC^*_{\omega_1}(\gamma)$ for γ as in Definition 4.3 implies that every first order structure $\mathcal{M}$ with a countable language and universe γ has an elementary substructure $\mathcal{N}$ of size $\aleph_1$ such that $\mathcal{N} \cap \omega_1$ is countable. It implies $\mathcal{S}^\gamma$ as well, and in fact it implies the strong form of $\mathcal{S}^\gamma$ saying that for every club $E \subseteq [\gamma]^{\aleph_0}$ there is a club $C \subseteq \omega_1$ such that $\{ot(X) : X \in E, X \cap \omega_1 = \nu\}$ includes a club for every $\nu \in C$. I do not know if this strong form of $\mathcal{S}^\gamma$ is equivalent to $\mathcal{S}^\gamma$ or not.

Note that $CC^*_{\omega_1}(\kappa)$ has an effect on cardinal arithmetic for successor cardinals κ. Specifically, if $\kappa = \lambda^+$ and $CC^*_{\omega_1}(\kappa)$ holds, then $\kappa^{\aleph_0} = \kappa$. To see this, fix a surjective map $e_\alpha : \lambda \longrightarrow \alpha$ for every $\alpha < \kappa$. By $CC^*_{\omega_1}(\kappa)$, there is a club E of $[\kappa]^{\aleph_0}$ such that for each $X \in E$, $X = N \cap \kappa$ for some countable elementary subtructure N of some large enough H_θ which contains $(e_\alpha)_{\alpha < \kappa}$ and, moreover, X can be end–extended with an element of E. It follows that each element of E is of the form $e_\alpha``\xi$ (for some $\alpha < \kappa$ and $\xi < \omega_1$), so that $|E| = \kappa$. But by a result in [8] every club of $[\kappa]^{\aleph_0}$ has size $\kappa^{\aleph_0}$.

THEOREM 4.6 (Asperó). *Suppose κ is a cardinal carrying a κ–complete ideal $\mathcal{I}$ such that player II has a winning strategy in $G'_{\mathcal{I}}$. Then, $CC^*_{\omega_1}(\kappa)$. Hence, if, in addition, BPSR holds, then so does ψ_{AC}.*

PROOF. Fix $F : [\kappa]^{<\omega} \longrightarrow \kappa$. Let θ be a large enough cardinal as in Lemma 4.4. It will be enough to prove that every countable substructure N of H_θ containing F, $\mathcal{I}$, and a winning strategy σ for player II in the game $G'_{\mathcal{I}}$ can be extended to another such structure N' such that $N \cap \kappa$ is a proper initial segment of $N' \cap \kappa$. So let G be an $(N, \mathcal{I}^+)$–generic set whose intersection is nonempty (such a G exists by Lemma 4.4) and let N' be the collection of all $f(\eta)$, where f is any function in N whose domain is κ and $\eta := \min \bigcap G$. By κ–completeness of $\mathcal{I}$, $\eta \geq sup(N \cap \kappa)$. As the identity function on κ is in N_ν, $\eta \in N'$, and it is also clear that $N \subseteq N'$. Now let α be some ordinal in $N' \cap \eta$. There is some function f in N such that $\alpha = f(\eta)$. It follows that $\{\xi < \kappa : f(\xi) < \xi\}$ is in G. By Lemma 4.5, we may assume that $\mathcal{I}$ is normal. Hence, by the $(N, \mathcal{I}^+)$–genericity of G we get some $\beta \in \kappa \cap N$ such that $f^{-1}(\beta) \in G$. But then, $\alpha = \beta$. There remains to prove $N' \preccurlyeq H_\theta$. Let π_0 be the collapsing function of N. Let $\overline{G} = \pi_0``G$. Then, $\overline{G}$ is $(\pi_0(\mathcal{I})^+)^{\overline{N}}$–generic over $\overline{N} := \pi_0``N$. Let M and j be the corresponding generic ultrapower of $\overline{N}$ and generic elementary embedding $j : \overline{N} \longrightarrow M$, respectively. We may assume that M is a transitive subclass of $\overline{N}[\overline{G}]$.

CLAIM 4.7. N' is collapsible and, letting π_1 be its collapsing function, $M = \pi_1``N'$ and $j(\pi_0(x)) = \pi_1(x)$ for every $x \in N$.

PROOF. Given functions f and g with domain κ, $[\pi_0(f)]_{\overline{G}} \in [\pi_0(g)]_{\overline{G}}$ iff $\{\xi < \kappa : f(\xi) \in g(\xi)\} \in G$ iff $f(\eta) \in g(\eta)$. This shows that the function sending $f(\eta)$ to $[\pi_0(f)]_{\overline{G}}$ (where f is a function in N with domain κ) is an isomorphism between N' and M, i.e. it is the transitive collapse π_1 of N'. Finally, for every $x \in N$, $j(\pi_0(x))$ is $[\langle \pi_0(x) : \xi < \pi_0(\kappa)\rangle]_{\overline{G}}$, but this is $\pi_1(c_x(\eta))$, where c_x is the x–constant function with domain κ, i.e., $j(\pi_0(x)) = \pi_1(x)$.   ⊣

Now let $\varphi(x_0, \ldots, x_{n-1})$ be a formula, let $f_0, \ldots, f_{n-1}$ be functions in N with domain κ and suppose $N' \models \varphi(f_0(\eta), \ldots, f_{n-1}(\eta))$. Then,

$$M \models \varphi\left([\pi_0(f_0)]_{\overline{G}}, \ldots, [\pi_0(f_{n-1})]_{\overline{G}}\right),$$

i.e., $\{\xi < \kappa : H_\theta \models \varphi(f_0(\xi), \ldots, f_{n-1}(\xi))\}$ belongs to G. But then, by the definition of η, $H_\theta \models \varphi(f_0(\eta), \ldots, f_{n-1}(\eta))$. This shows $N' \preccurlyeq H_\theta$.   ⊣

I will finish this section by mentioning a result which does not quite involve ψ_{AC}, but a local version of it. In early 2002 I considered the situation when forcing with $\mathcal{P}(\omega_1)/NS_{\omega_1}$ does not collapse ω_2. In February 2002 I proved that in this case the following weak form of $\mathcal{S}^{\omega_2}$ holds:

There is a stationary $S \subseteq \omega_1$ and $Z \subseteq \omega_1$ such that both $\mathcal{S}^{\omega_2}_{S',Z}$ and $\mathcal{S}^{\omega_2}_{S',\omega_1 \setminus Z}$ are stationary subsets of $[\omega_2]^{\aleph_0}$ for every stationary $S' \subseteq S$.

By the considerations in Section 2, this is clearly enough to obtain, in the presence of $BPSR$, the following local form of ψ_{AC} which suffices to imply the existence of a well–order of $\mathcal{P}(\omega_1)$ definable over H_{ω_2}.

DEFINITION 4.4. $\psi_{AC,local}$ is the statement that there are $S, Z \subseteq \omega_1$ such that S is stationary and such that for every stationary $S' \subseteq S$, if $S \setminus S'$ is also stationary, then there is an ordinal δ, a surjection $e : \omega_1 \longrightarrow \delta$ and a club $C \subseteq \omega_1$ such that

$$S' \cap C \subseteq \{v < \omega_1 : ot(e``v) \in Z\}$$

and

$$(S \setminus S') \cap C \subseteq \{v < \omega_1 : ot(e``v) \notin Z\}.$$

That this form of ψ_{AC} yields a coding of subsets of ω_1 by ordinals less than ω_2 can be established by essentially the same argument as in the corresponding proof for ψ_{AC}.

THEOREM 4.8 (Asperó). *Suppose that forcing with $\mathcal{P}(\omega_1)/NS_{\omega_1}$ does not collapse ω_2. Then there are stationary subsets S and T of ω_1 such that $\mathcal{S}^{\omega_2}_{S',T}$ and $\mathcal{S}^{\omega_2}_{S',\omega_1 \setminus T}$ are stationary subsets of $[\omega_2]^{\aleph_0}$ for every stationary $S' \subseteq S$.*

PROOF. Assume otherwise. Then, for all stationary $S, T \subseteq \omega_1$ there is some stationary $S' \subseteq S$ such that either $\mathcal{S}^{\omega_2}_{S',T}$ is nonstationary or $\mathcal{S}^{\omega_2}_{S',\omega_1 \setminus T}$ is nonstationary. Let $F : [\omega_2]^{<\omega} \longrightarrow \omega_2$ be a function witnessing this. This means that either there is no countable $X \subseteq \omega_2$ closed under F which is in $\mathcal{S}^{\omega_2}_{S',T}$, or else there is no countable $X \subseteq \omega_2$ closed under F which is in $\mathcal{S}^{\omega_2}_{S',\omega_1 \setminus T}$. Letting D be the club of all ordinals $\alpha < \omega_2$ which are closed under F, it is easy to see that either

$$S' \Vdash_{NS^+_{\omega_1}} D \cap j(T) = \emptyset$$

(in the first case), or

$$S' \Vdash_{NS_{\omega_1}^+} D \subseteq j(T)$$

(in the second case).

But then, fixing T, the collection of stationary sets deciding whether $D \subseteq j(T)$ or $D \cap j(T) = \emptyset$ for some club $D \subseteq \omega_2$ in the ground model is dense in $NS_{\omega_1}^+$. Hence, $NS_{\omega_1}^+$ forces that

$$\dot{\mathcal{U}} := \{Z \in \mathcal{P}(\omega_1)^V : D \subseteq j(Z) \text{ for some club } D \subseteq \omega_2^V \text{ in } V\}$$

is a V–ultrafilter over ω_1^V, meaning that $\dot{\mathcal{U}} \subseteq \mathcal{P}(\omega_1)^V$ is a non–principal filter which measures all subsets of ω_1^V in V.

Let $\bar{r} = \langle r_\xi : \xi < \omega_1 \rangle$ be a sequence in V of pairwise distinct elements of the Cantor space 2^ω and let G be $NS_{\omega_1}^+$–generic over V. Then, for every $n < \omega$ there is a unique $\varepsilon_n \in 2$ such that $\{\xi < \omega_1^V : r_\xi(n) = \varepsilon_n\} \in \dot{\mathcal{U}}[G]$, i.e. such that $D_n \subseteq \{\alpha < j(\omega_1^V) : j(\bar{r})(\alpha)(n) = \varepsilon_n\}$ for a club $D_n \subseteq \omega_2^V$ in V. From this I will derive a contradiction using the fact that $\omega_1^{V[G]} = \omega_2^V = j(\omega_1^V)$. All D_n are clubs of $\omega_1^{V[G]}$. Therefore we can take distinct α, $\beta \in \bigcap_n D_n$. It follows that $j(\bar{r})(\alpha)(n) = \varepsilon_n = j(\bar{r})(\beta)(n)$ for all n, which is a contradiction since $j(\bar{r})$ is a sequence of pairwise distinct reals. $\dashv$

COROLLARY 4.9. *Suppose that forcing with $NS_{\omega_1}^+$ does not collapse ω_2. If BPSR holds, then $2^{\aleph_1} = \aleph_2$. Hence, NS_{ω_1} is presaturated.*

PROOF. Recall that NS_{ω_1} is presaturated if and only if NS_{ω_1} is precipitous and forcing with $NS_{\omega_1}^+$ does not collapse ω_2. That NS_{ω_1} is precipitous if $NS_{\omega_1}^+$ does not collapse ω_2 and $2^{\aleph_1} = \aleph_2$ is a result from [8]. $\dashv$

I am unaware of any large cardinal consequence of $\psi_{AC,local}$.

§5. Result not involving ψ_{AC}.

In the first part of this section I will focus on a couple of results involving some form of generic absoluteness, together with some anti-large cardinal hypothesis. Then I will turn to the developments that took place in early 2002 after Theorem 4.8 was proved.

I proved the following result in spring 1999 [3].

THEOREM 5.1 (Asperó). *Suppose BMM holds and $X^\sharp$ fails to exist for some $X \subseteq \omega_1$. Then $2^{\aleph_0} = \aleph_2$. In fact, for every stationary $S \subseteq \omega_1$ there are ordinals $\xi < \alpha < \delta < \omega_2$ and $\beta < \delta$ and a continuous and increasing sequence $(X_\nu)_{\nu<\omega_1}$ of countable subsets of δ such that $\delta = \bigcup_\nu X_\nu$ and, given any ν, $X_\nu \cap \omega_1 \in S$ if and only if $cf^{L[X]}(sup(X_\nu)) = \alpha$ and, letting C be the $<_{L[X]}$–least cofinal $D \subseteq sup(X_\nu)$ in $L[X]$ in order type α, the ξ-th member of C is β.*

The proof of this result uses Jensen's Covering Lemma.

The additional hypotheses in Theorem 4.1 are in terms of the existence of (an inner model of) large cardinals, whereas the extra assumption in Theorem 5.1 certainly is an anti-large cardinal hypothesis. These two results taken together — i.e., the fact that BMM implies $2^{\aleph_0} = \aleph_2$ in these two opposite circumstances — were seen at that time as a strong evidence for the conjecture that BMM implies $2^{\aleph_0} = \aleph_2$.

Next I will digress briefly to define a certain class of strengthenings of bounded forcing axioms, which first appeared in [1] (see also [3]). As it will become clear, these statements have been of relevance in the course of the present history.

DEFINITION 5.1. Let Γ be a class of posets in ZFC.[13]
A formula $\varphi(x)$ is a *provably Γ–persistent predicate* if and only if

$$ZFC \vdash \forall x \big[\varphi(x) \longrightarrow \big(\forall \mathbb{P} \in \Gamma, \Vdash_{\mathbb{P}} \varphi(\check{x})\big)\big].$$

My original motivation for Definition 5.1 came in spring 1999 from [29], where the notion of *provably ccc–persistent predicate* is defined. Σ_1 statements are always preserved upwards. Therefore, the following generic absoluteness principles involving classes of provably persistent predicates are strong versions of corresponding forms of $Abs(\Gamma)$.

DEFINITION 5.2.[14] Let Γ be a class of posets and let Σ be a class of formulas of the language of set theory. Then $BFA(\Gamma, \Sigma)$ denotes the following schema:
Let $\varphi(x)$ be a formula in Σ. Given $a \in H(\omega_2)$, if there is some $\mathbb{P}$ in Γ such that $\Vdash_{\mathbb{P}} \varphi(\check{a})$, then $\varphi(a)$.

Suppose Σ is the class of all provably Γ–persistent predicates. Then $BFA(\Gamma)'$ denotes $BFA(\Gamma, \Sigma)$ and, for every natural number $n \geq 1$, $BFA(\Gamma)'_n$ denotes $BFA(\Gamma, \Sigma \cap \Sigma_n)$, where Σ_n is the class of all Σ_n formulas of the language of set theory.

Note that, because of the undefinability of truth, one can only expect to express $BFA(\Gamma, \Sigma)$ as an axiom schema in general. However, due to the definability of the satisfaction relation for Σ_n formulas, if there is some $n < \omega$ such that $\Sigma \subseteq \Sigma_n$, then $BFA(\Gamma, \Sigma)$ can be expressed by a first order sentence.

If Γ is the class of stationary–set–preserving posets, we write BMM' and BMM'_n for $BFA(\Gamma)'$ and $BFA(\Gamma)'_n$, respectively.

Note that, whenever $\Gamma_0 \subseteq \Gamma_1$, the class of provably Γ_1–persistent predicates is included in the class of provably Γ_0–persistent predicates, and therefore $BFA(\Gamma_1)'$ $(BFA(\Gamma_1)'_n)$ does not necessarily imply $BBFA(\Gamma_0)$ $(BFA(\Gamma_0)'_n)$.

Note also that if Γ is a class of stationary–set–preserving posets, then $BFA(\Gamma)'_2$ implies that $(H_{\omega_2}, \in, NS_{\omega_1})$ is a Σ_1–elementary substructure of the corresponding structure $(H_{\omega_2}^{V^{\mathbb{P}}}, \in, NS_{\omega_1}^{V^{\mathbb{P}}})$ obtained after forcing with $\mathbb{P}$ for every $\mathbb{P}$ in Γ.

As an immediate corollary of Theorem 5.1, one obtains the following result (see [3]).

---

[13] Meaning that $ZFC \vdash \forall \mathbb{P} \in \Gamma (\mathbb{P}$ is a poset$)$
[14] J. D. Hamkins has defined related notions (see [18]).

COROLLARY 5.2. *Suppose* $X^\sharp$ *does not exist for some set of ordinals* X. *If* BMM_2' *holds, then* $2^{\aleph_0} = \aleph_2$.

Coll(ω_1, X) forces that there is a subset of ω_1 whose sharp does not exist. Since no poset can add the sharp of any set from the ground model, by BMM_2' there is such a subset of ω_1, and the conclusion follows from Theorem 5.1.

R. Schindler subsequently observed in 2001 that BMM_2' can be replaced by BMM in the above statement. Actually, he proved the following fact.

THEOREM 5.3 (Schindler). *Suppose* BMM *holds and* $\omega_1^{L[a]} < \omega_1$ *for every real* a. *Then* $X^\sharp$ *exists for every set of ordinals* X.

From Jensen's techniques for coding the universe with a real it follows that if $X^\sharp$ does not exist, then there is a set–sized stationary–set–preserving partial order which codes X into a real (this can be extracted from [26]). By adding some information to X if necessary, one may assume that $\omega_1 = \omega_1^{L[X]}$. It follows that this partial order forces the existence of a real r such that $\omega_1 = \omega_1^{L[r]}$, which is a Σ_1 statement with ω_1 as parameter. Therefore, such a real r exists in case BMM is true. Theorem 5.3 was also found, later and independently, by S. Friedman.

The following two–person game appears originally in [16] (see also [34]).

DEFINITION 5.3. Let $\lambda \geq \omega_2$ be a regular cardinal, let $v < \omega_1$ and let $h : [\lambda]^{<\omega} \longrightarrow \lambda$. Then $\mathcal{G}_{v,h}$ is the following game of length ω with players I and II:

At stage n, player I plays an interval J_n of ordinals in λ and an ordinal $\xi_n \in J_n$ and player II plays an ordinal $\mu_n \in \lambda$.

If $n \geq 1$, I is required to play J_n so that $inf(J_n) > \mu_{n-1}$.

Player I wins if and only if, letting $X = cl_h(\{\xi_n : n \in \omega\} \cup v)$, $X \subseteq v \cup \bigcup_{n \in \omega} J_n$ and $X \cap \omega_1 = v$.

It turns out (see [34]) that, given $h : [\lambda]^{<\omega} \longrightarrow \lambda$, there are club many $v < \omega_1$ such that player I has a winning strategy for $\mathcal{G}_{v,h}$.

In 1999 I found the following application of an idea — as far as I know, due to Todorčević — for coding a given real using a $\Diamond$–sequence $\overline{C}$ and the above games.

THEOREM 5.4 (Asperó-Todorčević). *Suppose the universe is a set-generic extension of* $L[X]$ *for some* $X \in H_{\omega_2}$ *(more generally, suppose that, for some* $X \in H_{\omega_2}$ *and some regular cardinal* $\lambda \geq \omega_2$, *every club of* λ *includes a club of* λ *belonging to* $L[X]$). *If* $(H_{\omega_2}, \in, NS_{\omega_1}) \preccurlyeq_{\Sigma_1} (H_{\omega_2}^{V^{\mathbb{P}}}, \in, NS_{\omega_1}^{V^{\mathbb{P}}})$ *for every proper poset* $\mathbb{P}$, *then* $2^{\aleph_0} = \aleph_2$.

The main idea for proving Theorem 5.4 is, given a real $a \subseteq \omega$, a unbounded in ω, to force the set $\{X \in [\lambda]^{\aleph_0} : [C_\alpha(n), C_\alpha(n+1)) \cap X \neq \emptyset$ iff $n \in a\}$ to be of size $\aleph_1$, where $(C_\alpha : \alpha < \lambda)$ is a $\Diamond_\lambda$–sequence in $L[X]$ and $(C_\alpha(n) :$

$n < \omega)$ is the increasing enumeration of C_α for each α (see [3] for the missing details).

Now we come to the last wave of results. The following lemma, which I found in March 2002, is of crucial importance in the proof of Theorem 5.6. Its proof originated in the argument for proving Theorem 4.8.

LEMMA 5.5 (Asperó). *Let* $\gamma = (2^{\aleph_1})^+$. *Consider the following clauses:*

(A) *For all* α, $\omega_2 \leq \alpha < \gamma$, *for all stationary* $S \subseteq \omega_1$ *and for all* $Z \subseteq \omega_1$ *there is some stationary* $S' \subseteq S$ *such that either*
   (a) $\mathcal{S}^\alpha_{S',Z} \cup \mathcal{S}^\alpha_{\omega_1 \setminus S', \omega_1}$ *includes a club of* $[\alpha]^{\aleph_0}$, *or*
   (b) $\mathcal{S}^\alpha_{S', \omega_1 \setminus Z} \cup \mathcal{S}^\alpha_{\omega_1 \setminus S', \omega_1}$ *includes a club of* $[\alpha]^{\aleph_0}$.
(B) *For all stationary* $S \subseteq \omega_1$ *and for all* $Z \subseteq \omega_1$ *there is a club* $D \subseteq \gamma$ *such that*
   (a) *either* $\mathcal{S}^\alpha_{S,Z}$ *is a stationary subset of* $[\alpha]^{\aleph_0}$ *for each* $\alpha \in D$ *of countable cofinality,*
   (b) *or* $\mathcal{S}^\alpha_{S, \omega_1 \setminus Z}$ *is a stationary subset of* $[\alpha]^{\aleph_0}$ *for each* $\alpha \in D$ *of countable cofinality.*

If either (A) *or* (B) *fails and BPSR holds, then* $2^{\aleph_1} = \aleph_2$ *and in fact there is a well–order of* $\mathcal{P}(\omega_1)$ *in order type* ω_2 *which is* Δ_1 *definable over* $(H_{\omega_2}, \in, NS_{\omega_1})$ *with parameters.*

PROOF. Let $\gamma = (2^{\aleph_1})^+$. Suppose (A) fails. Then there is some α, $\omega_2 \leq \alpha < \gamma$, a stationary set $S \subseteq \omega_1$ and $Z \subseteq \omega_1$ such that both $\mathcal{S}^\alpha_{S',Z}$ and $\mathcal{S}^\alpha_{S', \omega_1 \setminus Z}$ are stationary subsets of $[\alpha]^{\aleph_0}$ for each stationary $S' \subseteq S$. Hence, by *BPSR*, $\psi_{AC,local}$ holds.

Now suppose (B) fails. Then there is a stationary $S \subseteq \omega_1$ and $Z \subseteq \omega_1$ such that

(1) $A := \{\alpha < \gamma : cf(\alpha) = \omega \text{ and } \mathcal{S}^\alpha_{S,Z} \cup \mathcal{S}^\alpha_{\omega_1 \setminus S, \omega_1} \text{ includes a club of } [\alpha]^{\aleph_0}\}$ is a stationary subset of γ, and
(2) $B := \{\alpha < \gamma : cf(\alpha) = \omega \text{ and } \mathcal{S}^\alpha_{S, \omega_1 \setminus Z} \cup \mathcal{S}^\alpha_{\omega_1 \setminus S, \omega_1} \text{ includes a club of } [\alpha]^{\aleph_0}\}$ is a stationary subset of γ.

BPSR will ensure that a ϕ_{AC}–like statement holds which guarantees the existence of a coding, Δ_1 definable over $(H_{\omega_2}, \in, NS_{\omega_1})$ with a parameter, of subsets of ω_1. Let $(S_\xi)_{\xi < \omega_1}$ be a partition of ω_1 into stationary sets such that given any stationary $S' \subseteq \omega_1$ there is some ξ with $S' \cap S_\xi$ stationary. Fix $Y \subseteq \omega_1$. Then, since A and B are both stationary subsets of γ consisting of ordinals of countable cofinality, $S_Y := \{X \in [\gamma]^{\aleph_0} : \text{ if } X \cap \omega_1 \in \bigcup_{\xi \in Y} S_\xi, \text{ then } sup(X) \in A, \text{ and if } X \cap \omega_1 \notin \bigcup_{\xi \in Y} S_\xi, \text{ then } sup(X) \in B\}$ is a projective stationary subset of $[\gamma]^{\aleph_0}$ (see [13, Lemma 8]). Therefore, the following statement (∗) holds by *BPSR*.

$(*)$: There is an ordinal $\delta_Y < \omega_2$, a $\subseteq$–increasing and $\subseteq$–continuous sequence $\langle X_\nu : \nu < \omega_1 \rangle$ of countable subsets of δ_Y covering all δ_Y and a sequence $\langle e_\nu : \nu < \omega_1 \rangle$ such that for all $\nu < \omega_1$,

(α) e_ν is a surjection from ω_1 into $sup(X_\nu)$,

(β) if $X_\nu \cap \omega_1 \in \bigcup_{\xi \in Y} S_\xi$, then there is a club $C \subseteq \omega_1$ such that $ot(e_\nu``\zeta) \in Z$ for each $\zeta \in C \cap S$,

(γ) and if $X_\nu \cap \omega_1 \notin \bigcup_{\xi \in Y} S_\xi$, then there is a club $C \subseteq \omega_1$ such that $ot(e_\nu``\zeta) \notin Z$ for each $\zeta \in C \cap S$.

Finally, suppose Y and Y' are distinct subsets of ω_1 such that $\delta_Y = \delta_{Y'} = \delta$. Take clubs E_Y, $E_{Y'} \subseteq [\delta]^{\aleph_0}$ such that, given any $X \in E_Y$ and $X' \in E_{Y'}$, $X \cap \omega_1 \in \bigcup_{\xi \in Y} S_\xi$ if and only if $S \Vdash_{NS^+_{\omega_1}} sup(X) \in j(Z)$ and $X' \cap \omega_1 \in \bigcup_{\xi \in Y'} S_\xi$ if and only if $S \Vdash_{NS^+_{\omega_1}} sup(X') \in j(Z)$. Now, given any $\xi \in Y \triangle Y'$ and any $X \in E_Y \cap E_{Y'}$ such that $X \cap \omega_1 \in S_\xi$, we simultaneously have that $S \Vdash_{NS^+_{\omega_1}} sup(X) \in j(Z)$ and that $S \nVdash_{NS^+_{\omega_1}} sup(X) \in j(Z)$, which is a contradiction. $\dashv$

I obtained Theorems 5.6 and 5.8 in March 2002. The proof of Theorem 5.6 uses a sequence of distinct elements of 2^ω in very much the same way as the proof of Theorem 4.8 does.

THEOREM 5.6 (Asperó). *BMM implies $\mathfrak{d} = \aleph_2$. In fact, BMM implies that there is a scale $\mathcal{D}$ of order type ω_2 which is Σ_1 definable over $(H_{\omega_2}, \in, NS_{\omega_1})$.*

PROOF. Let $\gamma = (2^{\aleph_1})^+$. Suppose first that both (A) and (B) from Lemma 5.5 hold. Then, since $\gamma > 2^{\aleph_1}$ is a regular cardinal, from (B) it follows that there is a club $D \subseteq \gamma$ such that for all stationary $S \subseteq \omega_1$ and all $Z \subseteq \omega_1$, either $\mathcal{S}^\alpha_{S,Z}$ is stationary for all $\alpha \in D$ of countable cofinality, or else $\mathcal{S}^\alpha_{S,\omega_1 \setminus Z}$ is stationary for all $\alpha \in D$ of countable cofinality.

From (A) it follows — again since $\gamma > 2^{\aleph_1}$ — that for all stationary $S \subseteq \omega_1$ and all $Z \subseteq \omega_1$ there is some stationary $S' \subseteq S$ such that either $\{\alpha < \gamma : cf(\alpha) = \omega, \mathcal{S}^\alpha_{S',Z} \cup \mathcal{S}^\alpha_{\omega_1 \setminus S',\omega_1}$ includes a club of $[\alpha]^{\aleph_0}\}$ is a stationary subset of γ, or $\{\alpha < \gamma : cf(\alpha) = \omega, \mathcal{S}^\alpha_{S',\omega_1 \setminus Z} \cup \mathcal{S}^\alpha_{\omega_1 \setminus S',\omega_1}$ includes a club of $[\alpha]^{\aleph_0}\}$ is a stationary subset of γ.

Hence, for all stationary $S \subseteq \omega_1$ and all $Z \subseteq \omega_1$,

$(*)_{S,Z}$ there is some stationary $S' \subseteq S$ such that either

- $\mathcal{S}^\alpha_{S',Z} \cup \mathcal{S}^\alpha_{\omega_1 \setminus S',\omega_1}$ includes a club of $[\alpha]^{\aleph_0}$ for all $\alpha \in D$ of countable cofinality, or

- $\mathcal{S}^\alpha_{S',\omega_1 \setminus Z} \cup \mathcal{S}^\alpha_{\omega_1 \setminus S',\omega_1}$ includes a club of $[\alpha]^{\aleph_0}$ for all $\alpha \in D$ of countable cofinality.

As in the proof of Theorem 4.8, let $\bar{r} = (r_\xi)_{\xi < \omega_1}$ be a sequence of mutually distinct members of 2^ω. For every $n < \omega$ and every $\varepsilon < 2$ let $Z_{n,\varepsilon} = \{\xi < \omega_1 : r_\xi(n) = \varepsilon\}$. Fix $\alpha < \beta$ in D, $cf(\alpha) = cf(\beta) = \omega$. Now fix n and a stationary $\bar{S} \subseteq \omega_1$. Applying appropriate versions of $(*)_{S,Z}$ $n+1$ times, we obtain a stationary $S' \subseteq \overline{S}$ and a club $E \subseteq [\beta]^{\aleph_0}$ such that

$$\forall X \in E, X \cap \omega_1 \in S' \longrightarrow (\forall m \leq n)ot(X) \in Z_{m,0} \text{ iff } ot(X \cap \alpha) \in Z_{m,0},$$

i.e., $\Delta(r_X, r_{X \cap \alpha}) > n$.

Fix a partition $(S_n)_{n<\omega}$ of ω_1 into stationary sets. Pick a real $f \in \omega^\omega$. Then, from the considerations in the above paragraph it follows that

$$\mathcal{S}_f := \left\{ X \in [\beta]^{\aleph_0} : (\forall n)X \cap \omega_1 \in S_n \longrightarrow \Delta(r_X, r_{X \cap \alpha}) > f(n) \right\}$$

is a projective stationary subset of $[\beta]^{\aleph_0}$.

Hence, by applying BPSR to all $\mathcal{S}_f$'s we have that for every $f \in \omega^\omega$ there are ordinals $\delta'_f < \delta_f < \omega_2$ and a club $E_f \subseteq [\delta_f]^{\aleph_0}$ such that for every $X \in E_f$ and every n, if $X \cap \omega_1 \in S_n$, then $\Delta(r_X, r_{X \cap \delta'_f}) > f(n)$.

Given ordinals γ and γ' such that $\omega_1 \leq \gamma' < \gamma < \omega_2$, let $f_{\gamma,\gamma'}$ be the real defined as follows: Given a natural number n, $f_{\gamma,\gamma'}(n)$ is the least $m < \omega$ such that

$$\left\{ X \in [\gamma]^{\aleph_0} : X \cap \omega_1 \in S_n \text{ and } \Delta(r_X, r_{X \cap \gamma'}) = m \right\}$$

is a stationary subset of $[\gamma]^{\aleph_0}$.

Having fixed a Δ_1 enumeration $(\langle \gamma_\xi, \gamma'_\xi \rangle : \xi < \omega_2)$ of all pairs $\langle \gamma, \gamma' \rangle$ such that $\omega_1 \leq \gamma' < \gamma < \omega_2$, $\mathcal{D}$ will be an $<^*$–increasing sequence $\langle a_\xi : \xi < \omega_2 \rangle$, which will be defined by recursion as follows:

Take $\xi < \omega_2$ and suppose a_ζ has been defined for all $\zeta < \xi$. Then, $a_\xi = f_{\overline{\gamma},\overline{\gamma}'}$, where $\langle \overline{\gamma}, \overline{\gamma}' \rangle$ is the first pair $\langle \gamma, \gamma' \rangle$ such that $a_\zeta <^* f_{\gamma,\gamma'}$ for all $\zeta < \xi$ and $f_{\gamma_\xi,\gamma'_\xi} <^* f_{\gamma,\gamma'}$.

CLAIM 5.7. $\mathcal{D}$ is well defined and it is a scale.

PROOF. Since $\mathfrak{b} > \aleph_1$, for every $\xi < \omega_2$ there is some real f dominating $f_{\gamma_\xi,\gamma'_\xi}$ and all a_ζ $(\zeta < \xi)$. Then, $f(n) < f_{\delta_f,\delta'_f}(n)$ for all n. To see this, fix n. Since $A = \{ X \in [\delta_f]^{\aleph_0} : X \cap \omega_1 \in S_n \text{ and } \Delta(r_X, r_{X \cap \delta'_f}) = f_{\delta_f,\delta'_f}(n) \}$ is stationary, there is some $X \in E_f \cap A$. But then, $f(n) < f_{\delta_f,\delta'_f}(n)$. Hence, a_ξ is well defined. That $\mathcal{D}$ is a dominating family follows from its construction since every $f \in \omega^\omega$ is dominated by some $f_{\gamma,\gamma'}$. ⊣

Finally, since stationary subsets of $[\gamma]^{\aleph_0}$ and clubs of $[\gamma]^{\aleph_0}$ for $\gamma < \omega_2$ translate, respectively, into stationary subsets of ω_1 and clubs of ω_1, it is easy to see that $\mathcal{D}$ admits a Σ_1 definition over $(H_{\omega_2}, \in, NS_{\omega_1})$ with $(r_\xi)_{\xi<\omega_1}$ and $(S_n)_{n<\omega}$ as parameters.

If either (A) or (B) of Lemma 5.5 fail, then by that lemma there is an enumeration of ω^ω in length ω_2 which is Σ_1 definable over $(H_{\omega_2}, \in, NS_{\omega_1})$. Then, since $\mathfrak{b} = \aleph_2$, we may recursively build as in the paragraph before Claim 5.7, in a Σ_1 way over $(H_{\omega_2}, \in, NS_{\omega_1})$, an ω_2–sequence of reals whose range is dominating. ⊣

THEOREM 5.8 (Asperó). *Suppose* $(H_{\omega_2}, \in, NS_{\omega_1}) \preccurlyeq_{\Sigma_1} (H_{\omega_2}^{V^{\mathbb{P}}}, \in, NS_{\omega_1}^{V^{\mathbb{P}}})$ *for every stationary–set–preserving poset* $\mathbb{P}$. *Then* $2^{\aleph_0} = \aleph_2$.

PROOF. This follows closely the proof of Todorčević and Veličković that *PFA* implies $\mathfrak{b} = 2^{\aleph_0}$ (see [7, Theorem 3.16]). By Theorem 5.6, we may fix a scale — actually an $<^*$–unbounded sequence of reals suffices — $\mathcal{D} = \langle a_\xi : \xi < \omega_2 \rangle$ of order type ω_2. Moreover, we may assume that $\mathcal{D}$ is definable over $(H_{\omega_2}, \in, NS_{\omega_1})$ with a Σ_1 formula $\Phi(x, y)$, i.e., for all ξ and all r, $r = a_\xi$ if and only if $(H_{\omega_2}, \in, NS_{\omega_1}) \models \Phi(r, \xi)$.

I shall define a partition $O : [\omega_2]^2 \longrightarrow \omega$ such that given any real $f \in \omega^\omega$ we can find, using the stipulated generic absoluteness, an ordinal $\delta_f < \omega_2$ of uncountable cofinality which codes f, in the sense that there is a club $C \subseteq \delta_f$ of order type ω_1 and a decomposition $C = \bigcup_{i < \omega} X_i^n$ for each n such that $O(\xi' + n, \xi + n) = f(n)$ for all i and all $\xi' < \xi$ in X_i^n. The mapping $f \longrightarrow \delta_f$ will then be clearly one–to–one.

Let *osc* be Todorčević's *oscillating function*, i.e. *osc* is the mapping with domain $(\omega^{<\omega} \cup \omega^\omega) \times (\omega^{<\omega} \cup \omega^\omega)$ given by

$$osc(s, t) = \left| \left\{ n < \alpha : s(n) \le t(n) \text{ and } s(n + 1) > t(n + 1) \right\} \right|,$$

where $\alpha = \min\{|s|, |t|\} - 1$ if s or t are finite and $\alpha = \omega$ otherwise.

We also fix an enumeration $(\langle t_n, h_n \rangle : n < \omega)$ of all pairs $\langle t, h \rangle$ such that, for some $k < \omega$, $t \in (\omega^{<\omega})^k$ and $h : k \times k \longrightarrow \omega$, and such that the following holds: for all k, $t \in (\omega^{<\omega})^k$ and $h : k \times k \longrightarrow \omega$ there is some $l < \omega$ such that for all $i, j < k$, if $n = osc(t(i), t(j)) + l$, then $\langle t_n, h_n \rangle = \langle t, h \rangle$.

Pick $\xi' < \xi < \omega_2$. If there are no i, j such that $t_{osc(a_{\xi'}, a_\xi)}(i) \subseteq a_{\xi'}$ and $t_{osc(a_{\xi'}, a_\xi)}(j) \subseteq a_\xi$, then set $o(\xi', \xi) = 0$. Otherwise, let i_0, j_0 be minimal with these properties and set $o(\xi', \xi) = h_{osc(a_{\xi'}, a_\xi)}(i_0, j_0)$.

Now fix $f \in \omega^\omega$ and, working in $V^{Coll(\omega_1, \omega_2)}$, Let $C \subseteq \omega_2^V$ be a club of order type ω_1. The following claim is the key to proving that $\mathfrak{b} = 2^{\aleph_0}$ follows from *PFA*. It is proved using Todorčević's method of oscillating functions (see [30], Chapter 1 of [31], and also p. 51 of [9]).

CLAIM 5.9. Let $\mathbb{Q}_f$ be the finite support product of $\langle \mathbb{Q}_n^{f(n)} : n < \omega \rangle$, where $\mathbb{Q}_n^{f(n)} = (\overline{\mathbb{Q}}_n^{f(n)})^{<\omega}$ and $\overline{\mathbb{Q}}_n^{f(n)}$ is the set of all finite $f(n)$–homogeneous subsets of $\{\xi + n : \xi \in C\}$ under the coloring o ordered under inclusion. Then $\mathbb{Q}_f$ is ccc.

Now, in $V^{Coll(\omega_1, \omega_2) * \dot{\mathbb{Q}}_f}$ the following holds:

($*$) There is a club C of order type ω_1 of some ordinal $\delta < \omega_2$ and a sequence $\Sigma = \langle r_\xi : \xi < \delta \rangle$ such that $\Phi(r_\xi, \xi)$ for all $\xi < \delta$ and such that for every $n < \omega$ there is a decomposition $C = \bigcup_{i < \omega} X_i^n$ of C with the property that, for all i and all $\xi' < \xi$ in X_i^n, $o'(\xi' + n, \xi + n) = f(n)$ (where o' is defined as o with Σ instead of $\mathcal{D}$).

By our choice of Φ and since o is Δ_1 definable with $\mathcal{D}$ and $(\langle t_n, h_n \rangle : n < \omega)$ as parameters, $(*)$ can be expressed over $(H_{\omega_2}, \in, NS_{\omega_1})$ by means of a Σ_1 formula with parameters in H_{ω_2}. Since $\mathrm{Coll}(\omega_1, \omega_2) * \dot{\mathbb{Q}}_f$ is σ–closed $*$ ccc, $(*)$ holds. But then, since Φ defines $\mathcal{D}$, $\langle r_\xi : \xi < \delta \rangle = \mathcal{D} \upharpoonright \delta$ and $o' = o \upharpoonright [\delta]^2$. $\dashv$

Theorem 5.8 was originally stated in terms of BMM_2'. It was then pointed out by Todorčević that this principle can be replaced by the more natural generic absoluteness stipulated in the statement presented here.

The final development in this story is Todorčević's result that BMM alone implies $2^{\aleph_0} = \aleph_2$, which he obtained in April 2002. I will only give the main points in his argument, referring the reader to [32] for the missing details.

Using the Erdős–Rado partition relation $(2^{\aleph_1})^+ \longrightarrow (\omega_1)^2_{2^{\aleph_1}}$, Todorčević proved the existence, in ZFC, of a certain pair (α, β) of "canonical ordinals" below $(2^{\aleph_1})^+$. This completely general fact is the crucial tool in [32]. Specifically, he proved the following (remember the notation introduced in Section 3).

THEOREM 5.10 (Todorčević). *There are ordinals $\bar{\alpha} < \bar{\beta} < (2^{\aleph_1})^+$ such that for every sequence $(r_\xi)_{\xi < \omega_1}$ of mutually distinct elements of 2^ω and every stationary $S \subseteq \omega_1$, both*

$$\left\{ X \in \left[(2^{\aleph_1})^+ \right]^{\aleph_0} : X \cap \omega_1 \in S \text{ and } \Delta\big(r_{X \cap \bar{\alpha}}, r_{X \cap \bar{\beta}}\big) = \max \Delta\big(r_{X \cap \bar{\alpha}}, r_{X \cap \bar{\beta}}, r_X\big) \right\}$$

and

$$\left\{ X \in \left[(2^{\aleph_1})^+ \right]^{\aleph_0} : X \cap \omega_1 \in S \text{ and } \Delta\big(r_{X \cap \bar{\alpha}}, r_{X \cap \bar{\beta}}\big) = \min \Delta\big(r_{X \cap \bar{\alpha}}, r_{X \cap \bar{\beta}}, r_X\big) \right\}$$

are stationary subsets of $[(2^{\aleph_1})^+]^{\aleph_0}$.

From this, it immediately follows that for every pair of ordinals $\bar{\alpha} < \bar{\beta}$ as given by the theorem, every sequence $(r_\xi)_{\xi < \omega_1}$ of pairwise distinct elements of 2^ω and every set $Y \subseteq \omega_1$,

$$\left\{ X \in \left[(2^{\aleph_1})^+ \right]^{\aleph_0} : X \cap \omega_1 \in Y \text{ iff } \Delta\big(r_{X \cap \bar{\alpha}}, r_{X \cap \bar{\beta}}\big) = \max \Delta\big(r_{X \cap \bar{\alpha}}, r_{X \cap \bar{\beta}}, r_X\big) \right\}$$

is a projective stationary subset of $[(2^{\aleph_1})^+]^{\aleph_0}$.

Now, a standard application of $BPSR$ to the above set gives the following.

THEOREM 5.11 (Todorčević). *BMM implies θ_{AC}. Thus, in particular it implies that there is a well–order of the reals in length ω_2 which is Δ_1 definable over $(H_{\omega_2}, \in, NS_{\omega_1})$ with parameters.*

REFERENCES

[1] D. ASPERÓ, *Bounded Forcing Axioms and the Continuum*, Ph.D. thesis, U. Barcelona, Barcelona, 2000.

[2] ———, *A maximal bounded forcing axiom*, *The Journal of Symbolic Logic*, vol. 67 (2002), no. 1, pp. 130–142.

[3] ———, *Bounded forcing axioms and the size of the continuum*, *Logic Colloquium 2000*, Lecture Notes in Logic, vol. 19, ASL, Urbana, IL, 2005, pp. 211–227.

[4] ———, *On a convenient property for* $[\gamma]^{\aleph_0}$, Submitted.

[5] D. ASPERÓ and J. BAGARIA, *Bounded forcing axioms and the continuum*, **Annals of Pure and Applied Logic**, vol. 109 (2001), no. 3, pp. 179–203.

[6] D. ASPERÓ and P. WELCH, *Bounded Martin's maximum, weak Erdős cardinals, and* ψ_{AC}, **The Journal of Symbolic Logic**, vol. 67 (2002), no. 3, pp. 1141–1152.

[7] J. BAGARIA, *Bounded forcing axioms as principles of generic absoluteness*, **Archive for Mathematical Logic**, vol. 39 (2000), no. 6, pp. 393–401.

[8] J. BAUMGARTNER and A. TAYLOR, *Saturation properties of ideals in generic extensions. I*, **Transactions of the American Mathematical Society**, vol. 270 (1982), no. 2, pp. 557–574.

[9] M. BEKKALI, *Topics in Set Theory*, Lecture Notes in Mathematics, vol. 1476, Springer-Verlag, Berlin, 1991.

[10] O. DEISER and H. D. DONDER, *Canonical functions, non-regular ultrafilters and Ulam's problem on* ω_1, **The Journal of Symbolic Logic**, vol. 68 (2003), no. 3, pp. 713–739.

[11] K. J. DEVLIN and S. SHELAH, *A weak version of* $\Diamond$ *which follows from* $2^{\aleph_0} < 2^{\aleph_1}$, **Israel Journal of Mathematics**, vol. 29 (1978), no. 2-3, pp. 239–247.

[12] Q. FENG and T. JECH, *Projective stationary sets and a strong reflection principle*, **Journal of the London Mathematical Society. Second Series**, vol. 58 (1998), no. 2, pp. 271–283.

[13] M. FOREMAN, M. MAGIDOR, and S. SHELAH, *Martin's maximum, saturated ideals, and nonregular ultrafilters. I*, **Annals of Mathematics. Second Series**, vol. 127 (1988), no. 1, pp. 1–47.

[14] S. FUCHINO, *On potential embedding and versions of Martin's axiom*, **Notre Dame Journal of Formal Logic**, vol. 33 (1992), no. 4, pp. 481–492.

[15] F. GALVIN, T. JECH, and M. MAGIDOR, *An ideal game*, **The Journal of Symbolic Logic**, vol. 43 (1978), no. 2, pp. 284–292.

[16] M. GITIK, *Nonsplitting subset of* $\mathcal{P}_\kappa(\kappa^+)$, **The Journal of Symbolic Logic**, vol. 50 (1985), no. 4, pp. 881–894.

[17] M. GOLDSTERN and S. SHELAH, *The bounded proper forcing axiom*, **The Journal of Symbolic Logic**, vol. 60 (1995), no. 1, pp. 58–73.

[18] J. D. HAMKINS, *A simple maximality principle*, **The Journal of Symbolic Logic**, vol. 68 (2003), no. 2, pp. 527–550.

[19] T. JECH, *Set Theory*, Perspectives in Mathematical Logic, Springer-Verlag, Berlin, 1997.

[20] T. JECH and S. SHELAH, *A note on canonical functions*, **Israel Journal of Mathematics**, vol. 68 (1989), no. 3, pp. 376–380.

[21] D. KUEKER, *Countable approximations and Löwenheim-Skolem theorems*, **Annals of Pure and Applied Logic**, vol. 11 (1977), no. 1, pp. 57–103.

[22] K. KUNEN, *Set Theory: An Introduction to Independence Proofs*, Studies in Logic and the Foundations of Mathematics, vol. 102, North-Holland Publishing Co., Amsterdam, 1980.

[23] P. LARSON, *The size of* $\tilde{T}$, **Archive for Mathematical Logic**, vol. 39 (2000), no. 7, pp. 541–568.

[24] D. MARTIN and R. SOLOVAY, *Internal Cohen extensions*, **Annals of Pure and Applied Logic**, vol. 2 (1970), no. 2, pp. 143–178.

[25] T. MIYAMOTO, *A note on weak segments of PFA*, **Proceedings of the Sixth Asian Logic Conference (Beijing, 1996)**, World Sci. Publishing, River Edge, NJ, 1998, pp. 175–197.

[26] R. SCHINDLER, *Coding into K by reasonable forcing*, **Transactions of the American Mathematical Society**, vol. 353 (2001), no. 2, pp. 479–489.

[27] S. SHELAH, *Semiproper forcing axiom implies Martin maximum but not PFA$^+$*, **The Journal of Symbolic Logic**, vol. 52 (1987), no. 2, pp. 360–367.

[28] ———, *Forcing axiom failure for* $\lambda > \aleph_1$, [Sh 784], Available at http://arxiv.org/abs/math.LO/0112286.

[29] J. STAVI and J. VÄÄNÄNEN, *Reflection principles for the continuum*, **Logic and Algebra**, Contemporary Mathematics, vol. 302, AMS, Providence, RI, 2002, pp. 59–84.

[30] S. Todorčević, *Oscillations of real numbers*, **Logic Colloquium '86 (Hull, 1986)**, Studies in Logic and the Foundations of Mathematics, vol. 124, North-Holland, Amsterdam, 1988, pp. 325–331.

[31] ———, **Partition Problems in Topology**, Contemporary Mathematics, vol. 84, American Mathematical Society, Providence, RI, 1989.

[32] ———, *Generic absoluteness and the continuum*, **Mathematical Research Letters**, vol. 9 (2002), no. 4, pp. 465–471.

[33] ———, *Localized reflection and fragments of PFA*, **Set Theory (Piscataway, NJ, 1999)**, DIMACS Series, vol. 58, AMS, Providence, RI, 2002, pp. 135–148.

[34] B. Veličković, *Forcing axioms and stationary sets*, **Advances in Mathematics**, vol. 94 (1992), no. 2, pp. 256–284.

[35] P. D. Welch, *On unfoldable cardinals, ω-closed cardinals, and the beginning of the inner model hierarchy*, **Archive for Mathematical Logic**, vol. 43 (2004), no. 4, pp. 443–458.

[36] H. Woodin, **The Axiom of Determinacy, Forcing Axioms, and the Nonstationary Ideal**, de Gruyter Series in Logic and its Applications, vol. 1, Walter de Gruyter & Co., Berlin, 1999.

UNIVERSITY OF BRISTOL
DEPARTMENT OF MATHEMATICS
UNIVERSITY WALK
BRISTOL, BS8 1TW, UK
E-mail: David.Aspero@bris.ac.uk

AXIOMS OF GENERIC ABSOLUTENESS

JOAN BAGARIA

Abstract. We give a unified presentation of the set-theoretic axioms of generic absoluteness, we survey the known results regarding their consistency strength and some of their consequences as well as their relationship to other kinds of set-theoretic axioms, and we provide a list of the main open problems.

§1. Introduction. A common theme of Set Theory, after the discovery by Cohen in 1963 of the method of *forcing* for building models of ZFC, has been *how to get rid of forcing*. The forcing technique has proved to be, over the last 40 years, an extremely powerful and flexible tool for building models of ZFC with the most varied properties, thereby proving the independence from the ZFC axioms of a large amount of mathematical statements. Spurred by this array of independence results, it has become a challenge for Set Theory to discover new axioms that would eliminate the relativity of the truth of mathematical statements with respect to different models of ZFC obtained by forcing. The ultimate axiom of this sort would be to require that V is elementary equivalent to $V^{\mathbb{P}}$ for every forcing notion $\mathbb{P}$. But this is clearly impossible, since we can force incompatible statements, for instance, the Continuum Hypothesis and its negation. So, one has to restrict either the class of statements that are absolute between V and its generic extensions, or the class of forcing extensions, or both. But how do we choose between two forceable statements that cannot be both simultaneously true? The answer is that we do not choose, that is, we do not discriminate against statements of the same logical complexity. We want all forceable statements of the same logical complexity to be true. This is always the case for Σ_1 statements, i.e., formulas of the language of Set Theory which, in normal form, have only existential quantifiers. But for formulas of

2000 *Mathematics Subject Classification.* 03E65, 03E35, 03E15, 03E50, 03E55.
Key words and phrases. Generic absoluteness; Axioms of Set Theory; Forcing; Projective posets; Bounded forcing axioms.
Research partially supported by the research project BFM2002-03236 of the Spanish Ministry of Science and Technology, and by the research project 2002SGR 00126 of the Departament d'Universitats, Recerca i Societat de la Informació of the Generalitat de Catalunya.

Logic Colloquium '02
Edited by Z. Chatzidakis, P. Koepke, and W. Pohlers
Lecture Notes in Logic, 27
© 2006, ASSOCIATION FOR SYMBOLIC LOGIC

28

the next higher level of complexity, the Σ_2, requiring that all those that are forceable are true, is false (both the Continuum Hypothesis and its negation are Σ_2). Thus, we must abandon the ultimate axiom and settle for weaker forms. Namely, we may require that some definable subclass W of V, which is seen as an approximation to V, is sufficiently elementarily equivalent to W as computed in some generic extensions of V. Whenever W is a substructure of W as computed in the generic extensions, then one can even require that W be an elementary substructure of the generic W. That is, we may have elements of the ground-model W as parameters. Many axioms of this sort have been studied in the literature and are currently one of the main topics in the foundations of Set Theory. These are the axioms of *generic absoluteness*, which may also be called axioms of *generic invariance*, that is, they assert that whatever can be forced is true, subject only to the restrictions that are strictly necessary for them to be consistent with the axioms of ZFC.

We give here a unified presentation of the axioms of generic absoluteness. We survey the known results regarding their consistency strength and their relationship to the axioms of large cardinals and the bounded forcing axioms. We illustrate some of their consequences, specially in Descriptive Set Theory, and we provide a list of the main open questions.

Our notation and basic definitions are standard, as in [25]. V is the class of all sets. If $\mathbb{P}$ is a forcing notion, i.e., a partial ordering, in V, then $V^{\mathbb{P}}$ is the Boolean-valued model corresponding to the Boolean completion of $\mathbb{P}$. In this survey we will only consider forcing notions that are *sets*, as opposed to *class* forcing notions. A formula of the first-order language of set theory is Σ_0 if all its quantifiers are bounded. For $n > 0$, a formula is Σ_n if, in its normal form, begins with a sequence of n blocs of unbounded quantifiers of the same kind, starting with a block of existential quantifiers, and followed by a Σ_0 formula. L is the constructible universe. For an infinite cardinal κ, $H(\kappa)$ is the set of all sets whose transitive closure has cardinality $< \kappa$. For a transitive set X, $L(X)$ is the smallest transitive model of ZF that contains all the ordinals and X. $L(\mathbb{R})$ is the smallest transitive model of ZF that contains all ordinals and all the reals. If x is a real number, $L[x]$ is the least transitive model of ZFC that contains all the ordinals and x.

I want to thank David Asperó, Sy Friedman, Ralf Schindler, and Hugh Woodin for their comments on the first versions of this paper. I am also grateful to S. Friedman for sending me some of his unpublished work on these topics. Finally, I thank the members of the Organizing and Scientific Committee of the Colloquium Logicum 2002 for inviting me to Münster.

§2. A general framework for axioms of generic absoluteness. An axiom of generic absoluteness is given by requiring that some definable subclass W of V is sufficiently elementarily equivalent to W, as computed in certain generic extensions of V. That is, for some class of sentences Φ and all forcing notions

$\mathbb{P}$ belonging to some class of forcing notions Γ,

$$W^V \equiv_\Phi W^{V^{\mathbb{P}}}$$

read: W^V is Φ-elementarily equivalent to $W^{V^{\mathbb{P}}}$. Namely, for all $\varphi \in \Phi$ and all $\mathbb{P} \in \Gamma$,

$$W^V \models \varphi \text{ iff } W^{V^{\mathbb{P}}} \models \varphi.$$

Thus, we have three variables:

- The subclass W of V.
- The class Φ of sentences.
- The class Γ of forcing notions.

By combining these three variables we obtain the different sorts of axioms of generic absoluteness. In general, the larger the classes W, Φ, and Γ, the stronger the axiom. But there is always a trade-off between them. For instance, as we will see, in the extreme case when one of the three variables is maximal, e.g., when $W = V$, Φ is the class of all sentences, or Γ is the class of all forcing notions, then the other two must be very small for the axiom to be consistent with ZFC.

To unify our presentation we will use the following notation for axioms of generic absoluteness: $\mathcal{A}(W, \Phi, \Gamma)$ is the assertion that W is Φ-elementarily equivalent to $W^{V^{\mathbb{P}}}$ for all $\mathbb{P} \in \Gamma$.

Since we want $\mathcal{A}(W, \Phi, \Gamma)$ to be an axiom, hence a sentence in the first-order language of Set Theory, W, Φ, and Γ must be definable classes.

Notice that if $\Phi \subseteq \Phi'$ and $\Gamma \subseteq \Gamma'$, then $\mathcal{A}(W, \Phi', \Gamma')$ implies $\mathcal{A}(W, \Phi, \Gamma)$. Also notice that $\mathcal{A}(W, \Phi, \Gamma)$ is equivalent to $\mathcal{A}(W, \bar{\Phi}, \Gamma)$, where $\bar{\Phi}$ is the closure of Φ under finite Boolean combinations. Thus, for instance, $\mathcal{A}(W, \Sigma_n, \Gamma)$ is equivalent to $\mathcal{A}(W, \Pi_n, \Gamma)$.

An important case is when Φ is a class of sentences with parameters. In this case, for $\mathcal{A}(W, \Phi, \Gamma)$ to make sense, the parameters must belong to $W \cap W^{V^{\mathbb{P}}}$, for all $\mathbb{P} \in \Gamma$. In particular, if $\Phi = \Sigma_n(W)$, i.e., Φ is the class of all Σ_n sentences with parameters from W, and $W \subseteq W^{V^{\mathbb{P}}}$ for all $\mathbb{P} \in \Gamma$, then $\mathcal{A}(W, \Phi, \Gamma)$ is just the assertion that W is a Σ_n-elementary substructure of $W^{V^{\mathbb{P}}}$, for all $\mathbb{P} \in \Gamma$. As is customary, instead of $\Sigma_n(W)$ we will write $\underset{\sim}{\Sigma}_n$. If the class of parameters is some X properly contained in W, then we write $\Sigma_n(X)$ for the class of Σ_n sentences with parameters from X.

We write Σ_ω for the class of all sentences of the first-order language of set theory, and $\underset{\sim}{\Sigma}_\omega$ for the class of all such sentences with parameters in W. If $\Phi = \Sigma_\omega$, then $\mathcal{A}(W, \Phi, \Gamma)$ should be regarded as an axiom schema, that is, as $\mathcal{A}(W, \Sigma_n, \Gamma)$ for each $n \in \omega$. Similarly if $\Phi = \underset{\sim}{\Sigma}_\omega$.

If Γ contains only one element, $\mathbb{P}$, then we will write $\mathcal{A}(W, \Phi, \mathbb{P})$, instead of $\mathcal{A}(W, \Phi, \Gamma)$. If Γ is the class of all set-forcing notions, then we just write $\mathcal{A}(W, \Phi)$.

§3. Some trivial cases. Let us first consider some forms of $\mathcal{A}(W, \Phi, \Gamma)$ that are either provable or refutable in ZFC. So, as axioms, they are of no interest, since they are either trivial or false.

3.1. $\Phi \subseteq \Sigma_0$: In this case, $\mathcal{A}(W, \Phi, \Gamma)$ holds for all transitive W and all Γ such that $W^{\widetilde{V}}$ is contained in $W^{V^{\mathbb{P}}}$ for all $\mathbb{P} \in \Gamma$.

3.2. $\Phi \subseteq \Sigma_1(H(\omega_1))$: By the Levy-Shoenfield absoluteness theorem (see [25]), we have that $\mathcal{A}(W, \Phi, \Gamma)$ holds for every transitive model W of a weak fragment of ZF that contains the parameters of Φ, and all Γ, provided W^V is contained in $W^{V^{\mathbb{P}}}$ for all $\mathbb{P} \in \Gamma$. In particular, $\mathcal{A}(H(\omega_1), \underset{\sim}{\Sigma_1})$, $\mathcal{A}(H(\kappa), \Sigma_1(H(\omega_1)))$, $\kappa > \omega_1$, and $\mathcal{A}(V, \Sigma_1(H(\omega_1)))$ all hold.

3.3. $W = V$: A proper class W, different from V, can never be an elementary substructure of V, since otherwise, by elementarity, for every ordinal α, $V_\alpha^W = V_\alpha$, and so $W = V$. In particular, $\mathcal{A}(W, \underset{\sim}{\Sigma_1}, \{\mathbb{P}\})$ fails for any nontrivial $\mathbb{P}$, i.e., any $\mathbb{P}$ that adds some new set.

3.4. $W = L$: For every forcing notion $\mathbb{P}$, $L^V = L^{V^{\mathbb{P}}}$. Hence, in this case $\mathcal{A}(W, \Phi)$ holds for all Φ.

3.5. $W = H(\omega)$: For every forcing notion $\mathbb{P}$, $H(\omega)^V = H(\omega)^{V^{\mathbb{P}}}$. Hence, $\mathcal{A}(W, \Phi)$ is true for all Φ.

§4. A natural class of axioms. A family of natural axioms of generic absoluteness is obtained when:

(1) $W = H(\kappa)$ or $W = L(H(\kappa))$, for some definable uncountable cardinal κ;

(2) Φ is the class of Σ_n sentences, some $n \in \omega$, or the class Σ_ω of all sentences, with or without parameters from W; and

(3) Γ is a definable class of posets which contains at least a non-trivial element.

Why the $H(\kappa)$ and not the V_α, α an ordinal? On the one hand, for a regular cardinal κ, $H(\kappa)$ is a model of ZF minus the Power-set axiom, and so it satisfies Replacement, thus being a better model in the sense of forcing. For instance, if $\mathbb{P} \in H(\kappa)$, then, a filter $G \subseteq \mathbb{P}$ is generic over V iff it is generic over $H(\kappa)$. Moreover, for $\Phi = \underset{\sim}{\Sigma_1}$, which, as we will see, is one of the most relevant cases, we have that if $\kappa < \lambda$, then $\mathcal{A}(H(\lambda), \underset{\sim}{\Sigma_1}, \Gamma)$ implies $\mathcal{A}(H(\kappa), \underset{\sim}{\Sigma_1}, \Gamma)$.

Of particular interest, besides the class of all posets, are the classes of posets that have been extensively studied in the literature and form an increasing chain, namely, the ccc posets, the proper and semi-proper posets, the posets that preserve stationary subsets of ω_1, and the posets that preserve ω_1 (see [25] for the definitions). To make the notation more readable, we shall write *ccc* for the class of ccc posets, *Proper* for the class of proper posets, *Semi-proper* for the class of semi-proper posets, *Stat-pres* for the class of posets that preserve stationary subsets of ω_1, and *ω_1-pres* for the class of posets that preserve ω_1. We have:

$$ccc \subset Proper \subset Semi\text{-}proper \subset Stat\text{-}pres \subset \omega_1\text{-}pres.$$

Of interest are also some subclasses of the ccc posets, namely, the classes of posets which are σ-centered (a poset is σ-centered if it can be partitioned into countably many classes so that each class is finite-wise compatible), the σ-linked (a poset is σ-linked if it can be partitioned into countably many classes so that each class is pair-wise compatible), the Knaster (a poset has the Knaster property if every uncountable set contains an uncountable subset of pair-wise compatible elements), and the Productive-ccc (a poset is productive-ccc if its product with any ccc poset is also ccc). We will write *Prod-ccc* and *Knaster* for the classes of productive-ccc posets and Knaster posets, respectively. These classes also form a chain:

$$\sigma\text{-}centered \subset \sigma\text{-}linked \subset Knaster \subset Prod\text{-}ccc \subset ccc.$$

Another classification of forcing notions is obtained with regard to their definability. Both interesting and natural are the axioms $\mathcal{A}(H(\kappa), \Phi, \Gamma)$, where Γ consists of those posets in one of the classes above that are definable (Σ_n or Π_n definable, some n) in $H(\kappa)$, with or without parameters. An important example is the class of projective posets, namely, those definable in $H(\omega_1)$ with parameters. As usual, that a poset is, say, Σ_n-definable in $H(\omega_1)$, means that the set, the ordering relation, and the incompatibility relation are all Σ_n-definable in $H(\omega_1)$. Similarly for the Π_n-definable posets.

Of special interest are the axioms of the form $\mathcal{A}(H(\omega_2), \Phi, \Gamma)$, both for their consequences and for being equivalent to the Bounded Forcing Axioms (see 6.4 below): Given a partial ordering $\mathbb{P}$, the *Bounded Forcing Axiom for $\mathbb{P}$*, in short $BFA(\mathbb{P})$, is the following statement:

For every collection $\{I_\alpha : \alpha < \omega_1\}$ of maximal antichains of $\mathbf{B} =_{df} r.o.(\mathbb{P}) \setminus \{0\}$, each of size $\leq \omega_1$, there exists a filter $G \subseteq \mathbf{B}$ such that for every α, $I_\alpha \cap G \neq \emptyset$.

For a class of posets Γ, $BFA(\Gamma)$ is the statement that for every $\mathbb{P} \in \Gamma$, $BFA(\mathbb{P})$.

MA_{ω_1}, Martin's axiom for ω_1, is $BFA(ccc)$. BPFA, the bounded proper forcing axiom, is $BFA(Proper)$. BSPFA, the bounded semi-proper forcing axiom, is $BFA(Semi\text{-}proper)$. Finally, BMM, the bounded Martin's maximum, is $BFA(Stat\text{-}pres)$.

Thus, in this paper we will concentrate on axioms of the form $\mathcal{A}(W, \Phi, \Gamma)$, where $W = H(\kappa)$ or $L(H(\kappa))$, where $\Phi = \Sigma_n, \underset{\sim}{\Sigma}_n, \Sigma_\omega$, or $\underset{\sim}{\Sigma}_\omega$, and where Γ is one of the classes of forcing notions considered above.

We shall begin by looking at the smallest non-trivial W, namely, $H(\omega_1)$.

§5. $W = H(\omega_1)$. For every forcing notion $\mathbb{P}$, $H(\omega_1) \subseteq H(\omega_1)^{V^{\mathbb{P}}}$. Hence, from 3.2 above, the first interesting case is when $\Phi = \Sigma_2$.

5.1. $\Phi = \Sigma_2$: $\mathcal{A}(H(\omega_1), \Sigma_2)$ fails in L, since saying that there exists a non-constructible real is Σ_2, hence it is not provable in ZFC. However,

THEOREM 5.1. [29, 41] *If $X^\sharp$ exists for every set X, then $\mathcal{A}(H(\omega_1), \underset{\sim}{\Sigma}_2)$ holds.*

The following theorem follows from a result of Feng-Magidor-Woodin [17], and was proved independently by S. Friedman (see [13]).

THEOREM 5.2.
1. $\mathcal{A}(H(\omega_1), \Sigma_2)$ *is equiconsistent with ZFC.*
2. $\mathcal{A}(H(\omega_1), \underset{\sim}{\Sigma}_2)$ *is equiconsistent with the existence of a Σ_2-reflecting cardinal. i.e., a regular cardinal κ such that $V_\kappa \preccurlyeq_{\Sigma_2} V$.*

The proof of the Theorem above actually shows that $\mathcal{A}(H(\omega_1), \underset{\sim}{\Sigma}_2)$ implies that ω_1 is a Σ_2-reflecting cardinal in $L[x]$, for every real x. This has recently been improved by S. Friedman:

THEOREM 5.3. [18] $\mathcal{A}(H(\omega_1), \Sigma_2, Stat\text{-}pres)$ *implies that ω_1 is a Σ_2-reflecting cardinal in $L[x]$, for every real x. Hence, by 5.2, $\mathcal{A}(H(\omega_1), \underset{\sim}{\Sigma}_2, Stat\text{-}pres)$ is equiconsistent with the existence of a Σ_2-reflecting cardinal.*

In particular, Friedman shows, using R. Schindler's *faster reshaping forcing* that $\mathcal{A}(H(\omega_1), \underset{\sim}{\Sigma}_2, Stat\text{-}pres)$ implies that ω_1 is inaccessible in $L[x]$, for every real x.

It follows immediately from [19] and [9] that $\mathcal{A}(H(\omega_1), \Sigma_2, Semi\text{-}proper)$ does not imply that ω_1^L is countable. Actually, as S. Friedman [18] has recently observed, $\mathcal{A}(H(\omega_1), \underset{\sim}{\Sigma}_2, Semi\text{-}proper)$ is equiconsistent with ZFC. (This contradicts the statement of Theorem 2 of [13], the proof of which actually shows that if $\mathcal{A}(H(\omega_1), \underset{\sim}{\Sigma}_2, Proper)$ holds after forcing with a certain proper poset, then either ω_1 is Mahlo in L or ω_2 is inaccessible in L.)

As we will see in Section 6 below, Bounded Forcing Axioms are actually equivalent to axioms of generic $\underset{\sim}{\Sigma}_1$-absoluteness for $H(\omega_2)$ [8, 9]. But they also imply generic $\underset{\sim}{\Sigma}_2$-absoluteness for $H(\omega_1)$. Namely,

THEOREM 5.4. [8, 32] MA_{ω_1} *implies $\mathcal{A}(H(\omega_1), \underset{\sim}{\Sigma}_2, ccc)$.*

More generally,

THEOREM 5.5. [8]
1. *BPFA implies $\mathcal{A}(H(\omega_1), \underset{\sim}{\Sigma}_2, Proper)$.*
2. *BSPFA implies $\mathcal{A}(H(\omega_1), \underset{\sim}{\Sigma}_2, Semi\text{-}proper)$.*
3. *BMM implies $\mathcal{A}(H(\omega_1), \underset{\sim}{\Sigma}_2, Stat\text{-}pres)$.*

That the last four implications cannot be reversed can be easily seen, as S. Friedman has pointed out, by noticing that all axioms of the form $\mathcal{A}(H(\omega_1), \underset{\sim}{\Sigma}_n, \Gamma)$ are preserved after collapsing the continuum to ω_1 by σ-closed forcing. Hence, they are all consistent with CH, and so they do not imply any of the bounded forcing axioms.

It is also worth mentioning the following surprising result of S. Friedman:

THEOREM 5.6. [18] *Let θ be the statement that every subset of ω_1 is constructible from a real, that is, for every $X \subseteq \omega_1$ there exists $x \subseteq \omega$ with $X \in L[x]$.*
Suppose that ω_1 is not weakly-compact in $L[x]$ for some $x \subseteq \omega$. Then,

1. MA_{ω_1} *is equivalent to* $\mathcal{A}(H(\omega_1), \Sigma_2, ccc)$ *plus* θ.
2. *BPFA is equivalent to* $\mathcal{A}(H(\omega_1), \Sigma_2, Proper)$ *plus* θ.
3. *BSPFA is equivalent to* $\mathcal{A}(H(\omega_1), \Sigma_2, Semi\text{-}proper)$ *plus* θ.

Let us point out that while BSPFA is consistent with $\omega_1^L = \omega_1$ [19], BMM implies that ω_1 is weakly-compact in $L[x]$ for every $x \subseteq \omega$ (see Theorem 6.6 below).

For many of the ccc and proper forcing notions $\mathbb{P}$ associated to regularity properties of sets of reals, like Lebesgue measurability, the Baire property, or the Ramsey property, there are interesting characterizations of the axioms $\mathcal{A}(H(\omega_1), \Sigma_2, \mathbb{P})$ in terms of these regularity properties holding for the projective classes of Δ_2^1 and Σ_2^1 sets. Analogous characterizations also hold for the axioms $\mathcal{A}(H(\omega_1), \Sigma_2, \widetilde{\mathbb{P}})$ and the Kleene classes Δ_2^1 and Σ_2^1. We state here only the parametrized forms. For instance, from [7] we have:

Let *Cohen* be the poset for adding a Cohen real, and let *Random* be the poset for adding a random real. Then,

- $\mathcal{A}(H(\omega_1), \Sigma_2, Cohen)$ is equivalent to the statement that every Δ_2^1 set of reals has the property of Baire.
- $\mathcal{A}(H(\omega_1), \Sigma_2, Random)$ is equivalent to the statement that every Δ_2^1 set of reals is Lebesgue measurable.

The following are due to H. Judah [14, 27]: Let *Amoeba* be the amoeba poset for measure, let *Amoeba-category* be the amoeba poset for category, and let *Hechler* be the Hechler forcing for adding a dominating real. Then

- $\mathcal{A}(H(\omega_1), \Sigma_2, Hechler)$ and $\mathcal{A}(H(\omega_1), \Sigma_2, Amoeba\text{-}category)$ are both equivalent to the statement that every Σ_2^1 set of reals has the property of Baire.
- $\mathcal{A}(H(\omega_1), \Sigma_2, Amoeba)$ is equivalent to the statement that every Σ_2^1 set of reals is Lebesgue measurable.

Similar results for the Mathias forcing were obtained by Halbeisen-Judah [20]. If *Mathias* is the Mathias forcing, then:

- $\mathcal{A}(H(\omega_1), \Sigma_2, Mathias)$ is equivalent to the statement that every Σ_2^1 set of reals is Ramsey.

It should be noted that all these forcing notions are proper and Borel, i.e., the set of conditions is a Borel set, and both the ordering and the incompatibility relation are Borel subsets of the plane. The following is a long-standing open question:

QUESTION 5.7. Does $\mathcal{A}(H(\omega_1), \underset{\sim}{\Sigma}_\omega, \Gamma)$, for Γ the class of Borel ccc forcing notions, imply that every projective set of real numbers is Lebesgue measurable?

The most general result along these lines is the following theorem of Feng-Magidor-Woodin:

THEOREM 5.8. [17] $\mathcal{A}(H(\omega_1), \underset{\sim}{\Sigma}_2)$ *is equivalent to the statement that every* $\underset{\sim}{\Delta}^1_2$ *set of reals is universally Baire.*

Let us now consider the next level of complexity of Φ.

5.2. $\Phi = \Sigma_3$: We start with the following version of a result from [7], which uses a Lemma of H. Woodin from [41] to the effect that if X is an uncountable sequence of reals in V and c is Cohen-generic over V, then in $V[c]$ there is no random real over $L(X, c)$.

THEOREM 5.9. *Let ω_1-Random be the σ-linked forcing notion for adding ω_1 random reals.*

1. $\mathcal{A}(H(\omega_1), \Sigma_3, \{\omega_1\text{-}Random, Cohen\})$ *implies that ω_1 is inaccessible in* $L[x]$, *for every real x.*
2. $\mathcal{A}(H(\omega_1), \Sigma_2, Random)$ *plus* $\mathcal{A}(H(\omega_1), \Sigma_3, Cohen)$ *imply that ω_1 is inaccessible in $\widetilde{L}[x]$, for every real x.*

The point (for (1)) is that if $\omega_1^{L[x]} = \omega_1$, then we may add ω_1 random reals so that, in the forcing extension, for every real y there is a random real over $L(x, y)$. By $\mathcal{A}(H(\omega_1), \Sigma_3, \omega_1\text{-}Random)$ this holds in V, and by $\mathcal{A}(H(\omega_1), \Sigma_3, Cohen)$ it holds after adding a Cohen real c. So, in $V[c]$, we have that there is a random real over $L(x, c)$, contradicting the aforementioned result of Woodin.

The following results give the exact consistency strengths.

THEOREM 5.10. [17] *The following are equiconsistent:*

1. $\mathcal{A}(H(\omega_1), \Sigma_3)$.
2. *Every set has a sharp.*

THEOREM 5.11. [13] *The following are equiconsistent:*

1. $\mathcal{A}(H(\omega_1), \underset{\sim}{\Sigma}_3)$.
2. *Every set has a sharp and there exists a Σ_2-reflecting cardinal.*

R. Schindler [33] and, independently, S. Friedman [18], have shown that if for some set X, $X^\sharp$ does not exist, then there is a forcing notion that preserves stationary subsets of ω_1 and adds a real r such that, in the generic extension, $\omega_1^{L[r]} = \omega_1$. But then, since the sentence *There exists a real x such that* $\omega_1^{L[x]} = \omega_1$ is Σ_3 in $H(\omega_1)$, by $\mathcal{A}(H(\omega_1), \Sigma_3, Stat\text{-}pres)$ we have that there is such a real x in the ground model, contradicting Theorem 5.9. This shows that

THEOREM 5.12. *The following are equiconsistent*:

1. $\mathcal{A}(H(\omega_1), \Sigma_3, Stat\text{-}pres)$.
2. *Every set has a sharp.*

Hence, from 5.10, the axiom $\mathcal{A}(H(\omega_1), \Sigma_3, \omega_1\text{-}pres)$ is also equiconsistent with the existence of the sharp of every set.

The next theorem follows from some results due to Kunen and Harrington-Shelah [21] (see [11] and [13]):

THEOREM 5.13. *For all n such that $3 \leq n \leq \omega$, the following are equiconsistent*:

1. $\mathcal{A}(H(\omega_1), \Sigma_n, Knaster)$.
2. $\mathcal{A}(H(\omega_1), \underset{\sim}{\Sigma}_n, ccc)$.
3. *There exists a weakly compact cardinal.*

An argument of A. R. D. Mathias, also implicit in [26], is used in the following result to show that $\mathcal{A}(H(\omega_1), \Sigma_3, \sigma\text{-}centered)$, plus ω_1 is inaccessible in $L[x]$ for every real x, implies that ω_1 is a Mahlo cardinal in L.

THEOREM 5.14. [11] *For all n such that $3 \leq n \leq \omega$, the following are equiconsistent*:

1. $\mathcal{A}(H(\omega_1), \underset{\sim}{\Sigma}_n, \sigma\text{-}centered)$.
2. $\mathcal{A}(H(\omega_1), \underset{\sim}{\Sigma}_n, \sigma\text{-}linked)$.
3. *There exists a Mahlo cardinal.*

Let us recall from [12] that a ccc poset $\mathbb{P}$ is *strongly-$\underset{\sim}{\Sigma}_n$* if it is Σ_n-definable in $H(\omega_1)$ with parameters, and the predicate "x codes a maximal antichain of $\mathbb{P}$" is also Σ_n-definable in $H(\omega_1)$ with parameters. A projective poset $\mathbb{P}$ is *absolutely-ccc* if it is ccc in every inner model W of V which satisfies ZFC and contains the parameters of the definition of $\mathbb{P}$.

Let us also recall from [11] that a projective poset $\mathbb{P}$ is *strongly-proper* if for every countable transitive model N of a fragment of ZFC with the parameters of the definition of $\mathbb{P}$ in N and such that $(\mathbb{P}^N, \leq_{\mathbb{P}}^N, \perp_{\mathbb{P}}^N) \subseteq (\mathbb{P}, \leq_{\mathbb{P}}, \perp_{\mathbb{P}})$, and for every $p \in \mathbb{P}^N$, there is $q \leq p$ which is $(N, \mathbb{P})$-generic, i.e., if $N \models$ "A is a maximal antichain of $\mathbb{P}$", then A is predense below q.

Notice that if $\mathbb{P}$ is a projective poset, $N \preceq H(\lambda)$, and the parameters of the definition of $\mathbb{P}$ are in N, then a condition q is $(N, \mathbb{P})$-generic iff it is $(\bar{N}, \mathbb{P})$-generic, where $\bar{N}$ is the transitive collapse of N. Thus, a projective strongly-proper poset is proper.

From [7] (see 5.9 above) we have that just for $\Gamma = \{\omega_1\text{-}Random, Cohen\}$, the axiom $\mathcal{A}(H(\omega_1), \Sigma_3, \Gamma)$ already implies that ω_1 is inaccessible in $L[x]$, for every real x. Notice that both ω_1-*Random* and *Cohen* are Σ_1-definable in $H(\omega_1)$ ccc forcing notions. The sharpest result at the level of consistency strength of an inaccessible cardinal is the following:

THEOREM 5.15. [11, 12] *For every $3 \leq n \leq \omega$ the following are equiconsistent (modulo ZFC)*:

1. $\mathcal{A}(H(\omega_1), \Sigma_n, \{\omega_1\text{-}Random, Cohen\})$.
2. $\mathcal{A}(H(\omega_1), \underset{\sim}{\Sigma}_n, \Gamma)$, where Γ is the class of posets that are absolutely-ccc and strongly-$\underset{\sim}{\Sigma}_2$.
3. $\mathcal{A}(H(\omega_1), \underset{\sim}{\Sigma}_n, \Gamma)$, where Γ is the class of strongly-proper posets that are Σ_2 definable in $H(\omega_1)$ with parameters.
4. There exists an inaccessible cardinal.

This result is optimal, for there exists a, provably in ZFC, ccc poset $\mathbb{P}$ which is both Σ_2 and Π_2 definable in $H(\omega_1)$, without parameters, and for which the axiom $\mathcal{A}(H(\omega_1), \Sigma_3, \mathbb{P})$ fails if ω_1 is not a Π_1-Mahlo cardinal in L (see [12]). A regular cardinal κ is Σ_n-Mahlo (Π_n-Mahlo) if every club subset of κ that is Σ_n-definable (Π_n-definable) in $H(\kappa)$ contains an inaccessible cardinal. Every Π_1-Mahlo cardinal is an inaccessible limit of inaccessible cardinals.

The general equiconsistency result for proper forcing notions is due to R. Schindler:

THEOREM 5.16. [34, 35] For every $3 \leq n \leq \omega$ the following are equiconsistent:

1. $\mathcal{A}(H(\omega_1), \Sigma_n, Proper)$.
2. $\mathcal{A}(H(\omega_1), \underset{\sim}{\Sigma}_n, Proper)$.
3. There exists a remarkable cardinal.

For the definition of remarkable cardinal see [34]. The proof of the Theorem actually shows that $\mathcal{A}(H(\omega_1), \Sigma_3, Proper)$ implies that ω_1 is a remarkable cardinal in L.

The following are the main open questions about the consistency strength of these axioms:

QUESTIONS 5.17. What is the exact consistency strength of

1. $\mathcal{A}(H(\omega_1), \Sigma_3, \omega_1\text{-}pres)$?
2. $\mathcal{A}(H(\omega_1), \underset{\sim}{\Sigma}_3, Stat\text{-}pres)$?
3. $\mathcal{A}(H(\omega_1), \underset{\sim}{\Sigma}_3, Semi\text{-}proper)$?
4. $\mathcal{A}(H(\omega_1), \underset{\sim}{\Sigma}_3, Semi\text{-}proper)$?

The next theorem generalizes 5.4 and 5.5:

THEOREM 5.18. [9] Assume that $x^\sharp$ exists for every real x and that the second uniform indiscernible is $< \omega_2$. Then,

1. MA_{ω_1} implies $\mathcal{A}(H(\omega_1), \Sigma_3, ccc)$.
2. BPFA implies $\mathcal{A}(H(\omega_1), \underset{\sim}{\Sigma}_3, Proper)$.
3. BSPFA implies $\mathcal{A}(H(\omega_1), \underset{\sim}{\Sigma}_3, Semi\text{-}proper)$.
4. BMM implies $\mathcal{A}(H(\omega_1), \underset{\sim}{\Sigma}_3, Stat\text{-}pres)$.

There are few results on the consequences of axioms of the form $\mathcal{A}(H(\omega_1), \underset{\sim}{\Sigma}_3, \Gamma)$ in Descriptive Set Theory. In particular there are no known equivalences with regularity properties of projective sets of reals, as in the case of the

Σ_2-absoluteness axioms considered in the last section. However, we do have the following result of H. Judah (see [7]):

THEOREM 5.19. 1. $A(H(\omega_1), \underset{\sim}{\Sigma}_3, \{Amoeba\text{-}category, Cohen\})$ *implies that every* $\underset{\sim}{\Delta}^1_3$ *set of reals has the property of Baire.*

2. $\tilde{A}(H(\omega_1), \underset{\sim}{\Sigma}_3, \{Amoeba, Random\})$ *implies that every* $\underset{\sim}{\Delta}^1_3$ *set of reals is Lebesgue measurable.*

J. Brendle proved a similar result for the Ramsey property:

THEOREM 5.20. [15] $A(H(\omega_1), \underset{\sim}{\Sigma}_3, Mathias)$ *implies that every* $\underset{\sim}{\Sigma}^1_3$ *set of reals is Ramsey.*

It is also shown in [15] that $A(H(\omega_1), \underset{\sim}{\Sigma}_3, Mathias)$ implies that ω_1 is inaccessible in $L[x]$, for every real x.

An analogous result also holds for Hechler forcing [16], namely, $A(H(\omega_1), \underset{\sim}{\Sigma}_3, Hechler)$ implies that ω_1 is inaccessible in $L[x]$, for every real x.

The main general result is the following:

THEOREM 5.21. [17] $A(H(\omega_1), \underset{\sim}{\Sigma}_3)$ *implies that every* $\underset{\sim}{\Delta}^1_3$ *set of reals is universally Baire.*

It is also shown in [17] that the converse does not hold.

5.3. $\Phi = \Sigma_n, 4 \leq n \leq \omega$. A generalization of an argument of Woodin [41], yields the following result from [23]:

THEOREM 5.22. $A(H(\omega_1), \underset{\sim}{\Sigma}_4)$ *implies that for every set* X, $X^\dagger$ *exists.*

The general equiconsistency result is due to Hauser-Woodin (see [23]):

THEOREM 5.23. *The following are equiconsistent*:

1. $A(H(\omega_1), \Sigma_\omega)$.
2. $A(H(\omega_1), \underset{\sim}{\Sigma}_\omega)$.
3. *There are infinitely-many strong cardinals.*

A proper class of Woodin cardinals suffices to imply full absoluteness for $H(\omega_1)$ under all set-forcing extensions. Indeed, the following theorem of Woodin is a consequence of the results proved in [42] (see also [25]).

THEOREM 5.24. *If there is a proper class of Woodin cardinals, then* $A(H(\omega_1), \underset{\sim}{\Sigma}_\omega)$.

The following is a natural open question from [17]:

QUESTION 5.25. Does $A(H(\omega_1), \underset{\sim}{\Sigma}_\omega)$ imply that every projective set of reals is universally Baire?

For the class of projective posets we have the following results: recall from [12] that a regular cardinal κ is Σ_ω-Mahlo if every club subset of κ that is definable (with parameters) in $H(\kappa)$ contains an inaccessible cardinal. If κ is Mahlo, then the set of Σ_ω-Mahlo cardinals is stationary on κ (see [12]).

THEOREM 5.26. [12] *Let Γ be the class of projective posets, i.e., posets that are definable with parameters in $H(\omega_1)$. The following are equiconsistent*:

1. $\mathcal{A}(H(\omega_1), \underset{\sim}{\Sigma}_\omega, \Gamma \cap \text{absolutely-ccc})$.
2. *There exists an Σ_ω-Mahlo cardinal*.

THEOREM 5.27. [11] *For Γ the class of projective posets, $CON(ZFC + \text{There exists an } \Sigma_\omega\text{-Mahlo cardinal})$ implies $CON(ZFC + \mathcal{A}(H(\omega_1), \underset{\sim}{\Sigma}_\omega, \Gamma \cap \text{strongly-proper}))$.*

Whether a Σ_ω-Mahlo cardinal is necessary for the consistency of $\mathcal{A}(H(\omega_1), \underset{\sim}{\Sigma}_\omega, \Gamma \cap \text{strongly-proper})$, is still open.

Other open questions are the following:

QUESTIONS 5.28.
1. What is the consistency strength of $\mathcal{A}(H(\omega_1), \Sigma_\omega, \omega_1\text{-pres})$?
2. What is the consistency strength of $\mathcal{A}(H(\omega_1), \Sigma_\omega, Stat\text{-pres})$?

   The same questions for sentences with parameters, i.e., for $\Phi = \underset{\sim}{\Sigma}_\omega$, are also open. They are also open for projective posets in each one of the corresponding classes of posets. They are also open for $\Phi = \underset{\sim}{\Sigma}_n$, $n \geq 4$.
3. What is the exact consistency strength of $\mathcal{A}(H(\omega_1), \underset{\sim}{\Sigma}_n)$ for $n \geq 4$?

   K. Hauser (unpublished) showed that this consistency strength is at least that of $n - 3$ strong cardinals and at most that of $n - 3$ strong cardinals with a Σ_2-reflecting cardinal above them.

§6. $W = H(\omega_2)$. The axiom $\mathcal{A}(H(\omega_2), \Sigma_2, \Gamma)$ is false for most classes of forcing notions. Even $\mathcal{A}(H(\omega_2), \Sigma_2, \sigma\text{-centered})$ is false if CH fails. Indeed, by adding ω_1 Cohen reals, a σ-centered forcing notion, one adds a Luzin set, that is, an uncountable set of reals that intersects every meager set in at most a countable set. Notice that saying that there exists a Luzin set is a Σ_2 statement in $H(\omega_2)$. But then we may iterate in length the continuum *Amoeba-category*, a σ-centered forcing notion, so that in the generic extension every set of size ω_1 is meager. Since any iteration of *Amoeba-category* with finite support is Knaster, the argument shows that $\mathcal{A}(H(\omega_2), \Sigma_2, Knaster)$ is false. Note that the argument also shows that $\mathcal{A}(H(\omega_2), \underset{\sim}{\Sigma}_2, \sigma\text{-centered})$ is false, since given any set of reals in $H(\omega_2)$ we can force with *Amoeba-category* to make it meager.

6.1. $\Phi = \Sigma_1$: Since $\mathcal{A}(H(\omega_2), \Sigma_1)$ holds (see 3.2), the interesting case is with parameters.

Notice that $\mathcal{A}(H(\omega_2), \underset{\sim}{\Sigma}_1, \mathbb{P})$ implies the negation of CH, for any $\mathbb{P}$ that adds a real number.

For ccc posets, the axiom turns out to be equivalent to Martin's axiom for ω_1, as given in the following result due independently to Stavi-Väänänen [37], and [8]:

THEOREM 6.1. *The following are equivalent*:

1. $\mathcal{A}(H(\omega_2), \underset{\sim}{\Sigma}_1, ccc)$.
2. MA_{ω_1}.

In fact, we have that for every ccc poset $\mathbb{P}$, $\mathcal{A}(H(\omega_2), \underset{\sim}{\Sigma}_1, \mathbb{P})$ is equivalent to $MA_{\omega_1}(\mathbb{P})$.

Since, as it is well-known, Martin's axiom is consistent relative to ZFC, we have as a corollary of the Theorem above that so is the axiom $\mathcal{A}(H(\omega_2), \underset{\sim}{\Sigma}_1, ccc)$.

For some particular forcing notions we have nice characterizations of the corresponding axioms, for instance, from 6.1 and some basic results on the additivity of the ideals of null and meager sets of reals (see [7]), we have the following:

THEOREM 6.2.

1. $\mathcal{A}(H(\omega_2), \underset{\sim}{\Sigma}_1, Amoeba)$ *is equivalent to the* ω_1-*additivity of the Lebesgue measure*.
2. $\mathcal{A}(H(\omega_2), \underset{\sim}{\Sigma}_1, Amoeba\text{-}category)$ *is equivalent to the* ω_1-*additivity of the Baire property*.

From [9] and [19], we have an exact consistency result for the classes of proper and semi-proper posets:

THEOREM 6.3. *The following are equiconsistent*:

1. $\mathcal{A}(H(\omega_2), \underset{\sim}{\Sigma}_1, Proper)$.
2. $\mathcal{A}(H(\omega_2), \underset{\sim}{\Sigma}_1, Semi\text{-}proper)$.
3. *There exists a* Σ_2-*reflecting cardinal*.

This is a consequence of the following result, which generalizes 6.1, together with the results of [19] on the consistency strength of *BPFA* and *BSPFA*. It shows that the bounded forcing axioms are, in fact, equivalent to axioms of generic absoluteness for $H(\omega_2)$, thus revealing them as natural axioms of set Theory.

THEOREM 6.4. [9]

1. *The following are equivalent*:
   (a) $\mathcal{A}(H(\omega_2), \underset{\sim}{\Sigma}_1, Proper)$.
   (b) *BPFA*.
2. *The following are equivalent*:
   (a) $\mathcal{A}(H(\omega_2), \underset{\sim}{\Sigma}_1, Semi\text{-}proper)$.
   (b) *BSPFA*.
3. *The following are equivalent*:
   (a) $\mathcal{A}(H(\omega_2), \underset{\sim}{\Sigma}_1, Stat\text{-}pres)$.
   (b) *BMM*.

Asperó-Welch [6] produced a model of BSPFA where BMM fails, assuming the consistency of a large-cardinal notion slightly weaker than an ω_1-Erdös cardinal. This was later improved by Asperó [4] by constructing such a model from the optimal large cardinal assumption, namely, the existence of a Σ_2-reflecting cardinal. Also, Schindler has shown, starting from cardinals $\kappa < \lambda < \mu < \nu$ such that κ is remarkable, λ is Σ_2-reflecting, μ is Woodin, and ν is measurable, that forcing with the Levy collapse of κ to ω_1, and then forcing BPFA with a proper forcing in V_λ, one obtains a model in which BSPFA fails.

$\mathcal{A}(H(\omega_2), \underset{\sim}{\Sigma}_1, \mathbb{P})$ is clearly inconsistent with ZFC for any $\mathbb{P}$ that collapses ω_1. Further, the axiom $\mathcal{A}(H(\omega_2), \underset{\sim}{\Sigma}_1, \omega_1\text{-}pres)$ is also inconsistent with ZFC. For if S is a stationary and co-stationary subset of ω_1, then, by Baumgartner-Harrington-Kleinberg [22], we can add a club $C \subseteq S$ while preserving ω_1. But then the axiom would imply that such a club exists in the ground model, and so the complement of S is not stationary.

A natural question is: what is the the maximal class Γ for which $\mathcal{A}(H(\omega_2), \underset{\sim}{\Sigma}_1, \Gamma)$ is consistent? This has been answered by D. Asperó and H. Woodin in the following ways. On the one hand, Asperó proved the following:

THEOREM 6.5. [2] *Let Γ be the class of all posets $\mathbb{P}$ such that for every set X of cardinality $\aleph_1$ of stationary subsets of ω_1 there is a condition $p \in \mathbb{P}$ such that p forces that S is stationary for every $S \in X$. The axiom $\mathcal{A}(H(\omega_2), \underset{\sim}{\Sigma}_1, \Gamma)$ is maximal, i.e., if $\mathbb{P} \notin \Gamma$, then $\mathcal{A}(H(\omega_2), \underset{\sim}{\Sigma}_1, \mathbb{P})$ fails. The axiom can be forced assuming the existence of a Σ_2-reflecting cardinal which is the limit of strongly compact cardinals.*

On the other hand, Woodin [43] provides a fine analysis of the axioms of the form $\mathcal{A}(H(\omega_2), \underset{\sim}{\Sigma}_1, \Gamma)$, assuming the axiom $(*)$ (see [43]), plus that every set X belongs to a model with a Woodin cardinal above X, an assumption which is, consistency-wise, weaker that the existence of a proper class of Woodin cardinals, hence consistency-wise much weaker than the large-cardinal assumption in Asperó's Theorem above. Woodin shows that for every $A \in H(\omega_2)$ and every Π_1 sentence, with A as a parameter, that holds in V, there is a ω_1-sequence of stationary subsets of ω_1 such that any forcing notion that forces the negation of the sentence must destroy one of the stationary sets in the sequence. Thus, this yields a stronger form of Asperó's result, under a, consistency-wise, much weaker hypothesis. However, while Asperó's axiom can be forced assuming large cardinals, this is not known to be the case for Woodin's $(*)$ axiom.

Numerous consequences, mostly combinatorial, of the axioms BPFA, BSPFA, and BMM are known (see [5] and [40]).

Woodin [43] showed that if one assumes the existence of a measurable cardinal, or that the non-stationary ideal on ω_1 is precipitous, then *BMM* implies a combinatorial principle, called ψ_{AC}, which in turn implies that there is a well-ordering of the reals in length ω_2 which is definable, with parameters, in $H(\omega_2)$, and hence $\mathfrak{c} = \aleph_2$. Further, Asperó [1] showed that if BMM holds and for some $x \in H(\omega_2)$, $x^\sharp$ does not exist, then also $\mathfrak{c} = \aleph_2$. Improving on Woodin's result, Asperó-Welch [6] showed that if there exists a ω_1-Erdös cardinal, then BMM implies ψ_{AC}. More consequences of BMM were obtained by Asperó [3], for instance, he showed that BMM implies that the dominating number is $\aleph_2$. Finally, in a truly remarkable result, Todorcevic [39] has recently shown that BMM implies that there is a well-ordering of the reals in length ω_2 which is definable, with parameters, in $H(\omega_2)$, and hence $\mathfrak{c} = \aleph_2$. For this reason alone BMM deserves a detailed study. Unfortunately, its consistency strength is not known, and, while it may even imply that every projective set is determined, it is not even known whether it implies that every projective set of reals numbers is Lebesgue measurable. A few months ago R. Schindler proved the following:

THEOREM 6.6 (R. Schindler). *BMM implies that for every set X there is an inner model with a strong cardinal containing X.*

Thus, in particular, BMM implies that for every set X, $X^\sharp$ exists. In his Ph. D. Thesis, G. Hjorth [24] proved that Martin's axiom plus the existence of the sharp of every real number imply that every Σ_3^1 set of reals is Lebesgue measurable. It follows that BMM implies that every $\underset{\sim}{\Sigma}_3^1$ set of reals is Lebesgue measurable. The upper bound on the consistency strenght of BMM is given by Woodin [43], where he shows that a model with a proper class of Woodin cardinals suffices to obtain a model of BMM.

Showing that Bounded Forcing Axioms imply that the size of the continuum is $\aleph_2$ requires some method for coding reals by ordinals less than ω_2. Two such methods were devised by Woodin and Todorcevic, respectively, in their proofs that BMM (plus a measurable cardinal in the case of Woodin) implies $\mathfrak{c} = \aleph_2$. Very recently, Justin T. Moore has discovered a new coding method which yields yet a further improvement on the aforementioned chain of results of Woodin, Asperó, Asperó-Welch, and Todorcevic:

THEOREM 6.7. [30] *BPFA implies that there is a well-ordering of the reals in length ω_2 which is definable, with parameters, in $H(\omega_2)$, and hence $\mathfrak{c} = \aleph_2$.*

QUESTIONS 6.8.
1. Let Γ be as in Theorem 6.5. Is $\mathcal{A}(H(\omega_2), \underset{\sim}{\Sigma}_1, \Gamma)$ equivalent to the seemingly weaker BMM? (Under MA_{ω_1} there are posets in Γ that do not preserve stationary subsets of ω_1 (see [2]).
2. Is the axiom $\mathcal{A}(H(\omega_2), \Sigma_2, \sigma\text{-}linked)$ consistent?
3. What is the consistency strength of BMM?
4. let σ-closed $*$ ccc be the class of forcing notions consisting of an iteration of a σ-closed poset followed by a ccc poset. Such posets are proper. Does $\mathcal{A}(H(\omega_2), \Sigma_2, \sigma\text{-closed} * \text{ccc})$ imply $\mathfrak{c} = \aleph_2$?

§7. $W = H(\kappa), \kappa \geq \omega_3$. The arguments from the beginning of last section show that $\mathcal{A}(H(\kappa), \underset{\sim}{\Sigma}_2, \sigma\text{-centered})$ is false, even for the σ-centered posets that belong to $H(\omega_3)$.

The axiom $\mathcal{A}(H(\kappa), \Sigma_1)$ holds (see 3.2). So, the only interesting case is for $\Phi = \underset{\sim}{\Sigma}_1$.

From [8] we have that $\mathcal{A}(H(\omega_3), \underset{\sim}{\Sigma}_1, ccc)$ is equivalent to MA_{ω_2}.

However, $\mathcal{A}(H(\kappa), \underset{\sim}{\Sigma}_1, Proper)$ is false, all $\kappa \geq \omega_3$. Indeed, let $\mathbb{P}$ be the forcing notion for collapsing ω_2 to ω_1 with countable conditions. This is a σ-closed, hence proper, forcing notion. But $\mathcal{A}(H(\kappa), \underset{\sim}{\Sigma}_1, \mathbb{P})$ implies that ω_2 and ω_1 have the same cardinality.

$\mathcal{A}(H(\kappa), \underset{\sim}{\Sigma}_1, ccc)$ is easily seen to be false for $\kappa > \mathfrak{c}$. For $\kappa < \mathfrak{c}$, we have from [8] that $\mathcal{A}(H(\kappa), \underset{\sim}{\Sigma}_1, ccc)$ is equivalent to MA_κ, and $\mathcal{A}(H(\mathfrak{c}), \underset{\sim}{\Sigma}_1, ccc)$ is equivalent to MA.

§8. $W = L(H(\omega_1))$. Since every element of $H(\omega_1)$ can be easily coded by a real number, we have that $L(H(\omega_1)) = L(\mathbb{R})$, the smallest inner model of ZF that contains all the real numbers.

In general, $L(\mathbb{R}) \not\subseteq L(\mathbb{R})^{V^{\mathbb{P}}}$. For instance, by adding ω_1 Cohen reals over L, call this model V, and then yet one more Cohen real c over V, we have that $\mathbb{R}^V \notin L(\mathbb{R})^{V[c]}$. But even if we have $L(\mathbb{R}) \subseteq L(\mathbb{R})^{V^{\mathbb{P}}}$, we cannot have $L(\mathbb{R}) \preccurlyeq_{\Sigma_1} L(\mathbb{R})^{V^{\mathbb{P}}}$ if $\mathbb{P}$ adds some real. So, the most we can hope for is $\mathcal{A}(L(\mathbb{R}), \Sigma, \Gamma)$, i.e., without parameters, or, if we want parameters, we have to restrict them to ordinals and reals. So, whenever we write $\underset{\sim}{\Sigma}_n$ or $\underset{\sim}{\Sigma}_\omega$, we mean $\Sigma_n(OR \cup \mathbb{R})$ and $\Sigma_\omega(OR \cup \mathbb{R})$, respectively.

Notice that $\mathcal{A}(L(\mathbb{R}), \underset{\sim}{\Sigma}_\omega, \Gamma)$ means that for every $\mathbb{P} \in \Gamma$ there is an elementary embedding

$$j : L(\mathbb{R}) \longrightarrow L(\mathbb{R})^{V^{\mathbb{P}}}$$

that is the identity on the ordinals and, of course, on the reals.

The following theorem of H. Woodin (see [43]) shows that under large cardinals the theory of $L(\mathbb{R})$, with real parameters, is generically absolute.

THEOREM 8.1. *If there is a proper class of Woodin cardinals, then $\mathcal{A}(L(\mathbb{R}), \Sigma_\omega(\mathbb{R}))$ holds.*

Woodin and, independently, Steel, have shown that the consistency strength of $\mathcal{A}(L(\mathbb{R}), \Sigma_\omega(\mathbb{R}))$ is roughly that of the existence of infinitely-many Woodin cardinals (see [38]). They have also shown that, actually, $\mathcal{A}(L(\mathbb{R}), \Sigma_\omega(\mathbb{R}))$ implies that the Axiom of Determinacy, AD, holds in $L(\mathbb{R})$.

It follows from results of Shelah and Woodin that $\mathcal{A}(L(\mathbb{R}), \underset{\sim}{\Sigma}_1, Semi\text{-}proper)$ is false, assuming the existence of large cardinals. For if, for instance, there is a supercompact cardinal, then Shelah has shown that one can force, by a semi-proper forcing notion that collapses the supercompact cardinal to ω_2, the non-stationary ideal on ω_1 to be saturated (see [25]) (a Woodin cardinal actually suffices for this (Shelah [36])). Since, in addition, in this model every set has a sharp, it follows from results of Woodin [43] that $L(\mathbb{R})$ computes correctly ω_2. Let $\alpha = \omega_2^V$. Then the Σ_2 sentence with α as a parameter, which states that α is a cardinal greater or equal than ω_2, is true in the $L(\mathbb{R})$ of the ground model V, but false in the $L(\mathbb{R})$ of the generic extension.

However, I. Neeman and J. Zapletal [31] have shown that $\mathcal{A}(L(\mathbb{R}), \underset{\sim}{\Sigma}_\omega,$ *Proper*) follows from large cardinals. Namely,

THEOREM 8.2. *If δ is a weakly-compact Woodin cardinal, then $\mathcal{A}(L(\mathbb{R}), \underset{\sim}{\Sigma}_\omega, \mathbb{P})$ holds for every proper poset $\mathbb{P} \in V_\delta$. Hence, $\mathcal{A}(L(\mathbb{R}), \underset{\sim}{\Sigma}_\omega, Proper)$ follows from the existence of a proper class of weakly-compact Woodin cardinals.*

In fact, Woodin [43] shows that the axiom $\mathcal{A}(L(\mathbb{R}), \underset{\sim}{\Sigma}_\omega, \Gamma)$ holds for a class Γ which is larger that *Proper*, assuming only the existence of a proper class of Woodin cardinals, and gives some applications.

The consistency strength of $\mathcal{A}(L(\mathbb{R}), \underset{\sim}{\Sigma}_\omega, Proper)$ is rather low. Namely, as a consequence of [35] one has that:

THEOREM 8.3. *The following are equiconsistent:*

1. $\mathcal{A}(L(\mathbb{R}), \underset{\sim}{\Sigma}_\omega, Proper)$.
2. *There exists a remarkable cardinal.*

For ccc forcing notions we have the following exact equiconsistency results. The first one follows from work of Kunen and Harrington-Shelah [21] (see [11]).

THEOREM 8.4. *The following are equiconsistent:*

1. $\mathcal{A}(L(\mathbb{R}), \underset{\sim}{\Sigma}_\omega, Knaster)$.
2. $\mathcal{A}(L(\mathbb{R}), \underset{\sim}{\Sigma}_\omega, ccc)$.
3. *There exists a weakly compact cardinal.*

For σ-linked forcing notions a Mahlo cardinal suffices:

THEOREM 8.5. [11] *The following are equiconsistent:*

1. $\mathcal{A}(L(\mathbb{R}), \underset{\sim}{\Sigma}_\omega, \sigma\text{-}centered)$.
2. $\mathcal{A}(L(\mathbb{R}), \underset{\sim}{\Sigma}_\omega, \sigma\text{-}linked)$.
3. *There exists a Mahlo cardinal.*

And when restricting to projective forcing notions we have the following:

THEOREM 8.6. [12] *Let Γ be the class of projective posets. The following are equiconsistent:*

1. $\mathcal{A}(L(\mathbb{R}), \underset{\sim}{\Sigma}_\omega, \Gamma \cap absolutely - ccc)$.
2. *There exists an Σ_ω-Mahlo cardinal.*

THEOREM 8.7. [11] *For Γ the class of projective posets, $CON(ZFC + There exists an \Sigma_\omega$-Mahlo cardinal$)$ implies $CON(ZFC + \mathcal{A}(L(\mathbb{R}), \underset{\sim}{\Sigma}_\omega, \Gamma \cap strongly\text{-}proper))$.*

It is open whether a Σ_ω-Mahlo cardinal is necessary for the consistency of $\mathcal{A}(L(\mathbb{R}), \underset{\sim}{\Sigma}_\omega, \Gamma \cap strongly\text{-}proper)$.

We finally look at the general projective ccc case:

Let us recall from [10] that if κ is a cardinal and $n \in \omega$, we say that κ is Σ_n-*weakly compact* (Σ_n-w.c., for short) iff κ is inaccessible and for every $R \subseteq V_\kappa$ which is definable by a Σ_n formula (with parameters) over V_κ and every Π_1^1 sentence Φ, if

$$\langle V_\kappa, \in, R \rangle \models \Phi$$

then there is $\alpha < \kappa$ such that

$$\langle V_\alpha, \in, R \cap V_\alpha \rangle \models \Phi.$$

That is, κ *reflects* Π_1^1 *sentences with* Σ_n *predicates.*

Also, recall from [28] that a cardinal κ is Σ_ω-*weakly compact* (Σ_ω-w.c., for short), iff κ is Σ_n-w.c. for every $n \in \omega$.

THEOREM 8.8. [10] *Let* Γ_n *be the class of ccc posets that are* Σ_n *or* Π_n *definable in* $H(\omega_1)$ *with parameters. The following are equiconsistent*:

1. $\mathcal{A}(L(\mathbb{R}), \underset{\sim}{\Sigma}_\omega, \Gamma_n)$.
2. *There exists a* Σ_n-*w.c. cardinal.*

COROLLARY 8.9. [10] *Let* Γ *be the class of projective ccc forcing notions. The following are equiconsistent*:

1. $\mathcal{A}(L(\mathbb{R}), \underset{\sim}{\Sigma}_\omega, \Gamma)$.
2. *There exists a* Σ_ω-*w.c. cardinal.*

§9. $W = L(H(\omega_2))$. Notice that for every sentence φ, $H(\omega_2) \models \varphi$ iff $L(H(\omega_2))$ satisfies the Σ_2 sentence $\exists x(x = H(\omega_2) \wedge x \models \varphi)$. Thus, if Φ contains Σ_2, then $\mathcal{A}(L(H(\omega_2)), \Phi, \Gamma)$ implies $\mathcal{A}(H(\omega_2), \Phi, \Gamma)$. Hence, the negative results for $\mathcal{A}(H(\omega_2), \Phi, \Gamma)$ from Section 6 apply also to $\mathcal{A}(L(H(\omega_2)), \Phi, \Gamma)$. For instance, $\mathcal{A}(L(H(\omega_2)), \Sigma_2, ccc)$ and $\mathcal{A}(L(H(\omega_2)), \underset{\sim}{\Sigma}_2, \sigma\text{-}centered)$ are false.

Also, we have that $\mathcal{A}(L(H(\omega_2)), \underset{\sim}{\Sigma}_1, Proper)$, is false. For if we collapse ω_2 to ω_1 by σ-closed (hence proper) forcing, then in the $H(\omega_2)$ of the generic extension there is an injection of ω_2^V into ω_1. But saying that there is such an injection is Σ_1 in the parameters ω_2^V and ω_1. Hence, the axiom implies that such an injection exists in V, which is impossible.

Similar considerations apply to the case $W = L(H(\omega_3))$, etc.

QUESTION 9.1. Is $\mathcal{A}(L(H(\omega_2)), \underset{\sim}{\Sigma}_1, ccc)$ consistent?

REFERENCES

[1] DAVID ASPERÓ, *Bounded Forcing Axioms and The Continuum*, Ph.D. thesis, Universitat de Barcelona, 2002.

[2] ———, *A maximal bounded forcing axiom*, *The Journal of Symbolic Logic*, vol. 67 (2002), no. 1, pp. 130–142.

[3] ———, *Generic absoluteness for* Σ_1 *formulas and the continuum problem*, Submitted, 2002.

[4] ———, *On a convenient property about* $[\gamma]^{\aleph_0}$, Submitted, 2002.

[5] DAVID ASPERÓ and JOAN BAGARIA, *Bounded forcing axioms and the continuum*, *Annals of Pure and Applied Logic*, vol. 109 (2001), pp. 179–203.

[6] DAVID ASPERÓ and PHILIP WELCH, *Bounded Martin's Maximum, weak Erdös cardinals, and ψ_{AC}*, **The Journal of Symbolic Logic**, vol. 67 (2002), no. 3, pp. 1141–1152.

[7] JOAN BAGARIA, *Definable Forcing and Regularity Properties of Projective Sets of Reals*, Ph.D. thesis, University of California, Berkeley. 1991.

[8] ———, *A characterization of Martin's axiom in terms of absoluteness*, **The Journal of Symbolic Logic**, vol. 62 (1997), pp. 366–372.

[9] ———, *Bounded forcing axioms as principles of generic absoluteness*, **Archive for Mathematical Logic**, vol. 39 (2000), pp. 393–401.

[10] JOAN BAGARIA and ROGER BOSCH, *Generic absoluteness under projective forcing*, to appear.

[11] ———, *Proper forcing extensions and Solovay models*, **Archive for Mathematical Logic**, vol. 43 (2004), no. 6, pp. 739–750.

[12] ———, *Solovay models and forcing extensions*, **The Journal of Symbolic Logic**, vol. 69 (2004), no. 3, pp. 742–766.

[13] JOAN BAGARIA and SY D. FRIEDMAN, *Generic absoluteness*, **Annals of Pure and Applied Logic**, vol. 108 (2001), no. 1-3, pp. 3–13.

[14] TOMEK BARTOSZYŃSKI and HAIM JUDAH, **Set Theory, On The Structure of The Real Line**, A. K. Peters, 1995.

[15] JÖRG BRENDLE, *Combinatorial properties of classical forcing notions*, **Annals of Pure and Applied Logic**, vol. 73 (1995), pp. 143–170.

[16] JÖRG BRENDLE, HAIM JUDAH, and SAHARON SHELAH, *Combinatorial properties of Hechler forcing*, **Annals of Pure and Applied Logic**, vol. 59 (1992), pp. 185–199.

[17] QI FENG, MENACHEM MAGIDOR. and W. HUGH WOODIN, *Universally Baire sets of reals*, **Set Theory of The Continuum** (H. Judah, W. Just, and H. Woodin, editors), MSRI, Springer, 1992, pp. 203–242.

[18] SY D. FRIEDMAN, *Generic Σ_3^1 absoluteness*, **The Journal of Symbolic Logic**, vol. 69 (2004), no. 1, pp. 73–80.

[19] MARTIN GOLDSTERN and SAHARON SHELAH, *The bounded proper forcing axiom*, **The Journal of Symbolic Logic**, vol. 60 (1995), no. 1, pp. 58–73.

[20] LORENZ HALBEISEN and HAIM JUDAH, *Mathias absoluteness and the Ramsey property*, **The Journal of Symbolic Logic**, vol. 61 (1996), no. 1, pp. 177–194.

[21] LEO HARRINGTON and SAHARON SHELAH, *Some exact equiconsistency results in set theory*, **Notre Dame Journal of Formal Logic**, vol. 26 (1985), pp. 178–187.

[22] LEO A. HARRINGTON, JAMES E. BAUMGARTNER, and E. M. KLEINBERG, *Adding a closed unbounded set*, **The Journal of Symbolic Logic**, vol. 41 (1976), no. 2, pp. 481–482.

[23] KAI HAUSER, **The Consistency Strenght of Projective Absoluteness**, Habilitationschrift, Heidelberg, 1993.

[24] GREG HJORTH, **The Influence of u_2**, Ph.D. thesis, University of California, Berkeley, 1993.

[25] THOMAS JECH, **Set Theory. The Third Millenium Edition, Revised and Expanded**, Springer-Verlag, 2002.

[26] RONALD JENSEN and ROBERT M. SOLOVAY, *Some applications of almost disjoint sets*, **Mathematical Logic and Foundations of Set Theory** (Y. Bar-Hillel, editor), North-Holland, 1970, pp. 84–104.

[27] HAIM JUDAH, *Absoluteness for projective sets*, **Logic Colloquium '90** (J. Oikkonen and Väänänen, editors), Logic Colloquium '90, Lecture Notes in Logic, vol. 2, Springer, 1993, pp. 145–154.

[28] AMIR LESHEM, *On the consistency of the definable tree property on $\aleph_1$*, **The Journal of Symbolic Logic**, vol. 65 (2000), pp. 1204–1214.

[29] DONALD A. MARTIN and ROBERT SOLOVAY, *A basis theorem for Σ_3^1 sets of reals*, **Annals of Mathematics**, vol. 89 (1969), pp. 138–160.

[30] JUSTIN T. MOORE, *Proper forcing, the continuum, and set mapping reflection*, Preprint.

[31] ITAY NEEMAN and JINDŘICH ZAPLETAL, *Proper forcings and absoluteness in I*ℝ, *Commentationes Mathematicae Universitatis Carolinae*, vol. 39 (1998), no. 2, pp. 281–301.

[32] ANDREZJ ROSLANOWSKI and SAHARON SHELAH, *Simple forcing notions and forcing axioms*, *The Journal of Symbolic Logic*, vol. 62 (1997), no. 4, pp. 1297–1314.

[33] RALF SCHINDLER, *Coding into K by reasonable forcing*, *Transactions of the American Mathematical Society*, vol. 353 (2000), pp. 479–489.

[34] ——, *Proper forcing and remarkable cardinals*, *The Bulletin of Symbolic Logic*, vol. 6 (2000), pp. 176–184.

[35] ——, *Proper forcing and remarkable cardinals II*, *The Journal of Symbolic Logic*, vol. 66 (2001), pp. 1481–1492.

[36] SAHARON SHELAH, *Iterated forcing and normal ideals on* ω_1, *Israel Journal of Mathematics*, vol. 60 (1987), pp. 345–380.

[37] JONATHAN STAVI and JOUKO VÄÄNÄNEN, *Reflection principles for the continuum*, *Logic and Algebra* (Providence, RI), Contemporary Mathematics, vol. 302, AMS, 2002, pp. 59–84.

[38] JOHN R. STEEL, *Core models with more Woodin cardinals*, *The Journal of Symbolic Logic*, vol. 67 (2002), no. 3, pp. 1197–1226.

[39] STEVO TODORČEVIĆ, *Generic absoluteness and the continuum*, *Mathematical Research Letters*, vol. 9 (2002), pp. 1–7.

[40] ——, *Localized reflection and fragments of PFA*, *DIMACS Series*, vol. 58 (2002), pp. 135–148.

[41] W. HUGH WOODIN, *On the consistency strength of projective uniformization*, *Proc. of the Herbrand Symposium, Logic Colloquium '81* (J. Stern, editor), Logic Colloquium '81, North-Holland, 1982, pp. 365–384.

[42] ——, *Supercompact cardinals, sets of reals, and weakly homogeneous trees*, *Proceedings of the National Academy of Sciences of the United States of America*, vol. 85 (1988), no. 18, pp. 6587–6591.

[43] ——, *The Axiom of Determinacy, Forcing Axioms and the Nonstationary Ideal*, de Gruyter Series in Logic and Its Applications, vol. 1, Walter de Gruyter, 1999.

ICREA (INSTITUCIÓ CATALANA DE RECERCA I ESTUDIS AVANÇATS)
and
DEPARTAMENT DE LÒGICA, HISTÒRIA I FILOSOFIA DE LA CIENCIA
UNIVERSITAT DE BARCELONA
BALDIRI REIXAC, S/N
08028 BARCELONA, CATALONIA (SPAIN)
E-mail: joan.bagaria@icrea.es

GENERALISED DYNAMIC ORDINALS — UNIVERSAL MEASURES FOR IMPLICIT COMPUTATIONAL COMPLEXITY

ARNOLD BECKMANN

Abstract. We extend the definition of dynamic ordinals to *generalised dynamic ordinals*. We compute generalised dynamic ordinals of all fragments of relativised bounded arithmetic by utilising methods from Boolean complexity theory, similar to Krajíček in [14]. We indicate the role of generalised dynamic ordinals as universal measures for implicit computational complexity. I.e., we describe the connections between generalised dynamic ordinals and witness oracle Turing machines for bounded arithmetic theories. In particular, through the determination of generalised dynamic ordinals we re-obtain well-known independence results for relativised bounded arithmetic theories.

§1. Introduction. Implicit computational complexity denotes the collection of approaches to computational complexity which define and classify the complexity of computations without direct reference to an underlying machine model. These approaches are formal systems which cover a wide range, including applicative functional programming languages, linear logic, bounded arithmetic and finite model theory (cf. [17]). In this paper, we contribute to the idea of characterising the computational complexity of such formal systems by universal measures, such that the formal systems describe exactly the same complexity class, if and only if they agree in their universal measures. In general, we aim at connections which can be represented as follows:

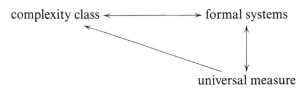

2000 *Mathematics Subject Classification.* 03F20; 03F30, 68Q15, 68R99.

Key words and phrases. Bounded arithmetic; Dynamic ordinals; Universal measures; Witness oracle Turing machines; Implicit computational complexity; Independence results; Håstad's Switching Lemmas; Cut-reduction by switching.

This work has been supported by a Marie Curie Individual Fellowship #HPMF-CT-2000-00803 from the European Commission.

Logic Colloquium '02
Edited by Z. Chatzidakis, P. Koepke, and W. Pohlers
Lecture Notes in Logic, 27

Many formal systems admit such kind of universal measures. For example, in case of "strong" implicit computational complexity, e.g. for number-theoretic functions which are computable by primitive recursive functionals in finite types, so-called proof-theoretic ordinals have proven useful as universal measures of proof and computation (and also consistency) strength (cf. [19]). With respect to our general picture this situation can be represented as follows:

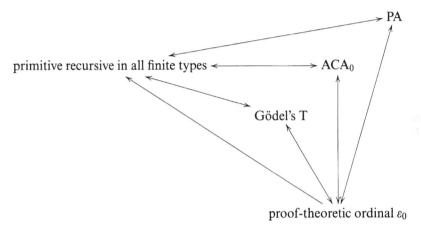

In this paper, we will focus on weak, also called low-level, complexity classes, i.e. complexity classes below EXPTIME. We will approach the general idea of finding universal measures by doing a case study for a particular framework of weak implicit computational complexity called bounded arithmetic. We already argued in [3] that so-called dynamic ordinals can be viewed as universal measures for *some* fragments of bounded arithmetic and corresponding bounded witness oracle Turing machine classes. In this paper, we will extend this project by defining and computing generalised dynamic ordinals and indicating their role as universal measures for *all* bounded arithmetic theories.

Bounded arithmetic theories are logical theories of arithmetic given as restrictions of Peano arithmetic. Quantification and induction are restricted ("bounded") in such a manner that complexity-theoretic classes can be closely tied to provability in these theories. A hierarchy of bounded formulas, Σ_i^b, and of theories $S_2^1 \subseteq T_2^1 \subseteq S_2^2 \subseteq T_2^2 \subseteq S_2^3 \ldots$ has been defined by Buss [6]. The class of predicates definable by Σ_i^b (or Π_i^b) formulas is precisely the class of predicates in the ith level Σ_i^p (respectively Π_i^p) of the polynomial time hierarchy. The Σ_i^b-definable functions of S_2^i are precisely the functions which are polynomial time computable with an oracle from Σ_{i-1}^p (cf. [6]). Krajíček [13] has characterised the Σ_{i+1}^b-definable multivalued functions of S_2^i as $FP^{\Sigma_i^b}(\text{wit}, O(\log n))$. Here, $FP^{\Sigma_i^b}(\text{wit}, O(\log n))$ denotes the class of multivalued functions computable by a polytime Σ_i^b-witness oracle Turing machine with the number of queries bounded by $O(\log n)$, see Section 3 for a precise definition. These results are extended and generalised by Pollett [20] to all bounded arithmetic theories.

It is an open problem of bounded arithmetic whether the hierarchy of theories collapses. This problem is connected with the open problem in complexity theory whether the polynomial time hierarchy PH collapses — the P=?NP problem is a sub-problem of this — in the following way: The hierarchy of bounded arithmetic theories collapses, if and only if the polynomial time hierarchy collapses provably in bounded arithmetic (cf. [16, 8, 23]). The case of relativised complexity classes and theories is completely different. The existence of an oracle A is proven in [1, 22, 11], such that the polynomial time hierarchy in this oracle PH^A does not collapse, hence in particular $P^A \neq NP^A$ holds. Building on this one can show $T_2^i(\alpha) \neq S_2^{i+1}(\alpha)$ [16]. Here, the relativised theories $S_2^i(\alpha)$ and $T_2^i(\alpha)$ result from S_2^i, and T_2^i respectively, by adding a free set variable α and the relation symbol $\in$. Similarly also, $S_2^i(\alpha) \neq T_2^i(\alpha)$ is proven in [13], and separation results for further relativised theories (dubbed $\Sigma_n^b(\alpha)$-L^mIND) are proven in [20]. Independently of these, and with completely different methods (see below), we have shown separation results for theories of relativised bounded arithmetic in [2, 4]. Despite all answers in the relativised case, all separation questions continue to be open for theories without set parameters.

The above mentioned alternative approach to the study of relativised bounded arithmetic theories is called dynamic ordinal analysis [2, 4]. Inspired from proof-theoretic ordinal analysis, which has its origin in Gentzen's consistency proof for PA, the proof theoretic strengths of bounded arithmetic theories are characterised by so-called dynamic ordinals. The dynamic ordinals $DO(T(\alpha))$ for some relativised bounded arithmetic theories $T(\alpha)$ have been defined and computed in [2, 4]. $DO(T(\alpha))$ is a set of unary number-theoretic functions, which characterises the amount of $\Pi_1^b(\alpha)$-order-induction provable in $T(\alpha)$. In [3], we have described how this fits into our general program on finding universal measure by connecting dynamic ordinals with witness oracle computations. The above mentioned characterisation of definable multivalued functions of higher bounded arithmetic theories suggests the following definition of generalised dynamic ordinals (for more details on this motivation see the discussion in [3]): The i-th generalised dynamic ordinal $DO_i(T(\alpha))$ of a relativised theory of bounded arithmetic $T(\alpha)$ characterises the amount of $\Pi_i^b(\alpha)$-order-induction provable in $T(\alpha)$ (thus, the usual dynamic ordinal is just the first generalised dynamic ordinal).

In this paper, we will define and compute generalised dynamic ordinals for all bounded arithmetic theories. This computation utilises methods from Boolean complexity like Håstad's Switching Lemmas [11, 10] to obtain a special cut-elimination technique, which we denote by "cut-reduction by switching". Krajíček has been the first utilising such methods from Boolean complexity to reduce the complexity of propositional proofs [14], and Buss and Krajíček successfully adapted these methods to reduce the oracle complexity of witnessing arguments [9]. Cut-reduction by switching will be formulated as a cut-elimination method. Usual cut-elimination procedures (like Gentzen or Tait style cut-elimination) eliminate outermost connectives of cut-formulas first. In general, the cost of applying such cut-elimination techniques is an exponential blow-up of certain parameters of derivations like their height. This blow up would destroy the computation of generalised dynamic ordinals. But still, the computation needs a reduction of the complexity of cut-formulas. Cut-reduction by switching will reduce cuts "inside-out", but will leave the proof-skeleton unchanged, e.g. the height of the derivation will remain the same. The price will be that not only the cut-formulas are reduced, but also the formula which is derived. This can be addressed by well-known utilisations of so-called Sipser functions [14, 9], again originating from Boolean complexity [11, 10].

Our results will be, that for all $i > 0$, the generalised dynamic ordinals are computed to

$$\mathrm{DO}_i\left(\mathrm{T}_2^i(\alpha)\right) = \left\{\lambda n.2^{|n|^c} : c \text{ a number}\right\},$$

$$\mathrm{DO}_i\left(\mathrm{S}_2^i(\alpha)\right) = \left\{\lambda n.|n|^c : c \text{ a number}\right\},$$

$$\mathrm{DO}_i\left(\mathrm{sR}_2^{i+1}(\alpha)\right) = \left\{\lambda n.2^{||n||^c} : c \text{ a number}\right\},$$

and more generally for $m > 0$

$$\mathrm{DO}_i\left(\Sigma_{i+m-1}^b(\alpha)\text{-}\mathrm{L}^m\mathrm{IND}\right) = \left\{\lambda n.2_m\left(c \cdot (|n|_{m+1})\right) : c \text{ a number}\right\}.$$

In particular we re-obtain the above mentioned separations of bounded arithmetic theories by dynamic ordinal analysis. We have displayed this situation in Fig. 1. The method of proving separations by dynamic ordinal analysis heavily differs from the above mentioned separation results, as no characterisations of definable functions and no oracle constructions are involved. In the figure, we mean with $S <_i T$ that the theories S and T are separated by a $\forall\Sigma_i^b$-sentence and that S is included in the consequences of T; with $S \not\subseteq_i T$ that S is separated from T by a $\forall\Sigma_i^b$-sentence, but not necessarily included; and with $S \preceq_i T$ that T is $\forall\Sigma_i^b$-conservative over S. The conservation results displayed in the figure have been proven by Buss [7].

Furthermore, we obtain the following connections between the i-th dynamic ordinal of relativised bounded arithmetic theories and the Σ_{i+1}^b-definable multivalued functions of their unrelativised companions. For $i > 0$ and for T from the following infinite list of theories

$$\mathrm{T}_2^i, \mathrm{S}_2^{i+1}, \mathrm{S}_2^i, \mathrm{sR}_2^{i+1}, \text{ and } \Sigma_{i+m-1}^b\text{-}\mathrm{L}^m\mathrm{IND} \text{ for all } m > 0,$$

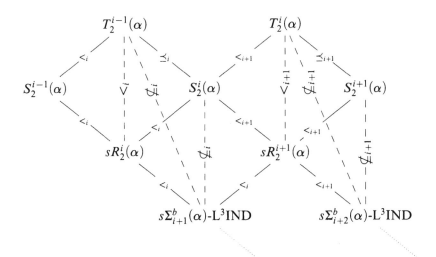

FIGURE 1. Independence results obtained by dynamic
ordinal analysis

we obtain:

> A multivalued function f is Σ_{i+1}^b-definable in T, if and only if f is computable by some polytime Σ_i^b-witness oracle Turing machine with the number of queries bounded by $\log(\mathrm{DO}_i(T(\alpha)))$.

This indicates that dynamic ordinals do in fact also characterise the *computational complexity* of bounded arithmetic theories. What is still missing is an intrinsic insight into this connection; this is work in progress.

The paper is organised as follows. In the following section we review the definition of bounded arithmetic theories. In Section 3 we define witness oracle Turing machines and review results characterising definable multivalued functions of bounded arithmetic theories by witness oracle Turing machines. The fourth section summarises definition of and results on dynamic ordinals, and gives the definition as well as lower bounds of the i-th generalised dynamic ordinal. In Section 5 we introduce a version of Gentzen's propositional proof system LK and prove basic properties like usual cut-elimination, and review translations from bounded arithmetic to LK. Section 6 introduces cut-reduction by switching and sketches of proof of this, which will be used in Section 7 to prove lower bounds to the height of derivations of the order induction principle. The results from Sections 5 to 7 have meanwhile been subject of a technical report [5]. In Section 8 we utilise the lower bound on derivation heights proved in Section 7 to compute the missing upper bounds on dynamic ordinals, which in turn will be used to obtain the connections between generalised dynamic ordinals and witness oracle Turing machines.

§2. Bounded arithmetic.

Let $\mathbb{N}$ denote the set of non-negative integers $0, 1, 2, \ldots$.

Bounded arithmetic can be formulated as the fragment $I\Delta_0 + \Omega_1$ of Peano arithmetic in which induction is restricted to bounded formulas and Ω_1 expresses a growth rate strictly smaller than exponentiation, namely that $2^{|x|^2}$ exists for all x. Here, $|x|$ denotes the length of the binary representation of x, i.e. an integer valued logarithm of x. The same fragment is obtained by extending the language of Peano arithmetic, and we will follow this approach first given by Buss, cf. [6]. Let us recall some definitions.

The language of bounded arithmetic[1] $\mathcal{L}_{BA}$ consists of function symbols 0 (zero), S (successor), + (addition), · (multiplication), $|x|$ (binary length), $\lfloor \frac{1}{2}x \rfloor$ (binary shift right), $x \# y$ (smash, $x\#y := 2^{|x| \cdot |y|}$), $x \dot{-} y$ (arithmetical subtraction), $\mathrm{MSP}(x, i)$ (Most Significant Part) and $\mathrm{LSP}(x, i)$ (Less Significant Part), and relation symbols = (equality) and ≤ (less than or equal). The meaning of MSP and LSP as number-theoretic functions is uniquely determined by stipulating that

$$x = \mathrm{MSP}(x, i) \cdot 2^i + \mathrm{LSP}(x, i), \quad \mathrm{LSP}(x, i) < 2^i$$

holds for all x and i. Restricted exponentiation $2^{\min(x, |y|)}$ can be defined by

$$2^{\min(x, |y|)} = \mathrm{MSP}\left(y \# 1, |y| \dot{-} x\right),$$

hence we can assume that restricted exponentiation is also part of our language $\mathcal{L}_{BA}$. We often write 2^t and mean $2^{\min(t, |x|)}$ if $t \leq |x|$ is clear from the context. Relativised bounded arithmetic is formulated in the language $\mathcal{L}_{BA}(\alpha)$ which is $\mathcal{L}_{BA}$ extended by one set variable α and the element relation symbol $\in$.

BASIC is a finite set of open axioms (cf. [6, 21, 12]) which axiomatises the non-logical symbols. When dealing with $\mathcal{L}_{BA}(\alpha)$ we assume that BASIC also contains the equality axioms for α.

Bounded quantifiers play an important role in bounded arithmetic. We abbreviate

$$(\forall x \leq t)A := (\forall x)(x \leq t \to A), \qquad (\exists x \leq t)A := (\exists x)(x \leq t \wedge A),$$

$$(\forall x < t)A := (\forall x \leq t)(t \not\leq x \to A), \quad (\exists x < t)A := (\exists x \leq t)(t \not\leq x \wedge A).$$

The quantifiers $(Qx \leq t)$, $(Qx < t)$, $Q \in \{\forall, \exists\}$, are called *bounded quanti-*

---

[1] For the sake of completeness we will fix a language for bounded arithmetic. In principle, our results can be obtained for any sufficient formulation, because they are stable under extending the language by arbitrary functions which have polynomial growth rate. See [4] for a discussion.

fiers. A bounded quantifier of the form $(Qx \leq |t|)$, $Q \in \{\forall, \exists\}$, is called a *sharply bounded quantifier*. A formula in which all quantifiers are (sharply) bounded is called a *(sharply) bounded formula*. Bounded formulas are stratified into levels:

i) $\Delta_0^b = \Sigma_0^b = \Pi_0^b$ is the set of all sharply bounded formulas.

ii) Σ_n^b-formulas are those which have a block of n alternating bounded quantifiers, starting with an existential one, in front of a sharply bounded kernel.

iii) Π_n^b is defined dually, i.e. the block of alternating quantifiers starts with a universal one.

In the relativised case $\Delta_0^b(\alpha)$, $\Sigma_n^b(\alpha)$, $\Pi_n^b(\alpha)$ are defined analogously.

Attention: In our definition, the class Σ_n^b consists only of *prenex*, also called *strict*, formulas. In other places in the literature like [6, 15], the definition of Σ_n^b is more liberal, and the class defined here is then denoted $s\Sigma_n^b$, where the "s" indicates "strict".

Induction is also stratified. Let $|x|_m$ denote the m-fold iteration of the binary length function, which can recursively be defined by $|x|_0 := x$ and $|x|_{m+1} := |(|x|_m)|$.

For Ψ is a set of $\mathcal{L}_{BA}$-formulas and m is a natural number, let Ψ-L^{m}IND denote the schema

$$\varphi(0) \wedge (\forall x < |t|_m)(\varphi(x) \rightarrow \varphi(Sx)) \rightarrow \varphi(|t|_m)$$

for all $\varphi \in \Psi$ and $\mathcal{L}_{BA}$-terms t.

For $m = 0$ this is the usual successor induction schema and will be denoted by Ψ-IND. In case $m = 1$ we often write Ψ-LIND.

The bounded arithmetic theories under consideration are given by

$$\text{BASIC} + \Sigma_n^b\text{-L}^m\text{IND}.$$

Usually we do not mention BASIC and simply call this theory Σ_n^b-L^{m}IND. Some of the theories have special names:

$$T_2^i := \Sigma_i^b\text{-IND},$$
$$S_2^i := \Sigma_i^b\text{-LIND},$$
$$sR_2^i := \Sigma_i^b\text{-L}^2\text{IND}.$$

For theories S, T let $S \subseteq T$ denote that all axioms in S are consequences of T. From the definition of the theories, the following two inclusions immediately follow:

$$\Sigma_n^b\text{-L}^{m+1}\text{IND} \subseteq \Sigma_n^b\text{-L}^m\text{IND},$$
$$\Sigma_n^b\text{-L}^m\text{IND} \subseteq \Sigma_{n+1}^b\text{-L}^m\text{IND}.$$

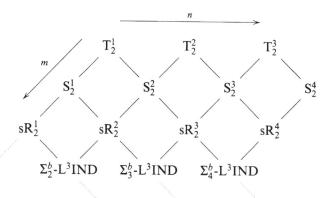

FIGURE 2. The theories Σ_n^b-L^mIND. Following any line rightwards takes one to super-theories. For example, $sR_2^1 \subseteq T_2^2$ following e.g. the path sR_2^1, S_2^1, T_2^1, S_2^2, T_2^2

A little bit more insight is needed to obtain

$$\Sigma_n^b\text{-}L^m\text{IND} \subseteq \Sigma_{n+1}^b\text{-}L^{m+1}\text{IND},$$

see [6] for a proof. Fig. 2 reflects the just obtained relations — going from left to right in the diagram means that the theory on the lefthand side of an edge is included in the theory on the righthand side. Similar definitions and results can be stated for relativised bounded arithmetic theories.

§3. **Witness oracle query complexity.** In this section we repeat the definition of witness oracle Turing machines and summarise how definable multivalued functions in bounded arithmetic theories are connected to witness oracle Turing machines.

A Turing machine with a witness oracle $Q(x) = (\exists y)R(x, y)$ is a Turing machine with a query tape for queries to Q that answers a query a as follows:

i) if $Q(a)$ holds, then it returns YES and some b such that $R(a, b)$ holds;
ii) if $\neg Q(a)$ holds, then it returns NO.

In general this type of Turing machines, called witness oracle Turing machines (WOTM), compute only *multivalued* functions rather than functions, as there may be multiple witnesses to affirmative oracle answers. A multivalued function is a relation $f \subseteq \mathbb{N} \times \mathbb{N}$ such that for all $x \in \mathbb{N}$ there exists some $y \in \mathbb{N}$ with $(x, y) \in f$. We express $(x, y) \in f$ as $f(x) = y$. A natural stratification of WOTMs, called bounded WOTMs, is obtained by bounding the number of oracle queries.

For Φ is a set of formulas, a multivalued function f is called Φ-definable in some theory T if there is a formula $\varphi(x, y)$ in Φ such that φ describes the graph of f and T proves the totality of f via φ, i.e.,

$$T \vdash (\forall x)(\exists y)\varphi(x, y),$$

$$\mathbb{N} \models (\forall x)(\forall y)[f(x) = y \leftrightarrow \varphi(x, y)].$$

Krajíček [13] has characterised the Σ^b_{i+1}-definable multivalued functions of T^i_2 and S^i_2 as $FP^{\Sigma^b_i}(\text{wit}, poly)$, and $FP^{\Sigma^b_i}(\text{wit}, O(\log n))$ respectively. $FP^{\Sigma^b_i}(\text{wit}, poly)$ and $FP^{\Sigma^b_i}(\text{wit}, O(\log n))$ are the classes of multivalued functions computable by a polynomial time WOTM which on inputs of length n uses fewer than respectively $n^{O(1)}$ and $O(\log n)$ witness queries to a Σ^b_i-oracle.

Pollett [20] obtains further relationships of definable multivalued functions and bounded polynomial time WOTM classes. The following version of bounded polynomial time WOTM classes goes back to [20].

DEFINITION 1. Let τ be a set of unary functions represented by terms in $\mathcal{L}_{BA}$. $FP^{\Sigma^b_i}(\text{wit}, \tau)$ is the class of multivalued functions computable by a polynomial time WOTM which on input x uses fewer than $l(t(x))$ witness queries to a Σ^b_i-oracle for some $l \in \tau$ and $\mathcal{L}_{BA}$-term t.

With id we denote the identity function, $\text{id}(n) = n$. The classes $FP^{\Sigma^b_i}(\text{wit}, poly)$ and $FP^{\Sigma^b_i}(\text{wit}, O(\log n))$ considered by Krajíček can be expressed as respectively $FP^{\Sigma^b_i}(\text{wit}, |\,\text{id}\,|)$ and $FP^{\Sigma^b_i}(\text{wit}, O(|\,\text{id}\,|_2))$ using the previous definition. The following characterisations of definable multivalued functions by bounded polynomial time WOTMs can be read of the results by Pollett [20], see [20] or [3] for more details. Let us remind that $2_m(x)$ and $|x|_m$ denote the m-fold iterations of the exponentiation function, respectively binary length function.

THEOREM 2 (Pollett [20]). Let $i \geq 0$ and $m \geq 1$.
A multivalued function f is Σ^b_{i+2}-definable in Σ^b_{m+i}-$L^m \text{IND}$, if and only if $f \in FP^{\Sigma^b_{i+1}}(\text{wit}, 2_{m-1}(O(|\,\text{id}\,|_{m+1})))$.

§4. Generalised dynamic ordinals. We start this section by repeating definitions of and results on dynamic ordinals for some fragments of bounded arithmetic from [4] and [3]. The underlying language will always be the language $\mathcal{L}_{BA}(\alpha)$ of relativised bounded arithmetic.

For $A(a)$ is a formula, let $S\text{Ind}(t, A)$ and $\mathcal{O}\text{Ind}(t, A)$ be defined by

$$S\text{Ind}(t, A) := A(0) \wedge (\forall x < t)(A(x) \rightarrow A(S\,x)) \rightarrow A(t),$$

$$\mathcal{O}\text{Ind}(t, A) := (\forall x < t)((\forall y < x)A(y) \rightarrow A(x)) \rightarrow (\forall x < t)A(x).$$

Order induction, denoted by $\mathcal{O}\text{Ind}$, applied to a formula A is logically equivalent to minimisation applied to the *negation* of A. It is well-known that over the base theory BASIC the schema Σ_i^b-IND is equivalent to minimisation for Σ_i^b-formulas which is equivalent (by coding one existential quantifier) to minimisation for Π_{i-1}^b-formulas [6, 15].

For Φ is a set of formulas, let $\mathcal{O}\mathrm{Ind}(t, \Phi)$ denote the schema of all instances $\mathcal{O}\mathrm{Ind}(t, A)$ for $A \in \Phi$ and $\mathcal{L}_{\mathrm{BA}}$-terms t. Similarly for $S\mathrm{Ind}$. When saying "let T be a theory" we always mean that T contains some weak base theory, say $S_2^0 \subseteq T$.

In [4] we have defined the dynamic ordinal $\mathrm{DO}(T)$ of a theory T by

$$\mathrm{DO}(T) := \{\lambda x.t \colon T \vdash (\forall x)\,\mathcal{O}\mathrm{Ind}\,(t, \Pi_1^b(\alpha))\}.$$

In this definition, it is understood that t ranges over $\mathcal{L}_{\mathrm{BA}}$-terms in which at most x occurs as a variable. Dynamic ordinals are sets of number theoretic functions, i.e. subsets of $^N\mathbb{N}$. Subsets of $^N\mathbb{N}$ can be arranged by eventual majorisability:

$$f \trianglelefteq g \;:\Longleftrightarrow\; g \text{ eventually majorises } f \;\Longleftrightarrow\; (\exists m)(\forall n \geq m)f(n) \leq g(n).$$

For subsets of number theoretic functions $D, E \subseteq {}^N\mathbb{N}$ we define

$$D \trianglelefteq E \;:\Longleftrightarrow\; (\forall f \in D)(\exists g \in E)f \trianglelefteq g,$$
$$D \equiv E \;:\Longleftrightarrow\; D \trianglelefteq E \& E \trianglelefteq D,$$
$$D \vartriangleleft E \;:\Longleftrightarrow\; D \trianglelefteq E \& E \ntrianglelefteq D.$$

$\trianglelefteq$ is a partial, transitive, reflexive ordering, $\vartriangleleft$ is a partial, transitive, irreflexive, not well-founded ordering, and $\equiv$ is an equivalence relation.

Using the big-O notation we will denote sets of unary number-theoretic functions in the following way:

$$f(O(g(\mathrm{id}))) := \{\lambda n.f(c \cdot g(n)) \colon c \in \mathbb{N}\}$$

for unary number-theoretic functions f and g.

The dynamic ordinals for certain bounded arithmetic theories are well established (cf. [4]):

$$\mathrm{DO}\left(T_2^1(\alpha)\right) \;\equiv\; 2_2\big(O(|\,\mathrm{id}\,|_2)\big) \;\equiv\; \mathrm{DO}\left(S_2^2(\alpha)\right),$$
$$\mathrm{DO}\left(S_2^1(\alpha)\right) \;\equiv\; 2_1\big(O(|\,\mathrm{id}\,|_2)\big),$$
$$\mathrm{DO}\left(sR_2^2(\alpha)\right) \;\equiv\; 2_2\big(O(|\,\mathrm{id}\,|_3)\big)$$

and more generally for $m > 0$

$$\mathrm{DO}\left(\Sigma_m^b(\alpha)\text{-}L^m\mathrm{Ind}\right) \;\equiv\; 2_m\big(O(|\,\mathrm{id}\,|_{m+1})\big).$$

In [3] we have described the following connections between the dynamic ordinal of some relativised bounded arithmetic theories and the Σ_2^b-definable multivalued functions of their unrelativised companions. For T from the following infinite list of theories

$$T_2^1, S_2^2, S_2^1, sR_2^2, \text{ and } \Sigma_m^b\text{-}L^m\mathrm{IND} \text{ for all } m > 0,$$

we obtain:

A multivalued function f is Σ_2^b-definable in T, if and only if f is computable by some polytime Σ_1^b-witness oracle Turing machine with the number of queries bounded by $\log(\mathrm{DO}(T(\alpha)))$.

Hence, the characterisation of definable multivalued functions of bounded arithmetic theories from the previous section suggests the following definition of generalised dynamic ordinals (see discussion in [3] for more details):

DEFINITION 3. The i-th generalised dynamic ordinal of an $\mathcal{L}_{\mathrm{BA}}(\alpha)$-theory T is defined by

$$\mathrm{DO}_i(T) := \left\{ \lambda x. t : T \vdash (\forall x)\, \mathcal{O}\mathrm{Ind}\left(t, \Pi_i^b(\alpha)\right) \right\}.$$

In this definition and in the next theorem, it is understood that t ranges over $\mathcal{L}_{\mathrm{BA}}$-terms in which at most x occurs as a variable. Observe that the previous definition of the dynamic ordinal of a theory T, $\mathrm{DO}(T)$, is the same as the first generalised dynamic ordinal of T, $\mathrm{DO}_1(T)$.

As generalised dynamic ordinals consist of terms in the language $\mathcal{L}_{\mathrm{BA}}$, a crude upper bound on generalised dynamic ordinals is always given by the growth rates of the functions representable by $\mathcal{L}_{\mathrm{BA}}$-terms:

$$\mathrm{DO}_i(T) \trianglelefteq 2^{|\mathrm{id}|^{O(1)}} = 2_2\big(O(|\mathrm{id}|_2)\big).$$

Generalised dynamic ordinals can also be characterised in terms of $\mathcal{S}\mathrm{Ind}$. In [4] we have shown that for sets of bounded formulas Φ which are closed under bounded universal quantification, we have $T \vdash \mathcal{O}\mathrm{Ind}(t, \Phi)$ if and only if $T \vdash \mathcal{S}\mathrm{Ind}(t, \Phi)$. Hence we have the following alternative characterisation of generalised dynamic ordinals:

COROLLARY 4. $\mathrm{DO}_i(T) = \{ \lambda x. t : T \vdash (\forall x)\, \mathcal{S}\mathrm{Ind}(t, \Pi_i^b(\alpha)) \}$.

In [4] we have shown that different dynamic ordinals imply a separation of the underlying theories. A similar property holds for generalised dynamic ordinals.

LEMMA 5. *Let S, T be two theories in the language of bounded arithmetic and assume $\mathrm{DO}_i(S) \neq \mathrm{DO}_i(T)$. Then S is separated from T by some $\forall \Sigma_{i+1}^b(\alpha)$-sentence.*

PROOF. Assume $f \in \mathrm{DO}_i(T) \setminus \mathrm{DO}_i(S)$. By the definition of generalised dynamic ordinals there is a term $t(x)$ and a $\Pi_i^b(\alpha)$-formula A such that $f(n) = t(n)$ and $T \vdash (\forall x)\, \mathcal{O}\mathrm{Ind}(t(x), A)$. But $f \notin \mathrm{DO}_i(S)$ implies $S \nvdash (\forall x)\, \mathcal{O}\mathrm{Ind}(t(x), A)$. Obviously, $\mathcal{O}\mathrm{Ind}(t(x), A) \in \Sigma_{i+1}^b(\alpha)$. ⊣

The language $\mathcal{L}_{\mathrm{BA}}$ includes the successor function, $+$ and $\cdot$, which enables us to speed-up induction polynomially. This has been carried out in [4] showing

THEOREM 6. [4, Theorem 9] $\Sigma_n^b\text{-}\mathrm{L}^m\mathrm{IND} \vdash \mathcal{O}\mathrm{Ind}(p(|x|_m), \Pi_n^b)$ *for polynomials p, if $m > 0$ or $n > 0$.*

Order induction for higher formula complexity is connected to order induction on larger orderings by speed-up techniques. The reader unfamiliar with such speedup techniques may consult [19, Chaper 15] for the transfinite case of speeding up induction in Peano arithmetic, or [2, Chaper 9] respectively [4, Section 3] for the adapted case to bounded arithmetic. The main ingredient which formalises this is the following jump set $\mathrm{Jp}(t, x, \alpha)$:

$$\{y \leq t : t \leq |x| \wedge (\forall z \leq 2^t)[z \subseteq \alpha \wedge z + 2^y \leq 2^t + 1 \rightarrow z + 2^y \subseteq \alpha]\}.$$

Here, $z \leq \alpha$ is an abbreviation for $\forall u < z\ u \in \alpha$. Iterations of Jp are defined by

$$\mathrm{Jp}_0(t, x, \alpha) = \alpha,$$
$$\mathrm{Jp}_{i+1}(t, x, \alpha) = \mathrm{Jp}\left(t, |x|_i, \mathrm{Jp}_i(t, x, \alpha)\right).$$

Let us remind that $|\cdot|_i$ denotes the i-fold iteration of $|\cdot|$, and that 2_m denotes the m-fold iteration of exponentiation. Using the iterated jump set we obtain the following connections:

THEOREM 7. [4, Corollary 15]

$$\mathrm{BASIC} \vdash t \leq |x|_m \rightarrow \left[\mathcal{O}\mathrm{Ind}\left(2_m(t), A\right) \leftrightarrow \mathcal{O}\mathrm{Ind}\left(t, \mathrm{Jp}_m(t, x, A)\right)\right].$$

PROOF IDEA. The direction from left to right of the equivalence follows directly. For the other direction we would have to prove the following statement, see [4, Section 3] for a definition of $\mathcal{O}\mathrm{Prog}$ and a proof of this:

$$\mathrm{BASIC} \vdash t \leq |x| \wedge \mathcal{O}\mathrm{Prog}(2^t, A) \rightarrow \mathcal{O}\mathrm{Prog}\left(t, \mathrm{Jp}(t, x, A)\right). \qquad \dashv$$

Concerning the complexity of the iterated jump we observe that

$$\mathrm{Jp}_n\left(t, x, \Pi_i^b\right) \subset \Pi_{n+i}^b$$

hence Theorem 6 and Theorem 7 together show the following Corollary, which has been formulated for the base case $i = 1$ in [4, Theorem 16].

COROLLARY 8. Let $0 \leq n < m$ or $n = m = 1$, let $i > 0$ and let c be some natural number, then $\Sigma_{n+i}^b\text{-}\mathrm{L}^m\mathrm{IND} \vdash \mathcal{O}\mathrm{Ind}(2_n(|x|_m^c), \Pi_i^b)$ and $\Sigma_{n+i}^b\text{-}\mathrm{L}^m\mathrm{IND} \vdash \mathcal{O}\mathrm{Ind}(2_{n+1}(c \cdot |x|_{m+1}), \Pi_i^b)$.

This establishes lower bounds on general dynamic ordinals. E.g., we obtain for $m > 0$:

$$\mathrm{DO}_{i+1}\left(\Sigma_{m+i}^b(\alpha)\text{-}\mathrm{L}^m\mathrm{Ind}\right) \trianglerighteq 2_m\left(O(|\mathrm{id}|_{m+1})\right).$$

For upper bounds we will utilise translations to propositional proof systems, which then will be studied proof-theoretically. But first we have to specify our favourite propositional proof system.

§5. The proof system Σ_i^{qp}-LK. In the following we give a natural modification of the definition of language and formulas of Gentzen's propositional proof system LK. In the way we will describe it here, it is sometimes attributed as "Tait-style". LK consists of constants $0, 1$, propositional variables $p_0, p_1, p_2, \ldots$ (also called atoms; we may use $x, y, \ldots$ as meta-symbols for variables), the connectives negation $\neg$, conjunction $\bigwedge$ and disjunction $\bigvee$ (conjunction and disjunction are both of unbounded finite arity), and auxiliary symbols like parentheses. Formulas are defined inductively: constants, atoms and negated atoms are formulas (they are called literals), and if Φ is a finite set of formulas, then $\bigwedge \Phi$ and $\bigvee \Phi$ are formulas, too. In general, negation is defined as an operation according to the de Morgan laws, i.e., $\neg\varphi$ denotes the formula obtained from φ by interchanging $\bigwedge$ and $\bigvee$, 0 and 1, and atoms and their negations. The *logical depth*, or just *depth*, $\mathrm{dp}(\varphi)$ of a formula φ, is the maximal nesting of $\bigwedge$ and $\bigvee$ in it. In particular, constants and atoms have depth 0, the depths of φ and $\neg\varphi$ are equal, and $\mathrm{dp}(\bigvee \Phi)$ equals $1 + \max\{\mathrm{dp}(\varphi) : \varphi \in \Phi\}$.

In our setting, *cedents* $\Gamma, \Delta, \ldots$ are finite *sets* of formulas, not *sequences* as in [14], and the meaning of a cedent Γ is $\bigvee \Gamma$. Γ is called a tautological cedent if $\bigvee \Gamma$ is a tautology. We often abuse notation by writing Γ, φ or $\Gamma \vee \varphi$ instead of $\Gamma \cup \{\varphi\}$, or by writing $\varphi_1, \ldots, \varphi_k$ instead of $\{\varphi_1, \ldots, \varphi_k\}$.

Our version of LK does not have structural rules as special inferences, they will be obtained as derivable rules. LK consists of four inference rules: initial cedent rule, introduction rules for $\bigwedge$ and $\bigvee$, and cut-rule. We are going to define $\mathcal{C}$-LK where $\mathcal{C}$ denotes the set of permissable cut formulas.

DEFINITION 9. We inductively define that Γ is $\mathcal{C}$-LK provable with height η, in symbols $\vdash_{\mathcal{C}}^{\eta} \Gamma$, for Γ a cedent, $\mathcal{C}$ a set of formulas and $\eta \in \mathbb{N}$. $\vdash_{\mathcal{C}}^{\eta} \Gamma$ holds if and only if one of the following four conditions is fulfilled:

(Init): Γ is an initial cedent, i.e. $1 \in \Gamma$, or $x, \neg x \in \Gamma$ for some variable x.

($\bigwedge$): There is some $\bigwedge \Phi \in \Gamma$ and $\eta' < \eta$ such that $\vdash_{\mathcal{C}}^{\eta'} \Gamma, \varphi$ for all $\varphi \in \Phi$.

($\bigvee$): There is some $\bigvee \Phi \in \Gamma$ and $\varphi \in \Phi$ and $\eta' < \eta$ such that $\vdash_{\mathcal{C}}^{\eta'} \Gamma, \varphi$.

(Cut): There is some $\varphi \in \mathcal{C}$ and $\eta' < \eta$ such that $\vdash_{\mathcal{C}}^{\eta'} \Gamma, \varphi$ and $\vdash_{\mathcal{C}}^{\eta'} \Gamma, \neg\varphi$.

The formula φ in the application of the cut-rule is called the *cut-formula* of this inference.

In order to make our definition of Σ_i^{qp}-LK precise we have to define a fine structure on constant depth formulas (cf. [14]).

DEFINITION 10. Let S, t, i be in $\mathbb{N}$. We inductively define $\varphi \in \Sigma_i^{S,t}$ by the following clauses:

i) $\varphi \in \Sigma_0^{S,t}$ if and only if φ is a $\bigwedge$ or $\bigvee$ of at most t many literals.

ii) $\varphi \in \Sigma_{i+1}^{S,t}$ if and only if φ is in $\Sigma_i^{S,t} \cup \Pi_i^{S,t}$, or it has the form of a $\bigvee$ of at most S many formulas from $\Sigma_i^{S,t} \cup \Pi_i^{S,t}$.

iii) A formula is in $\Pi_i^{S,t}$, if and only if its negation is in $\Sigma_i^{S,t}$.

Now we are prepared to say what we mean by Σ_i^{qp}-LK. Here and in the following, the superscript "qp" stands for "quasi-polynomial".

DEFINITION 11. Let $i \in \mathbb{N}$, let $f, \eta : \mathbb{N} \to \mathbb{N}$ be functions, and let $(\Gamma_n)_n$ be a sequence of tautological cedents.

We say that $(\Gamma_n)_n$ is (i, f)-LK (or (Σ_i, f)-LK) provable with height η, if and only if there is a sequence of subsets $\mathcal{C}_n \subseteq \Sigma_i^{f(n), \log(f(n))}$ of cardinality bounded by $f(n)$ such that Γ_n is $\mathcal{C}_n$-LK provable of height $\eta(n)$.

Then Σ_i^{qp}-LK denotes $(\Sigma_i, 2^{(\log n)^{O(1)}})$-LK, i.e. $(\Gamma_n)_n$ is Σ_i^{qp}-LK provable with height η iff there is a $c \in \mathbb{N}$ such that $(\Gamma_n)_n$ is $(i, 2^{(\log n)^c})$-LK provable with height η.

We will often abuse notation and write Γ_n is Σ_i^{qp}-LK provable with height $\eta(n)$, instead of $(\Gamma_n)_n$ is Σ_i^{qp}-LK provable with height η.

In the previous definition, we included a strange looking condition that the number of distinct cut-formulas is bounded, too. The reason is technical: during the computation of dynamic ordinals we will encounter LK derivations whose heights grow stronger than poly-logarithmically. For derivations with poly-logarithmic height we would not need such a condition as it would be fulfilled implicitly: a derivation, in which all cut-formulas are in $\Sigma_i^{2^{(\log n)^c}, (\log n)^c}$, has the property that the fan-in to each node in the derivation tree is bounded by $2^{(\log n)^c}$. Hence, if the height of such a derivation tree is bounded by $(\log n)^c$, then the number of nodes in the tree is bounded by $2^{(\log n)^{2c}}$, and thus the number of different cut-formulas in the derivation has the same bound. This argument obviously fails if the heights of the derivation trees grow stronger than poly-logarithmically. Now, having a quasi-polynomial upper bound on the number of different cut-formulas is essential for the cut-reduction by switching method needed to compute the generalised dynamic ordinals. It should be said at this point, that this condition will be fulfilled by translations of bounded arithmetic derivations, hence we obtain a stronger result this way.

Structural rules are not included in the definition of LK. They can be obtained as derivable rules which is stated in the next proposition. It is readily proven by induction on η.

PROPOSITION 12 (Structural Rule). *Assume $\eta \leq \eta', \mathcal{C} \subseteq \mathcal{C}'$ and $\Gamma \subseteq \Gamma'$, then $\vdash_{\mathcal{C}}^{\eta} \Gamma$ implies $\vdash_{\mathcal{C}'}^{\eta'} \Gamma'$.*

The following propositions on $\bigwedge$-Inversion and $\bigvee$-Exportation are readily proven by induction on η.

PROPOSITION 13 ($\bigwedge$-Inversion). *Assume $\vdash_{\mathcal{C}}^{\eta} \Gamma, \bigwedge \Phi$, then $\vdash_{\mathcal{C}}^{\eta} \Gamma, \varphi$ holds for all $\varphi \in \Phi$.*

PROPOSITION 14 ($\bigvee$-Exportation). *Suppose* $\vdash^{\eta}_{C} \Gamma, \bigvee \Phi$ *holds, then* $\vdash^{\eta}_{C} \Gamma, \Phi$.

The proof of the next Lemma and Proposition follows the same pattern as the standard one which can be found e.g. in [2, 4].

LEMMA 15 (Cut-Elimination Lemma). *Let* $\varphi \in \Sigma^{S,t}_{i+1}$ *and* $C \subseteq \Sigma^{S,t}_i$ *such that* C *includes all* $\Sigma^{S,t}_i$*-sub-formulas and all negations of* $\Pi^{S,t}_i$*-sub-formulas of* φ. *If* $\vdash^{\eta_0}_{C} \Gamma, \varphi$ *and* $\vdash^{\eta_1}_{C} \Delta, \neg\varphi$, *then* $\vdash^{\eta_0+\eta_1}_{C} \Gamma, \Delta$.

PROPOSITION 16 (Cut-Elimination Theorem). *Let* $C \subseteq \Sigma^{S,t}_{i+1}$ *be closed under sub-formulas and let* $C' := C \cap (\Sigma^{S,t}_i \cup \Pi^{S,t}_i)$. *Then* $\vdash^{\eta}_{C} \Gamma$ *implies* $\vdash^{2^{\eta}}_{C'} \Gamma$.

We repeat the translation (also called embedding) of provability in $S^i_2(\alpha)$, $T^i_2(\alpha)$, and more general of $\Sigma^b_i(\alpha)$-L^{m+1}IND, to LK from [2, 4]. Let $\log^{(k)}(n)$ be the k-times iterated logarithm applied to n, and $2_k(n)$ the k-times iterated exponentiation applied to n.

There exists a canonical translation due to Paris and Wilkie [18] from the language of bounded arithmetic to the language of LK (see [15, 9.1.1], or [2, 4]). Let φ be a formula in the language of bounded arithmetic in which no individual (i.e. first order) variable occurs free — we call such a formula (first order) closed. Then $\llbracket \varphi \rrbracket$ denotes the translation of φ to the language of LK, which for example maps the atom $\alpha(t)$, for t a closed term of value $m_t \in \mathbb{N}$, to the propositional variable p_{m_t}, and bounded quantifiers to connectives $\bigwedge$, respectively $\bigvee$, e.g. $\llbracket (\forall x \leq t)\varphi(x) \rrbracket = \bigwedge_{j \leq m_t} \llbracket \varphi(j) \rrbracket$. It follows that a formula $\varphi(x)$ from $\Sigma^b_i(\alpha)$ (with x being the only variable occurring free in φ) translates to $(\llbracket \varphi(n) \rrbracket)_n$ in Σ^{qp}_i, i.e. there is some $c \in \mathbb{N}$ such that $\llbracket \varphi(n) \rrbracket \in \Sigma^{2^{(\log n)^c},(\log n)^c}_i$ for all $n \in \mathbb{N}$.

REMARK 17. The last statement is not totally correct in the way it is stated, because, Δ^b_0 formulas may have unbounded depth. If we want to be really precise we could proceed as follows: First we define a more restricted version of Σ^b_i and Π^b_i. E.g., we define $\hat{\Sigma}^b_1$ formulas to be of the form: one bounded existential quantifier followed by disjunctions of one sharply bounded universal quantifier followed by conjunctions of atomic formulas. It can be shown in weak theories of Bounded Arithmetic that Σ^b_1 formulas are equivalent to $\hat{\Sigma}^b_1$ formulas. Hence we obtain an equivalent definition of our theories using the more restricted classes. Second, we let $\hat{\Sigma}^{S,t}_i$ be the stratification of propositional formulas corresponding to the restrictions $\hat{\Sigma}^b_i$. E.g., $\hat{\Sigma}^{S,t}_0$ are the formulas having at most two levels of fan-in t disjunctions. Then we obviously have that $\hat{\Sigma}^b_i$ formulas translate into $\hat{\Sigma}^{qp}_i$-formulas. Now the translation of proofs in bounded arithmetic which we are going to describe will produce $\hat{\Sigma}^{qp}_i$-LK-derivations, which finally can be transformed into Σ^{qp}_i-LK-derivations by merging two levels of connectives of the same type using

$$\vdash^{h}_{C} \Gamma, \bigvee_{i<n}(\bigvee\Phi_i) \implies \vdash^{h}_{C} \Gamma, \bigvee(\bigcup_{i<n}\Phi_i)$$

and

$$\vdash^{h}_{C} \Gamma, \bigwedge_{i<n}(\bigwedge \Phi_i) \implies \vdash^{h}_{C} \Gamma, \bigwedge(\bigcup_{i<n} \Phi_i).$$

For notational simplicity we will assume in this Section that Δ_0^b formula have the form one sharply bounded quantifier followed by an atomic formula. Assuming this, the above statement is correct.

The same proofs as in [2, 4] also show the following Theorem, which is here formulated using Σ_i^{qp}-LK.

THEOREM 18. *Let $\varphi(x)$ be a formula in the language of bounded arithmetic, in which at most the variable x occurs free.*

i) *If $S_2^i(\alpha) \vdash \varphi(x)$, then $[\![\varphi(n)]\!]$ has some Σ_i^{qp}-LK derivation of height $O(\log^{(2)} n)$.*

ii) *If $T_2^i(\alpha) \vdash \varphi(x)$, then $[\![\varphi(n)]\!]$ has some Σ_i^{qp}-LK derivation of height $(\log n)^{O(1)}$.*

iii) *If $\Sigma_i^b(\alpha)$-L^mIND $\vdash \varphi(x)$, then $[\![\varphi(n)]\!]$ has some Σ_i^{qp}-LK derivation of height $O(\log^{(m+1)} n)$.*

Combining this Theorem with the Cut-Elimination Theorem we obtain

COROLLARY 19. *Let $\varphi(x)$ be a formula in the language of bounded arithmetic, in which at most the variable x occurs free.*

i) *If $T_2^i(\alpha) \vdash \varphi(x)$ or $S_2^{i+1}(\alpha) \vdash \varphi(x)$, then $[\![\varphi(n)]\!]$ is Σ_i^{qp}-LK provable with height $(\log n)^{O(1)}$. In this case we say that $[\![\varphi(n)]\!]$ is poly-logarithmic-height restricted Σ_i^{qp}-LK provable.*

ii) *If $\Sigma_{m+i+1}^b(\alpha)$-L^{m+1}IND $\vdash \varphi(x)$, then $[\![\varphi(n)]\!]$ is Σ_i^{qp}-LK provable with height $2_m((\log^{(m+1)} n)^{O(1)})$. In this case we say that $[\![\varphi(n)]\!]$ is $2_m((\log^{(m+1)} n)^{O(1)})$-height restricted Σ_i^{qp}-LK provable.*

Height restricted proof systems have been subject of a technical report [5], which in particular covers the content of the next two sections. It is the second part of the previous Corollary where the technical condition discussed after Definition 11 comes into play. For example, if $m = 1$ then the resulting heights are of size $2^{(\log \log n)^c}$ which grow stronger than poly-logarithmically in general. Hence, having the additional restriction on the number of cut-formulas is a proper assertion, which is fulfilled as we are considering translations of bounded arithmetic derivations.

§6. **Cut-reduction by switching.** Usual cut-elimination procedures (like Gentzen or Tait style cut-elimination) eliminate outermost connectives of cut-formulas first. In general, the cost of applying such cut-elimination techniques is an exponential blow-up of certain parameters of derivations like their heights, as seen in the previous section. Later we want to show that the translations of the order induction principles need certain heights of LK-proofs. Our lower bounds technique will only work if the heights of the proofs grow sub-linear. Thus, in order to reduce the degree of cut formulas in the derivations in Corollary 19 we cannot apply the Cut-Elimination Theorem any further, as this would result in upper bounds on heights which grow too fast.

At this point, the elimination of cuts, which is necessary in our proof of lower bounds, needs a different cut-elimination technique which we call cut-reduction by switching. It relies on methods from Boolean complexity, i.e. Håstad's Switching Lemmas [11, 10]. In [14] such Boolean complexity techniques are successfully applied to reduce the complexity of Σ_i^{qp}-LK refutations. We will follow [9] where the same approach is used to reduce the complexity of oracle computations related to definable functions in bounded arithmetic. Cut-reduction by switching will reduce cuts "inside-out", but will leave the proof-skeleton unchanged, e.g. the heights will remain the same. The price will be that not only the cut-formulas are reduced, but also the formula which is derived. The idea is to find a so-called restriction (i.e. a partial substitution of propositional variables by truth values) for a given derivation of a formula φ such that after applying that restriction to the proof, cut-formulas are sufficiently reduced but the restriction of φ is sufficiently meaningful.

In order to formulate cut-reduction by switching, we need some notation. Our logarithms are always base 2.

(1) Fix $m \geq 1$, $i \geq 0$. Let $[m]$ denote the set $\{0, \ldots, m - 1\}$. For $x, y_1, \ldots, y_i \in \mathbb{N}$ let $p_{x,y_1,\ldots,y_i}$ be a Boolean variable, and let

$$B_i(m) = \{ p_{x,y_1,\ldots,y_i} : x, y_1, \ldots, y_i < m \}.$$

The cardinality of $B_i(m)$ is m^{i+1}. We shall henceforth use $\vec{y}$ as an abbreviation of $y_1, \ldots, y_i$ or $y_1, \ldots, y_{i-1}$, depending on the context it occurs. Note that $B_0(m)$ is the set of variables p_x with $x < m$.

(2) A propositional formula is Σ_1^t, if and only if it is a disjunction of conjunctions of at most t literals, i.e. if it is in $\Sigma_1^{S,t}$ for some S. A propositional formula is Π_1^t if and only if its negation is Σ_1^t, and it is Δ_1^t if and only if it is equivalent to both Σ_1^t and Π_1^t. A formula φ is *hereditarily* Δ_1^t, denoted by $\varphi \in \underline{\Delta}_1^t$, if and only if every sub-formula of φ is Δ_1^t. We inductively define for $i \geq 0$:

$$\varphi \in \Pi_i^{S,t} \Longleftrightarrow \neg\varphi \in \Sigma_i^{S,t},$$

$$\varphi \in \Sigma_0^{S,t} \Longleftrightarrow \varphi \in \underline{\Delta}_1^t,$$

$$\varphi \in \Sigma_1^{S,t} \Longleftrightarrow \varphi \equiv \bigvee_{j<w}\varphi_j \text{ and } \varphi_j \in \underline{\Delta}_1^t \text{ for all } j < w,$$

$$\varphi \in \Sigma_{i+2}^{S,t} \Longleftrightarrow \varphi \equiv \bigvee_{j<w}\varphi_j \text{ and } w \leq S \text{ and } \varphi_j \in \Pi_{i+1}^{S,t} \text{ for all } j < w.$$

Observe that for the definition of $\Sigma_1^{S,t}$, we do *not* assume $w \leq S$!

(3) We define for $x < m$ some general $\Sigma_i^{m,1}$-formulas $D_{i,m}(x)$ in m^i variables from $B_i(m)$. They compute so-called Sipser functions [10] and are defined by

$$D_{i,m}(x) \quad = \quad \bigwedge_{y_1 < m} \bigvee_{y_2 < m} \cdots \quad Q^{i-1}_{y_{i-1} < m} \quad Q^i_{y_i < m} \quad p_{x,\vec{y}}$$

where either Q^{i-1} or Q^i is $\bigwedge$, depending on whether i is even or odd, respectively, and the other is $\bigvee$.

(4) We are now ready to formulate cut-reduction by switching. The notation $B[p_x \leftarrow \varphi_x : x \in M]$ denotes the result of simultaneously replacing variable p_x by formula φ_x for all $x \in M$.

THEOREM 20 (Cut-Reduction by Switching). *Let $i \in \mathbb{N}$ and $\varepsilon \in \mathbb{R}$ with $i \geq 1$ and $0 < \varepsilon < \frac{1}{2}$. Let $M \subseteq \mathbb{N}$ be some infinite set. For $m \in M$, let $\eta_m \in \mathbb{N}$, $t = t(m) = m^{\frac{1}{2}-\varepsilon}$, $S = S(m) = 2^t$, B_m a formula with variables in $B_0(m)$, and $C_m \subset \Sigma_i^{S,t}$ with $|C_m| \leq S$. Furthermore, assume that $B_m[p_x \leftarrow D_{i,m}(x) : x < m]$ is C_m-LK provable with height η_m.*

Then, for all $m \in M$ which are sufficiently large, there is some $Q \subset [m]$ such that

i) $|[m] \setminus Q| \geq \sqrt{m \cdot \log m}$;
ii) $B_m[p_x \leftarrow 0 : x \in Q]$ *is Δ_1^t-LK provable with height η_m.*

We now sketch the proof of this Theorem. We go on introducing notation.
(5) Let $i, m \geq 1$. We have already defined sets $B_i(m)$ of propositional variables. They are partitioned into blocks via

$$(B_i(m))_{(x,y_1,\ldots,y_{i-1})} := \{ p_{x,y_1,\ldots,y_{i-1},z} : z < m \}$$

for $(x, y_1, \ldots, y_{i-1}) \in [m]^i$.

(6) A restriction ρ on $B_i(m)$ is a map going from $B_i(m)$ to $\{0, 1, *\}$:

$$\rho : B_i(m) \longrightarrow \{0, 1, *\}.$$

We should think of $\rho(p) = 0$ or $\rho(p) = 1$ as p is *replaced by* 0 or 1 respectively, and of $\rho(p) = *$ as p is *left unchanged*. Alternatively, we can think of ρ as a partial map going from $B_i(m)$ to $\{0, 1\}$.

(7) The probability space $\mathbb{R}_{i,m}^+(q)$ of restrictions ρ for $0 < q < 1$ is given as follows. Let $x < m$, $\vec{y} \in [m]^{i-1}$ and $y_i < m$.

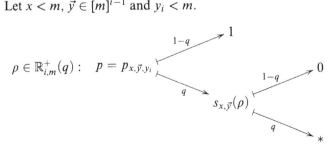

Meaning: first choose $s_{x,\bar{y}}$ such that $s_{x,\bar{y}} = *$ with probability q and $s_{x,\bar{y}} = 0$ with probability $1 - q$; then choose $\rho(p)$ such that $\rho(p) = s_{x,\bar{y}}$ with probability q and $\rho(p) = 1$ with probability $1 - q$.

Define $\mathbb{R}^-_{i,m}(q)$ by interchanging 0 and 1:

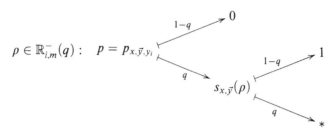

$$\rho \in \mathbb{R}^-_{i,m}(q): \quad p = p_{x,\bar{y},y_i}$$

(8) Let $\rho \in \mathbb{R}^+_{i,m}(q)$. We define a transformation $\lceil_{g\rho}$ which maps formulas with variables in $B_i(m)$ to formulas with variables in $B_{i-1}(m)$:

i) Apply ρ.
ii) Assign 1 to every $p_{x,\bar{y},z}$ with $\rho(p_{x,\bar{y},z}) = *$ such that there is some $z < z' < m$ with $\rho(p_{x,\bar{y},z'}) = *$. I.e., all but one variable in a block are touched.
iii) Rename each $p_{x,\bar{y},z}$ by $p_{x,\bar{y}}$.

For $\rho \in \mathbb{R}^-_{i,m}(q)$ replace 1 by 0.

(9) The following lemma is Håstad's second switching lemma, see [10].

LEMMA 21 (Håstad [10]). *Let* $i \geq 1$ *and* $v \in \{+, -\}$. *Let* φ *be a* $\Sigma^{S,t}_{i+1}$-*formula with variables from* $B_i(m)$ *and* $0 < q < 1$. *Then*

$$\Pr_{\rho \in \mathbb{R}^v_{i,m}(q)} \left[\varphi\lceil_{g\rho} \notin \Sigma^{S,t}_i \right] \leq S^i \cdot (6qt)^t.$$

I.e., the probability of a randomly chosen ρ *from* $\mathbb{R}^v_{i,m}(q)$ *that the formula* $\varphi\lceil_{g\rho}$ *is not equivalent to some* $\Sigma^{S,t}_i$-*formula is at most* $S^i \cdot (6qt)^t$.

(10) For the following inductive proof, the previously defined Sipser functions $D_{i,m}(x)$ have to be modified. We define $\bar{D}_{i,m}(x)$ for every $x < m$ with variables from $B_i(m)$. They compute modified Sipser functions (cf. [10, 9]) and are defined by

$$\bar{D}_{i,m}(x) = \bigwedge_{y_1 < m} \bigvee_{y_2 < m} \cdots Q^{i-1}_{y_{i-1} < m} \underset{y_i < \sqrt{\frac{1}{2}(i+1)m \log m}}{Q^i} p_{x,\bar{y}}$$

where either Q^{i-1} or Q^i is $\bigwedge$, depending on whether i is even or odd, respectively, and the other is $\bigvee$. Note that for distinct x, the formulas $\bar{D}_{i,m}(x)$ contain distinct propositional variables.

(11) The next lemma is also due to Håstad [10]. We repeat essentially the version stated by Buss and Krajíček [9].

We say that a formula φ contains formula ψ, written as $\psi \subseteq \varphi$, if by renaming and/or erasing some variables, we can transform φ into ψ.

LEMMA 22. Let m be big (i.e. $m \geq 10^{30}$), $i \geq 1$, $\overline{m} := \sqrt{\frac{1}{2}(i+1)m \log m}$, $q := \sqrt{\frac{2(i+1)\log m}{m}}$ and assume $q \leq \frac{1}{5}$. Then the following holds:

i) Assume $i \geq 2$ and let $v(i) = +$ or $v(i) = -$ if i is odd or even respectively. For all $x < m$:

$$\Pr_{\rho \in \mathbb{R}_{i,m}^{v(i)}(q)} \left[\bar{D}_{i-1,m}(x) \not\subseteq \bar{D}_{i,m}(x)\restriction_{g\rho} \right] \leq \frac{1}{3}m^{-2}.$$

I.e., the probability of a randomly chosen ρ from $\mathbb{R}_{i,m}^{v(i)}(q)$ that the formula $\bar{D}_{i,m}(x)\restriction_{g\rho}$ does not contain $\bar{D}_{i-1,m}(x)$ is at most $\frac{1}{3}m^{-2}$.

ii) For $i = 1$ we have for all $x < m$:

$$\Pr_{\rho \in \mathbb{R}_{1,m}^{+}(q)} \left[\bar{D}_{1,m}(x)\restriction_{g\rho} = 1 \right] \leq \frac{1}{6}m^{-2}.$$

I.e., the probability of a randomly chosen ρ from $\mathbb{R}_{1,m}^{+}(q)$ that the formula $\bar{D}_{1,m}(x)$ is transformed to 1 by $\restriction_{g\rho}$ is at most $\frac{1}{6}m^{-2}$.
For $R \subseteq [m]$ with $|R| \geq m$ we have

$$\Pr_{\rho \in \mathbb{R}_{1,m}^{+}(q)} \left[|\{x \in R : s_x(\rho) = *\}| \geq \frac{1}{2}q \cdot |R| \right] \geq 1 - \frac{1}{6}m^{-2}.$$

I.e., the probability of a randomly chosen ρ from $\mathbb{R}_{1,m}^{+}(q)$ that for at least an $\frac{1}{2}q$-fraction of R the corresponding variables p_x are left unchanged by ρ (i.e. are assigned $*$) is at least $1 - \frac{1}{6}m^{-2}$.

(12) Utilising this we obtain the following lemmas which immediately proof our Cut-Reduction by Switching Theorem 20. For the rest of this section fix $\varepsilon \in \mathbb{R}$ with $0 < \varepsilon < \frac{1}{2}$. Fix some infinite set $M \subseteq \mathbb{N}$. For $m \in M$, let $t = t(m) = m^{\frac{1}{2}-\varepsilon}$, $S = S(m) = 2^t$, and B_m a formula with variables in $B_0(m)$.

LEMMA 23. Let $i \geq 1$, let $f : \mathbb{N} \to \mathbb{N}$ be some function, and let $\mathcal{C}_m \subseteq \Sigma_{i+1}^{S,t}$ be given such that $|\mathcal{C}_m| \leq S$ and $B_m[p_x \leftarrow \bar{D}_{i+1,m}(x): x < m]$ is $\mathcal{C}_m$-LK provable with height $f(m)$ for all $m \in M$.

Then, for $m \in M$ sufficiently large, there is some $\mathcal{C}'_m \subset \Sigma_i^{S,t}$ such that $|\mathcal{C}'_m| \leq S$ and $B_m[p_x \leftarrow \bar{D}_{i,m}(x): x < m]$ is $\mathcal{C}'_m$-LK provable with height $f(m)$.

LEMMA 24. Let $f : \mathbb{N} \to \mathbb{N}$ be some function, and let $\mathcal{C}_m \subset \Sigma_1^{S,t}$ be given such that $|\mathcal{C}_m| \leq S$ and $B_m[p_x \leftarrow \bar{D}_{1,m}(x): x < m]$ is $\mathcal{C}_m$-LK provable with height $f(m)$ for all $m \in M$.

Then, for $m \in M$ sufficiently large, there is some $Q = Q_m \subseteq [m]$ such that

i) $|[m] \setminus Q| \geq \sqrt{m \cdot \log m}$;
ii) $B_m[p_x \leftarrow 0: x \in Q]$ is $\boldsymbol{\Delta}_1^t$-LK provable with height $f(m)$.

§7. **Lower bounds on heights of Σ_i^{qp}-LK-proofs of order induction.** In this section we will prove lower bounds on heights of Σ_i^{qp}-LK proofs of the order induction principle for some particular Σ_i^{qp}-property (given by the Sipser functions $D_{i,m}(x)$). This will be obtained by applying the Cut-Reduction by Switching Theorem from the previous Section and the lower bound theorem for Δ_1^i-resolution proofs of the order induction principle to be proven next. This lower bound is also called "Boundedness Theorem" in the setting of ordinal analysis.

The order induction principle $\mathcal{O}\mathrm{Ind}(m)$ is given by the formula

$$\mathcal{O}\mathrm{Ind}(m) := \bigwedge_{x<m}\left(\left(\bigwedge_{y<x}p_y\right) \to p_x\right) \to \bigwedge_{x<m}p_x$$

(of course $A \to B$ is an abbreviation of $\bigvee\{\neg A, B\}$). The meaning is easily understood if we consider its contraposition which expresses minimisation: if some variables among $p_0, \ldots, p_{m-1}$ are false then there is one with minimal index. It is the translation of our previously defined $\mathcal{L}_{BA}$-formula $\mathcal{O}\mathrm{Ind}(m, \alpha)$ to LK.

THEOREM 25 (Boundedness). $\left|\frac{\eta}{\Delta_1^i}\right.$ $\mathcal{O}\mathrm{Ind}(n) \Rightarrow n \leq \eta \cdot t$.

We will give a detailed proof of this Theorem in the next subsection. But before we do this we utilise the Boundedness Theorem. The complexity of the order induction principle is extended by replacing variables p_x by the Sipser function $D_{i,m}(x)$ from the previous section. The next theorem states the lower bound for Σ_i^{qp}-LK derivations of the extended order induction principle. With $\mathrm{rng}(f)$ we will denote the range of a number-theoretic function f.

THEOREM 26. *Let $i \in \mathbb{N}$ with $i \geq 1$. Let $f, \eta : \mathbb{N} \to \mathbb{N}$ be some number-theoretic functions such that $\eta(n) = (\log n)^{\Omega(1)}$. Assume that $\mathcal{O}\mathrm{Ind}(f(n))$ $[p_x \leftarrow \neg D_{i,f(n)}(x): x < f(n)]$ is Σ_i^{qp}-LK provable with height $\eta(n)$. Then, $\eta(n) = f(n)^{\Omega(1)}$, or, equivalently, $f(n) = \eta(n)^{O(1)}$.*

PROOF. Assume for the sake of contradiction that the assumptions of the Theorem are satisfied, but $f(n) \neq \eta(n)^{O(1)}$. In particular, $\mathrm{rng}(f)$ must be unbounded.

By assumption, we have that $\mathcal{O}\mathrm{Ind}(f(n))[p_x \leftarrow \neg D_{i,f(n)}(x): x < f(n)]$ is Σ_i^{qp}-LK provable with height $\eta(n)$. This means that there is some $c \in \mathbb{N}$ such that for all $n \in \mathbb{N}$ we can fix a set of formulas $\bar{C}_n$ with the following properties: let $\bar{t} = \bar{t}(n) = (\log n)^c$ and $\bar{S} = \bar{S}(n) = 2^{\bar{t}}$, then $\bar{C}_n \subseteq \Sigma_i^{\bar{S},\bar{t}}$, $|\bar{C}_n| \leq \bar{S}$ and $\mathcal{O}\mathrm{Ind}(f(n))[p_x \leftarrow \neg D_{i,f(n)}(x): x < f(n)]$ is $\bar{C}_n$-LK provable with height $\eta(n)$. By assumption $\log n = \eta(n)^{O(1)}$, hence there is some d and $N_0 \in \mathbb{N}$ such that $\log n \leq \eta(n)^d$ for $n \geq N_0$. W.l.o.g., we can choose $c \cdot d > 1$.

We will construct some infinite subset M of $\mathrm{rng}(f)$ which can be used to apply the Cut-Elimination by Switching Theorem. Let m_0 be any number. We will construct some $m_1 > m_0$ which we will put into the set M. Fix some $n_0 \geq N_0$ with $(\log n_0)^{4c} \geq m_0$. As $f(n) \neq \eta(n)^{O(1)}$ there must be some $n_1 > n_0$ satisfying $f(n_1) > \eta(n_1)^{d \cdot 4 \cdot c}$. Let $m_1 := f(n_1)$, then $m_1 > \eta(n_1)^{d \cdot 4 \cdot c} \geq (\log n_1)^{4c} \geq m_0$. Thus $(\log n_1)^c \leq m_1^{\frac{1}{4}}$ and $m_0 < m_1$. Hence $\bar{C}_{n_1} \subseteq \Sigma_i^{S,t}$ and $|\bar{C}_{n_1}| \leq S$ for $t := m_1^{\frac{1}{4}}$ and $S := 2^t$. Put m_1 into the set M and define $\mathcal{C}_{m_1} := \bar{C}_{n_1}$ and $\eta_{m_1} := \eta(n_1)$. Go on defining $m_2, m_3, \ldots$ in the same fashion.

Then, the prerequisites of the Cut-Reduction by Switching Theorem are satisfied, and we obtain some large $m \in M$, some set $Q \subset [m]$ not too big (i.e. $|[m] \setminus Q| \geq \sqrt{m \cdot \log m} \geq \sqrt{m}$) and a Δ_1^t-LK derivation of $\mathcal{O}\mathrm{Ind}(m)[p_x \leftarrow 1 : x \in Q]$ of height η_m. By pruning and renaming of variables this can be transformed into a Δ_1^t-LK derivation of $\mathcal{O}\mathrm{Ind}(m - |Q|)$ of height η_m, hence the Boundedness Theorem yields $m - |Q| \leq \eta_m \cdot t = \eta_m \cdot m^{\frac{1}{4}}$, which together with the largeness condition on Q rewrites to $\eta_m \geq m^{\frac{1}{4}}$. By construction of M there is some n such that $\eta(n) = \eta_m$ and $m = f(n) > \eta(n)^{d \cdot 4 \cdot c}$, contradicting the previously obtained $m \leq \eta(n)^4$, as $c \cdot d > 1$. $\dashv$

7.1. The proof of the Boundedness Theorem. For this subsection we fix $t \in \mathbb{N}$, $t \geq 1$. By $\overset{\eta}{\underset{\bullet}{\vdash}} \varphi$ we denote that φ is Δ_1^t-LK provable with height η. A formula φ will always be one in the language of LK. We want to prove the Boundedness Theorem, i.e.

$$\overset{\eta}{\underset{\bullet}{\vdash}} \mathcal{O}\mathrm{Ind}(n) \implies n \leq \eta \cdot t.$$

Before we can do this we first have to fix some suitable notation.

Let φ be an LK-formula. For a set $M \subseteq \mathbb{N}$ we define $\varphi[M]$ to be the result of replacing p_i by 1 if $i \in M$, and by 0 if $i \notin M$. Then let $M \vDash \varphi$ if and only if $\varphi[M]$ is true.

For two sets $M^+, M^- \subseteq \mathbb{N}$ we define $[M^+, M^-]$ to be the set of all subsets M of $\mathbb{N}$ that contain M^+ but are disjoint from M^-:

$$[M^+, M^-] := \{ M : M^+ \subseteq M \subseteq \mathbb{N} \setminus M^- \}.$$

DEFINITION 27. For a formula φ and a truth value $v \in \{0, 1\}$ we define that (M^+, M^-) fixes φ to v, if and only if M^+ and M^- are disjoint subsets of $\mathbb{N}$ (this implies $[M^+, M^-] \neq \emptyset$) and the truth of φ is fixed on $[M^+, M^-]$ to v, i.e. $\varphi[M] = v$ for all $M \in [M^+, M^-]$. We say that (M^+, M^-) fixes φ, if and only if (M^+, M^-) fixes φ to some truth value $v \in \{0, 1\}$.

A true Δ_1^t-formula φ can always be fixed to 1 by a pair M^+, M^- which is small, i.e. the cardinality of M^+ and M^- together is bounded by t, denoted by $|M^+| + |M^-| \leq t$. In addition, M^+, M^- can be chosen to respect any given satisfying assignment of φ:

LEMMA 28. Let $\varphi \in \Delta_1^t$ and $M_0 \subseteq \mathbb{N}$ such that $M_0 \vDash \varphi$. Then there are $M^+ \subseteq M_0$ and $M^- \subseteq \mathbb{N}$ satisfying $|M^+| + |M^-| \leq t$, $M_0 \cap M^- = \emptyset$ and (M^+, M^-) fixes φ to 1.

PROOF. The assumption $\varphi \in \Delta_1^t$ particularly implies $\varphi \in \Delta_1^t$. Hence, φ is equivalent to some $\bigvee_{x<S} \bigwedge_{y<t} \theta_{xy}$ for some S and some literals θ_{xy}. From the assumption $M_0 \vDash \varphi$ it follows that there is some $x_0 < S$ such that $M_0 \vDash \bigwedge_{y<t} \theta_{x_0 y}$. Fix such an $x_0 < S$. Let

$$M^+ := \{i : \theta_{x_0 y} = p_i \text{ for some } y < t\},$$
$$M^- := \{i : \theta_{x_0 y} = \neg p_i \text{ for some } y < t\}.$$

Then the assertion follows. ⊣

The following Lemma is the main technical part for proving the Boundedness Theorem 25. Let

$$\mathcal{O}\mathrm{Prog}(m) := \bigwedge_{x<m} \left(\left(\bigwedge_{y<x} p_y \right) \to p_x \right)$$

hence $\mathcal{O}\mathrm{Ind}(m)$ has the form $\neg \mathcal{O}\mathrm{Prog}(m) \vee \bigwedge_{x<m} p_x$.

LEMMA 29. $\vdash_\bullet^\eta \neg \mathcal{O}\mathrm{Prog}(n), p_m \Rightarrow m < \eta \cdot t$.

PROOF OF THE BOUNDEDNESS THEOREM 25. Assume $\vdash_\bullet^\eta \mathcal{O}\mathrm{Ind}(n)$. By applying first $\bigvee$-Exportation and then $\bigwedge$-Inversion from Section 5 we obtain $\vdash_\bullet^\eta \neg \mathcal{O}\mathrm{Prog}(n), p_{n-1}$. Hence, the above Lemma shows $n - 1 < \eta \cdot t$ and the assertion follows. ⊣

PROOF OF THE ABOVE LEMMA. Assume for the sake of contradiction that

$$\vdash_\bullet^\eta \neg \mathcal{O}\mathrm{Prog}(n), p_m \quad \text{and} \quad \eta \cdot t \leq m.$$

For a finite set $M \subseteq \mathbb{N}$ let $\overline{\mathrm{en}}_M$ denote the enumeration function of $\mathbb{N} \setminus M$. Let $\mathcal{R}^\gamma(M)$ be the set $\{a : a < \overline{\mathrm{en}}_M(\gamma)\} \cup M$.

We will construct by recursion on l sets $\Delta_l \subseteq \Delta_1^t$, $M_l^+, M_l^- \subseteq \mathbb{N}$ for $l = \eta, \ldots, 0$ satisfying the property $\mathcal{G}(l, \Delta_l, M_l^+, M_l^-)$ given by

  i) $\vdash_\bullet^l \neg \mathcal{O}\mathrm{Prog}(n), \Delta_l$.
  ii) $|M_l^+| + |M_l^-| \leq t \cdot (\eta - l)$.
  iii) all $\varphi \in \Delta_l$, which are not variables, are fixed by (M_l^+, M_l^-) to 0.
  iv) $\mathcal{R}^{l \cdot t}(M_l^+) \nvDash \Delta_l$.
  v) $\mathcal{R}^{l \cdot t}(M_l^+) \cap M_l^- = \emptyset$.

For $l = 0$ the assertion follows. Because, if we have constructed $\Delta_0 \subseteq \Delta_1^t$, $M_0^+, M_0^- \subseteq \mathbb{N}$ which satisfy $\mathcal{G}(0, \Delta_0, M_0^+, M_0^-)$, then $\mathcal{G}(0, \Delta_0, M_0^+, M_0^-)$ i) shows $\vdash_\bullet^0 \neg \mathcal{O}\mathrm{Prog}(n), \Delta_0$, hence Δ_0 must be an axiom. But this contradicts $\mathcal{G}(0, \Delta_0, M_0^+, M_0^-)$ iv) and the assertion follows.

We now prove the assertion by backwards-induction from $l = \eta$ to 0. To start the induction for $l = \eta$ let $\Delta_\eta := \{p_m\}$ and $M_\eta^+ := M_\eta^- := \emptyset$. Then $\mathcal{G}(\eta, \Delta_\eta, M_\eta^+, M_\eta^-)$ i), ii), iii), v) immediately follow. For $\mathcal{G}(\eta, \Delta_\eta, M_\eta^+, M_\eta^-)$ iv) observe that $\overline{\mathrm{en}}_\emptyset(\eta \cdot t) = \eta \cdot t \leq m$, hence $m \notin \mathcal{R}^{\eta \cdot t}(\emptyset)$.

For the induction step $l+1 \rightsquigarrow l$ assume that we have constructed $\Delta_{l+1} \subseteq \Delta_1^t$, $M_{l+1}^+, M_{l+1}^- \subseteq \mathbb{N}$ satisfying $\mathcal{G}(l+1, \Delta_{l+1}, M_{l+1}^+, M_{l+1}^-)$. We will consider the last inference in $\mathcal{G}(l+1, \Delta_{l+1}, M_{l+1}^+, M_{l+1}^-)$ i) which leads to $\vdash_{\bullet}^{l+1} \neg \mathcal{O}\mathrm{Prog}(n), \Delta_{l+1}$. Let $\mathcal{R}^*$ abbreviate $\mathcal{R}^{(l+1)\cdot t}(M_{l+1}^+)$. In order to simplify sub-cases, we first argue that it is enough to find some $\psi \in \Delta_1^t$ and $M^+, M^- \subseteq \mathbb{N}$ satisfying the following property:

I) $\vdash_{\bullet}^{l} \neg \mathcal{O}\mathrm{Prog}(n), \Delta_{l+1}, \psi$.
II) (M^+, M^-) fixes ψ to 0.
III) $|M^+| + |M^-| \leq t$.
IV) $M^+ \subseteq \mathcal{R}^*$.
V) $\mathcal{R}^* \cap M^- = \emptyset$.

Then, $\Delta_l := \Delta_{l+1} \cup \{\psi\}$, $M_l^+ := M_{l+1}^+ \cup M^+$, $M_l^- := M_{l+1}^- \cup M^-$ will satisfy property $\mathcal{G}(l, \Delta_l, M_l^+, M_l^-)$, because $\mathcal{G}(l, \Delta_l, M_l^+, M_l^-)$ i) and ii) are obvious; and for $\mathcal{G}(l, \Delta_l, M_l^+, M_l^-)$ iii), iv) and v) we observe

A) $\mathcal{R}^{l\cdot t}(M_l^+) \subseteq \mathcal{R}^{l\cdot t+t}(M_{l+1}^+) \cup M^+ = \mathcal{R}^*$. This follows, because $\overline{\mathrm{en}}_{M \cup \{a\}} (\gamma) \leq \overline{\mathrm{en}}_M (\gamma + 1)$, hence $\mathcal{R}^\gamma(M \cup \{a\}) \subseteq \mathcal{R}^{\gamma+1}(M) \cup \{a\}$.
B) V) and the induction hypothesis $\mathcal{G}(l+1, \Delta_{l+1}, M_{l+1}^+, M_{l+1}^-)$ v) imply $\mathcal{R}^* \cap M_l^- = \emptyset$, hence $\mathcal{G}(l, \Delta_l, M_l^+, M_l^-)$ v) follows using A).
C) A) and B) show $\emptyset \neq [M_l^+, M_l^-]$. By construction $[M_l^+, M_l^-] \subseteq [M_{l+1}^+, M_{l+1}^-]$, hence (M_l^+, M_l^-) fixes all $\varphi \in \Delta_{l+1}$ which are not variables, to 0. Furthermore, $[M_l^+, M_l^-] \subseteq [M^+, M^-]$, hence II) implies that ψ is fixed to 0 by (M_l^+, M_l^-). Thus, $\mathcal{G}(l, \Delta_l, M_l^+, M_l^-)$ iii) follows.
D) Utilising B) and A) we obtain $\mathcal{R}^{l\cdot t}(M_l^+), \mathcal{R}^* \in [M_l^+, M_l^-]$ hence $\mathcal{G}(l+1, \Delta_{l+1}, M_{l+1}^+, M_{l+1}^-)$ iv) shows that (M_l^+, M_l^-) fixes every formula in Δ_{l+1} to 0. In particular, $\mathcal{R}^{l\cdot t}(M_l^+) \not\vDash \Delta_{l+1}$, which shows $\mathcal{G}(l, \Delta_l, M_l^+, M_l^-)$ iv).

Now we distinguish sub-cases according to the last inference which leads to $\vdash_{\bullet}^{l+1} \neg \mathcal{O}\mathrm{Prog}(n), \Delta_{l+1}$. In the sub-cases, we either construct ψ, M^+, M^- satisfying I) to V), or we directly construct Δ_l, M_l^+, M_l^- satisfying $\mathcal{G}(l, \Delta_l, M_l^+, M_l^-)$, depending which is easier.

($\bigwedge$): There is some $\varphi = \bigwedge_{j<J} \varphi_j \in \Delta_{l+1}$ such that $\vdash_{\bullet}^{l} \neg \mathcal{O}\mathrm{Prog}(n), \Delta_{l+1}, \varphi_j$ for all $j < J$. By induction hypothesis $\mathcal{G}(l+1, \Delta_{l+1}, M_{l+1}^+, M_{l+1}^-)$ iv) we have that $\mathcal{R}^* \not\vDash \varphi$. Thus, there is some $j_0 < J$ such that $\mathcal{R}^* \not\vDash \varphi_{j_0}$.
Let $\psi := \varphi_{j_0}$, then $\psi \in \Delta_1^t$ [$\Rightarrow$ I)]. By Lemma 28 there are some $M^+ \subseteq \mathcal{R}^*$ [$\Rightarrow$ IV)] and $M^- \subseteq \mathbb{N}$ such that $\mathcal{R}^* \cap M^- = \emptyset$ [$\Rightarrow$ V)], $|M^+| + |M^-| \leq t$ [$\Rightarrow$ III)] and (M^+, M^-) fixes ψ to 0 [$\Rightarrow$ II)].

($\bigvee$): The first sub case is that $\neg \mathcal{O}\mathrm{Ind}(n)$ is not the main formula of the inference. Then, there is some $\varphi = \bigvee_{j<J} \varphi_J \in \Delta_{l+1}$ such that $\vdash_{\bullet}^{l} \neg \mathcal{O}\mathrm{Prog}(n), \Delta_{l+1}, \varphi_{j_0}$ for some $j_0 < J$. By induction hypothesis $\mathcal{G}(l+1, \Delta_{l+1}, M_{l+1}^+, M_{l+1}^-)$ iv) we have that $\mathcal{R}^* \not\vDash \varphi$, thus also $\mathcal{R}^* \not\vDash \varphi_{j_0}$. Now the same argumentation as in the $\bigwedge$-case can be applied.

Now assume that the main formula is $\neg \mathcal{O}\mathrm{Ind}(n)$. Then, there is some $x < n$ such that

$$\vdash^{ll}_{\bullet} \neg \mathcal{O}\mathrm{Prog}(n), \Delta_{l+1}, \left(\bigwedge_{y<x} p_y\right) \wedge \neg p_x.$$

A) Assume, there is some $y < x$ such that $y \notin \mathcal{R}^{l \cdot t}(M^+_{l+1})$. By $\bigwedge$-Inversion we obtain $\vdash^{ll}_{\bullet} \neg \mathcal{O}\mathrm{Prog}(n), \Delta_{l+1}, p_y$. Let $\Delta_l := \Delta_{l+1}, p_y$ $[\Rightarrow \mathcal{G}(l, \Delta_l, M^+_l, M^-_l) \ \mathrm{i})]$, $M^+_l := M^+_{l+1}$ and $M^-_l := M^-_{l+1}$ $[\Rightarrow \mathcal{G}(l, \Delta_l, M^+_l, M^-_l) \ \mathrm{ii})$, $\mathrm{iii})]$. Now $\mathcal{R}^{l \cdot t}(M^+_l) \subseteq \mathcal{R}^*[\Rightarrow \mathcal{G}(l, \Delta_l, M^+_l, M^-_l) \ \mathrm{v})]$, hence, using the assumption $y \notin \mathcal{R}^{l \cdot t}(M^+_l)$, we obtain $\mathcal{R}^{l \cdot t}(M^+_l) \nvDash \Delta_l$ $[\Rightarrow \mathcal{G}(l, \Delta_l, M^+_l, M^-_l) \ \mathrm{iv})]$.

B) Now assume that A) does not hold, hence $y \in \mathcal{R}^{l \cdot t+1}(M^+_{l+1})$ for all $y < x$. This implies $x \in \mathcal{R}^{l \cdot t+1}(M^+_{l+1}) \subseteq \mathcal{R}^*$. Because, $\overline{\mathrm{en}}_M(\gamma) \notin \mathcal{R}^\gamma(M)$, hence $y \in \mathcal{R}^\gamma(M)$ for all $y < x$ implies $\overline{\mathrm{en}}_M(\gamma) \geq x$, hence $\overline{\mathrm{en}}_M(\gamma + 1) > x$ and in sequel $x \in \mathcal{R}^{\gamma+1}(M)$.
By $\bigwedge$-Inversion we obtain $\vdash^{ll}_{\bullet} \neg \mathcal{O}\mathrm{Prog}(n), \Delta_{l+1}, \neg p_x$. Let $\psi := \neg p_x$ $[\Rightarrow \mathrm{I})]$, $M^+ := \{x\}$ and $M^- := \emptyset$ $[\Rightarrow \mathrm{II})$, $\mathrm{III})$, $\mathrm{IV})$, $\mathrm{V})]$.

(Cut): There is some $\varphi \in \Delta^t_1$ such that $\vdash^{ll}_{\bullet} \neg \mathcal{O}\mathrm{Prog}(n), \Delta_{l+1}, \varphi$ and $\vdash^{ll}_{\bullet} \neg \mathcal{O}\mathrm{Prog}(n), \Delta_{l+1}, \neg \varphi$. W.l.o.g. we may assume $\mathcal{R}^* \nvDash \varphi$. The same argumentation as in the $\bigwedge$-case yields the assertion. $\dashv$

§8. Generalised dynamic ordinals revisited.

We have now collected all tools which are needed to compute the missing upper bounds on generalised dynamic ordinals. Our strategy will be to translate a $\Sigma^b_{m+i}(\alpha)$-$\mathrm{L}^m\mathrm{IND}$-proof of $\mathcal{O}\mathrm{Ind}(t(x), \Pi^b_i)$ to Σ^{qp}_i-LK, and then use the result on lower bounds of the order induction principle, Theorem 26, to obtain tight upper bounds on generalised dynamic ordinals.

For the following considerations fix some $i, m \in \mathbb{N}$ with $m > 0$ and some

$$f \in \mathrm{DO}_{i+1} \left(\Sigma^b_{m+i}(\alpha)\text{-}\mathrm{L}^m\mathrm{IND} \right).$$

First, we define some general $\Pi^b_i(\alpha)$-formula $A^{\alpha,i}(a, x)$, which is translated by the Paris-Wilkie translation to the Sipser function $D_{i,a}$ as defined before. $A^{\alpha,i}(a, x)$ is given by the formula

$$(\forall y_1 < a)\,(\exists y_2 < a) \cdots (Q^{i-1} y_{i-1} < a)\,(Q^i y_i < a)\, \alpha(\langle x, y_1, \ldots, y_i \rangle)$$

where either Q^{i-1} or Q^i is $\forall$, depending on whether i is even or odd, respectively, and the other is $\exists$. Here, $\langle z_1, \ldots, z_j \rangle$ denotes some sequence coding function expressible in the language of bounded arithmetic. Then, the definition of DO_{i+1} yields that there is some term t such that $f = \lambda x.t(x)$ and

$$\Sigma^b_{m+i}(\alpha)\text{-}\mathrm{L}^m\mathrm{IND} \vdash (\forall x)\,\mathcal{O}\mathrm{Ind}\left(t, \neg A^{\alpha,i}(t, \cdot)\right).$$

Utilising Theorem 19 shows that there is some $c \geq 1$ such that eventually

$$[\![\mathcal{O}\mathrm{Ind}(t(n), \neg A^{\alpha,i}(t(n), \cdot))]\!]$$

is Σ_i^{qp}-LK provable with height $2_{m-1}((\log^{(m)} n)^c)$. By identifying $p_{\langle x, \vec{y} \rangle}$ with $p_{x,\vec{y}}$, we see that these derivations transform to Σ_i^{qp}-LK proofs of

$$\mathcal{O}\text{Ind}(t(n))\big[p_x \leftarrow \neg D_{i,t(n)}(x): x < t(n)\big]$$

of height $2_{m-1}((\log^{(m)} n)^c)$.

Now we are in the situation that we can apply the Lower Bound Theorem 26, because $2_{m-1}((\log^{(m)} n)^c) = (\log n)^{\Omega(1)}$. Hence, we obtain that

$$f(n) = t(n) = 2_{m-1}\big((\log^{(m)} n)^c\big)^{O(1)} = 2_m\big(O\big(\log^{(m+1)} n\big)\big).$$

Together with Corollary 8, this shows:

THEOREM 30. *Let* $m > 0$. *The* $i + 1$ *generalised dynamic ordinal of the theory* $\Sigma_{m+i}^b(\alpha)$-L^mIND *can be described as*:

$$\text{DO}_{i+1}\big(\Sigma_{m+i}^b(\alpha)\text{-}L^m\text{Ind}\big) \equiv 2_m\big(O(|\,\text{id}\,|_{m+1})\big).$$

Hence we have for $i > 0$:

$$
\begin{aligned}
\text{DO}_i(T_2^i(\alpha)) &\equiv 2_2(O(|\,\text{id}\,|_2)) &\equiv \text{DO}_i(S_2^{i+1}(\alpha)), \\
\text{DO}_i(S_2^i(\alpha)) &\equiv 2_1(O(|\,\text{id}\,|_2)), \\
\text{DO}_i(sR_2^{i+1}(\alpha)) &\equiv 2_2(O(|\,\text{id}\,|_3)).
\end{aligned}
$$

We also compare definable multivalued functions of unrelativised theories with the generalised dynamic ordinals of their relativised companions.

THEOREM 31. *Let* $i \geq 0$. *For any theory* T *from the infinite list*
$$T_2^{i+1}, S_2^{i+2}, S_2^{i+1}, sR_2^{i+2}(= \Sigma_{i+2}^b\text{-}L^2\text{IND}), \Sigma_{i+3}^b\text{-}L^3\text{IND}, \ldots$$
we have:

A multivalued function f *is* Σ_{i+2}^b-*definable in* T, *if and only if*
$$f \in \text{FP}^{\Sigma_{i+1}^b}(\text{wit}, \log(\text{DO}_{i+1}(T(\alpha)))).$$

This indicates that generalised dynamic ordinals do in fact also characterise the *computational complexity* of bounded arithmetic theories.

REFERENCES

[1] THEODORE BAKER, JOHN GILL, and ROBERT SOLOVAY, *Relativizations of the* $\mathcal{P} = ?\mathcal{NP}$ *question*, **SIAM Journal on Computing**, vol. 4 (1975), no. 4, pp. 431–442.

[2] ARNOLD BECKMANN, *Seperating Fragments of Bounded Predicative Arithmetic*, Ph.D. thesis, Westfälische Wilhelms-Universität, Münster, 1996.

[3] ———, *A note on universal measures for weak implicit computational complexity*, **Logic for Programming, Artificial Intelligence, and Reasoning** (Matthias Baaz and Andrei Voronkov, editors), Lecture Notes in Computer Science, vol. 2514, Springer, Berlin, 2002, pp. 53–67.

[4] ———, *Dynamic ordinal analysis*, **Archive for Mathematical Logic**, vol. 42 (2003), no. 4, pp. 303–334.

[5] ———, *Height restricted constant depth LK*, Research Note Report TR03-034, Electronic Colloquium on Computational Complexity, 2003, http://www.eccc.uni-trier.de/eccc-reports/2003/TR03-034/.

[6] SAMUEL R. BUSS, *Bounded Arithmetic*, Studies in Proof Theory. Lecture Notes, vol. 3, Bibliopolis, Naples, 1986.

[7] ———, *Axiomatizations and conservation results for fragments of bounded arithmetic*, **Logic and Computation (Pittsburgh, PA, 1987)** (Wilfried Sieg, editor), Contemporary Mathematics, vol. 106, AMS, Providence, RI, 1990, pp. 57–84.

[8] ———, *Relating the bounded arithmetic and polynomial time hierarchies*, **Annals of Pure and Applied Logic**, vol. 75 (1995), no. 1-2, pp. 67–77.

[9] SAMUEL R. BUSS and JAN KRAJÍČEK, *An application of Boolean complexity to separation problems in bounded arithmetic*, **Proceedings of the London Mathematical Society. Third Series**, vol. 69 (1994), no. 1, pp. 1–21.

[10] JOHAN HÅSTAD, *Almost optimal lower bounds for small depth circuits*, **Randomness and Computation**, vol. 5 (1969), pp. 143–170.

[11] ———, **Computational Limitations of Small Depth Circuits**, MIT Press, Cambridge, MA, 1987.

[12] JAN JOHANNSEN, *A note on sharply bounded arithmetic*, **Archive for Mathematical Logic**, vol. 33 (1994), no. 2, pp. 159–165.

[13] JAN KRAJÍČEK, *Fragments of bounded arithmetic and bounded query classes*, **Transactions of the American Mathematical Society**, vol. 338 (1993), no. 2, pp. 587–598.

[14] ———, *Lower bounds to the size of constant-depth propositional proofs*, **The Journal of Symbolic Logic**, vol. 59 (1994), no. 1, pp. 73–86.

[15] ———, **Bounded Arithmetic, Propositional Logic, and Complexity Theory**, Cambridge University Press, Cambridge, 1995.

[16] JAN KRAJÍČEK, PAVEL PUDLÁK, and GAISI TAKEUTI, *Bounded arithmetic and the polynomial hierarchy*, **Annals of Pure and Applied Logic**, vol. 52 (1991), no. 1-2, pp. 143–153.

[17] DANIEL LEIVANT, *Substructural termination proofs and feasibility certification*, **Proceedings of the 3rd Workshop on Implicit Computational Complexity (Aarhus)**, 2001, pp. 75–91.

[18] JEFF B. PARIS and ALEX J. WILKIE, *Counting problems in bounded arithmetic*, **Methods in Mathematical Logic (Caracas, 1983)** (Carlos Augusto di Prisco, editor), Lecture Notes in Mathematics, vol. 1130, Springer, Berlin, 1985, pp. 317–340.

[19] WOLFRAM POHLERS, **Proof Theory: An introduction**, Lecture Notes in Mathematics, vol. 1407, Springer-Verlag, Berlin, 1989.

[20] CHRIS POLLETT, *Structure and definability in general bounded arithmetic theories*, **Annals of Pure and Applied Logic**, vol. 100 (1999), no. 1-3, pp. 189–245.

[21] GAISI TAKEUTI, RSUV *isomorphisms*, **Arithmetic, Proof Theory, and Computational Complexity (Prague, 1991)** (P. Clote and J. Krajíček, editors), Oxford Logic Guides, vol. 23, Oxford Univ. Press, New York, 1993, pp. 364–386.

[22] ANDREW C. YAO, *Separating the polynomial-time hierarchy by oracles*, **Proc. 26th Ann. IEEE Symp. on Foundations of Computer Science** (Robert E. Tarjan, editor), 1985, pp. 1–10.

[23] DOMENICO ZAMBELLA, *Notes on polynomially bounded arithmetic*, **The Journal of Symbolic Logic**, vol. 61 (1996), no. 3, pp. 942–966.

UNIVERSITY OF WALES SWANSEA
SINGLETON PARK
SWANSEA SA2 8PP, UK
E-mail: A.Beckmann@swansea.ac.uk

THE WORM PRINCIPLE

LEV D. BEKLEMISHEV

Abstract. In [4] a new approach to proof-theoretic analysis of Peano arithmetic was suggested based on the notion of graded provability algebra. Here we read off a version of independent combinatorial Hydra battle principle from graded provability algebra. This allows for simple independence proofs of both principles based on provability algebraic methods.

§1. Introduction. This paper is a companion to [4] where a proof-theoretic analysis of Peano arithmetic based on the concept of *graded provability algebra* is given. A particular feature of this approach was that a system of ordinal notation for ε_0 naturally emerges from the closed fragment of a certain decidable propositional modal logic. This unusual view of ε_0 suggests to have a closer look at some traditional applications of proof-theoretic analysis such as combinatorial independence results.

Here we present a simple statement of combinatorial nature that is independent of Peano arithmetic and is motivated by the provability algebraic view of ε_0. The principle asserts the termination of a certain combinatorial game similar to the well-known *Hydra battle* of L. Kirby and J. Paris [11] which we call the *Worm battle*. In fact, modulo some details, the Hydra principle and the Worm principle turn out to be mutually translatable and one can easily infer the independence of the one from that of the other.[1] Thus, we essentially provide an alternative proof of the Kirby–Paris independence result. However, the Worm principle also has some independent interest because of the naturality of its provability algebraic interpretation.

Incidentally, L. Carlucci brought to our attention a paper by M. Hamano and M. Okada [10] where a very similar principle is derived from a restricted version of the much stronger *Buchholz hydra battle* principle [6]. The authors call it *one-dimensional version* of Buchholz hydra battle and also establish its intertranslatability with the usual Hydra battle. This might indicate some

---

Supported by the Russian Foundation for Basic Research and the Council for Support of Leading Scientific Schools.

[1]This direct translatability has been noticed in private correspondence independently by A. Weiermann, G. Lee and L. Carlucci.

Logic Colloquium '02
Edited by Z. Chatzidakis, P. Koepke, and W. Pohlers
Lecture Notes in Logic, 27

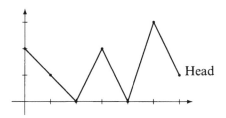

FIGURE 1. A Worm

(as yet unclear) relationships between our provability algebraic approach and the much stronger systems of ordinal notations.

We tried to make the present paper possibly self-contained. So, we reproduce main ingredients of the graded provability algebra approach at the beginning of the paper. There is essentially only one result — the reduction property of graded provability algebras — for the proof of which we refer the reader to [4, 3]. The results of this paper were presented in tutorial lectures at the European Logic Colloquium in Münster, Germany, August 2002.

§2. The Worm principle. The game deals with objects called *worms*. A worm is a just a finite function $f : [0, n] \to \mathbb{N}$. Worms can be specified as lists of natural numbers $w = (f(0), f(1), \ldots, f(n))$. For example, $w = 2102031$ is a worm (where we omit commas assuming all elements are <10). $f(n)$ is called the *head* of the worm. The empty worm is denoted by $\varnothing$.

Now we describe the rules of the game. Informally, the battle starts with an arbitrary worm and at each step the head of the worm is being affected so that it decreases by 1. In response the worm grows according to the two simple rules below. Unlike the original Hydra battle, the Worm battle is fully deterministic.

Formally, we specify a function $\mathrm{next}(w, m)$, where $w = (f(0), f(1), \ldots, f(n))$ is a worm and m is a step of the game:

1. If $f(n) = 0$ then $\mathrm{next}(w, m) := (f(0), \ldots, f(n-1))$. In this case the head of the worm is cut away.

2. If $f(n) > 0$ let $k := \max_{i<n} f(i) < f(n)$.

The worm w (with the head decreased by 1) is then the concatenation of two parts, the *good*[2] part $r := (f(0), \ldots, f(k))$, and the *bad* part $s := (f(k+1), \ldots, f(n-1), f(n) - 1)$. We define

$$\mathrm{next}(w, m) := r * \underbrace{s * s * \cdots * s}_{m+1 \text{ times}}.$$

Now let $w_0 := w$ and $w_{n+1} := \mathrm{next}(w_n, n+1)$.

---

[2]This part can also be empty.

As a typical example consider the worm $w = 2102031$ depicted in Figure 1. At first step we obtain $k = 4$; $r = 21020$; $s = 30$; $next(w, 1) = 210203030$. Then the game proceeds as follows:

$$w_0 = 2102031$$
$$w_1 = 210203030$$
$$w_2 = 21020303$$
$$w_3 = 21020302222$$
$$w_4 = 2102030222122212221222122212221$$
$$w_5 = 2102030(22212221222122212220)^6$$
$$\cdots$$

Notice that w_n is defined by primitive recursion. In fact, w_n is an elementary function of n and (the code of) w. This can be seen from the estimate

$$|w_n| \le (n + 2)! \cdot |w_0|$$

showing that the length of a worm grows only elementarily in the course of the game. Also notice that the maximal size of the elements of the worm can only decrease. This allows to write out a Δ_0-formula in three variables stating $w_n = u$.

The intended true PA-unprovable principle asserts that **E**very **W**orm eventually **D**ies:

$$\text{EWD} :\iff \forall w \exists n \; w_n = \varnothing.$$

THEOREM 1. EWD *is true but unprovable in* PA. *In fact*, EWD *is equivalent to* 1-*consistency of* PA *within* EA.

The notion of 1-consistency is introduced in the next section.

§3. *n***-provability and *n*-consistency.** As our basic fragment of arithmetic we take *elementary arithmetic* EA. The precise formulation of EA is not important, for definiteness we specify the language of EA as that of Peano arithmetic augmented by a symbol exp for the function 2^x and a symbol $\le$. Δ_0-formulas in the language of EA are those with all quantifier occurrences bounded by terms. Π_n- and Σ_n-formulas are obtained from Δ_0 by adding a quantifier prefix in the standard way. Axioms of EA consist of some minimal set of open defining axioms for all the symbols of the language and the induction schema for Δ_0-formulas. Peano arithmetic PA can be obtained from EA by adding the full induction schema. We shall also use an extension of EA by an axiom stating that the superexponentiation function $\lambda x \cdot \exp^{(x)}(x)$ is total, denoted EA^+. EA and EA^+ are finitely axiomatizable fragments of primitive recursive arithmetic PRA.

Let $Th_{\Pi_n}(\mathbb{N})$ denote the set of all true arithmetical Π_n-sentences. A theory T is called *n-consistent* if $T + Th_{\Pi_n}(\mathbb{N})$ is consistent. Let $n\text{-Con}(T)$ be the natural Π_{n+1}-formula expressing the n-consistency of T. $\langle n \rangle_T \varphi$ stands for n-Con$(T + \varphi)$.

The formula $[n]_T \varphi := \neg \langle n \rangle_T \neg \varphi$ formalizes the n-provability of φ in T. Thus, $[n]_T \varphi$ asserts that φ is provable from the axioms of T and some true Π_n-sentences. For $n = 0$ these concepts coincide with the usual Gödel's consistency assertion and provability predicate for T. We also write $\Box$ for $[0]$ and $\Diamond$ for $\langle 0 \rangle$.

Properties of the n-provability are very similar to those of the usual provability predicate. First of all, the n-provability predicate $[n]_T$ satisfies Bernays–Löb derivability conditions.

PROPOSITION 1.

(i) $T \vdash \varphi \Rightarrow \mathsf{EA} \vdash [n]_T \varphi$;

(ii) $\mathsf{EA} \vdash [n]_T(\varphi \to \psi) \to ([n]_T \varphi \to [n]_T \psi)$;

(iii) $\mathsf{EA} \vdash [n]_T \varphi \to [n]_T [n]_T \varphi$.

The last condition is a consequence of a more general fact known as provable Σ_{n+1}-completeness.

PROPOSITION 2. *For any Σ_{n+1}-formula $\sigma(x_1, \ldots, x_k)$, with exactly the variables $x_1, \ldots, x_k$ free,*

$$\mathsf{EA} \vdash \sigma(x_1, \ldots, x_k) \longrightarrow [n]_T \sigma(\dot{x}_1, \ldots, \dot{x}_k).$$

Here $\sigma(\dot{x}_1, \ldots, \dot{x}_k)$ under $[n]_T$ denotes the definable Kalmar elementary term for the function that, given a tuple $n_1, \ldots, n_k$, outputs the code $\ulcorner \sigma(\bar{n}_1, \ldots, \bar{n}_k) \urcorner$ of the result of substitution of the numerals $\bar{n}_1, \ldots, \bar{n}_k$ for variables $x_1, \ldots, x_k$ in σ.

As a standard consequence of the derivability conditions and the arithmetical fixed point lemma we obtain formalized *Löb's Theorem*.

PROPOSITION 3. *For any formula φ,*

$$\mathsf{EA} \vdash [n]_T([n]_T \varphi \to \varphi) \longleftrightarrow [n]_T \varphi.$$

Another useful lemma shows that n-consistency assertions are equivalent to the uniform reflection principles for T. Let $\mathsf{RFN}_{\Pi_{n+1}}(T)$ denote the schema

$$\{ \forall x (\Box_T \varphi(\dot{x}) \longrightarrow \varphi(x)) : \varphi \in \Pi_{n+1} \}.$$

PROPOSITION 4. *Over EA,*

$$n\text{-Con}(T) \iff \mathsf{RFN}_{\Pi_{n+1}}(T).$$

PROOF. ($\Rightarrow$) If $\varphi(x) \in \Pi_{n+1}$, then $\neg \varphi(x)$ implies $[n]_T \neg \varphi(\dot{x})$ by Σ_{n+1}-completeness. Therefore $\Box_T \varphi(\dot{x})$ implies $[n]_T(\varphi(\dot{x}) \wedge \neg \varphi(\dot{x}))$, that is, $[n]_T \bot$.

($\Leftarrow$) If $[n]_T \bot$, then for some true $\pi \in \Pi_n$, $\Box_T \neg \pi$ by formalized Deduction theorem. Let $\varphi(x) \in \Pi_n$ be a truth-definition for Π_n-formulas so that

EA proves:

$$\text{For each } \psi \in \Pi_n, \quad \text{EA} \vdash \psi \longleftrightarrow \varphi(\ulcorner \psi \urcorner).$$

We have $\Box_T \neg \varphi(\ulcorner \pi \urcorner)$ but $\varphi(\ulcorner \pi \urcorner)$. ⊣

§4. Graded provability algebras. Let T be a theory containing EA. Since the formulas $[0]_T, [1]_T, \ldots$ satisfy Bernays–Löb derivability conditions, all of them correctly define operators

$$\varphi \longmapsto [n]_T \varphi$$

acting on the Lindenbaum boolean algebra $\mathcal{B}_T$ of T. (Here and below we freely identify formulas of T and the corresponding elements of $\mathcal{B}_T$.) The enriched structure $\mathcal{M}_T^\infty = (\mathcal{B}_T, [0]_T, [1]_T, \ldots)$ is called the *graded provability algebra of* T.

Terms of this algebra correspond to propositional *polymodal formulas*, that is, the formulas built up from propositional variables and $\bot, \top$ by boolean connectives and $[0]_T, [1]_T, \ldots$. The identities of $\mathcal{M}_T^\infty$ are exactly characterized by the following system **GLP** due to G. Japaridze ([8, 5]).

Axioms: (i) Boolean tautologies;
(ii) $[n](\varphi \to \psi) \to ([n]\varphi \to [n]\psi)$;
(iii) $[n]([n]\varphi \to \varphi) \to [n]\varphi$;
(iv) $[m]\varphi \to [n]\varphi$, for $m \le n$;
(v) $\langle m \rangle \varphi \to [n]\langle m \rangle \varphi$, for $m < n$.

Rules: modus ponens, $\varphi \vdash [n]\varphi$.

This is expressed by the arithmetical completeness theorem for **GLP**.

PROPOSITION 5 (Japaridze). *For any sound theory T containing* EA,

$$\boldsymbol{GLP} \vdash \varphi(\vec{x}) \iff \mathcal{M}_T^\infty \vDash \forall \vec{x} \, (\varphi(\vec{x}) = \top).$$

We note that the soundness ($\Rightarrow$) follows immediately from Propositions 1–3. We will essentially only rely on the soundness part in this paper.

The graded provability algebra of T provides a kind of big, universal structure where all the extensions of T formulated in the arithmetical language 'live in'. Any arithmetical theory extending T is embeddable as a filter into the Lindenbaum algebra of T. In particular, fragments of PA above EA can be viewed as particular filters in $\mathcal{M}_{\text{EA}}^\infty$. However, in order that the machinery of provability algebras be applicable to these theories, the structure $\mathcal{M}_{\text{EA}}^\infty$ has to 'see' these filters, in other words, they have to be, in some sense, nicely definable in the structure $\mathcal{M}_{\text{EA}}^\infty$.

This is what we have to know about PA (see [14, 15]):

PROPOSITION 6. *The following relationships hold provably in* EA:

(i) $I\Sigma_n \equiv \text{EA} + \langle n+1 \rangle_{\text{EA}} \top$, *for* $n \ge 1$;
(ii) $\text{PA} \equiv \text{EA} + \{\langle n \rangle_{\text{EA}} \top : n < \omega\}$.

When reasoning in provability algebras we shall often confuse extensions of T and filters in $\mathcal{M}_T^\infty$. $U \subseteq_n V$ denotes the Π_{n+1}-conservativity of U over V, i.e., for every $\pi \in \Pi_n$ such that $U \vdash \pi$ we have $V \vdash \pi$. $U \equiv_n V$ means $U \subseteq_n V$ and $V \subseteq_n U$. The same notation is also applied to arbitrary *sets* of elements of $\mathcal{M}_T^\infty$ and means the corresponding relation between theories/filters axiomatized by those sets.

The following crucial property of graded provability algebras is proved in [3, 4].

PROPOSITION 7 (Reduction). *Assume T is a Π_{n+2}-axiomatized theory containing* EA. *Then for all $\varphi \in \mathcal{M}_T^\infty$ the following holds in $\mathcal{M}_T^\infty$ (provably in* EA$^+$):

$$\{\langle n + 1 \rangle_T \varphi\} \equiv_n \{Q_k^n(\varphi) : k < \omega\},$$

where

$$Q_0^n(\varphi) = \langle n \rangle_T \varphi,$$
$$Q_{k+1}^n(\varphi) = \langle n \rangle_T (Q_k^n(\varphi) \wedge \varphi).$$

Thus, the filter generated by all Π_{n+1}-consequences of an element $\langle n + 1 \rangle_T \varphi \in \mathcal{M}_T^\infty$ of complexity Π_{n+2} can be generated by specific Π_{n+1}-elements $Q_k^n(\varphi)$. It is important that these elements are definable by terms in the language of $\mathcal{M}_T^\infty$. We call this property of $\mathcal{M}_T^\infty$ *reduction property*.

We conclude this section with a corollary of the reduction property concerning the n-consistency orderings. The *n-consistency ordering* $<_n$ on $\mathcal{M}_T^\infty$ is defined by

$$\psi <_n \varphi \iff T \vdash \varphi \longrightarrow \langle n \rangle_T \psi.$$

Clearly, $<_n$ is transitive and, by Löb's theorem, irreflexive on $\mathcal{M}_T^\infty \setminus \{\bot\}$.

For $\alpha = \langle n + 1 \rangle_T \varphi$ we define $\alpha[\![k]\!] := Q_k^n(\varphi)$. The following corollary tells that the limit of the sequence $\alpha[\![k]\!]$ (in the sense of the ordering $<_n$ on $\mathcal{M}_T^\infty$) is α.

COROLLARY 8. *Assume $\mathcal{M}_T^\infty$ satisfies the reduction property. If $\psi <_n \alpha$, then $\exists k : \psi <_n \alpha[\![k]\!]$. Hence $\alpha[\![0]\!] <_n \alpha[\![1]\!] <_n \cdots \longrightarrow \alpha$.*

PROOF. It is obvious that $\alpha[\![k]\!] <_n \alpha[\![k + 1]\!]$ for any k. Likewise, a simple induction on k shows $\alpha[\![k]\!] <_n \alpha$, for any k.

Now, if $T \vdash \alpha \to \langle n \rangle_T \psi$, then by the reduction property $\langle n \rangle_T \psi$ belongs to the filter generated by $\{\alpha[\![k]\!] : k < \omega\}$. Hence, $T \vdash \alpha[\![k]\!] \to \langle n \rangle_T \psi$, for some k. $\dashv$

As a word of caution we remark that the sequence $\alpha[\![k]\!]$ is not uniquely determined by the element $\alpha \in \mathcal{M}_T^\infty$. Different choices of φ such that $\alpha = \langle n + 1 \rangle_T \varphi$ can yield different sequences.

§5. An ordinal notation system for ε_0. Work in the closed fragment of **GLP**. Recall that by Japaridze's theorem, for any closed modal formula φ,

$$(*) \qquad\qquad GLP \vdash \varphi \iff T \vdash \varphi^*,$$

where φ^* denotes the interpretation of φ in $\mathcal{M}_T^\infty$. This means that the Lindenbaum algebra $\mathcal{M}_0^\infty$ of the closed fragment of **GLP** is canonically embeddable into $\mathcal{M}_T^\infty$, if T is sound. We shall consider a certain substructure of $\mathcal{M}_0^\infty$ as our ordinal notation system for ε_0.

Let S be the set of formulas generated from $\top$ by $\langle 0 \rangle, \langle 1 \rangle, \dots$. An element of S typically has the form

$$\alpha = \langle n_1 \rangle \langle n_2 \rangle \dots \langle n_k \rangle \top.$$

We identify such elements with words in the alphabet of natural numbers

$$\alpha = n_1 n_2 \dots n_k.$$

The empty word $\varnothing$ is identified with $\top$. S_n denotes the restriction of S to the alphabet $\{n, n+1, \dots\}$.

We define the relation $<_n$ on S in analogy with that on the algebra $\mathcal{M}_T^\infty$:

$$\beta <_n \alpha \iff GLP \vdash \alpha \longrightarrow \langle n \rangle \beta.$$

By $(*)$ we have, for sound T,

$$\beta <_n \alpha \iff \mathcal{M}_T^\infty \vDash \beta^* <_n \alpha^*.$$

PROPOSITION 9. $(S_n, <_n)$ *is well-founded of height ε_0. Modulo provable equivalence in* **GLP** *the ordering is linear.*

We shall use the elements of S as our codes for the ordinals below ε_0. Recall that **GLP** is elementary decidable, so the ordering $(S_n, <_n)$ is elementary decidable, too.

The proof of this theorem is elaborated in [4], but it is not really needed for the treatment of the Worm principle below. To help the reader's intuition we only give the easy correspondence between S and the ordinals below ε_0.

Define $o(0^k) = k$. If $\alpha = \alpha_1 0 \alpha_2 0 \cdots 0 \alpha_n$, where all $\alpha_i \in S_1$ and not all of them empty, then

$$o(\alpha) = \omega^{o(\alpha_n^-)} + \cdots + \omega^{o(\alpha_1^-)},$$

where β^- is obtained from $\beta \in S_1$ by replacing every letter $m+1$ by m.

Notice that $o : S \to \varepsilon_0$ is obviously onto. Proposition 9 (for the case $n = 0$) can be now more explicitly formulated as follows: For all $\alpha, \beta \in S$,

$$GLP \vdash \alpha \longleftrightarrow \beta \quad \text{iff } o(\alpha) = o(\beta);$$
$$GLP \vdash \beta \longrightarrow \Diamond \alpha \quad \text{iff } o(\alpha) < o(\beta).$$

EXAMPLE 10. $o(2101) = \omega^{o(0)} + \omega^{o(10)} = \omega + \omega^{\omega^0 + \omega^1} = \omega^\omega$. Accordingly, we have

$$GLP \vdash 2101 \longleftrightarrow (21 \wedge 01) \longleftrightarrow 21 \longleftrightarrow 2.$$

The first equivalence here follows from Lemma 11(ii) below.

§6. Reading off the fundamental sequences. Reduction property provides the fundamental sequences for certain elements of $\mathcal{M}_T^\infty$. We want to use the same fundamental sequences in our ordinal notation system. We only have to establish that modulo GLP the set S is closed under the operation $\alpha \mapsto \alpha[\![n]\!]$.

In [4] it is shown that S is closed under conjunction. $\wedge$ together with the operators $\langle n \rangle$ is sufficient to build the formulas Q_k which describe the fundamental sequences. Here we prefer to take a shortcut which also allows to explicitly calculate the fundamental sequences.

LEMMA 11. *Some derivations in* GLP:

(i) *If* $m < n$, *then* $\vdash \langle n \rangle \varphi \wedge \langle m \rangle \psi \leftrightarrow \langle n \rangle (\varphi \wedge \langle m \rangle \psi)$;
(ii) *If* $\alpha \in S_{n+1}$, *then* $\vdash \alpha \wedge n\beta \leftrightarrow \alpha n\beta$.
(iii) *If* $m \leq n$, *then* $\vdash nm\alpha \rightarrow m\alpha$.

PROOF. Statement (i):

$$GLP \vdash \langle n \rangle \varphi \wedge \langle m \rangle \psi \longrightarrow [n]\langle m \rangle \psi \quad \text{by Axiom (v)}$$
$$\longrightarrow \langle n \rangle (\varphi \wedge \langle m \rangle \psi).$$

Statement (ii) follows by repeated application of (i). Statement (iii) is theorem $[m]\varphi \rightarrow [m][m]\varphi$ of GLP.   ⊣

LEMMA 12.   (i) *If* $\alpha = \langle n + 1 \rangle \gamma$ *with* $\gamma \in S_{n+1}$, *then for all* k, $GLP \vdash \alpha[\![k]\!] \leftrightarrow (n\gamma)^{k+1}$.

(ii) *If* $\alpha = \langle n + 1 \rangle \gamma m\beta$, *where* $\gamma \in S_{n+1}$ *and* $m \leq n$, *then for any* k, $GLP \vdash \alpha[\![k]\!] \leftrightarrow (n\gamma)^{k+1}m\beta$.

PROOF. (i) We argue by induction on k. For $k = 0$ we have $\alpha[\![0]\!] = \langle n \rangle \gamma \in S$. For the induction step we have

$$GLP \vdash \alpha[\![k + 1]\!] \longleftrightarrow \langle n \rangle (\gamma \wedge (n\gamma)^{k+1})$$
$$\longleftrightarrow \langle n \rangle (\gamma (n\gamma)^{k+1}) \quad \text{by Lemma 11(ii)}$$
$$\longleftrightarrow (n\gamma)^{k+2}.$$

(ii) Similarly, for the induction step we have

$$GLP \vdash \alpha[\![k + 1]\!]$$
$$\longleftrightarrow \langle n \rangle (\gamma m\beta \wedge (n\gamma)^{k+1}m\beta)$$
$$\longleftrightarrow \langle n \rangle (\gamma (m\beta \wedge (n\gamma)^{k+1}m\beta)) \quad \text{by Lemma 11(i)}$$
$$\longleftrightarrow (n\gamma)^{k+2}m\beta \quad \text{by Lemma 11(iii)}. \qquad \dashv$$

§7. **Two soundness proofs.** In this section we give two proofs of truth of EWD. The first proof is a standard one and is based on ordinal assignments and transfinite induction on ε_0 as the only principle that goes beyond EA.

The second proof is more interesting in that it does not use the well-foundedness of ε_0. Instead, it relies directly on the 1-consistency assumption for PA and uses the reasoning in graded provability algebras.

7.1. First proof. For any worm w define an element $i(w) \in S$ by writing the word w in the reverse order. Observe that by Lemma 12 we always have $i(\text{next}(w, n)) = i(w)\llbracket n \rrbracket$, if w ends with $m > 0$. After all, the definition of $\text{next}(w, n)$ was read off from that of $\alpha\llbracket n \rrbracket$ in this way.

Recall the ordinal assignment $o : S \to \varepsilon_0$ defined in Section 5. It induces an assignment of ordinals to worms by $o(w) := o(i(w))$.

Let w_n be the worm game sequence starting from w and let $\alpha_n := i(w_n)$ be the corresponding sequence of elements of S. If w_n ends with 0 we obviously have $o(w_{n+1}) < o(w_n)$. Otherwise we have:

$$o(w_{n+1}) = o(\text{next}(w_n, n + 1)) = o(\alpha_n\llbracket n + 1 \rrbracket) < o(\alpha_n) = o(w_n).$$

This proves $o(w_n)$ to be a strictly decreasing sequence of ordinals, which contradicts the well-foundedness of ε_0.

Notice that this proof uses Proposition 9 to ensure that $o(\alpha\llbracket n \rrbracket) < o(\alpha)$. Alternatively, one can check this directly on the basis of the definition of $o(\alpha)$. We omit the details.

In contrast, our second proof does not rely on ordinals. In a sense, the use of transfinite induction is replaced by the use of Löb's theorem (and the 1-consistency assumption for PA).

7.2. Second proof. We prove EA + 1-Con(PA) $\vdash$ EWD. Now we shall interpret worms as elements of a graded provability algebra.

We work in $\mathcal{M}_{EA}^{\infty}$. For any $\alpha \in S$ we denote by α^* its arithmetical interpretation in $\mathcal{M}_{EA}^{\infty}$. Define $w^* := (i(w)^+)^*$, where α^+ means increasing every element of α by 1.

Thus, for example, $(103)^* = \langle 4 \rangle_{EA} \langle 1 \rangle_{EA} \langle 2 \rangle_{EA} \top$. In the following reasoning we omit the subscript $_{EA}$ everywhere without causing confusion.

LEMMA 13. *For any w, PA $\vdash w^*$.*

PROOF. We argue by induction on $|w|$. If $w = vn$ and $m >$ any letter in w, then

$$EA \vdash v^* \wedge \langle m + 1 \rangle \top \longrightarrow \langle m + 1 \rangle v^* \quad \text{by Lemma 11} \longrightarrow \langle n + 1 \rangle v^*.$$

By Proposition 6 and the induction hypothesis

$$PA \vdash v^* \wedge \langle m + 1 \rangle \top,$$

so PA $\vdash \langle n + 1 \rangle v^*$, which yields the induction step. $\dashv$

LEMMA 14. *For any w,*

$$\text{EA} \vdash \forall n \, (w_n \neq \varnothing \longrightarrow \Box(w_n^\star \longrightarrow \langle 1\rangle w_{n+1}^\star)).$$

PROOF. It is sufficient to prove

(1) $\forall w \neq \varnothing \, \forall n \, \text{EA} \vdash w^\star \longrightarrow \langle 1\rangle \text{next}(w,n)^\star$

by an argument formalizable in EA.

Let $\alpha = i(w)$. If α begins with 0, the claim is obvious. If α begins with $k+1$, by Lemma 12 $\alpha[\![n]\!] = i(\text{next}(w,n))$. An easy induction on n yields:

$$\forall n \, \textbf{\textit{GLP}} \vdash \alpha \longrightarrow \Diamond \alpha[\![n]\!].$$

The axioms of $\textbf{\textit{GLP}}$ are stable under $(\cdot)^+$, that is, if $\textbf{\textit{GLP}} \vdash \varphi$, then $\textbf{\textit{GLP}} \vdash \varphi^+$. Hence we obtain

$$\forall n \, \textbf{\textit{GLP}} \vdash \alpha^+ \longrightarrow \langle 1\rangle \alpha[\![n]\!]^+.$$

This yields (1) by the easy soundness part of Japaridze's theorem.   ⊣

The following argument uses the same idea as Solovay's proof of the fact that there do not exist provably uniformly descending hierarchies of consistency assertions.

LEMMA 15. $\text{EA} \vdash \langle 1\rangle w_0^\star \to \exists n \, w_n = \varnothing$.

PROOF. We prove $\forall n \, w_n \neq \varnothing \to \forall n \, [1]\neg w_n^\star$ essentially using Löb's principle.

$$\begin{aligned}
\text{EA} \vdash \forall n \, w_n \neq \varnothing \wedge [1]\forall n[1]\neg w_n^\star &\longrightarrow [1]\forall n[1]\neg w_{n+1}^\star \\
&\longrightarrow \forall n[1][1]\neg w_{n+1}^\star \\
&\longrightarrow \forall n[1]\neg w_n^\star \quad \text{by Lemma 14.}
\end{aligned}$$

$$\begin{aligned}
\text{EA} \vdash [1]\forall n \, w_n \neq \varnothing &\longrightarrow [1]([1]\forall n[1]\neg w_n^\star \longrightarrow \forall n[1]\neg w_n^\star) \\
&\longrightarrow [1]\forall n[1]\neg w_n^\star \quad \text{by Löb.}
\end{aligned}$$

$$\begin{aligned}
\text{EA} \vdash \forall n \, w_n \neq \varnothing &\longrightarrow [1]\forall n \, w_n \neq \varnothing \quad \text{by } \Sigma_2\text{-completeness} \\
&\longrightarrow [1]\forall n[1]\neg w_n^\star \\
&\longrightarrow \forall n[1]\neg w_n^\star \\
&\longrightarrow [1]\neg w_0^\star,
\end{aligned}$$

as required.   ⊣

We conclude the proof of Theorem 1. From Lemmas 13 and 15 we obtain

$$\begin{aligned}
\text{PA} &\vdash \langle 1\rangle w^\star, \\
\text{EA} &\vdash \langle 1\rangle w^\star \longrightarrow \exists n \, w_n = \varnothing.
\end{aligned}$$

Hence, $\forall w$ PA $\vdash \exists n \, w_n = \varnothing$. This proof is formalizable in EA, so 1-Con(PA) implies $\forall w \exists n \, w_n = \varnothing$ by Σ_1-reflection.

§8. **Provably total computable functions.** For the independence proof of the Worm principle we need to know a few non-PA-specific facts concerning provably total computable functions. Essentially, we will need Proposition 20 below. The material of this section is mostly folklore. The reader can find some additional details in [2].

Recall that a function $f : \mathbb{N}^k \to \mathbb{N}$ is called *provably total computable in T*, if for some Σ_1-formula $\varphi(\vec{x}, y)$ there holds:

(i) $f(\vec{x}) = y \iff \mathbb{N} \vDash \varphi(\vec{x}, y)$;
(ii) $T \vdash \forall \vec{x} \exists y \varphi(\vec{x}, y)$.

$\mathcal{F}(T)$ denotes the class of all provably total computable functions in T. $\mathcal{F}(EA)$ is known to coincide with the *(Kalmar) elementary functions* $\mathcal{E}$. The class $\mathcal{E}$ is defined as the closure of $0, 1, +, \cdot, 2^x$, projection functions and the characteristic function of $\leq$ by composition and bounded recursion, that is, by primitive recursion with the restriction that the resulting function is bounded by some previously generated function. Thus, it is easy to see that any elementary function is bounded by some fixed iterate of 2^x.

For T containing EA, the class $\mathcal{F}(T)$ contains $\mathcal{E}$ and is closed under composition, but generally not under the bounded recursion. Also notice that $\mathcal{F}(T)$ only depends on the set of Π_2-consequences of T. Hence, if T is Π_2-conservative over U, then $\mathcal{F}(T) \subseteq \mathcal{F}(U)$.

For many natural theories T the classes $\mathcal{F}(T)$ have been characterized recursion-theoretically. For example, by a well-known result of C. Parsons [17] and independently of G. Mints [16], $\mathcal{F}(I\Sigma_1)$ coincides with the class of primitive recursive functions. On the other hand, already W. Ackermann [1] and G. Kreisel [12] established that the class $\mathcal{F}(PA)$ coincides with the class of $<\varepsilon_0$-recursive functions. Later some other characterizations of this class have been obtained. In particular, H. Schwichtenberg and S. Wainer (see [18, 7]) characterized $\mathcal{F}(PA)$ by a hierarchy of functions very close to those introduced below.

The notion of provably total computable function is tightly related to that of 1-consistency. In order to explain this relationship we consider the corresponding *programs*, or *indices* of such functions. Fix some natural coding of Turing machines and a Σ_1-formula $\varphi_e(x) = y$ expressing the statement that the Turing machine coded by e on input x halts and outputs y. Usually, one represents φ using Kleene's T-predicate as follows:

$$\varphi_e(x) = y \longleftrightarrow \exists z \, (T(e, x, z) \wedge U(z) = y),$$

where formula $T \in \Delta_0$ and the function U is elementary. Recall that $T(e, x, z)$ expresses that z is a full protocol of a terminating computation of machine e on input x.

DEFINITION 16. Let T be an elementary presented theory. A number e is a T-index, if $e = \langle e_1, e_2 \rangle$ where

- e_1 codes a Turing machine;
- e_2 codes a T-proof of $\forall x \exists y \varphi_{\bar{e}_1}(x) = y$.

With this indexing of provably total computable functions a universal function ψ^T is associated:

$$\psi_e^T(x) := \begin{cases} \varphi_{e_1}(x), & \text{if } e = \langle e_1, e_2 \rangle \text{ is a } T\text{-index;} \\ 0, & \text{otherwise.} \end{cases}$$

The usual diagonalization argument shows that ψ^T, as a function of arguments e and x, does not belong to $\mathcal{F}(T)$. Therefore, the statement of its totality delivers an independent principle for T.

LEMMA 17. $\text{EA} \vdash \forall e, x \exists y \, \psi_e(x) = y \leftrightarrow \text{RFN}_{\Pi_2}(T)$.

PROOF. The totality of ψ is expressed by the formula:

$$(2) \qquad \forall e_1, e_2, x \, (\text{Prf}_T(e_2, \ulcorner \forall x \exists y \varphi_{\bar{e}_1}(x) = y \urcorner) \longrightarrow \exists y \varphi_{e_1}(x) = y).$$

Any Π_2-sentence is equivalent to the one of the form $\forall x \exists y \varphi_{\bar{e}_1}(x) = y$ for a suitable index e_1, so (2) is equivalent to $\text{RFN}_{\Pi_2}(T)$.   ⊣

If $f(\vec{x})$ is a function whose graph is definable, let $f\!\downarrow$ denote the formula $\forall \vec{x} \exists y f(\vec{x}) = y$. The following basic result almost immediately follows from Herbrand's theorem (cf. [2] for details).

PROPOSITION 18. *Suppose the graph of f is elementary. Then $g \in \mathcal{F}(\text{EA} + f\!\downarrow)$ iff g can be obtained from elementary functions and f by composition.*

We denote by $\mathbf{C}(f)$ the closure of $\mathcal{E} \cup \{f\}$ under composition. We can define the *jump* $\mathcal{F}(T)'$ of the class of provably total computable functions of T as $\mathbf{C}(\psi^T)$. From Lemma 17 we now obtain

COROLLARY 19. *If T is a Σ_1-sound theory, then $\mathcal{F}(\text{EA} + \langle 1 \rangle_T \top) = \mathcal{F}(T)'$.*

PROOF. The graph of ψ^T is not elementary. Consider the function $\tilde{\psi}^T$ such that $\tilde{\psi}_e^T(x)$ encodes the full protocol of the computation of $\psi_e(x)$, in other words

$$\tilde{\psi}_e^T(x) := \begin{cases} \mu z.\text{T}(e_1, x, z), & \text{if } e = \langle e_1, e_2 \rangle \text{ is a } T\text{-index;} \\ 0, & \text{otherwise.} \end{cases}$$

It is easy to see that $\tilde{\psi}^T$ has an elementary graph, $\mathbf{C}(\psi^T) = \mathbf{C}(\tilde{\psi}^T)$ and

$$\text{EA} \vdash \tilde{\psi}^T\!\downarrow \longleftrightarrow \psi^T\!\downarrow.$$

Now apply Proposition 18 to $f = \tilde{\psi}^T$.   ⊣

Proposition 18 leads to an alternative 'proofs-free' indexing of functions in $\mathcal{F}(T)$ for $T = \text{EA} + f\!\downarrow$. Terms composed of elementary functions and a function symbol f have a natural Gödel numbering. These numbers can be considered as codes of provably total programs. So, with this Gödel numbering we can associate another universal function $\theta_e^T(x)$ that computes the value of the term with index e on x:

$$\theta_e^T(x) := \begin{cases} t[f,x], & \text{if } e = \ulcorner t \urcorner; \\ 0, & \text{otherwise.} \end{cases}$$

The two kinds of indexing are equivalent provably in EA^+, because the Herbrand theorem is verifiable in EA^+ and allows to extract an explicit term from a proof of totality of a computable function. The converse is also true: for any $\mathbf{C}(f)$-term t one can elementarily construct a Kleene index e_1 of the function $\lambda x.t[f,x]$ and a T-proof e_2 of its totality.

LEMMA 20. *Suppose f has an elementary graph and*

(i) $\text{EA} \vdash \forall x \, f(x) > 2^x$;

(ii) $\text{EA} \vdash \forall x, y \, (x \le y \to f(x) \le f(y))$.

Then $\text{EA} \vdash \lambda x.f^{(x)}(x)\downarrow \leftrightarrow \langle 1 \rangle_{\text{EA}} f \downarrow$.

Here and below we assume that the Δ_0-definition of the graph of $\lambda x.f^{(x)}(x)$ is constructed from the Δ_0-definition of the graph of f in a natural way. (Using provable monotonicity of f this is easy.)

PROOF. Let $T := \text{EA} + f \downarrow$. By Lemma 17, $\langle 1 \rangle_{\text{EA}} f \downarrow$ is EA-equivalent to $\psi^T \downarrow$. The formula $\psi^T \downarrow$, in particular, implies EA^+ and hence $\theta^T \downarrow$. Vice versa, $\theta^T \downarrow$ also implies EA^+ and hence $\psi^T \downarrow$. So, it is sufficient to show that $\lambda x.f^{(x)}(x)\downarrow$ is equivalent to the totality of θ^T.

Clearly, if θ is total, then for every k the function $f^{(k)}$ is also total. Indeed, $f^{(k)} \in \mathbf{C}(f)$ and the Gödel number of $f^{(k)}$ is obtained elementarily from k. We can evaluate it using θ, so $\lambda x.f^{(x)}(x)$ is total.

In the other direction, we use the monotonicity of f. Under the given assumptions, every term $g \in \mathbf{C}(f)$ can be majorized by a fixed iterate of the function f:

$$\text{EA} \vdash \forall x \, (g(x) \le f^{(k)}(x)).$$

The number k can be computed elementarily from the Gödel number of g, say, by a function $j(e)$.

Assume $\lambda x.f^{(x)}(x)$ is total. To show that, for any e and x, the value $\theta_e^T(x)$ is defined, consider the value $f^{(j(e))}(x)$. This value is smaller than $f^{(z)}(z)$, where $z := \max(j(e), x)$, hence it is defined. Therefore, the yet smaller value $\theta_e^T(x)$ is also defined. ⊣

EXAMPLE 21. The classes $\mathcal{E} \subseteq \mathcal{E}' \subseteq \mathcal{E}'' \subseteq \cdots$ form the so-called *Grzegorczyk hierarchy* [9]. It is well-known that the union of this hierarchy coincides with the class of primitive recursive functions (see also [18]). From the previous proposition we conclude that the class $\mathcal{E}^{(n)}$ coincides with $\mathbf{C}(F_n)$, where

$$F_0(x) := 2^x + 1; \qquad F_{n+1}(x) := F_n^{(x)}(x).$$

The functions F_n are all primitive recursive and their graphs are elementary definable. The extension of EA by the axioms $F_n\!\downarrow$ for all $n \geq 1$ is an alternative axiomatization of the *primitive recursive arithmetic* PRA.

Define $T_\omega^n := T + \{\langle n \rangle_T^k \top : k < \omega\}$. Repeated application of Lemma 20 yields the following well-known fact.

PROPOSITION 22.   (i) $\mathsf{EA}_\omega^1 \equiv \mathsf{PRA}$.

(ii) $\mathcal{F}(\mathsf{EA}_\omega^1)$ *is the class of primitive recursive functions.*

§9. Independence of the Worm principle. In this section we let $w[\![n]\!] := \mathrm{next}(w, n)$ and

$$w[\![n \ldots n + k]\!] := w[\![n]\!][\![n + 1]\!] \ldots [\![n + k]\!].$$

We introduce an analogue of the Hardy functions as follows. Let $h_w(n)$ be the smallest k such that $w[\![n \ldots n + k]\!] = \varnothing$.

We need some nice properties of h established within EA. We notice that the function

$$W(w, n, k) := w[\![n \ldots n + k]\!]$$

is elementary, because it can be defined by bounded recursion on k, similarly to the worm sequence w_n. This gives a natural elementary representation of the relation $h_w(n) = k$ in EA.

The following notion will be used to establish the monotonicity of the h functions. Let $v \trianglelefteq u$ iff $v = u[\![0]\!][\![0]\!] \ldots [\![0]\!]$. This essentially means that v is an initial segment of u except possibly for the last letter, which should be not larger than the corresponding letter in u.

LEMMA 23. *If $h_w(m)$ is defined and $u \trianglelefteq w$, then*

$$\exists k \; w[\![m \ldots m + k]\!] = u.$$

PROOF. The n-th letter in w can only change if all letters to the right of it are deleted. So, if w rewrites to $\varnothing$, it cannot possibly miss the u state.   ⊣

Here and below the assumption "$h_w(n)$ *is defined*" is needed to formalize the argument inside EA, because EA may not be able to verify that h_w is a total function. In the standard model of arithmetic this assumption is, of course, superfluous, as we know that EWD is true.

COROLLARY 24. *If $h_w(n)$ is defined, then $\forall m \leq n \exists k \; w[\![n \ldots n+k]\!] = w[\![m]\!]$.*

LEMMA 25. *If $v \trianglelefteq u$ and $x \leq y$ then $h_v(x) \leq h_u(y)$.*

PROOF. Repeating Corollary 24 obtain $s_0, s_1, \ldots$ such that

$$u[\![y \ldots y + s_0]\!] = v[\![x]\!]$$
$$u[\![y \ldots y + s_0 + s_1]\!] = v[\![x]\!][\![x + 1]\!]$$
$$\vdots$$

Therefore, all the steps of the rewrite sequence for v occur in the rewrite sequence for u.   ⊣

LEMMA 26. $h_{u0v}(n) = h_u(n + h_v(n) + 2) + h_v(n) + 1 > h_u(h_v(n))$.

PROOF. Nothing can happen to the 0 between u and v until the v part is eliminated. So, the worm $u0v$ first rewrites to $u0$ in $h_v(n)$ steps and then to $\varnothing$.   ⊣

COROLLARY 27. If $w \in S_1$, then $h_{w1}(n) > h_w^{(n)}(n)$.

PROOF. Observe that $w1[\![n]\!] = w0w0\ldots w0$.   ⊣

Now we can formulate the main lemma. As usual, $h_w\!\downarrow$ denotes the formula $\forall x \exists y\, h_w(x) = y$. We also let $w^* := \alpha^*$, where $\alpha = i(w)$.

LEMMA 28. $EA \vdash \forall w \in S_1\, (h_{1111w}\!\downarrow \rightarrow \langle 1 \rangle w^*)$.

Using this lemma we can easily give

PROOF OF THE INDEPENDENCE OF EWD.

$$EA \vdash \forall w \exists n\, w_n = \varnothing \longrightarrow \forall w \in S_1\, h_w\!\downarrow$$
$$\longrightarrow \forall n\, \langle 1 \rangle \langle n \rangle \top$$
$$\longrightarrow \text{1-Con(PA)}.$$

Here, there first implication holds, because for every worm w and a number x we can find another worm $w' := w0^x$ such that $w'[\![0\ldots x-1]\!] = w$. So, w' dies iff $h_w(x)$ is defined. The third implication holds by the formalization of Proposition 6.   ⊣

REMARK 29. In the proof below we essentially use Proposition 20. Notice that by Corollary 27 we have $h_{111}(x) > 2^x$, therefore the same inequality holds for the function $h_{111w}(x)$, where w is any worm. This is the only reason why we inserted 1111 in front of w in the formulation of the main lemma. In fact, a somewhat sharper statement is obtained if one defines

$$w" := \begin{cases} w, & \text{if } w \text{ begins with } m > 1; \\ 1111w, & \text{otherwise.} \end{cases}$$

Then one can prove in EA that $\forall w \in S_1\, (h_{w"}\!\downarrow \rightarrow \langle 1 \rangle w^*)$.

PROOF OF LEMMA 28. By Löb we can use as an additional assumption

$$A := \forall w \in S_1\, [1](h_{1111w}\!\downarrow \longrightarrow \langle 1 \rangle w^*).$$

Indeed, if we prove that $EA \vdash A \rightarrow \forall w \in S_1\, (h_{1111w}\!\downarrow \rightarrow \langle 1 \rangle w^*)$, then by Löb's theorem the statement of the lemma will also be provable in EA.

The proof relies on the reduction property of $\mathcal{M}_{EA}^\infty$. More precisely, we shall use the following corollary.

COROLLARY 30. Suppose $\alpha \in S_1$ begins with $m > 1$. Then

$$EA^+ \vdash \langle 1 \rangle \alpha^* \longleftrightarrow \forall n \langle 1 \rangle \alpha[\![n]\!]^*.$$

PROOF. In $\mathcal{M}_{\mathrm{EA}}^{\infty}$, by the reduction property, $\alpha^* \equiv_1 \{\alpha[\![n]\!]^* : n < \omega\}$. Therefore, formalizably in EA^+, α^* proves a false Σ_1-sentence iff $\alpha[\![n]\!]^*$ does, for some n. In other words,

$$\mathrm{EA}^+ \vdash \langle 1 \rangle \alpha^* \longleftrightarrow \forall n \langle 1 \rangle \alpha[\![n]\!]^*. \qquad \dashv$$

To prove Lemma 28 we reason in EA and consider two cases according to the last symbol in $1111w$.

If $1111w = v1$, then $h_{v1}\!\downarrow$ implies $\lambda x.h_v^{(x)}(x)\!\downarrow$, by Lemma 27.

The function h_v is increasing, has an elementary graph and grows at least exponentially. So, if $w = \varnothing$, the claim is obvious: $h_{1111}\!\downarrow$ implies the totality of superexponentiation and hence $\langle 1 \rangle\top$. If w is nonempty, $v = 1111v_0$ and we reason as follows:

$$\lambda x.h_v^{(x)}(x)\!\downarrow \longrightarrow \langle 1 \rangle h_v\!\downarrow \quad \text{by Proposition 20}$$
$$\longrightarrow \langle 1 \rangle\langle 1 \rangle v_0^* \quad \text{by the assumption } A$$
$$\longrightarrow \langle 1 \rangle w^*.$$

If $1111w = v$ ends with $m > 1$, then $v[\![n]\!] = 1111(w[\![n]\!])$. We have

$$h_v\!\downarrow \longrightarrow \lambda x.h_{v[x]}(x+1)\!\downarrow$$
$$\longrightarrow \forall n\, h_{v[\![n]\!]}\!\downarrow.$$

This follows by formalizing the following argument.

Consider any n and x. We have to show that $h_{v[\![n]\!]}(x)$ is defined. If $x \leq n$, then $h_{v[\![n]\!]}(x) \leq h_{v[\![n]\!]}(n+1)$. If $x \geq n$, then $h_{v[\![n]\!]}(x) \leq h_{v[x]}(x+1)$. In both cases we know that the larger value is defined, hence so is the smaller value.

Further we obtain

$$\forall n\, h_{v[\![n+1]\!]}\!\downarrow \longrightarrow \forall n\, h_{v[\![n]\!]1}\!\downarrow \quad \text{as } v[\![n]\!]1 \unlhd v[\![n+1]\!]$$
$$\longrightarrow \forall n\, \langle 1 \rangle h_{v[\![n]\!]}\!\downarrow \quad \text{as before}$$
$$\longrightarrow \forall n\, \langle 1 \rangle\langle 1 \rangle w[\![n]\!]^* \quad \text{by } A$$
$$\longrightarrow \forall n\, \langle 1 \rangle w[\![n]\!]^* \quad \text{by } \Sigma_2\text{-completeness}$$
$$\longrightarrow \langle 1 \rangle w^* \quad \text{by Corollary 30.}$$

This ends the proofs of Lemma 28 and of Theorem 1. $\qquad \dashv$

From Lemma 28 we also obtain that the growth rate of the provably total computable functions of PA is bounded by the functions h_n, where n are just the single letter worms. (This, in a somewhat different manner, also implies the independence of the Worm principle.)

THEOREM 2. *If $f \in \mathcal{F}(\mathrm{PA})$ then for some n and almost all x, $f(x) \leq h_n(x)$.*

PROOF. First of all, we prove that

$$\mathrm{PA} \equiv_1 \mathrm{EA} + \{\langle 1 \rangle_{\mathrm{EA}} \langle n \rangle_{\mathrm{EA}}\top : n < \omega\}.$$

The inclusion $\supseteq$ follows from Lemma 13. For the opposite inclusion assume $\mathsf{PA} \vdash \pi$ with $\pi \in \Pi_2$. Then for some n we have $\mathsf{EA} + \langle n \rangle_{\mathsf{EA}} \top \vdash \pi$. Hence,

$$\mathsf{EA} \vdash \langle 1 \rangle_{\mathsf{EA}} \langle n \rangle_{\mathsf{EA}} \top \longrightarrow \langle 1 \rangle_{\mathsf{EA}} \pi \longrightarrow \pi, \quad \text{by } \Sigma_2\text{-completeness.}$$

This proves the claim. We conclude that if $f \in \mathcal{F}(\mathsf{PA})$ then

$$\mathsf{EA} + \langle 1 \rangle_{\mathsf{EA}} \langle n \rangle_{\mathsf{EA}} \top \vdash f{\downarrow},$$

for some n. By Lemma 28 it follows that

$$\mathsf{EA} + h_{1111n}{\downarrow} \vdash f{\downarrow}.$$

Lemma 18 now implies that f can be obtained by composition from elementary functions and h_{1111n}. Hence, by monotonicity, it is majorized by a fixed iterate of h_{1111n}, ergo by h_{1111n1} for almost all x.

Using a sharper version of Lemma 28 (see Remark 29) we can in the same way conclude that f is majorized by h_{n1} for almost all x, assuming $n > 1$. By Lemma 25 one also estimates $h_{n1}(x)$ by $h_{n+1}(x)$ from above. $\dashv$

§10. **Comparing ordinal notation systems.** It is interesting to compare the ordinal notation system for ε_0 studied in this paper and [4] with the standard one based on Cantor normal forms. Are they essentially the same?

An answer to this question is a matter of degree of precision. From a sufficiently general point of view they are equivalent. Certainly, the order relations are EA-provably elementary isomorphic.

However, in the proof-theoretic literature starting perhaps from [13] it has been argued that natural systems of ordinal notation should be perceived as orderings equipped with some additional operations. For the standard ordinal notation system one considers the operations $(0, +, \omega^x)$ as basic.[3] These operations are sufficient to generate all ordinal notations up to ε_0 as closed terms.

The ordinal notation system considered here can also be perceived as an ordered structure together with some additional operations. This is a many-one notation system, as different (**GLP**-equivalent) words from S can denote the same ordinal. It is possible to restrict the attention to the words in normal form (see [4]). Then one obtains a one-one ordinal notation system.

The operations come from the graded provability algebra and are essentially $\top$ (interpreted as 0) and $\langle n \rangle$, for each n. An additional operation that was used is conjunction $\wedge$. It is not, strictly speaking, needed to generate all necessary ordinal notations, but it is a natural operation of the Lindenbaum algebra and was useful in constructing the fundamental sequences. We notice that all

---

[3]Sometimes one also adds ordinal multiplication and/or suitable inverse operations to the basic list. Sometimes one also considers the single basic function $x + \omega^y$.

these operations respect **GLP**-equivalence, so they correctly define operations on ordinals (not just on their representations).

Let us examine how to express the operations of the two ordinal notation systems in terms of one another. The ordinal sum can be expressed as follows: for $\alpha, \beta \in S$,

$$\alpha + \beta = \begin{cases} \beta\alpha, & \exists k \ \beta = 0^k, \\ \beta 0\alpha, & \forall k \ \beta \neq 0^k. \end{cases}$$

The function ω^x corresponds to α^+ if $o(\alpha) = x$. Thus, the function $x + \omega^y$ is nicely expressible by $\alpha^+ 0\beta$, if $o(\alpha) = y$ and $o(\beta) = x$.

However, notice that $(\cdot)^+$ is, officially, not an operation of the structure. It is also easily seen that it is not expressible by any particular term. For that matter, the same problem is with $+$, because a function of two arguments cannot be expressed as a composition of unary functions. (Concatenation is *not* an operation of the structure!)

The opposite translation is also not quite straightforward. We have $o(\langle 0 \rangle \alpha) = o(\alpha) + 1$. However, for $n > 0$ the expressions become more complicated, e.g., if $o(\alpha) = \omega^{\alpha_k} \cdot n_k + \cdots + \omega^{\alpha_0} \cdot n_0$ in Cantor normal form, then

$$(3) \qquad o(\langle 1 \rangle \alpha) = \omega^{\alpha_k} \cdot n_k + \cdots + \omega^{\alpha_1} \cdot n_1 + \omega^{\alpha_0 + 1}.$$

(The latter expression is not in Cantor normal form in case $\alpha_1 = \alpha_0 + 1$.) The operation $\langle 2 \rangle$ works similarly but on the 'second floor' of the Cantor normal form expression, that is, in this case on the normal form of the ordinal α_0:

$$o(\langle 2 \rangle \alpha) = \omega^{\alpha_k} \cdot n_k + \cdots + \omega^{\alpha_1} \cdot n_1 + \omega^{\langle 1 \rangle(\alpha_0)},$$

where the operation $\langle 1 \rangle$ on ordinals is computed by the right hand side of (3). The effect of $\langle n \rangle$ for $n > 2$ can be calculated similarly.

Conjunction $\wedge$ considered as an operation on ordinals is a bit puzzling. Of course, this operation satisfies all the identities of $\wedge$: it is symmetric, associative, idempotent. However, $x \wedge y$ does not always equal $\max(x, y)$, as one naturally expects. E.g.,

$$\omega \wedge (\omega + 1) = o(1 \wedge 01) = o(101) = \omega + \omega.$$

By Corollary 10 from [4] we know that, if both α and β are words from S_n and begin with n, then one of α and β implies the other. This means that in this case, indeed, $o(\alpha \wedge \beta) = \max(o(\alpha), o(\beta))$. We do not have a very nice expression for $o(\alpha) \wedge o(\beta)$, in general. There is a simple recursive algorithm of calculating it with Cantor normal forms, though (see [4, Lemma 11]).

Now we consider fundamental sequences. It turns out that the fundamental sequences $\alpha[\![n]\!]$ used in this paper are almost the same as those for the standard system of ordinal notation, which may even be a bit surprising taking into account all the differences discussed above. This was observed by A. Weiermann, G. Lee and L. Carlucci.

Let $\alpha = \alpha_0 0 \ldots 0\alpha_k$ with all $\alpha_i \in S_1$ and $o(\alpha) = \omega^{o(\alpha_k)} + \cdots + \omega^{o(\alpha_0)}$ in Cantor normal form. $o(\alpha)$ being a limit ordinal means that the word α_0 is not empty, hence $\alpha_0 = \langle m+1 \rangle \beta$ for some $\beta \in S_1$. By Lemma 12 the sequence $\alpha[\![n]\!]$ can be computed according to the two cases below.

(a) $m = 0$. Then $\alpha[\![n]\!] = (0\beta)^{n+1} 0\alpha_1 0 \cdots 0\alpha_k$, because $\beta \in S_1$ and is followed by 0 in α.

(b) $m > 0$. Then $\alpha[\![n]\!] = \alpha_0' 0\alpha_1 0 \cdots 0\alpha_k$, where $\alpha_0' = \alpha_0[\![n]\!] = ((m\beta^-)[\![n]\!])^+$. The latter equality holds because from Lemma 12 one immediately concludes that $\gamma^+[\![n]\!] = \gamma[\![n]\!]^+$ whenever γ begins with $m > 0$ and it makes sense to speak about $\gamma[\![n]\!]$.

This means that the fundamental sequence for an ordinal $\alpha = \omega^{\alpha_k} + \cdots + \omega^{\alpha_0}$ in Cantor normal form is as follows:

$$
\alpha[\![n]\!] = \begin{cases} \omega^{\alpha_k} + \cdots + \omega^{\alpha_1} + \omega^{\beta} \cdot (n+1) + 1, & \text{if } \alpha_0 = \beta + 1; \\ \omega^{\alpha_k} + \cdots + \omega^{\alpha_1} + \omega^{\alpha_0[\![n]\!]}, & \text{if } \alpha_0 \text{ is a limit ordinal.} \end{cases}
$$

So, the only difference with the standard fundamental sequences is the second "+1" in the first case.

There is a well-known exact correspondence between the standard fundamental sequences and a particular strategy for Hercules in his battle with Hydra: always chop off the rightmost head. This provides an immediate relationship between the Worms and Hydras.

Essentially, the Worm battle is slightly longer because it costs the Hercules additional steps each time to eliminate an extra new head of the Hydra sprouting from the grandfather of the head that has been cut away (this corresponds to the extra "+1"). But we can also estimate the worm function by the hydra function from above.

Let $\alpha[n]$ denote the standard fundamental sequences assignment and let $h_\alpha'(n)$ denote the minimal k such that $\alpha[n \ldots n + k] = \varnothing$. This essentially measures the length of the Hydra battle for the hydra associated with α. As before, $h_\alpha(n)$ stands for $h_w(n)$ where w is the unique worm in normal form corresponding to the ordinal $\alpha < \varepsilon_0$.

LEMMA 31. *For all ordinals α, for all n, $h_\alpha(n) \leq h_\alpha'(n+1)$.*

PROOF. First of all, observe that $\alpha[\![n]\!] \trianglelefteq \alpha[n+1]$, for any limit ordinal α. Here, the relation $\trianglelefteq$ on ordinals is induced by the corresponding relation on worms. This relationship is clear from the clauses (a) and (b) above.

To prove the lemma one argues by transfinite induction on α: $h_\alpha(n) = h_{\alpha[n]}(n + 1) \leq h_{\alpha[n+1]}(n + 1)$ by Lemma 25. Then $h_{\alpha[n+1]}(n + 1) \leq h'_{\alpha[n+1]}(n + 2) = h'_\alpha(n + 1)$, by the induction hypothesis.   ⊣

Therefore, by Theorem 2 the functions $h'_\alpha(n)$ majorize all provably total computable functions of PA which also implies the independence of the Hydra battle principle.

REFERENCES

[1] W. ACKERMANN, *Zur Widerspruchsfreiheit der Zahlentheorie*, **Mathematische Annalen**, vol. 117 (1940), pp. 162–194.

[2] L. D. BEKLEMISHEV, *Induction rules, reflection principles, and provably recursive functions*, **Annals of Pure and Applied Logic**, vol. 85 (1997), no. 3, pp. 193–242.

[3] ———, *Proof-theoretic analysis by iterated reflection*, **Archive for Mathematical Logic**, vol. 42 (2003), no. 6, pp. 515–552, DOI: 10.1007/s00153-002-0158-7.

[4] ———, *Provability algebras and proof-theoretic ordinals. I*, **Annals of Pure and Applied Logic**, vol. 128 (2004), no. 1-3, pp. 103–123.

[5] G. BOOLOS, *The Logic of Provability*, Cambridge University Press, Cambridge, 1993.

[6] W. BUCHHOLZ, *An independence result for* $(\Pi^1_1\text{-}CA)+BI$, **Annals of Pure and Applied Logic**, vol. 33 (1987), no. 2, pp. 131–155.

[7] W. BUCHHOLZ and S. WAINER, *Provably computable functions and the fast growing hierarchy*, **Logic and Combinatorics (Arcata, Calif., 1985)**, Contemporary Mathematics, vol. 65, AMS, Providence, RI, 1987, pp. 179–198.

[8] G. K. DZHAPARIDZE, *Polymodal provability logic*, **Intensional Logics and the Logical Structure of Theories (Telavi, 1985)**, "Metsniereba", Tbilisi, 1988, (Russian), pp. 16–48.

[9] A. GRZEGORCZYK, *Some classes of recursive functions*, **Rozprawy Matematyczne**, vol. 4 (1953), p. 46.

[10] M. HAMANO and M. OKADA, *A relationship among Gentzen's proof-reduction, Kirby-Paris' hydra game and Buchholz's hydra game*, **Mathematical Logic Quarterly**, vol. 43 (1997), no. 1, pp. 103–120.

[11] L. A. S. KIRBY and J. B. PARIS, *Accessible independence results for Peano arithmetic*, **The Bulletin of the London Mathematical Society**, vol. 14 (1982), no. 4, pp. 285–293.

[12] G. KREISEL, *On the interpretation of non-finitist proofs. II. Interpretation of number theory. Applications*, **The Journal of Symbolic Logic**, vol. 17 (1952), pp. 43–58.

[13] ———, *Wie die Beweistheorie zu ihren Ordinalzahlen kam und kommt*, **Jahresbericht der Deutschen Mathematiker-Vereinigung**, vol. 78 (1976/77), no. 4, pp. 177–223.

[14] G. KREISEL and A. LÉVY, *Reflection principles and their use for establishing the complexity of axiomatic systems*, **Zeitschrift für Mathematische Logik und Grundlagen der Mathematik**, vol. 14 (1968), pp. 97–142.

[15] D. LEIVANT, *The optimality of induction as an axiomatization of arithmetic*, **The Journal of Symbolic Logic**, vol. 48 (1983), no. 1, pp. 182–184.

[16] G. E. MINC, *Quantifier-free and one-quantifier systems*, **Zapiski Naučnyh Seminarov Leningradskogo Otdelenija Matematičeskogo Instituta im. V. A. Steklova Akademii Nauk SSSR (LOMI)**, vol. 20 (1971), pp. 115–133, In Russian.

[17] C. PARSONS, *On a number theoretic choice schema and its relation to induction*, **Intuitionism and Proof Theory (Proc. Conf., Buffalo, N.Y., 1968)** (A. Kino, J. Myhill, and R. E. Vessley, editors), North-Holland, Amsterdam, 1970, pp. 459–473.

[18] H. E. ROSE, **Subrecursion: Functions and Hierarchies**, Oxford Logic Guides, vol. 9, The Clarendon Press, Oxford University Press, New York, 1984.

DEPARTMENT OF PHILOSOPHY
UTRECHT UNIVERSITY
HEIDELBERGLAAN 8
3584 CS UTRECHT, THE NETHERLANDS
and
STEKLOV MATHEMATICAL INSTITUTE
GUBKINA 8
119991 MOSCOW, RUSSIAN FEDERATION
E-mail: Lev.Beklemishev@phil.uu.nl

"ONE IS A LONELY NUMBER": LOGIC AND COMMUNICATION

JOHAN VAN BENTHEM

Abstract. Logic is not just about single-agent notions like reasoning, or zero-agent notions like truth, but also about communication between two or more people. What we tell and ask each other can be just as logical as what we infer in Olympic solitude. We show how communication and other interactive phenomena can be studied systematically by merging epistemic and dynamic logic, leading to new types of question.

§1. Logic in a social setting.

1.1. Questions and answers. Consider the simplest type of communication: a question-answer episode between two agents. Here is an example. I approach you in a busy Roman street, A.D. 180, intent on rescuing my former general Maximus, now held as a gladiator, and ask:

Q   *Is this the road to the Colosseum?*

As a well-informed and helpful Roman citizen, you answer truly:

A   *Yes.*

This is the sort of thing that we all do competently millions of times in our lives. There is nothing to it. But what is going on? I learn the fact that this is the road to the Colosseum. But much more happens. By asking the question, I convey to you that I do not know the answer, and also, that I think it possible that you do. This information flows before you have said anything at all. Then, by answering, you do not just convey the topographical fact to me. You also bring it about that you know that I know, I know that you know I know, etc. This knowledge up to every finite depth of mutual reflection is called *common knowledge*. It involves a mixture of factual information and iterated information about what others know.

Logic Colloquium '02
Edited by Z. Chatzidakis, P. Koepke, and W. Pohlers
Lecture Notes in Logic, 27

These *epistemic overtones* concerning our mutual information are not mere side-effects. They may steer further concrete actions. Some bystanders' knowing that I know the road to Maximus' location may lead them to rush off and warn the Emperor Commodus about this inquisitive Dutch visitor — while my knowing that they know I know may lead me to prevent them from doing just that. So epistemic overtones are ubiquitous and important, and we are good at computing them! In particular, we are well-attuned to fine differences in group knowledge. Everyone's knowing individually that your partner is unfaithful is unpleasant, but shame explodes when you meet people and know they all know that they know.

This is just the tip of an iceberg. I have described one type of question, but there are others. If you are my student, you would not assume that my classroom question shows that I do not know the answer. It need not even convey that I think you know, since my purpose may be to expose your ignorance. Such phenomena have been studied from lots of angles. Philosophers of language have developed speech act theory, linguists study the semantics of questions, computer scientists study communication mechanisms, and game theorists have their signaling games. All these perspectives are important — but there is also a foothold for *logic*. This paper aims to show that communication is a typical arena for logical analysis. Logical models help in raising and solving basic issues not recognized before.

1.2. The puzzle of the Muddy Children. Subtleties of information flow are often high-lighted in puzzles, some with a long history of appeal to broad audiences. A perennial example are the Muddy Children:

> *After playing outside, two of three children have mud on their fore-heads. They all see the others, but not themselves, so they do not know their own status. Now their Father comes and says: "At least one of you is dirty". He then asks: "Does anyone know if he is dirty?" The children answer truthfully. As this question-answer episode repeats, what will happen?*

Nobody knows in the first round. But upon seeing this, the muddy children will both know in the second round, as each of them can argue as follows:

> *"If I were clean, the one dirty child I see would have seen only clean children around her, and so she would have known that she was dirty at once. But she did not. So I must be dirty, too!"*

This is symmetric for both muddy children — and therefore, both know in the second round. The third child knows it is clean one round later, after they have announced that. The puzzle is easily generalized to other numbers of clean and dirty children. Many variants are still emerging, as one can check by a simple Internet search.

Puzzles have a serious thrust, as they highlight subtle features of communication beyond simple questions and answers. E.g., consider the plausible putative *Learning Principle* stating that what we hear in public becomes common knowledge. This holds for announcing simple facts — such as the one in Tacitus that, long before international UN peace-keepers, German imperial guards already walked the streets of Rome. But the Principle is not valid in general! In the first round of Muddy Children, the muddy ones both announced the true fact that they did not know their status. But the result of that announcement was not that this ignorance became common knowledge. The announcement rather produced its own falsity, since the muddy children knew their status in the second round. Communicative acts involve *timing* and *information change*, and these may change truth values in complex ways. As we shall see, one of the virtues of logic is that it can help us keep all this straight.

1.3. Logical models of public communication. A logical description of our question-answer episode is easy to give. First, we need to picture the *relevant information* states, after that, we say how they are *updated*.

Answering a question. One initial information model for the group $\{Q, A\}$ of you and I has two states with "ϕ", "$\neg\phi$", with ϕ "this is the road to the Colosseum". We draw these states as points in a diagram. Also, we indicate agents' uncertainties between states. The labeled line shows that Q cannot distinguish between the two:

$$\phi \bullet \overset{Q}{\rule{2cm}{0.4pt}} \bigcirc \neg\phi$$

The black dot is an outside marker for the actual world where the agents live. There are no uncertainty lines for A. This reflects the fact that the Roman local A knows if this is the road to the Colosseum. But Q, though uninformed about the facts, sees that A knows in each eventuality, and hence he knows that A knows. This information about other's information is an excellent reason for asking a question.

Next, A's answer triggers an *update* of this information model. In this simple case, A's answer eliminates the option *not-ϕ*, thereby changing the initial situation into the following one-point diagram:

$$\phi \bullet$$

This picture has only one possible state of the world, where the proposition ϕ holds, and no uncertainty line for anyone. This indicates that ϕ is now common knowledge between you and me. Cognoscenti will recognize where we are heading. Information states are models for the modal logic S5 in its multi-agent version, and communication consists in actions which change such models. In what follows, we mean by "knowledge" only this much: "according to the agent's information".

Muddy Children: the movie. Here is a video of information updates for Muddy Children. States of the world assign **D** (dirty) or **C** (clean) to each child: 8 in total. In any of these, a child has one uncertainty. It knows about the others' faces, but cannot distinguish the state from one where its own **D/C** value is different:

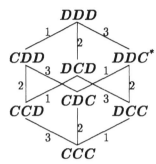

Updates start with the Father's elimination of the world **CCC**:

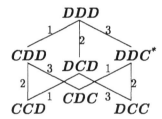

When no one knows his status, the bottom worlds disappear:

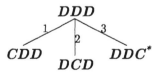

The final update is to

$$DDC^*$$

1.4. General communication. Update by elimination of worlds incompatible with a statement made publicly is a simple mechanism. Human communication in general is very complex, including many other propositional attitudes than knowledge, such as belief or doubt — and logically more challenging phenomena than announcing the truth, such as hiding and cheating. There are two main lines of research here. One is further in-depth analysis of public communication, which turns out to be a quite subtle affair. This will be the main topic of the present paper. The other direction is modeling more complex communicative actions, such as giving answers to questions which others do not hear, or which others overhear, etc. Natural language has a rich vocabulary for all kinds of attitudes toward information, speech acts, secrets, and so on — reflecting our natural proficiency with these. We will discuss more complex models briefly later in Section 9. Actually, this might seem a hopeless enterprise, as our behaviour is so diverse and open-ended. But fortunately, there exist simple realistic settings highlighting key aspects, viz. *games* that people engage in, which will also be discussed toward the end.

Some crucial references for this research program are [13, 40, 6, 14], and the extensive new version 2003 of the latter basic reference: [2]. Also well-worth reading is [41], which contains a mathematical analysis of all the subtle information passing moves in the well-known parlour game "Cluedo". The present paper builds on these references and others, while also including a number of unpublished results by the author over the past few years. Most ideas and results in what follows have been presented at public events since 1999, including the keynote lecture "Update Delights" at the 11th *ESSLLI* Summer School, Birmingham, 2000. An *ILLC* report version of this paper has been available on-line since 2002.

§2. **The basics of update logic.** The logic of public information update can largely be assembled from existing systems. We first survey basic epistemic logic and dynamic logic, and then discuss their combination.

2.1. Epistemic logic. *Language.* Epistemic logic has the following explicit notation for talking about knowledge of agents:

$$K_j \phi \qquad\qquad \text{agent } j \text{ knows that } \phi.$$

With such a symbolism, we can also analyse further patterns:

$\neg K_j \neg \phi \text{ (or } \Diamond_j \phi) \qquad$ agent j considers it *possible that* ϕ,
$K_j \phi \vee K_j \neg \phi \qquad\qquad$ agent j knows *if* ϕ,
$K_j \neg K_i \phi \qquad\qquad$ agent j knows that agent i does not know that ϕ.

E.g., in asking a "normal" question, Q conveys that he does not know if ϕ:

$$\neg K_Q \phi \wedge \neg K_Q \neg \phi$$

and also that he thinks that A might know:

$$\Diamond_Q (K_A \phi \vee K_A \neg \phi).$$

By answering affirmatively, A conveys that she knows that ϕ, but she also makes Q know that ϕ, that Q knows that A knows, and so on up to any finite iteration depth, leading to *common knowledge*, written as follows:

$$C_{\{Q,A\}} \phi.$$

Models. Models for this epistemic language are of the form

$$M = (S, \{\sim_j \mid j \in G\}, V)$$

with (a) S a set of worlds, (b) V a valuation function for proposition letters, and (c) for each agent $j \in G$, an *equivalence relation* $\sim_j$ relating worlds s to all worlds that j cannot distinguish from it. These models M may be viewed as collective information states.

Semantics. Next, in these models, an agent j *knows* exactly those propositions that are true in all worlds she cannot distinguish from the current one. That is:

$$M, s \models K_j \phi \quad \text{iff} \quad M, t \models \phi \quad \text{for all } t \text{ s.t. } s \sim_j t.$$

The related notation $\neg K_j \neg \phi$ or $\Diamond_j \phi$ works out to:

$$M, s \models \Diamond_j \phi \quad \text{iff} \quad M, t \models \phi \quad \text{for some } t \text{ s.t. } s \sim_j t.$$

In addition, there are several useful operators of "group knowledge":

Universal knowledge $E_G \phi$
This is just the conjunction of all formulas $K_j \phi$ for $j \in G$.

Common knowledge $C_G \phi$
This says at s that ϕ is true in every state reachable from s by finitely many $\sim$-steps for members of group G (e.g.,$\sim_1,\sim_2,\sim_1,\sim_3$).

Implicit knowledge $I_G \phi$
This says that ϕ is true in all states which are related to s via the intersection of all uncertainty relations $\sim_j$ for $j \in G$.

Logic. Information models validate an epistemic logic that can describe and automate reasoning with knowledge and ignorance of interacting agents. Here are its major validities:

$K_j(\phi \to \psi) \to (K_j \phi \to K_j \psi),$ *Knowledge Distribution,*
$K_j \phi \to \phi$ *Veridicality,*
$K_j \phi \to K_j K_j \phi$ *Positive Introspection,*
$\neg K_j \phi \to K_j \neg K_j \phi$ *Negative Introspection.*

The complete system is multi-S5, which serves in two different guises: describing the agents' own explicit reasoning, and describing our reasoning as theorists about them. And here are the required additional axioms for a complete logic with common knowledge:

$$C_G\phi \leftrightarrow \phi \wedge E_G C_G \phi \qquad \textit{Equilibrium Axiom,}$$
$$(\phi \wedge C_G(\phi \rightarrow E_G\phi)) \rightarrow C_G\phi \qquad \textit{Induction Axiom.}$$

This is the standard decidable version of epistemic logic.

2.2. Dynamic logic. The usual logic of knowledge by itself can only describe static snapshots of a communication sequence. To describe the updates themselves, we must add actions explicitly.

Language. The language now has formulas F and program expressions P on a par. We write this two-level syntax in abbreviated form with so-called Backus-Naur notation:

$$F := \textit{propositional atoms } p, q, r, \ldots \mid \neg F \mid (F \wedge F) \mid \Diamond_P F$$
$$P := \textit{basic actions } a, b, c, \ldots \mid (P; P) \mid (P \cup P) \mid P^* \mid (F)?$$

Semantics. This formalism is interpreted over polymodal models

$$M = \langle S, \{R_a \mid a \in A\}, V \rangle$$

which are intuitively seen as process graphs with states and possible basic transitions. The truth definition explains two notions in one recursion:

$$M, s \qquad \models \phi \qquad \qquad \phi \textit{ is true at state } s,$$
$$M, s_1, s_2 \models \pi \qquad \qquad \textit{the transition from } s_1 \textit{ to } s_2 \textit{ corresponds to}$$
$$\textit{a successful execution for the program } \pi.$$

Here are the inductive clauses:

▶ $\quad M, s \qquad \models p \qquad$ iff $\quad s \in V(p)$
$\quad M, s \qquad \models \neg\psi \qquad$ iff $\quad \textit{not } M, s \models \psi$
$\quad M, s \qquad \models \phi_1 \wedge \phi_2 \quad$ iff $\quad M, s \models \phi_1 \textit{ and } M, s \models \phi_2$
$\quad M, s \qquad \models \Diamond_\pi\phi \qquad$ iff $\quad \textit{for some } s' \textit{ with } M, s, s' \models \pi\text{:} \; M, s' \models \phi$

▶ $\quad M, s_1, s_2 \models a \qquad$ iff $\quad (s_1, s_2) \in R_a$
$\quad M, s_1, s_2 \models \pi_1; \pi_2 \quad$ iff $\quad \textit{there exists a } s_3 \textit{ with}$
$\qquad\qquad\qquad\qquad\qquad\qquad M, s_1, s_3 \models \pi_1 \textit{ and } M, s_3, s_2 \models \pi_2$
$\quad M, s_1, s_2 \models \pi_1 \cup \pi_2 \quad$ iff $\quad M, s_1, s_2 \models \pi_1 \textit{ or } M, s_1, s_2 \models \pi_2$
$\quad M, s_1, s_2 \models \pi^* \qquad$ iff $\quad \textit{some finite sequence of } \pi\text{-transitions}$
$\qquad\qquad\qquad\qquad\qquad\qquad \textit{in } M \textit{ connects } s_1 \textit{ with } s_2$
$\quad M, s_1, s_2 \models (\phi)? \qquad$ iff $\quad s_1 = s_2 \textit{ and } M, s_1 \models \phi.$

Thus, formulas have the usual Boolean operators, while the existential modality $\Diamond_\pi\phi$ is a weakest precondition true at only those states where program π can be performed to achieve the truth of ϕ. The program constructions are the usual regular operations of relational composition, Boolean choice, Kleene iteration, and tests for formulas. This system defines standard control operators on programs such as:

IF ε THEN π_1 ELSE π_2 $\quad$ $((\varepsilon)? \, ; \pi_1) \cup ((\neg \varepsilon)? \, ; \pi_2)$

WHILE ε DO π $\quad$ $((\varepsilon)? \, ; \pi)^* \, ; (\neg \varepsilon)?$

Logic. Dynamic logic expresses all of modal logic plus regular relational set algebra. Its complete set of validities is known (cf. [9]):

▶ All principles of the minimal modal logic for all modalities $\Box_\pi$
▶ Computation rules for weakest preconditions:

$$\Diamond_{\pi_1 ; \pi_2} \phi \;\leftrightarrow\; \Diamond_{\pi_1} \Diamond_{\pi_2} \phi.$$
$$\Diamond_{\pi_1 \cup \pi_2} \phi \;\leftrightarrow\; \Diamond_{\pi_1} \phi \vee \Diamond_{\pi_2} \phi.$$
$$\Diamond_{\phi?} \psi \;\leftrightarrow\; \phi \wedge \psi.$$
$$\Diamond_{\pi^*} \phi \;\leftrightarrow\; \phi \vee \Diamond_\pi \Diamond_{\pi^*} \phi.$$

▶ Induction Axiom: $(\phi \wedge \Box_{\pi^*}(\phi \to \Box_\pi \phi)) \to \Box_{\pi^*} \phi.$

Like basic epistemic logic, the logic of public announcement is decidable. This remains so with certain *extensions* of the basic language, such as the program construction $\cap$ of *intersection* — to which we will return below. Extended modal languages occur quite frequently in applications.

2.3. Dynamic epistemic logic. Analyzing communication requires a logic of knowledge in action, combining epistemic logic and dynamic logic. This may be done in at least two ways.

Abstract DEL. One can join the languages of epistemic and dynamic logic, and merge the signatures of their models. This yields abstract logics of knowledge and action: cf. [12] on planning, or [30] on imperfect information games. The general logic is the union of epistemic multi-S5 and dynamic logic. This is a good base for experimenting with further constraints. An example is agents having perfect memory for what went on in the course of communication (cf. [7]). This amounts to an extra commutation axiom

$$K_j \Box_a \phi \longrightarrow \Box_a K_j \phi.$$

Abstract *DEL* may be the best setting for general studies of communication.
Concrete update logic. More concretely, in Section 1, public announcement of a proposition ϕ changed the current epistemic model M, s, with actual world s to a new one as follows:

eliminate all worlds which currently do not satisfy ϕ.

Thus, we work in a universe whose states are epistemic models — either all of them or just some family — and basic actions are public announcements $A!$ of assertions A from the epistemic language. These actions are *partial functions*.

from	ϕ		*to*		
$\boldsymbol{M}, s$	s	$\neg\phi$	$\boldsymbol{M}\vert\phi, s$	s	

If A is true, then it can be truthfully announced with a unique update. From the standpoint of dynamic logic, this is just one concrete instance of abstract process models, plus epistemic extras. The appropriate logic has combined dynamic-epistemic assertions

$$\square_{A!}\ \phi\quad \text{\textit{``after truthful announcement of A, ϕ holds.''}}$$

The set-up of this system merges epistemic with dynamic logic, with some additions reflecting particulars of our update universe. There is a complete and decidable axiomatization [17, 6], with the following key axioms:

$$\diamond_{A!}\ p \quad\longleftrightarrow\quad A \wedge p\ \textit{for atomic facts } p,$$
$$\diamond_{A!}\ \neg\phi \quad\longleftrightarrow\quad A \wedge \neg\diamond_{A!}\ \phi,$$
$$\diamond_{A!}\ \phi \vee \psi \quad\longleftrightarrow\quad \diamond_{A!}\ \phi \vee \diamond_{A!}\ \psi,$$
$$\diamond_{A!}\ K_i\phi \quad\longleftrightarrow\quad A \wedge K_i(A \rightarrow \diamond_{A!}\ \phi).$$

Essentially, these compute preconditions $\diamond_{A!}\ \phi$ by *relativizing* the postcondition ϕ to A. The axioms can also be stated with the universal modal box, leading to versions like

$$\square_{A!}K_i\phi \longleftrightarrow A \longrightarrow K_i\square_{A!}\phi.$$

This is like the above law for Perfect Recall. As for common knowledge, the earlier epistemic language needs a little extension, with a *binary* version

$$C_G(A, \phi) \qquad \text{common knowledge of } \phi \text{ in the submodel defined by } A.$$

There is no definition for this in terms of just absolute common knowledge. Having added this feature, we can state the remaining reduction principle

$$\diamond_{A!}C_G\phi \longleftrightarrow A \wedge C_G(A, \diamond_{A!}\phi)$$

[38] develops this idea in detail. There are even further ways of combining knowledge and action. In particular, [26] defines a more thoroughly epistemized dynamic logic with transitions as objects in their own right.

DEL with program constructions. Public announcement is just one basic action. Conversation may involve more complex programming of what is said. Saying one thing after another amounts to program composition, choosing one's assertions involves choice, and Muddy Children even involved a guarded iteration:

WHILE "you don't know your status" DO "say so".

The basic logic of public update with the first two constructions is like its version with just basic announcements $A!$, because of the reduction axioms for composition and choice in dynamic logic. But with possible iteration of announcements, the system changes — and even loses its decidability [2].

§3. **Basic theory of information models.** *Special model classes.* Multi-S5 models can be quite complicated. But there are some subclasses of special interest. For instance, Muddy Children started with a full cube of 3-vectors, with accessibility given as the special equivalence relation

$X \sim_j Y$ which holds if the j-th coordinates are equal: $(X)_j = (Y)_j$.

Cube models are studied in algebraic logic [10] for their connections with assignment spaces over first-order models. But the subsequent Muddy Children updates led to submodels of such cubes. These already attain full epistemic generality [22]:

THEOREM 3.1. *Every multi-S5 model is representable up to bisimulation as a submodel of a full cube.*

Other special model classes arise in the study of card games [41] and games in general [21].

Bisimulation. Epistemic and dynamic logic are both standard modal logics (cf. [4]) with the following structural model comparison:

DEFINITION 3.2. A *bisimulation* between two models M, N is a binary relation $\equiv$ between their states m, n such that, whenever $m \equiv n$, then (a) m, n satisfy the same proposition letters, (b1) if mRm', then there exists a world n' with nRn' and $m' \equiv n'$, and (b2) the analogous "zigzag clause" holds in the opposite direction.

Our question-answer example has a bisimulation with this variant:

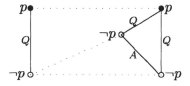

In a natural sense, these are two representations of the same information state. Bisimulation equivalence occurs naturally in update. Suppose the current model is like this, with the actual world indicated by the black dot:

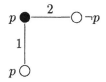

Note that all three worlds satisfy different epistemic formulas. Now, despite her uncertainty, *1* knows that *p*, and can say this — updating to the model

But this can be contracted via a bisimulation to the one-point model

In general, it is convenient to think of update steps with bisimulation contractions interleaved automatically.

Some basic results link bisimulation to truth of modal formulas. For convenience, we restrict attention to *finite models* — but this can be lifted.

Invariance and definability. Consider general models, or those of multi-S5.

LEMMA 3.3 (Invariance Lemma). *The following are equivalent*:

(a) M, s *and* N, t *are connected by a bisimulation*,
(b) M, s *and* N, t *satisfy the same modal formulas*.

Any model has a bisimilar *unraveled tree model*, but also a smallest *bisimulation contraction* satisfying the same modal formulas. Another useful tool are modal analogues of Scott-sentences for infinitary logic (cf. [3]):

LEMMA 3.4 (State Definition Lemma). *For each model* M, s *there is an epistemic formula* β *(involving common knowledge) such that the following assertions are equivalent*:

(a) $N, t \models \beta$,
(b) N, t *has a bisimulation* $\equiv$ *with* M, s *such that* $s \equiv t$.

PROOF. The version and proof given here are from [23, 25]. Consider any finite multi-S5 model M, s. This falls into a number of maximal "zones" of worlds that satisfy the same epistemic formulas in our language.

CLAIM 3.5. There exists a finite set of formulas ϕ_i ($1 \leq i \leq k$) such that (a) each world satisfies one of them, (b) no world satisfies two of them (i.e., they define a partition of the model), and (c) if two worlds satisfy the same formula ϕ_i, then they agree on all epistemic formulas.

To show this, take any world s, and find difference formulas $\delta^{s,t}$ between it and any t which does not satisfy the same epistemic formulas, where s satisfies $\delta^{s,t}$ while t does not. The conjunction of all $\delta^{s,t}$ is a formula ϕ_i true only in s and the worlds sharing its epistemic theory. We may assume that the ϕ_i also list all information about the proposition letters true and false throughout their partition zone. We also make a quick observation about uncertainty links between these zones:

\#   If any world satisfying ϕ_i is $\sim_a$-linked to a world satisfying ϕ_j,
then all worlds satisfying ϕ_i also satisfy $\Diamond_a \phi_j$.

Next take the following description $\beta_{M,s}$ of M, s:

(a)   all (negated) proposition letters true at s
plus the unique ϕ_i true at M, s.
  (b)   common knowledge for the whole group of
       (b1)   the disjunction of all ϕ_i,
       (b2)   all negations of conjunctions $\phi_i \wedge \phi_j$ $(i \neq j)$,
       (b3)   all implications $\phi_i \to \Diamond_a \phi_j$ for which situation \# occurs,
       (b4)   all implications $\phi_i \to \Box_a \vee \phi_j$ where the disjunction runs
              over all situations listed in the previous clause.

CLAIM 3.6. $M, s \models \beta_{M,s}$

CLAIM 3.7.   If $N, t \models \beta_{M,s}$, then there exists a bisimulation between N, t
and M, s.

To prove Claim 3.7, let N, t be any model for $\beta_{M,s}$. The ϕ_i partition N into
disjoint zones Z_i of worlds satisfying these formulas. Now relate all worlds in
such a zone to all worlds that satisfy ϕ_i in the model M. In particular, t gets
connected to s. We check that this connection is a bisimulation. The atomic
clause is clear from an earlier remark. But also, the zigzag clauses follow from
the given description. (a) Any $\sim_a$-successor step in M has been encoded in
a formula $\phi_i \to \Diamond_a \phi_j$ which holds everywhere in N, producing the required
successor there. (b) Conversely, if there is no $\sim_a$-successor in M, this shows
up in the limitative formula $\phi_i \to \Box_a \vee \phi_j$, which also holds in N, so that there
is no "excess" successor there either. This concludes the proof.   ⊣

The Invariance Lemma says that bisimulation has the right fit with the
modal language. The State Definition Lemma says that each semantic state
can be characterized by one epistemic formula. E.g., consider the two-world
model for our question-answer episode. Here is an epistemic formula which
defines its ϕ-state up to bisimulation:

$$\phi \wedge C_{\{Q,A\}}\big((K_A\phi \vee K_A \neg \phi) \wedge \neg K_Q \phi \wedge \neg K_Q \neg \phi\big)$$

This allows us to switch, in principle, between semantic accounts of information states as models M, s and syntactic ones in terms of complete defining
formulas. There is more to this than just technicality. For instance, syntactic
approaches have been dominant in related areas like belief revision theory,
where information states are not models but syntactic theories. It would be
good to systematically relate syntactic and semantic approaches to update,
but we shall stay semantic here.

Respectful and safe operations. The above also constrains epistemic update operations O. These should *respect bisimulation*:

If M, s and N, t are bisimilar, so are their values $O(M, s)$ and $O(N, t)$.

FACT 3.8. Public update respects bisimulation.

PROOF. Let $\equiv$ be a bisimulation between M, s and N, t. Consider their submodels $M|\phi, s$, $N|\phi, t$ after public update with ϕ. The restriction of $\equiv$ to these is still a bisimulation. Here is the zigzag clause. Suppose some world w has an $\sim_i$-successor v in $M|\phi, s$. This same v is still available in the other model: it remained in M since it satisfied ϕ. But then v also satisfied ϕ in N, t, because of the Invariance Lemma for the bisimulation $\equiv$ — and so it stayed in the updated model $N|\phi, t$, too.   ⊣

Many other proposed update operations respect bisimulations (cf. [8] on similar phenomena in process algebra). Finally, bisimulation also works for the full language of dynamic logic — but with a new twist [22]. Intertwined with invariance for formulas ϕ, one must show that the zigzag clauses go through for all regular program constructions: not just the atomic R_a, but each transition relation $[[\pi]]$:

FACT 3.9. Let $\equiv$ be a bisimulation between two models M, M', with $s \equiv s'$. Then it holds that

(i) s, s' verify the same formulas of propositional dynamic logic,
(ii) if $s[[\pi]]^M t$, then there exists t' with $s'[[\pi]]^{M'} t'$ and $s' \equiv t'$.

This observation motivates a new invariance for program operations:

DEFINITION 3.10. An operation $O(R_1, \ldots, R_n)$ on programs is *safe for bisimulation* if, whenever $\equiv$ is a relation of bisimulation between two models for their transition relations $R_1, \ldots, R_n$, then it is also a bisimulation for the defined relation $O(R_1, \ldots, R_n)$.

The core of the above program induction is that the three regular operations; $\cup$ * of PDL are safe for bisimulation. By contrast, the operation of program *intersection* is not safe:

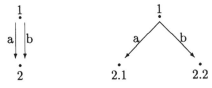

There is an obvious bisimulation here with respect to a, b — but the required zigzag clause fails for $R_a \cap R_b$.

After this technical extravaganza, we return to communication. In fact, the Muddy Children puzzle highlights a whole agenda of further questions. We already noted how its specific model sequence is characteristic for the field.

But in addition, it raises many further central issues, such as

(a)   the benefits of internal group communication,
(b)   the role of iterated assertion,
(c)   the interplay of update and inference in reasoning.

We will look into these as we go. But we start with an issue already noted: the putative "learning principle" that was refuted by Muddy Children.

§4. What do we learn from a statement? *Specifying speech acts.* Update logic is a sequel to speech act theories, which originated in philosophy, and then partly migrated to computer science (cf. [44]). Earlier accounts of speech acts often consist in formal specifications of preconditions and postconditions of successful assertions, questions, or commands. Some of these insights are quite valuable, such as those concerning assertoric force of assertions. E.g., in what follows, we will assume, in line with that tradition, that normal cooperative speakers may only utter statements which they know to be true. Even so, what guarantees that the specifications are correct? E.g., it has been said that answers to questions typically produce common knowledge of the answer.   But Muddy Children provided a counter-example to this naive "Learning Principle". Logical tools help us get clearer on pitfalls and solutions. The learning problem is a good example of this use.

Persistence under update. Public announcement of atomic facts p makes them common knowledge, and the same holds for other types of assertion. But, as we noted in Section 1, not all updates with ϕ result in common knowledge of ϕ! A simple counter-example is this. In our question-answer case, let A say truly

$$p \wedge \neg K_Q p \qquad \text{"}p\text{, but you don't know it."}$$

This utterance removes Q's lack of knowledge about the fact p, and thereby makes its own assertion false! Ordinary terminology is misleading here:

> *learning that* ϕ is ambiguous between: ϕ *was* the case, before the announcement, and ϕ *is* the case — after the announcement.

The explanation is that statements may change their truth value with update. For worlds surviving in the smaller model, factual properties do not change, but epistemic properties may. This raises a general logical issue of *persistence under update*:

> Which forms of epistemic assertion remain true at a world
> whenever other worlds are eliminated from the model?

These are epistemic assertions which, when publicly announced to a group, will always result in common knowledge. Examples are atomic facts p, and knowledge-free assertions generally, but also knowledge assertions $K_i p$ and ignorance assertions $\neg K_i p$.

New kinds of preservation result. Here is a relevant model-theoretic fact from modal logic (cf. [37]):

THEOREM 4.1. *The epistemic formulas without common knowledge that are preserved under submodels are precisely those definable using literals p, $\neg p$, conjunction, disjunction, and K_i-operators.*

Compare universal formulas in first-order logic, which are just those preserved under submodels. The obvious conjecture for the epistemic language with common knowledge would just add arbitrary formulas $C\phi$ as persistent forms. But this result is still open, as lifting first-order model theory to such non-first-order modal fixed-point languages seems non-trivial, even on a universe of finite models.

QUESTION. Which formulas of the full epistemic language with common knowledge are preserved under submodels?

In any case, what we need is not really full preservation under submodels, but rather preservation under "self-defined submodels":

When we *restrict* a model to those of its worlds which satisfy ϕ, then ϕ should hold throughout the remaining model, or in terms of an elegant validity: $\phi \rightarrow (\phi)^\phi$.

QUESTION. Which epistemic formulas imply their self-relativization?

Indeed, which *first-order* formulas are preserved in this self-fulfilling sense? Model-theoretic preservation questions of this special form seem new.

A non-issue? Many people find this particular issue annoying. Non-persistence seems a side-effect of infelicitous wording. E.g., when A said "p, but you don't know it", she should just have said "p", keeping her mouth shut about my mental state. Now, the Muddy Children brand of non-persistence is not as blatant as this. And in any case, dangers in timing aspects of what was true before and is true after an update are no more exotic than the acknowledged danger in computer science of confusing states of a process. Dynamic logics were developed precisely to keep track of that. But let's stop fencing: *can* we reword any message to make the non-persistence go away? An epistemic assertion A defines a set of worlds in the current model M. Can we always find an equivalent persistent definition? This would be easy if each world has a simple unique factual description, as is the case with hands in card games. But even without assuming this there is a method that works, at least locally:

FACT 4.2. In each model, every public announcement has a persistent equivalent leading to the same update.

PROOF. Without loss of generality, assume we are working only with bisimulation-contracted models which are also totally connected: no isolated components. Let w be the current world in model M. Let j publicly announce

A, updating to the submodel $M|A$ with domain $A^* = \{s \in M | M, s \models A\}$. If this is still M itself, then the announcement "True" is adequate, and persistent. Now suppose A^* is not the whole domain. Our persistent assertion consist of two disjuncts:

$$\Delta \vee \Sigma.$$

First we make Δ. Using the proof of the State Definition Lemma of Section 3, this is an epistemic definition for A^* in M formed by describing each world in it up to bisimulation, and then taking the disjunction of these.

Now for Σ. Again using the mentioned proof, write a formula which describes $M|A$ up to bisimulation. For concreteness, this had a common knowledge operator over a plain epistemic formula describing the pattern of states and links, true everywhere in the model $M|A$. No specific world description is appended, however.

Clearly $\Delta \vee \Sigma$ is common knowledge in $M|A$, because Σ is. But it also picks out the right worlds in M. First, any world in A^* satisfies its own disjunct of Δ. Conversely, suppose any world t in M satisfies $\Delta \vee \Sigma$. If it satisfies some disjunct of Δ, then it must then be in A^* by the bisimulation-minimality of the model. Otherwise, M, t satisfies Σ. But then by connectedness, every world in M satisfies Σ, and in particular, given the construction of Σ, there must be a bisimulation between M and $M|A$. But this contradicts the fact that the update was genuine. ⊣

Of course, this recipe for phrasing your assertions is ugly, and not recommended! Moreover, it is local to one model, and does not work uniformly. Recall that, depending on group size, muddy children may have to repeat the same ignorance statement any number of times before knowledge dawns. If there were one uniform persistent equivalent for that statement, the latter's announcement would always lead to common knowledge uniformly after some fixed finite stage.

§5. Internal communication in groups. *The best we can.* The muddy children might just tell each other what they see, and common knowledge of their situation is reached at once. The same holds for card players telling each other their hands. Of course, life is civilized precisely because we do not "tell it like it is". Even so, there is an issue of principle of what agents in a group can achieve by maximal communication. Consider two epistemic agents that find themselves in some collective information state M, at some actual situation s. They can tell each other things they know, thereby cutting down the model to smaller sizes. Suppose they wish to be maximally cooperative:

> What is the best correct information they can give each other via successive updates — and what does the resulting collective information state look like?

E.g., what is the best that can be achieved in the following model?

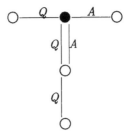

Geometrical intuition suggests that this must be:

This is correct! First, any sequence of mutual updates in a finite model must terminate in some minimal domain which can no longer be reduced. This is reached when everything each agent knows is already common knowledge: i.e., it holds in every world. But what is more, this minimal model is *unique*, and we may call it the "communicative core" of the initial model. Here is an explicit description, proved in [29]:

THEOREM 5.1. *Each model has a communicative core, viz. the set of worlds that are reachable from the actual world via all uncertainty links.*

PROOF. For convenience, consider a model with two agents only. The case with more than two agents is an easy generalization of the same technique. First, agents can reach this special set of worlds as follows. Without loss of generality, let all states t in the model satisfy a unique defining formula δ_t as in Section 3 — or obtained by an ad-hoc argument. Agent 1 now communicates all he knows by stating the *disjunction* $\vee\delta_t$ for all worlds t he considers indistinguishable from the actual one. This initial move cuts the model down to the actual world plus all its $\sim_1$-alternatives. Now there is a small technicality. The resulting model need no longer satisfy the above unique definability property. The update may have removed worlds that distinguished between otherwise similar options. But this is easy to remedy by taking the *bisimulation contraction*. Next, let 2 make a similar strongest assertion available to her. This cuts the model down to those worlds that are also $\sim_2$-accessible from the actual one. After that, everything any agent knows automatically becomes common knowledge, so further statements have no informative effect.

Next, suppose agents reach a state where further announcements have no effect. Then the following implications hold for all ϕ : $K_1\phi \rightarrow C_{\{1,2\}}\phi$, $K_2\phi \rightarrow C_{\{1,2\}}\phi$. Again using defining formulas, this means 1, 2 have the same alternative worlds. So, these form a *subset* of the above core. But in fact, all of it is preserved. An agent can only make statements that hold in all of its worlds, as it is included in his information set. Therefore, the whole core survives each episode of public update, and therefore, by induction, it survives all of them.   ⊣

A corollary of the preceding proof is this:

FACT 5.2. Agents need only 2 communication rounds to get to the core.

In particular, there is no need for repetitions by agents. E.g., let 1 truly say A (something he knows in the actual world), note the induced public update, and then say B (which he knows in the new state). Then he might just as well have asserted $A \wedge (B)^A$ straightaway: where $(B)^A$ is the relativization of B to A (cf. Section 6).

Incidentally, a two-step solution to the initial example of this section is the following rather existentialist conversation:

> Q sighs: "I don't know."
> A sighs: "I don't know either."

It does not matter if you forget the details here, because for our model, it also works in the opposite order.

The communicative core is the actual world plus every world connected to it by the intersection of all uncertainty relations. This is the range used in defining *implicit knowledge* for a group of agents in Section 2.1 Thus, maximal communication turns implicit knowledge of a group into common knowledge. As a slogan, this makes sense, but there are subtleties. It may be implicit knowledge that none of us know where the treasure is. But once the communicative core is all that is left, the location of the treasure may be common knowledge. Compare the difference between quantifier restriction and relativization. Implicit knowledge $I_G\phi$ looks only at worlds in the communicative core CC, but it then evaluates the formula ϕ from each world there in the whole model. By contrast, internal evaluation in the core is like evaluating totally relativized statements $(\phi)^{CC}$ in the model.

Another technicality is that the relevant *intersection* of relations, though keeping the logic decidable, is no longer safe for bisimulation in the sense of Section 4. Adding it to the language leads to a genuinely richer epistemic logic, for which some of the earlier model theory would have to be redone.

Planning assertions. The preceding topic really shows a shift in interest. Update logics can be used to analyze given assertions, but they can also be used to plan assertions meeting certain specifications. An example is the following puzzle from a mathematical Olympiad in Moscow (cf. [42]):

> 7 cards are distributed among A, B, C. A gets 3, B 3, C 1. How should A, B communicate publicly, in hearing of C, so that they find out the precise distribution of the cards while C does not?

There are solutions here — but their existence depends on the number of cards. This question may be seen as a generalization of the preceding one. How can a subgroup of all agents communicate maximally, while keeping the rest of the group as much in the dark as possible? Normally, this calls for hiding, but it is interesting to see — at least to politicians, or illicit lovers — that some of this can be achieved publicly. This sort of planning problem is only beginning to be studied.

Here we just observe an analogy with computer science. The dynamic epistemic logic of Section 2.3 is like well-known program logics manipulating correctness assertions

$$\phi \rightarrow \Box_{A!}\, \psi \qquad \text{if precondition } \phi \text{ holds, then saying } A \text{ always}$$
leads to a state where postcondition ψ holds.

Such triples may be looked at in different ways. Given an assertion, one can analyze its preconditions and postconditions, as we did for questions and answers. Or, given an assertion $A!$ plus a precondition ϕ we can look for their strongest postcondition ψ. E.g., with $\phi = True$, and A atomic, it is easy to show that the strongest postcondition is the common knowledge of A. This is program analysis. But there is also program *synthesis*. Given a precondition ϕ and a postcondition ψ, we can look for an assertion $A!$ guaranteeing the transition. Here the postcondition ψ may specify which agents are supposed to learn what, using assertions of the forms $K_j\alpha$ and $\neg K_j\beta$. Conversation planning is like this, determining what to say so that only the right people get the right information.

§6. **Public update as relativization.** The following technical intermezzo [27] joins forces with standard logic.

Semantic and syntactic relativization. Here is a simple observation. Announcing A amounts to a logical operation of *semantic relativization*

from a model M, s to the definable submodel $M|A$, s.

This explains all behaviour so far — while raising new questions. For a start, in the new model, we can again evaluate formulas that express knowledge and ignorance of agents, in the standard format $M|A, s \models \phi$. In standard logic, this may also be described via *syntactic relativization* of the formula ϕ by the update assertion A:

LEMMA 6.1 (Relativization Lemma). $M|A, s \models \phi$ iff $M, s \models (\phi)^A$.

This says we can either evaluate assertions in a relativized model, or equivalently, their relativized versions in the original model. For convenience, we assume that relativization is defined so that $(\phi)^A$ always implies A. For the basic epistemic language, this goes via the following recursion:

$$(p)^A = A \wedge p,$$
$$(\neg\phi)^A = A \wedge \neg(\phi)^A,$$
$$(\phi \vee \psi)^A = (\phi)^A \vee (\psi)^A,$$
$$(K_i\phi)^A = A \wedge K_i(A \longrightarrow (\phi)^A).$$

In this definition, one immediately recognizes the above axioms for public update. Whether this works entirely within the language of epistemic announcements depends on its strength. E.g., relativization was less straightforward with common knowledge, as no syntactic prefix "$A \to \ldots$" or "$A \wedge \ldots$" on absolute operators C_G does the job. But we saw how to extend epistemic logic with a binary restricted common knowledge operator. Actually, dynamic logic is better behaved in this respect.

FACT 6.2. Dynamic logic is closed under relativization.

PROOF. In line with the usual syntax of the system, we need a double recursion over formulas and programs. For formulas, the clauses are all as above, while we add

$$(\Box_\pi\phi)^A = \Box_{(\pi)^A}(\phi)^A.$$

For programs, here are the recursive clauses that do the job:

$$(R; S)^A = (R)^A; (S)^A,$$
$$(R \cup S)^A = (R)^A \cup (S)^A,$$
$$((\phi)?)^A = (A)?; (\phi)^A,$$
$$(\pi^*)^A = ((A)?; (\pi)^A)^*. \qquad \dashv$$

Now, common knowledge $C_G\phi$ may be viewed as a dynamic logic formula

$$\Box_{(\cup\{i | i \in G\})^*}\phi.$$

Therefore, we can get a natural relativization for epistemic logic by the above Fact, by borrowing a little syntax from dynamic logic.

General logic of relativization. Stripped of its motivation, update logic is an axiomatization of one model-theoretic operation, viz. relativization. There is nothing specifically modal about this. One could ask for a complete logic of relativizations $(\phi)^A$ in first-order logic, as done for *substitutions* $\Box_{t/x}\phi$ in [10].

QUESTION. What is the complete logic of the relativization operator in a standard first-order language?

At least we may observe that there is more than the axioms listed in Section 2.3. For instance, the following additional fact is easy to prove:

Associativity $((A)^B)^C$ is logically equivalent to $A^{((B)^C)}$.

In our update logic, performing two relativizations matches performing two consecutive updates. Thus Associativity amounts to the validity of

$$\Box_{A!;B!}\phi \longleftrightarrow \Box_{(\Box_{A!}B)!}\phi$$

Why was this not on the earlier list of the complete axiom system? The answer is a subtlety. That axiom system does indeed derive every valid formula. But it does so without being *substitution-closed*. In particular, the above basic axiom for atoms

$$\Diamond_{A!}p \longleftrightarrow A \wedge p$$

fails for arbitrary formulas ϕ. Define the *substitution core* of update logic as those valid schemata all of whose substitution instances are valid formulas. Associativity belongs to it, but it is not derivable schematically from the earlier axiom system.

QUESTION. Can one axiomatize the complete substitution core of public update logic?

There are also interestingly invalid principles, witness the discussion of persistence in Section 4. Announcing a true statement "p, but you don't know it" invalidates itself. More technically, even when $p \wedge \Diamond_1 \neg p$ holds, its self-relativization

$$(p \wedge \Diamond_1 \neg p)^{p \wedge \Diamond_1 \neg p} = p \wedge \Diamond_1 \neg p \wedge \Diamond_1(p \wedge \Diamond_1 \neg p \wedge p)$$

is a contradiction. Thus some assertions are self-refuting when announced, and the earlier-mentioned typographically pleasing principle is not part of a general logic of relativization:

$$\phi \longrightarrow (\phi)^\phi \qquad \text{holds only for special assertions } \phi.$$

We look at further issues in the logic of relativization in Section 7, including iterated announcement and its connections with fixed-point operators.

Excursion: richer systems of update. In standard logic, relativization often occurs together with other operations, such as *translation of predicates* — e.g., in the notion of *relative interpretation* of one theory into another. Likewise, the above connection extends to more sophisticated forms of epistemic update (cf. Section 9). For instance, when a group hears that a question is asked and answered, but only a subgroup gets the precise answer, we must use a new operation of *arrow elimination*, rather than world elimination. More precisely,

all arrows are removed for all members of that subgroup between those zones of the model that reflect different exhaustive answers.

Arrow elimination involves substitution of new accessibility relations for the current ones. E.g., when the question "ϕ?" is asked and answered, the uncertainty relations $\sim_i$ for agents i in the informed subgroup are replaced by the union of relations

$$(\phi)?; \sim_i; (\phi)? \cup (\neg\phi)?; \sim_i; (\neg\phi)?$$

This is a translation of the old binary relation $\sim_i$ into a new definable one.

Next on this road, there are more complex "product updates" — which correspond to those interpretations between theories which involve construction of new definable objects, like when we embed the rationals into the integers using ordered pairs. Axioms for update logics will then still axiomatize parts of the meta-theory of such general logical operations. Thus, progressively more complex notions of update correspond to more sophisticated theory relations from standard logic.

Finally, relativization suggests a slightly different view of eliminative update. So far, we discarded old information states. But now, we can keep the old information state, and perform "virtual update" via relativized assertions. Thus, the initial state already contains all possible future communicative developments. Another take on this at least keeps the old models around, doing updates with *memory*. There are also independent reasons for maintaining some past history in our logic, having to do with public updates which refer explicitly to the "epistemic past", such as:

"what you said, I knew already".

Section 9 below has more concrete examples.

§7. **Repeated announcement and limit behaviour.** *"Keep talking"*. In the Muddy Children scenario, an assertion of ignorance was repeated until it could no longer be made truly. In the given model, the statement was *self-defeating*: when repeated iteratively, it reaches a stage where it is not true anywhere. Of course, self-defeating ignorance statements lead to something good for us, viz. knowledge. There is also a counterpart to this limit behaviour: iterated announcement of *self-fulfilling* statements makes them common knowledge. This happened in one step with factual assertions and others in Section 4. Technically, this means that the process of repeatedly announcing A reaches a *fixed-point*. More subtle cases are discussed in [21], viz. repeated joint assertions of rationality by players in a strategic game, saying that one will only choose actions that may be best possible responses to what the others do. These may decrease the set of available strategy profiles until a "best zone" is reached consisting of either a Nash equilibrium, or at least some rationalizable profiles to be in.

Limits and fixed-points. Repeated announcement of rationality by two players *1, 2* has the following simple modal form, which we take for granted here without motivation:

$$JR: \qquad \Diamond_1 B_1 \wedge \Diamond_2 B_2.$$

Here the proposition letter B_i says that i's action in the current world is a *best response for i* to what the opponent is playing here. It can be shown that any finite game matrix has entries (worlds in the corresponding epistemic model) in a loop with alternating assertions B_i:

$$x_1 \models B_1 \sim_2 x_2 \models B_2 \sim_1 x_3 \models B_1 \sim_2 \cdots \sim_1 x_1 \models B_1.$$

Repeated announcement of joint rationality JR may keep removing worlds, as each round may remove worlds satisfying a B_i on which one conjunct depended. But clearly, whole loops of the kind described remain all the time, as they form a kind of mutual protection society. Thus, we have

FACT 7.1. Strong Rationality is self-fulfilling on finite game models.

The above suggests extending update logic with *fixed-point operators*. This is like extending modal logic to a μ-calculus, whose syntax defines smallest fixed-points $\mu p \bullet \phi(p)$ and greatest ones $\nu p \bullet \phi(p)$. [20] has details on this, [5] on more general fixed-point logics. We explore this a bit, as there are delicate issues involved (cf. again [21]). For a start, we have this

FACT 7.2. The stable set of worlds reached via repeated announcement of JR is defined inside the original full game model by the greatest fixed-point formula $\nu p \bullet (\Diamond_E (B_E \wedge p) \wedge \Diamond_A (B_A \wedge p))$.

Iterated announcement in dynamic logic. In any model M, we can keep announcing any formula ϕ, until we reach a fixed-point, empty or not:

$$\#(\phi, M).$$

E.g., *self-fulfilling* formulas ϕ in M become common knowledge in $\#(\phi, M)$:

$$\phi \longrightarrow (C_G \phi)^{\#(\phi, M)}.$$

What kind of fixed-point are we computing here? Technically $\#(\phi, M)$ arises by continued application of the following set function, taking intersections at limit ordinals:

$$F_{M,\phi}(X) = \{s \in X | M | X, s \models \phi\}$$

with $M|X$ the restriction of M to the set X.

This map F is *not monotone*, and the usual theory of fixed-points does not apply. The reason is the earlier fact that statements ϕ may change truth value when passing from M to submodels $M|X$. In particular, we do not recompute stages inside one unchanging model, as in the normal semantics of greatest fixed-point formulas $\nu p \bullet \phi(p)$, but in ever smaller models, changing the range of the modal operators. Thus we mix fixed-point computation with *relativization* (cf. Section 6). Despite *F*'s non-monotonicity, iterated announcement is a fixed-point procedure of sorts:

FACT 7.3. The iterated announcement limit is definable as a so-called inflationary fixed-point.

PROOF. Take any ϕ, and relativize it to a fresh proposition letter p, yielding $(\phi)^p$. Here p need not always occur positively (it becomes negative when relativizing positive K-operators). Now the earlier epistemic Relativization Lemma says that

$$M, s \models (\phi)^p \quad \text{iff} \quad M|X, s \models \phi.$$

Thus, the above definition of $F_{M,\phi}(X)$ as $\{s \in X | M|X, s \models \phi\}$ equals

$$\{s \in M | M, s \models (\phi)^p\} \cap X.$$

And this computes a greatest *inflationary fixed-point* [5]. ⊣

But then, why did iterated announcement of *JR* produce an ordinary greatest fixed-point in the epistemic μ-calculus? The above update map $F_{M,\phi}(X)$ is monotone with special sorts of formulas:

FACT 7.4. $F_{M,\phi}(X)$ is monotone for existential modal formulas.

The reason is that such formulas are preserved under model extensions, making their F monotone for set inclusion: cf. the related preservation issues in Section 4. As the epistemic μ-calculus is decidable, while its inflationary fixed-point extension is not, these issues of syntax matter to the complexity of reasoning about iterated announcement.

Excursion: comparing update sequences. Update logic is subtle, even here. What happens with different repeated announcements of rationality that players could make? [21] considers a weaker rationality assertion *WR* which follows from *JR*. Does this guarantee that their limits are included:

$$\#(SR, M) \subseteq \#(WR, M)?$$

The general answer is negative. Making weaker assertions repeatedly may lead to incomparable results. An example are the following formula ϕ and its consequence ψ:

$$\phi = p \wedge (\Diamond \neg p \longrightarrow \Diamond q),$$
$$\psi = \phi \vee (\neg p \wedge \neg q).$$

In the following epistemic model, the update sequence for ϕ stops in one step with the world *1*, whereas that for ψ runs: $\{1, 2, 3\}, \{1, 2\}, \{2\}$.

$$\begin{array}{ccc} 1 & \!\!\!\!\!-\!\!\!\!-\!\!\!\! & 2 \!\!\!\!-\!\!\!\!-\!\!\!\! 3 \\ p, \neg q & \neg p, \neg q & \neg p, q \end{array}$$

But sometimes, things work out.

FACT 7.5. If an *existential* ϕ implies ψ in M, then $\#(\phi, M) \subseteq \#(\psi, M)$.

PROOF. We always have the inclusion

$$T_\phi^\alpha(M) \subseteq T_\psi^\alpha(M).$$

The reason for this is the following implication:

$$\text{if } X \subseteq Y, \text{ then } F_{M,\phi}(X) \subseteq F_{M,\psi}(Y).$$

For, if $M|X, s \models \phi$ and $s \in X$, then $s \in Y$ and also $M|Y, s \models \phi$ — by the modal *existential* form of ϕ. But then we also have that $M|Y, s \models \psi$, by our valid implication. $\dashv$

One more type of fixed-point! Iterated announcement can be described by the finite iteration $*$ of dynamic logic (cf. Section 2.2). This extension is studied in [2], which shows that formulas of the form

$$\Diamond_{(A!)^*} C_G A$$

are not definable in the modal μ-calculus. Still, it is known that formulas

$$\Box_{(A!)^*} \phi$$

with program iteration are definable with greatest fixed-point operators

$$\nu p \bullet \phi \wedge \Box_{A!} p.$$

But these cannot be analyzed in the earlier style, as they involve relativizing p to A, rather than the more tractable A to p, as happened in our analysis of repeated announcement.

§8. **Inference versus update.** *Dynamic inference.* Standard epistemic logic describes inference in unchanging information models. But there is also a more lively notion following the dynamics of update (cf. [22]):

> Conclusion ϕ *follows dynamically* from premises $P_1, \ldots, P_k$ if after updating any information state with public announcements of the successive premises, all worlds in the end state satisfy ϕ.

Thus in dynamic-epistemic logic, the following implication must be valid:

$$\Box_{P_1!;\ldots;P_k!} C_G \phi.$$

This behaves differently from standard logic in its premise management:

Order of presentation matters
Conclusions from A, B need not be the same as from B, A:
witness $\neg Kp, p$ (consistent) versus $p, \neg Kp$ (inconsistent).

Multiplicity of occurrence matters
$\neg Kp \wedge p$ has different update effects from $(\neg Kp \wedge p) \wedge (\neg Kp \wedge p)$.

Adding premises can disturb conclusions
$\neg Kp$ implies $\neg Kp$ — but $\neg Kp, p$ does not imply $\neg Kp$.

By contrast, the structural rules of classical logic say that order, multiplicity, and overkill does not matter. Nevertheless, we have a description.

Structural rules and representation. [33] has three modified structural rules that are valid for dynamic inference as defined above:

Left Monotonicity	$X \Rightarrow A$ implies $B, X \Rightarrow A$,
Cautious Monotonicity	$X \Rightarrow A$ and $X, Y \Rightarrow B$ imply $X, A, Y \Rightarrow B$,
Left Cut	$X \Rightarrow A$ and $X, A, Y \Rightarrow B$ imply $X, Y \Rightarrow B$.

Moreover, the following completeness result holds:

THEOREM 8.1. *The structural properties of dynamic inference are axiomatized completely by the rules of Left Monotonicity, Cautious Monotonicity, and Left Cut.*

The core of the proof is a representation argument showing that any abstract finite tree model for modal logic can be represented up to bisimulation in the following form:

Worlds w go to a family of epistemic models M_w
Basic actions a go to suitable epistemic announcements (ϕ_a)!

This suggests that public update is a quite general process, which can encode arbitrary processes in the form of "conversation games".

Inference versus update. Here is a more general moral. Logic has two different inferential processes. The first is *ordinary inference*, related to implications $A \to B$. This stays inside one fixed model. The second process is a model-jumping *inference under relativization*, related to the earlier formulas $\Box_{A!} B$. Both are important, and update logics mix them.

Even so, one empirical issue remains. Muddy children *deduce* a solution: they did not draw update diagrams. What is going on inside our heads?

§9. The wider world.

As stated in Section 1, update analysis has two main directions: increased coverage by means of new models and mechanisms, and increased in-depth understanding of the logics that we already have. This paper has concentrated on the latter, hoping to show the interest of logical issues in communication. In this Section, the reader gets a lightning tour of what she has missed by taking this prong of the fork.

Keeping track of past assertions. Some puzzles involve reference to past states, with people saying things like "What you said did not surprise me" [11]. This says they knew at the previous state, calling for a further update there. To accomplish this, we need to maintain a stack of past updates, instead of just performing them and trashing previous stages. In the limit, as earlier, this richer framework might also include protocol information about the sort of communication we are in.

Privacy and hiding. The next stage of complexity beyond public communication involves hiding information, either willfully, or through partial observation. Here is about the simplest example, from [24]:

> *We have both just drawn a closed envelope. It is common knowledge between us that one envelope holds an invitation to a lecture on logic, the other to a wild night out in Amsterdam. We are both ignorant of the fate in store for us! Now I open my envelope, and read the contents, without showing them to you. Yours remains closed. Which information has passed exactly because of my action? I certainly know now which fate is in store for me. But you have also learnt something, viz.* that I know - *though not what I know. Likewise, I did not just learn what is in my envelope. I also learnt something about you, viz. that you know that I know. The latter fact has even become common knowledge between us. And so on. What is a general principle behind this?*

The initial information state is a familiar one of collective ignorance:

$$L\bullet \overset{me}{\underset{you}{\rule{2cm}{0.4pt}}} \bigcirc N$$

The intuitive update just removes my uncertainty link — while both worlds remain available, as they are needed to model your continued ignorance of the base fact:

$$L\bullet \underset{you}{\rule{2cm}{0.4pt}} \bigcirc N$$

Such updates occur frequently in card games, when players publicly show cards to some others, but not to all. But card updates can also blow up the size of a model. Suppose I opened my envelope, but you cannot tell if I read the card in it or not. Let us say that in fact, I did look. In that case, the intuitive update is to the model

$$
\begin{array}{ccc}
L\bigcirc & \overset{me}{\underset{you}{\rule{2cm}{0.4pt}}} & \bigcirc N \\
you\big| & & \big|you \\
L\bullet & \underset{you}{\rule{2cm}{0.4pt}} & \bigcirc N
\end{array}
$$

The road to this broader kind of update leads via the key publications [6, 14, 41]. The general idea is as follows. Complex communication involves two ingredients: a current information model M, and another epistemic model A of possible physical actions, which agents may not be able to distinguish. Moreover, these actions come with *preconditions* on their successful execution. E.g., truthful public announcement $A!$ can only happen in worlds where A holds. General update takes a *product* of the two models, giving a new information model $M \times A$ whose states (s, a) record new actions taken at s, provided the preconditions of a is satisfied in M, s. This may transform the old model M drastically. The basic epistemic stipulation is this. Uncertainty among new states can only come from existing uncertainty via indistinguishable actions:

$$(s, a) \sim_i (t, b) \quad \text{iff both } s \sim_i t \text{ and } a \sim_i b.$$

In the first card example, the actions were "read lecture", "read night out". Taking the proper preconditions into account and then computing the new uncertainties gives the correct updated model

$$read\ lecture\ \bullet\!\!\underset{you}{\rule{3cm}{0.4pt}}\!\!\bigcirc\ read\ night\ out$$

The second example involved a third action "do nothing" with precondition $\top$, which I can distinguish from the first two, but you cannot. Product update delivers a model with four worlds $(L, readL)$, $(L, do\ nothing)$, $(N, read\ N)$, $(N, do\ nothing)$ — with the agents' uncertainties as computed by product update exactly as those of the earlier intuitive diagram.

Clearly, truth values of epistemic propositions can change drastically in product update. Dynamic-epistemic logic now gets very exciting, involving combining epistemic formulas true in M and epistemic information about A expressed in a suitable language. Many of the concerns for public update in this paper will return in more sophisticated versions. Here is one intriguing example of a new issue concerning general update — for a much broader list of new open problems in update logic, cf. [34].

Evolution in the Update Universe. Unlike updates for public announcement, product update can blow up the size of the initial input model M. Here is a simple illustration of this possible blow-up of information models.

EXAMPLE 1. Suppose a public announcement of the true fact P takes place in a group $\{1, 2\}$, but 2 is not sure whether it was an announcement of P, or just some identity action Id which could happen anywhere. In that case, a two-world model with P and $\neg P$ turns into a three-world model with states $(p, "P!")$, (p, Id) and $(\neg P, Id)$.

A typical example of such blow-up occurs again in *games*. Players start from an initial situation M, say a deal of cards, and the action model A contains all possible moves that they have — with preconditions restricting when these are available at players' turns. The game tree consists of all possible evolutions through the nodes in a tree model. Such structures may be defined more precisely as follows:

DEFINITION 9.1. Let an initial epistemic model M be given, and also an action model A. Then the *update evolution model* $TREE(M, A)$ is the infinite epistemic model consisting of disjoint copies of all successive product update layers

$$M, MxA, (MxA)xA, \ldots$$

[30] gives necessary and sufficient conditions on agents' abilities under which an arbitrary extensive game with imperfect information can be represented in the format $TREE(M, A)$. Next, as we have seen, the horizontal stages of $TREE(M, A)$ can grow in size all the time. This blow-up by product update is noticeable with sophisticated communication moves in parlour games such as "Cluedo" [41]. But often, this infinity seems spurious — as happens in many games, where complexity of information can grow in mid-play, but then starts decreasing again toward the end game. The counter-acting force is epistemic *bisimulation contraction*.

EXAMPLE 2. Stabilization under bisimulation. Consider a model with two worlds P, $\neg P$, between wich agent 1 is uncertain, though 2 is not. The actual world has P.

Now a true announcement of P takes place, and agent 1 hears this. But agent 2 thinks the announcement might just be a statement "True" which could hold anywhere. The action model for this scenario looks as follows:

The next two levels of $TREE(A, A)$ then look as follows:

There is an *epistemic bisimulation* between the last two levels, linking the lower three worlds to the left with the single world $(P, P!)$ in MxA. Thus, $(MxA)xA$ is bisimilar with MxA, and the potentially infinite iteration in the update evolution remains finite modulo bisimulation.

Indeed, the original version of this paper stated a *"Finite Evolution Conjecture"* saying that, starting from a given finite M and A, the model $TREE(M, A)$ always remains finite modulo bisimulation. This would imply that some levels in the tree must be bisimilar, say at depth k and $l > k$ — after which, infinite "oscillation" follows. But this was refuted in [19], which shows that the Conjecture fails in some models with two $S5$-agents. Nevertheless, the Conjecture holds for many special cases of such models, e.g., when the epistemic accessibility relations for all agents in the model A are linearly ordered by inclusion. [19] uses finite pebble games over M and A to determine when $TREE(M, A)$ is finite modulo bisimulation. This is the case iff the "responding player" in the pebble game has a winning strategy. The computational complexity of having a finite update evolution modulo bisimulation is still unknown. Thus large classes of communicative scenarios might still satisfy the Finite Evolution Conjecture — e.g., those corresponding to most parlour games.

§10. General communication. *Complexity and diversity.* General tools like product update, single-step or iterated, can chart many varieties of communication, and their broad patterns. In particular, one then finds natural *thresholds* in types of communication. One such threshold occurs when passing from mere partial information to misleading, lying and cheating. In principle, product update also describes the latter, but there is a lot of fine-structure yet to be understood. A more technological challenge in this realm are hiding mechanisms, such as security protocols on the Internet.

Another source of variation are people's limitations, such as bounded memory or limited attention span. Epistemic updates can be performed by ideal agents, but the reality is, of course, that agents may have *bounded rationality*. This diversity of agents calls for variants of the above product update mechanism, which are found in [36].

Games and social software. Again, there is not just analysis, but also synthesis. Communication involves planning what we say, for a purpose. The proper broader setting for this are *games*, which involve preferences and strategies (cf. [1, 28, 32]). This is one instance of what has been called "social software" recently: the design of mechanisms satisfying epistemic specifications such as who gets to know what (cf. [15, 16]). For a very concrete example of social software (and hardware), just think of your *email*! When you announce a fact to one friend, and send some *cc*'s, the effect is public announcement. But if you use the "blind carbon-copy" button *bcc*, the new information state requires product update which may increase complexity. Handling email and the many communication options it offers is a logically highly non-trivial task once you start computing who is supposed to know what.

Communication channels. After all these sweeping vistas, we come down to earth with a simple example, again based on a puzzle. The 1998 "National Science Quiz" of the Dutch national research agency NWO [43] had the following question:

> Six people each know a secret. In one telephone call, two of them can share all secrets they have. What is the minimal number of calls that they have to make in order to ensure that all secrets become known to everyone?

The answers offered were: *7, 8, 9*. The correct one turns out to *8*. For N people, $2N - 4$ calls turns out to be optimal, a result which is not deep but surprisingly difficult to prove. The best algorithm notes that four people can share all secrets that they have in four steps:

1 calls 2, 3 calls 4, 1 calls 3, 2 calls 4.

So, single out any four people in the total group.

First let the other $N - 4$ call one of them, then let the four people share all secrets that they have, then let the $N - 4$ people call back to their informant.

The total number of calls will be $2N - 4$. Now, this clearly raises a general question. What happens to update logic when we make a further semantic parameter explicit, viz. the *communication network*? Our running example of public announcement presupposed some public broadcast system. The gossip puzzle assumes two-by-two telephone connections without conference calls. We can look for results linking up desired outcomes with properties of the network. E.g., it is easy to show that:

Universal knowledge of secrets can be achieved if and only if the network is *connected*: every two people must be connectible by some sequence of calls.

But there are many other intriguing phenomena. Suppose three generals with armies on different hilltops are planning a joint attack on the adversary in the plain below. They have completely reliable two-way telephone lines. One of them possesses some piece of information p which has to become common knowledge among them in order to execute a successful coordinated attack. Can they achieve this common knowledge of p? The answer is that it depends on the scenario, or *protocol*. But this involves genuine extensions of the update logic framework to a temporal setting [13, 31] — as well as structured groups with *channels* (cf. [18]).

If the generals only communicate secrets, even including information about all calls they made, then common knowledge is unattainable, just as in the more familiar two-generals problem with unreliable communication. Informally, there is always someone who is not sure that the last communication took place. More precisely, product update allowing for this uncertainty will leave at least one agent uncertainty chain from the actual world to a $\neg p$-world, preventing common knowledge. But what about general A phoning general B, sharing the information, and telling him that he will call C, tell him about this whole conversation, including the promise to call him? This is like mediators going back and forth between estranged parties. Can this produce common knowledge? Again, it depends on whether agents are sure that promises are carried out. If they are, then a scenario arises with actions and observations where product update will indeed deliver common knowledge.

We leave matters at this informal stage here. Our aim in this excursion has merely been to show that update logic fits naturally with other aspects of communication, such as the availability and reliability of channels.

§11. Logic and communication. Traditionally, logic is about reasoning. If I want to find out something, I sit back in my chair, close my eyes, and think. Of course, I might also just go out, and ask someone, but this seems like cheating. At the University of Groningen, we once did a seminar reading Newton's two great works: *Principia Mathematica*, and the *Optics*. The first was fine: pure deduction, and facts only admitted when they do not spoil the show. But the *Optics* was a shock, for being so terribly unprincipled! Its essential axioms even include some brute facts, for which you have to go out on a sunny day, and see what light does on when falling on prisms or films. For Newton, what we can observe is as hard a fact as what we can deduce. The same is true for ordinary life: questions to nature, or other knowledgeable sources such as people, provide hard information. And the general point of this paper is that logic has a lot to say about this, too. One can see this as an extension of the agenda, and it certainly is. But eventually, it may also have repercussions for the original heartland. Say, what would be the crucial desirable meta-properties of first-order logic when we add the analysis of communication as one of its core tasks?

I will not elaborate on this, as this paper has already taken up too much of the reader's time. And in any case, the time for mere communication has passed. Now that you have told me about the road to the Colosseum, the time has come for me to use this knowledge, and act according to my soldier's pledge, made long ago.

§12. Added in print. This paper was written in 2002, after some earlier years as a "traveling lecture" — and it appeared as an *ILLC* Tech Report in Amsterdam. In the meantime, various new developments have taken place, including a better insight into the Update Universe [19], solutions to some of the problems mentioned here (cf. the new survey [34]), uniform axiomatizations of dynamic epistemic logics with common knowledge in terms of reduction axioms [38], and a better view of fruitful connections with temporal logic and game theory [35]. We thank many of the authors mentioned in this paper for their comments on various drafts through the years. Also, the first textbook on dynamic epistemic logic is preparation, cf. [39].

REFERENCES

[1] A. BALTAG, *Logics for insecure communication*, **Proceedings TARK VIII**, Morgan Kaufmann, Los Altos, 2001, pp. 111–121.

[2] A. BALTAG, L. MOSS, and S. SOLECKI, *Updated version of "The Logic of Public Announcements, Common Knowledge and Private Suspicions, Proceedings TARK 1998"*, 2003, Department

of Cognitive Science, Indiana University, Bloomington and Department of Computing, Oxford University.

[3] J. BARWISE and L. MOSS, *Vicious Circles*, CSLI Publications, Stanford, 1997.

[4] P. BLACKBURN, M. DE RIJKE, and Y. VENEMA, *Modal Logic*, Cambridge University Press, Cambridge, 2001.

[5] H.-D. EBBINGHAUS and J. FLUM, *Finite Model Theory*, Springer, Berlin, 1995.

[6] J. GERBRANDY, *Bisimulations on Planet Kripke*, Technical Report DS-1999-01, Institute for Logic, Language and Computation, University of Amsterdam, 1999.

[7] J. HALPERN and M. VARDI, *The complexity of reasoning about knowledge and time*, **Journal of Computer and Systems Science**, vol. 38 (1989), no. 1, pp. 195–237.

[8] M. HOLLENBERG, *Logic and Bisimulation*, vol. XIV, Publications Zeno Institute of Philosophy, Zeno Institute of Philosophy, University of Utrecht, 1998.

[9] D. KOZEN, D. HAREL, and J. TIURYN, *Dynamic Logic*, MIT Press, Cambridge (Mass.), 2000.

[10] M. MARX and Y. VENEMA, *Multi-Dimensional Modal Logic*, Kluwer Academic Publishers, Dordrecht, 1997.

[11] J. MCCARTHY, *Two Puzzles About Knowledge*, 2002.

[12] B. MOORE, *A Formal Theory of Knowledge and Action*, research report, SRI International, Menlo Park, 1985.

[13] Y. MOSES, M. VARDI, R. FAGIN, and J. HALPERN, *Reasoning About Knowledge*, MIT Press, Cambridge (Mass.), 1995.

[14] L. MOSS, S. SOLECKI, and A. BALTAG, *The logic of public announcements, common knowledge and private suspicions*, **Proceedings TARK 1998**, Morgan Kaufmann Publishers, Los Altos, 1998, pp. 43–56.

[15] R. PARIKH, *Social software*, **Synthese**, vol. 132 (2002), pp. 187–211.

[16] M. PAULY, *Logic for Social Software*, Dissertation DS-2001-10, Institute for Logic, Language and Computation, University of Amsterdam, 2001.

[17] J. PLAZA, *Logics of public communications*, **Proceedings of the 4th International Symposium on Methodologies for Intelligent Systems** (M. L. Emrich et al., editors), North-Holland, 1989, pp. 201–216.

[18] F. ROELOFSEN, *Logical Perspectives on Distributed Knowledge and its Dynamics*, Master of Logic Thesis, University of Amsterdam, 2005.

[19] T. SADZIK, *Dynamic Epistemic Update System*, 2004, Graduate School of Business, Stanford University.

[20] C. STIRLING, *Bisimulation, modal logic, and model checking games*, **Logic Journal of the IGPL. Interest Group in Pure and Applied Logics**, vol. 7 (1999), no. 1, pp. 103–124, Special issue on Temporal Logic.

[21] J. VAN BENTHEM, *Rational Dynamics*, invited lecture, LOGAMAS workshop, department of computer science, University of Liverpool. In S. Vannucci, ed., Logic, Game Theory and Social Choice III, University of Siena, department of political economy, 19-23. To appear in International Journal of Game Theory.

[22] ———, *Exploring Logical Dynamics*, CSLI Publications, Stanford, 1996.

[23] ———, *Dynamic Bits and Pieces*, Report LP-97-01, Institute for Logic, Language and Computation, University of Amsterdam, 1997.

[24] ———, *Annual Report Spinoza Award Project "Logic in Action"*, Technical report, Institute for Logic, Language and Computation, University of Amsterdam, 1998.

[25] ———, *Dynamic Odds and Ends*, Report ML-98-08, Institute for Logic, Language and Computation, University of Amsterdam, 1998.

[26] ———, *Radical Epistemic Dynamic Logic*, note for course "Logic in Games", Institute for Logic, Language and Computation, University of Amsterdam, 1999.

[27] ——, *Update as Relativization*, manuscript, Institute for Logic, Language and Computation, University of Amsterdam, 1999.

[28] ——, *Logic in Games*, lecture notes, Institute for Logic, Language and Computation, University of Amsterdam, 1999–2002, http://staff.science.uva.nl/~johan.

[29] ——, *Update Delights*, invited lecture, ESSLLI Summer School, Birmingham, and manuscript, Institute for Logic, Language and Computation, University of Amsterdam, 2000.

[30] ——, *Games in dynamic epistemic logic*, **Bulletin of Economic Research**, vol. 53 (2001), no. 4, pp. 219–248.

[31] ——, *Logic and The Dynamics of Information*, 2003, in L. Floridi, ed., Minds and Machines 13:4, 503–519.

[32] ——, *Rational dynamics and epistemic logic in games*, **Logic, Game Theory and Social Choice III** (S. Vannucci, editor), Department of political economy, University of Siena, 2003, pp. 19–23.

[33] ——, *Structural properties of dynamic reasoning*, **The Dynamic Turn** (J. Peregrin, editor), Kluwer Academic Publishers, Amsterdam, 2003, pp. 15–31.

[34] ——, *Open problems in update logic*, **Mathematical Problems from Applied Logic** (N. Rozhkowska, editor), Plenum Publishers, Russian Academy of Sciences, Novosibirsk, 2005.

[35] ——, *Update and Revision in Games*, 2005, paper for APA-ASL Meeting, San Francisco. Continuation of "A Mini-Guide to Logic in Action" Philosophical Researches 2004, Supplement, 21–30, Chinese Academy of Sciences, Beijing.

[36] J. VAN BENTHEM and F. LIU, *Diversity of logical agents in games*, **Philosophia Scientiae**, vol. 8 (2004), no. 2, pp. 163–178.

[37] J. VAN BENTHEM, I. NÉMETI, and H. ANDRÉKA, *Modal logics and bounded fragments of predicate logic*, **Journal of Philosophical Logic**, vol. 27 (1998), no. 3, pp. 217–274.

[38] J. VAN BENTHEM, J. VAN EIJCK, and B. KOOI, *A Logic for Communication and Change*, **Proceedings TARK 10, Singapore** (R. van der Meyden, editor), ILLC & CWI Amsterdam, Philosophy Department, Groningen, 2005, pp. 253–261.

[39] W. VAN DER HOEK, B. KOOI, and H. VAN DITMARSCH, *Dynamic Epistemic Logic*, Kluwer-Springer Academic Publishers, Dordrecht, to appear.

[40] R. VAN DER MEYDEN, *Common knowledge and update in finite environments*, **Information and Computation**, vol. 140 (1998), no. 2, pp. 115–157.

[41] H. VAN DITMARSCH, *Knowledge Games*, Dissertation DS-2000-06, Institute for Logic, Language and Computation, University of Amsterdam, 2000.

[42] ——, *Keeping Secrets With Public Communication*, Technical Report, Department of Computer Science, University of Otago, 2002.

[43] NATIONALE WETENSCHAPSQUIZ, *Gossip Puzzle*, 1998, http://www.nwo.nl/.

[44] M. WOOLDRIDGE, *An Introduction to Multi-Agent Systems*, John Wiley, Colchester, 2002.

INSTITUTE FOR LOGIC
  LANGUAGE AND COMPUTATION
  UNIVERSITY OF AMSTERDAM
    PLANTAGE MUIDERGRACHT 24
      1018 TV AMSTERDAM, THE NETHERLANDS
  and
  DEPARTMENT OF PHILOSOPHY
  STANFORD UNIVERSITY
    STANFORD, CALIFORNIA 94305, USA
E-mail: johan@science.uva.nl

COMPUTABLE VERSIONS OF THE UNIFORM
BOUNDEDNESS THEOREM

VASCO BRATTKA

Abstract. We investigate the computable content of the Uniform Boundedness Theorem and of the closely related Banach-Steinhaus Theorem. The Uniform Boundedness Theorem states that a pointwise bounded sequence of bounded linear operators on Banach spaces is also uniformly bounded. But, given the sequence, can we also *effectively* find the uniform bound? It turns out that the answer depends on how the sequence is "given". If it is just given with respect to the compact open topology (i.e. if just a sequence of "programs" is given), then we cannot even compute an upper bound of the uniform bound in general. If, however, the pointwise bounds are available as additional input information, then we can effectively compute an upper bound of the uniform bound. Additionally, we prove an effective version of the contraposition of the Uniform Boundedness Theorem: given a sequence of linear bounded operators which is not uniformly bounded, we can effectively find a witness for the fact that the sequence is not pointwise bounded. As an easy application of this theorem we obtain a computable function whose Fourier series does not converge.

§1. Introduction. In this paper we want to study the computational content of some theorems of functional analysis. The Uniform Boundedness Theorem is one of the central theorems of functional analysis and it has first been published in Banach's thesis [1].

THEOREM 1.1 (Uniform Boundedness Theorem). *Let X be a Banach space, Y a normed space and let $(T_i)_{i \in \mathbb{N}}$ be a sequence of bounded linear operators $T_i : X \to Y$. If $\{\|T_i x\| : i \in \mathbb{N}\}$ is bounded for each $x \in X$, then $\{\|T_i\| : i \in \mathbb{N}\}$ is bounded.*

Here, $\|T_i\| := \sup_{\|x\| \leq 1} \|T_i x\|$ denotes the *bound* of T_i. Roughly speaking, the Uniform Boundedness Theorem states that each pointwise bounded sequence of linear bounded operators is also uniformly bounded. But, given a pointwise bounded sequence $(T_i)_{i \in \mathbb{N}}$ of linear bounded operators, can we also *effectively* find the uniform bound? This will be one of the main questions

2000 *Mathematics Subject Classification.* 03F60, 03D45, 46S30.

Key words and phrases. Computable functional analysis, Effective representations.

A preliminary extended abstract version of this paper has been published as [7].

Work partially supported by DFG Grant BR 1807/4-1.

Logic Colloquium '02
Edited by Z. Chatzidakis, P. Koepke, and W. Pohlers
Lecture Notes in Logic, 27
© 2006, ASSOCIATION FOR SYMBOLIC LOGIC

studied in this paper and we will see that the answer depends on how the sequence is "given":

(1) If $(T_i)_{i \in \mathbb{N}}$ is available as a point in $C(X, Y)^{\mathbb{N}}$, then *arbitrary lower bounds* of the uniform bound can be determined, but in general no upper bound of the uniform bound.

(2) If $(T_i)_{i \in \mathbb{N}}$ is available as a point in $C(X, \mathcal{B}(\mathbb{N}, Y))$, then, additionally, *some upper bound* of the uniform bound can be determined, but in general not the uniform bound itself.

Besides these versions of the Uniform Boundedness Theorem we will also study computable versions of the contraposition of the theorem. It turns out that given a sequence $(T_i)_{i \in \mathbb{N}}$ of linear bounded operators which is not uniformly bounded (i.e. $(\|T_i\|)_{i \in \mathbb{N}}$ is not bounded), we can effectively find a point $x \in X$ such that $(\|T_i x\|)_{i \in \mathbb{N}}$ is not bounded. This corresponds to versions of the theorem which are known in constructive analysis [4].

As a second theorem we will study the Banach-Steinhaus Theorem which is closely related to the Uniform Boundedness Theorem and it has first been published by Banach and Steinhaus in [3].

THEOREM 1.2 (Banach-Steinhaus Theorem). *Let X be a Banach space, let Y be a normed space and $(T_i)_{i \in \mathbb{N}}$ a sequence of linear and bounded operators $T_i : X \to Y$ which converges pointwise. Then by $Tx := \lim_{n \to \infty} T_n x$ a linear and bounded operator $T : X \to Y$ is defined.*

Additionally, $\|T\| \leq \sup_{n \in \mathbb{N}} \|T_n\|$ holds in the situation of the theorem. Proofs of the classical versions of these theorems can be found in standard textbooks on functional analysis, see e.g. [14]. From the computational point of view these theorems are interesting, since their classical proofs rely more or less on the Baire Category Theorem and therefore they count as "non-constructive".

We will study these theorems from the point of view of computable analysis, which is the Turing machine based theory of computability on real numbers and other topological spaces. Pioneering work on this theory has been presented by Turing [29], Banach and Mazur [2], Lacombe [20] and Grzegorczyk [15]. Recent monographs have been published by Pour-El and Richards [24], Ko [18] and Weihrauch [31]. Certain aspects of computable functional analysis have already been studied by several authors, see for instance [22, 33, 34, 32].

We close the introduction with a short survey on the organisation of this paper. In the following section we will present some preliminaries from computable analysis. In Section 3 we discuss computable metric spaces and computable Banach spaces and in Section 4 we shortly present some results on effective continuity which we will use in the following. Section 5 and 6 are devoted to different computable versions of the Uniform Boundedness Theorem. In Section 7 we apply a computable version of the contraposition of the Uniform Boundedness Theorem in order to construct a computable function whose Fourier series does not converge. In Section 8 we discuss a computable version of the Banach-Steinhaus Theorem. Some further proofs and details are included in [5]. In the Conclusions we briefly compare our results with known results from Bishop's school of constructive analysis [4] and with Simpson's approach to reverse mathematics [26].

§2. **Preliminaries from computable analysis.** In this section we briefly summarize some notions of Weihrauch's representation based approach to computable analysis. For details the reader is refered to [31]. The basic idea of this approach is to represent infinite objects like real numbers, functions or sets, by infinite strings over some alphabet Σ (which should at least contain the symbols 0 and 1). Thus, a *representation* of a set X is a surjective mapping $\delta :\subseteq \Sigma^\omega \to X$ and in this situation we will call (X, δ) a *represented space*. Here Σ^ω denotes the set of infinite sequences over Σ and the inclusion symbol is used to indicate that the mapping might be partial. If we have two represented spaces, then we can define the notion of a computable function, as illustrated by the diagram in Figure 1.

DEFINITION 2.1 (Computable function). Let (X, δ) and (Y, δ') be represented spaces. A function $f :\subseteq X \to Y$ is called (δ, δ')-*computable*, if there exists some computable function $F :\subseteq \Sigma^\omega \to \Sigma^\omega$ such that $\delta' F(p) = f\delta(p)$ for all $p \in \text{dom}(f\delta)$.

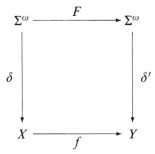

FIGURE 1. Computability of a function $f :\subseteq X \to Y$.

Of course, we have to define computability of functions $F :\subseteq \Sigma^\omega \to \Sigma^\omega$ to make this definition complete, but this can be done via Turing machines: F is computable if there exists some Turing machine, which computes infinitely long and transforms each sequence $p \in \text{dom}(F)$, written on the input tape, into the corresponding sequence $F(p)$, written on the one-way output tape. Later on, we will also need computable multi-valued operations $f :\subseteq X \rightrightarrows Y$, which are defined analogously to computable functions by substituting $\delta' F(p) \in f\delta(p)$ for the equation in Definition 2.1 above. If the represented spaces are fixed or clear from the context, then we will simply call a function or operation f *computable*.

For the comparison of representations it will be useful to have the notion of *reducibility* of representations. If δ, δ' are both representations of a set X, then δ is called *reducible* to δ', $\delta \leq \delta'$ in symbols, if there exists a computable function $F :\subseteq \Sigma^\omega \to \Sigma^\omega$ such that $\delta(p) = \delta' F(p)$ for all $p \in \text{dom}(\delta)$. Obviously, $\delta \leq \delta'$ holds, if and only if the identity id $: X \to X$ is (δ, δ')–computable. Moreover, δ and δ' are called *equivalent*, $\delta \equiv \delta'$ in symbols, if $\delta \leq \delta'$ and $\delta' \leq \delta$.

Analogously to the notion of computability we can define the notion of (δ, δ')–*continuity* for single- and multi-valued operations, by substituting a continuous function $F :\subseteq \Sigma^\omega \to \Sigma^\omega$ for the computable function F in the definitions above. On Σ^ω we use the *Cantor topology*, which is simply the product topology of the discrete topology on Σ. The corresponding reducibility will be called *continuous reducibility* and we will use the symbols $\leq_t$ and $\equiv_t$ in this case. Again we will simply say that the corresponding function is *continuous*, if the representations are fixed or clear from the context. If not mentioned otherwise, we will always assume that a represented space is endowed with the final topology induced by its representation.

This will lead to no confusion with the ordinary topological notion of continuity, as long as we are dealing with *admissible* representations. A representation δ of a topological space X is called *admissible*, if δ is maximal among all continuous representations δ' of X, i.e. if $\delta' \leq_t \delta$ holds for all continuous representations δ' of X. If δ, δ' are admissible representations of topological spaces X, Y, then a function $f :\subseteq X \to Y$ is (δ, δ')–continuous, if and only if it is sequentially continuous, cf. [25, 8].

Given a represented space (X, δ), we will occasionally use the notions of a *computable sequence* and a *computable point*. A *computable sequence* is a computable function $f : \mathbb{N} \to X$, where we assume that $\mathbb{N} = \{0, 1, 2, \dots\}$ is represented by $\delta_\mathbb{N}(1^n 0^\omega) := n$ and a point $x \in X$ is called *computable*, if there is a constant computable sequence with value x (or, equivalently, if there is a computable p such that $x = \delta(p)$).

Given two represented spaces (X, δ) and (Y, δ'), there is a canonical representation $[\delta, \delta']$ of $X \times Y$ and a representation $[\delta \to \delta']$ of certain functions $f : X \to Y$. If δ, δ' are *admissible* representations of sequential topological spaces, then $[\delta \to \delta']$ is actually a representation of the set $\mathcal{C}(X, Y)$ of continuous functions $f : X \to Y$. If $Y = \mathbb{R}$, then we write for short $\mathcal{C}(X) := \mathcal{C}(X, \mathbb{R})$. The function space representation can be characterized by the fact that it admits evaluation and type conversion.

PROPOSITION 2.2 (Evaluation and type conversion). *Let* $(X, \delta), (Y, \delta')$ *be admissibly represented sequential topological spaces and let* (Z, δ'') *be a represented space.*

(1) (*Evaluation*) ev $: \mathcal{C}(X, Y) \times X \to Y, (f, x) \mapsto f(x)$ *is* $([[\delta \to \delta'], \delta], \delta')$–*computable,*

(2) (*Type conversion*) $f : Z \times X \to Y$, is $([\delta'', \delta], \delta')$–*computable, if and only if the function* $\check{f} : Z \to C(X, Y)$, *defined by* $\check{f}(z)(x) := f(z, x)$ *is* $(\delta'', [\delta \to \delta'])$–*computable*.

The proof of this proposition is based on a version of smn– and utm–Theorem, see [31, 25]. If (X, δ), (Y, δ') are admissibly represented sequential topological spaces, then we will always assume that $C(X, Y)$ is represented by $[\delta \to \delta']$ in the following. We can conclude by evaluation and type conversion that the computable points in the space $(C(X, Y), [\delta \to \delta'])$ are just the (δ, δ')–computable functions $f : X \to Y$. If (X, δ) is a represented space, then we will always assume that the set of sequences $X^{\mathbb{N}}$ is represented by $\delta^{\mathbb{N}} := [\delta_{\mathbb{N}} \to \delta]$. The computable points in $(X^{\mathbb{N}}, \delta^{\mathbb{N}})$ are just the computable sequences in (X, δ). Moreover, we assume that X^n is always represented by δ^n, which can be defined inductively by $\delta^1 := \delta$ and $\delta^{n+1} := [\delta^n, \delta]$.

§3. **Computable metric and Banach spaces.** In this section we will briefly discuss computable metric spaces and computable Banach spaces. The notion of a computable Banach space will be the central notion for all following results. Computable metric spaces have been used in the literature at least since Lacombe [21]. Restricted to computable points they have also been studied by various authors [19, 23, 27, 30]. We consider computable metric spaces as special separable metric spaces but on all points and not only restricted to computable points. Pour-El and Richards have introduced a closely related axiomatic characterization of sequential computability structures for Banach spaces [24] which has been extended to metric spaces by Mori, Tsujii, and Yasugi [33].

Before we start with the definition of computable metric spaces we mention that we will denote *open balls* of a metric space (X, d) by $B(x, \varepsilon) := \{y \in X : d(x, y) < \varepsilon\}$ for all $x \in X$, $\varepsilon \geq 0$ and correspondingly *closed balls* by $\overline{B}(x, \varepsilon) := \{y \in X : d(x, y) \leq \varepsilon\}$. Occasionally, we denote complements of sets $A \subseteq X$ by $A^c := X \setminus A$.

DEFINITION 3.1 (Computable metric space). A tuple (X, d, α) is called a *computable metric space*, if

(1) $d : X \times X \to \mathbb{R}$ is a metric on X,
(2) $\alpha : \mathbb{N} \to X$ is a sequence which is dense in X,
(3) $d \circ (\alpha \times \alpha) : \mathbb{N}^2 \to \mathbb{R}$ is a computable (double) sequence in $\mathbb{R}$.

Here, we tacitly assume that the reader is familiar with the notion of a computable sequence of reals, but we will come back to that point below. Obviously, any computable metric space is separable. Occasionally, we will say for short that X is a *computable metric space*. The *Cauchy representation* $\delta_X :\subseteq \Sigma^{\omega} \to X$ of a computable metric space (X, d, α) is defined by

$$\delta_X\left(01^{n_0+1}01^{n_1+1}01^{n_2+1}\ldots\right) := \lim_{i \to \infty} \alpha(n_i)$$

for all $n_i \in \mathbb{N}$ such that $(\alpha(n_i))_{i \in \mathbb{N}}$ converges and $d(\alpha(n_i), \alpha(n_j)) \leq 2^{-i}$ for all $j > i$ (and δ_X undefined for all other input sequences). In the following we tacitly assume that computable metric spaces are represented by their Cauchy representations. If X is a computable metric space, then it is easy to see that $d : X \times X \to \mathbb{R}$ becomes computable [8]. All Cauchy representations are admissible with respect to the corresponding metric topology.

An important computable metric space is $(\mathbb{R}, d_{\mathbb{R}}, \alpha_{\mathbb{R}})$ with the Euclidean metric $d_{\mathbb{R}}(x, y) := |x - y|$ and some standard numbering of the rational numbers $\mathbb{Q}$, as $\alpha_{\mathbb{R}}\langle i, j, k \rangle := (i - j)/(k + 1)$. Here, we use the definition $\langle i, j \rangle := 1/2(i + j)(i + j + 1) + j$ for *Cantor pairs* and this definition can be extended inductively to finite tuples. Similarly, we can define $\langle p, q \rangle \in \Sigma^\omega$ for sequences $p, q \in \Sigma^\omega$. For short we will occasionally write $\overline{k} := \alpha_{\mathbb{R}}(k)$. In the following we assume that $\mathbb{R}$ is endowed with the Cauchy representation $\delta_{\mathbb{R}}$ induced by the computable metric space given above. This representation of $\mathbb{R}$ can also be defined, if $(\mathbb{R}, d_{\mathbb{R}}, \alpha_{\mathbb{R}})$ just fulfills (1) and (2) of the definition above and this leads to a definition of computable real number sequences without circularity. Occasionally, we will also use the represented space $(\mathbb{Q}, \delta_{\mathbb{Q}})$ of rational numbers with $\delta_{\mathbb{Q}}(1^n 0^\omega) := \alpha_{\mathbb{R}}(n) = \overline{n}$.

Many important representations can be deduced from computable metric spaces, but we will also need some ad hoc defined representations. For instance, we will use two further representations $\rho_<, \rho_>$ of the real numbers, which correspond to weaker information on the represented real numbers. Here

$$\rho_< \left(01^{n_0+1} 01^{n_1+1} 01^{n_2+1} \ldots \right) = x :\Longleftrightarrow \{ q \in \mathbb{Q} : q < x \} = \{ \overline{n_i} : i \in \mathbb{N} \}$$

and $\rho_<$ is undefined for all other sequences. Thus, $\rho_<(p) = x$, if p is a list of all rational numbers smaller than x. Analogously, $\rho_>$ is defined with ">" instead of "<". We write $\mathbb{R}_< = (\mathbb{R}, \rho_<)$ and $\mathbb{R}_> = (\mathbb{R}, \rho_>)$ for the corresponding represented spaces. The computable numbers in $\mathbb{R}_<$ are called *left-computable* real numbers and the computable numbers in $\mathbb{R}_>$ *right-computable* real numbers. The representations $\rho_<$ and $\rho_>$ are admissible with respect to the *lower* and *upper topology* on $\mathbb{R}$, which are induced by the open intervals (q, ∞) and $(-\infty, q)$, respectively. Yet another representation $\rho_<^*$ of the real numbers can be defined by

$$\rho_<^* \left(01^{n+1} p \right) = x :\Longleftrightarrow \rho_<(p) = x \leq n.$$

Thus, $\rho_<^*(q) = x$, if q contains some upper bound $n \geq x$ and a list of all rational numbers smaller than x. We will write $\mathbb{R}_<^* = (\mathbb{R}, \rho_<^*)$ for the corresponding represented space. Continuity with respect to $\mathbb{R}_<^*$ will always be understood as $\rho_<^*$–continuity.

Computationally, we do not have to distinguish the complex numbers $\mathbb{C}$ from $\mathbb{R}^2$. Thus, we can directly define a representation of $\mathbb{C}$ by $\delta_{\mathbb{C}} := \delta_{\mathbb{R}}^2$. If $z = a + ib \in \mathbb{C}$, then we denote by $\overline{z} := a - ib \in \mathbb{C}$ the *conjugate complex number* and by $|z| := \sqrt{a^2 + b^2}$ the *absolute value* of z. Alternatively to the ad hoc definition of $\delta_{\mathbb{C}}$, we could consider $\delta_{\mathbb{C}}$ as Cauchy representation of a computable metric space $(\mathbb{C}, d_{\mathbb{C}}, \alpha_{\mathbb{C}})$, where $\alpha_{\mathbb{C}}$ is a numbering of $\mathbb{Q}[i]$, defined by $\alpha_{\mathbb{C}}\langle n, k \rangle := \overline{n} + \overline{k}i$ and $d_{\mathbb{C}}(w, z) := |w - z|$ is the Euclidean metric on $\mathbb{C}$. The corresponding Cauchy representation is equivalent to $\delta_{\mathbb{R}}^2$. In the following we will consider vector spaces over $\mathbb{R}$, as well as over $\mathbb{C}$. We will use the notation $\mathbb{F}$ for a field which always might be replaced by both, $\mathbb{R}$ or $\mathbb{C}$. Correspondingly, we use the notation $(\mathbb{F}, d_{\mathbb{F}}, \alpha_{\mathbb{F}})$ for a computable metric space which might be replaced by both computable metric spaces $(\mathbb{R}, d_{\mathbb{R}}, \alpha_{\mathbb{R}})$, $(\mathbb{C}, d_{\mathbb{C}}, \alpha_{\mathbb{C}})$ defined above. We will also use the notation $Q_{\mathbb{F}} = \text{range}(\alpha_{\mathbb{F}})$, i.e. $Q_{\mathbb{R}} = \mathbb{Q}$ and $Q_{\mathbb{C}} = \mathbb{Q}[i]$.

For the definition of a computable Banach space it is helpful to have the notion of a computable vector space which we will define next.

DEFINITION 3.2 (Computable vector space). A represented space (X, δ), where $(X, +, \cdot, 0)$ is a vector space over $\mathbb{F}$, is called a *computable vector space* (over $\mathbb{F}$), if the following conditions hold:

(1) $+ : X \times X \to X, (x, y) \mapsto x + y$ is computable,
(2) $\cdot : \mathbb{F} \times X \to X, (a, x) \mapsto a \cdot x$ is computable,
(3) $0 \in X$ is a computable point.

If (X, δ) is a computable vector space over $\mathbb{F}$, then $(\mathbb{F}, \delta_{\mathbb{F}})$, (X^n, δ^n) and $(X^{\mathbb{N}}, \delta^{\mathbb{N}})$ are computable vector spaces over $\mathbb{F}$. If, additionally, (X, δ), (Y, δ') are admissibly represented second countable T_0–spaces, then the function space $(\mathcal{C}(Y, X), [\delta' \to \delta])$ is a computable vector space over $\mathbb{F}$. Here we tacitly assume that the vector space operations on product, sequence and function spaces are defined componentwise. The proof for the function space is a straightforward application of evaluation and type conversion. The central definition for the present investigation will be the notion of a computable Banach space. Here, a *fundamental sequence* $e : \mathbb{N} \to X$ is a sequence whose linear span is dense in X.

DEFINITION 3.3 (Computable normed space). A tuple $(X, \| \ \|, e)$ is called a *computable normed space*, if

(1) $(X, \| \ \|)$ is a normed space with a fundamental sequence $e : \mathbb{N} \to X$,
(2) (X, d, α_e) with $\alpha_e \langle k, \langle n_0, \ldots, n_k \rangle \rangle := \sum_{i=0}^{k} \alpha_{\mathbb{F}}(n_i) e_i$ and $d(x, y) := \|x - y\|$ is a computable metric space,
(3) (X, δ_X) with Cauchy representation δ_X is a computable vector space.

If in the situation of the definition the underlying space $(X, \| \ \|)$ is even a Banach space, i.e. if (X, d) is a complete metric space, then $(X, \| \ \|, e)$ is called a *computable Banach space*. If the norm and the fundamental sequence are clear from the context or locally irrelevant, we will say for short that X is a *computable normed space* or a *computable Banach space*. We will always assume that computable normed spaces are represented by their Cauchy representations, which are admissible with respect to the norm topology. If X is a computable normed space, then $\| \ \| : X \to \mathbb{R}$ is a computable function. Of course, all computable Banach spaces are separable. In the following proposition a number of computable Banach spaces are defined.

PROPOSITION 3.4 (Computable Banach spaces). *Let* $p \in \mathbb{R}$ *be a computable real number with* $1 \leq p < \infty$ *and let* $a < b$ *be computable real numbers. The following spaces are computable Banach spaces over* $\mathbb{F}$.

(1) $(\mathbb{F}^n, \| \ \|_p, e)$ *and* $(\mathbb{F}^n, \| \ \|_\infty, e)$ *with*

- $\|(x_1, x_2, \ldots, x_n)\|_p := \sqrt[p]{\sum_{k=1}^{n} |x_k|^p}$ *and*

  $\|(x_1, x_2, \ldots, x_n)\|_\infty := \max_{k=1,\ldots,n} |x_k|$,
- $e_i = e(i) = (e_{i1}, e_{i2}, \ldots, e_{in})$ *with* $e_{ik} := \begin{cases} 1 & \text{if } i = k, \\ 0 & \text{else.} \end{cases}$

(2) $(\ell_p, \| \ \|_p, e)$ *with*
- $\ell_p := \{x \in \mathbb{F}^{\mathbb{N}} : \|x\|_p < \infty\}$,
- $\|(x_k)_{k \in \mathbb{N}}\|_p := \sqrt[p]{\sum_{k=0}^{\infty} |x_k|^p}$,
- $e_i = e(i) = (e_{ik})_{k \in \mathbb{N}}$ *with* $e_{ik} := \begin{cases} 1 & \text{if } i = k, \\ 0 & \text{else.} \end{cases}$

(3) $(\mathcal{C}[a, b], \| \ \|, e)$ *with*
- $\mathcal{C}[a, b] := \{f : [a, b] \to \mathbb{R} : f \ continuous\}$,
- $\|f\| := \max_{t \in [a,b]} |f(t)|$,
- $e_i(t) = e(i)(t) = t^i$.

We leave it to the reader to check that these spaces are actually computable Banach spaces. If not stated differently, then we will assume that $(\mathbb{F}^n, \| \ \|)$ is endowed with the maximum norm $\| \ \| = \| \ \|_\infty$. It is known that the Cauchy representation $\delta_{\mathcal{C}[a,b]}$ of $\mathcal{C}[a, b] = \mathcal{C}([a, b], \mathbb{R})$ is equivalent to $[\delta_{[a,b]} \to \delta_{\mathbb{R}}]$, where $\delta_{[a,b]}$ denotes the restriction of $\delta_{\mathbb{R}}$ to $[a, b]$ (cf. [31, Lemma 6.1.10]). In the following we will occasionally utilize the sequence spaces ℓ_p to construct counterexamples. Since we also want to use some non-separable normed spaces (which cannot be computable by definition) we give an ad hoc definition for representations of such spaces. Especially, we will deal with the space of bounded sequences.

DEFINITION 3.5 (Space of bounded sequences). Let $(X, \| \ \|)$ be a computable normed space. Let $\mathcal{B}(\mathbb{N}, X) := \{x \in X^{\mathbb{N}} : \|x\| < \infty\}$ be endowed with the supremum norm $\|(x_k)_{k \in \mathbb{N}}\| := \sup_{k \in \mathbb{N}} \|x_k\|$ and the representation $\delta_{\mathcal{B}(\mathbb{N},X)}$, defined by

$$\delta_{\mathcal{B}(\mathbb{N},X)}\langle p, q \rangle = x : \iff \delta_X^{\mathbb{N}}(p) = x \text{ and } \delta_{\mathbb{R}}(q) = \|x\|.$$

One can prove that this space is a computable normed space in a generalized sense, see [9]. For the following we assume that $\mathcal{B}(\mathbb{N}, X)$ is endowed with the sequentialization of the weakest topology which contains the subtopology on $\mathcal{B}(\mathbb{N}, X)$ of the product topology on $X^{\mathbb{N}}$ and which makes the norm $\| \ \| : \mathcal{B}(\mathbb{N}, X) \to \mathbb{R}$ continuous. We will call this topology τ the *weak$^=$ topology* on $\mathcal{B}(\mathbb{N}, X)$. It follows from results of Schröder [25] that $\delta_{\mathcal{B}(\mathbb{N},X)}$ is admissible with respect to the weak$^=$ topology, see also [9, 11].

§4. **Effective continuity.** In this section we want to present some results on effective continuity and hyperspaces which we will use to prove our main results on the Uniform Boundedness Theorem. Especially, we will use representations of the hyperspace $\mathcal{O}(X)$ of open subsets and the hyperspace $\mathcal{A}(X)$ of closed subsets of X. Such representations have been studied in the Euclidean case in [12, 31] and for the metric case in [10].

DEFINITION 4.1 (Hyperspace of open subsets). Let (X, d, α) be a computable metric space. We endow $\mathcal{O}(X) := \{U \subseteq X : U \text{ open}\}$ with the representation $\delta_{\mathcal{O}(X)}$, defined by

$$\delta_{\mathcal{O}(X)}\left(01^{\langle n_0, k_0 \rangle + 1}01^{\langle n_1, k_1 \rangle + 1}01^{\langle n_2, k_2 \rangle + 1} \ldots\right) := \bigcup_{i=0}^{\infty} B\left(\alpha(n_i), \overline{k_i}\right).$$

As a first result we will prove that given an open set $U \in \mathcal{O}(X)$ and a point $x \in U$, we can effectively find some neighborhood of x which is included in U. This statement is made precise by the following lemma.

LEMMA 4.2. *Let (X, d, α) be a computable metric space. There exists a computable multi-valued operation $R :\subseteq X \times \mathcal{O}(X) \rightrightarrows \mathbb{N}$ such that for any open $U \subseteq X$ and $x \in U$ there exists some $k \in R(x, U)$; moreover $B(x, \overline{k}) \subseteq U$ and $\overline{k} > 0$ hold for all such k.*

PROOF. Given a sequence $\langle n_i, k_i \rangle_{i \in \mathbb{N}}$ such that $U = \bigcup_{i=0}^{\infty} B(\alpha(n_i), \overline{k_i})$ and a sequence $(m_i)_{i \in \mathbb{N}}$ such that $d(\alpha(m_i), \alpha(m_j)) \leq 2^{-j}$ for all $i > j$ and $x := \lim_{i \to \infty} \alpha(m_i) \in U$, there exist some numbers $i, j \in \mathbb{N}$ such that $d(\alpha(n_i), \alpha(m_j)) + 2^{-j} < \overline{k_i}$ and thus we can effectively find $i, j, k \in \mathbb{N}$ such that $d(\alpha(n_i), \alpha(m_j)) + 2^{-j} + \overline{k} < \overline{k_i}$ and $\overline{k} > 0$. Then $d(x, y) < \overline{k}$ implies

$$\begin{aligned}
d(\alpha(n_i), y) &\leq d(\alpha(n_i), \alpha(m_j)) + d(\alpha(m_j), x) + d(x, y) \\
&< d(\alpha(n_i), \alpha(m_j)) + 2^{-j} + \overline{k} \\
&< \overline{k_i}
\end{aligned}$$

for all $y \in X$ and thus $B(x, \overline{k}) \subseteq B(\alpha(n_i), \overline{k_i}) \subseteq U$. ⊣

The following result shows that we can represent open subsets by preimages of continuous functions. It is an effective version of the statement that open subsets of metric spaces coincide with the functional open subsets (the proof directly follows from [10, Theorem 3.10]).

PROPOSITION 4.3 (Functional open subsets). *Let X be a computable metric space. The map $Z : C(X) \to \mathcal{O}(X), f \mapsto X \setminus f^{-1}\{0\}$ is computable and admits a multi-valued computable right-inverse $\mathcal{O}(X) \rightrightarrows C(X)$.*

Now we are prepared to prove a theorem on effective continuity.

THEOREM 4.4 (Effective Continuity). *Let X, Y be computable metric spaces and let $T : X \to Y$ be a function. Then the following are equivalent:*

(1) $T : X \to Y$ *is computable,*
(2) $\mathcal{O}(T^{-1}) : \mathcal{O}(Y) \to \mathcal{O}(X), V \mapsto T^{-1}(V)$ *is well-defined and computable.*

PROOF. "(1)$\Longrightarrow$(2)" If $T : X \to Y$ is computable, then it is continuous and hence $\mathcal{O}(T^{-1})$ is well-defined. Given a function $f : Y \to \mathbb{R}$ such that $V = Y \setminus f^{-1}\{0\}$, we obtain $T^{-1}(V) = T^{-1}(Y \setminus f^{-1}\{0\}) = X \setminus (fT)^{-1}\{0\}$. Using Proposition 4.3 and the fact that the composition mapping $\circ : C(Y, \mathbb{R}) \times C(X, Y) \to C(X, \mathbb{R}), (f, T) \mapsto f \circ T$ is computable (which can be proved by evaluation and type conversion), we obtain that $\mathcal{O}(T^{-1})$ is computable.

"(2)$\Longrightarrow$(1)" We consider the computable metric spaces (X, d, α) and (Y, d', β). We note that T is continuous, if $\mathcal{O}(T^{-1})$ is well-defined. Given a Turing machine M which computes a realization of $\mathcal{O}(T^{-1})$, we construct a Turing machine M' which computes a realization of T. The machine M' with input $p \in \text{dom}(\delta_X)$ works in steps $k = 0, 1, 2, \ldots$ as follows. In step k machine M' simultaneously tests all values $n \in \mathbb{N}$ until some value is found with the following property: machine M with input $01^{\langle n,m \rangle+1}01^{\langle n,m \rangle+1}01^{\langle n,m \rangle+1}0 \ldots$ with $\overline{m} = 2^{-k-1}$ produces an output with subword $01^{\langle i,j \rangle+1}0$ such that $x = \delta_X(p) \in B(\alpha(i), \overline{j})$. As soon as such a subword is found, M' writes 01^{n+1} on the output tape and continues with the next step. We note that the property $\delta_X(p) \in B(\alpha(i), \overline{j})$ is only r.e. but not recursive. However, if M produces a suitable subword, then this can eventually be noticed by M'.

If this happens, i.e. if M' writes 01^{n+1} on its output tape, then we obtain $Tx \in B(\beta(n), 2^{-k-1})$ since $x \in B(\alpha(i), \overline{j}) \subseteq T^{-1}(B(\beta(n), \overline{m}))$. Moreover, M' actually produces an infinite output q, since for any $k \in \mathbb{N}$ there is some $n \in \mathbb{N}$ such that $Tx \in B(\beta(n), 2^{-k-1})$ and thus $x \in T^{-1}(B(\beta(n), 2^{-k-1}))$ and consequently M on input $01^{\langle n,m \rangle+1}01^{\langle n,m \rangle+1}0 \ldots$ with $\overline{m} = 2^{-k-1}$ has to produce some output with subword $01^{\langle i,j \rangle+1}0$ and $x \in B(\alpha(i), \overline{j})$. It follows $\delta_Y(q) = Tx$.   ⊣

We immediately obtain a uniform version of this theorem simply by using evaluation and type conversion (certain smn- and utm-Theorems, alternatively).

THEOREM 4.5 (Effective Continuity). *Let X, Y be computable metric spaces. Then the total map $\omega : T \mapsto \mathcal{O}(T^{-1})$, defined for all continuous $T : X \to Y$, is $([\delta_X \to \delta_Y], [\delta_{\mathcal{O}(Y)} \to \delta_{\mathcal{O}(X)}])$-computable and its partial inverse ω^{-1} is computable in the corresponding sense too.*

Finally, we briefly discuss the hyperspace $\mathcal{A}(X) := \{A \subseteq X : A \text{ closed}\}$ of closed subsets. Any representation of the hyperspace of open subsets induces a dual representation of the hyperspace of closed subsets. In this way we obtain a representation $\delta^>_{\mathcal{A}(X)}$, defined by $\delta^>_{\mathcal{A}(X)}(p) := X \setminus \delta_{\mathcal{O}(X)}(p)$. We will write $\mathcal{A}_>(X) := (\mathcal{A}(X), \delta^>_{\mathcal{A}(X)})$ for the corresponding represented space.

§5. The Uniform Boundedness Theorem.

In this section we will study computable versions of the Uniform Boundedness Theorem. The theorem states that each pointwise bounded sequence of bounded linear operators is also uniformly bounded, see Theorem 1.1. We start with an investigation of the bound $\|T\| := \sup_{x \in \overline{B}(0,1)} \|Tx\|$ of linear bounded operators $T : X \to Y$. As a first observation we note that the bound $\|T\|$ can be approximated from below and estimated from above, if it exists. This exactly means that it can be computed with respect to $\rho^*_<$.

THEOREM 5.1 (Bound). *Let X, Y be computable normed spaces. The partial map $\| \ \| :\subseteq \mathcal{C}(X, Y) \to \mathbb{R}^*_<, T \mapsto \|T\|$, defined for all linear bounded operators $T : X \to Y$, is computable.*

PROOF. The distance function $d_{\overline{B}(0,1)} : X \to \mathbb{R}, x \mapsto \inf_{y \in \overline{B}(0,1)} \|x - y\|$ of the closed unit ball $\overline{B}(0,1)$ is computable, since we obtain $d_{\overline{B}(0,1)}(x) = \max\{\|x\| - 1, 0\}$. Moreover, there exists some computable sequence $f : \mathbb{N} \to X$ such that range(f) is dense in $\overline{B}(0,1)$, see [10]. By continuity it follows $\|T\| = \sup_{x \in \overline{B}(0,1)} \|Tx\| = \sup_{n \in \mathbb{N}} \|Tf(n)\|$ for all linear bounded operators $T : X \to Y$. Since the norm of Y is computable and the supremum $\sup :\subseteq \mathbb{R}^\mathbb{N} \to \mathbb{R}_<$ is computable, we can conclude with the help of evaluation and type conversion that the map $\| \ \| :\subseteq \mathcal{C}(X, Y) \to \mathbb{R}_<$ is computable.

It remains to show that one can also determine an upper bound of $\|T\|$. Given a linear bounded operator $T : X \to Y$, we can effectively compute $T^{-1}B(0,1) \in \mathcal{O}(X)$ by Theorem 4.5 and the evaluation property since $B(0,1)$ is a computable point in $\mathcal{O}(X)$. By linearity of T, we obtain $0 \in T^{-1}B(0,1)$ and thus we can effectively find some rational number $r > 0$ such that $B(0,r) \subseteq T^{-1}B(0,1)$ by Lemma 4.2. But this implies $\|T\| \leq \frac{1}{r}$ by linearity of T. Thus, any $n \geq \frac{1}{r}$ is an appropriate result.   ⊣

As a corollary we immediately obtain that the bound of any computable linear operator is a left-computable real number.

COROLLARY 5.2. *Let X, Y be computable normed spaces. If $T : X \to Y$ is a computable linear operator, then $\|T\|$ is a left-computable real number.*

It is easy to see that in case of a finite-dimensional normed space X, the norm, considered as a function $\| \ \| :\subseteq \mathcal{C}(X, Y) \to \mathbb{R}$, is computable and thus $\|T\|$ is even a computable real number for any computable linear operator $T : X \to Y$ in this case, see [5]. On the other hand, one can prove that in the infinite-dimensional case the bound of a computable linear operator is not necessarily computable (not even for bijective operators on Hilbert spaces).

EXAMPLE 5.3. There exists some computable linear operator $T : \ell_2 \to \ell_2$ such that $\|T\|$ is not right-computable.

PROOF. Let $(a_n)_{n \in \mathbb{N}}$ be a positive and increasing computable sequence such that $a := \sup_{n \in \mathbb{N}} a_n \in \mathbb{R}$ exists but is not right-computable and define a linear bounded diagonal operator $T : \ell_2 \to \ell_2$ by $T(x_k)_{k \in \mathbb{N}} := (a_k x_k)_{k \in \mathbb{N}}$ for all $(x_k)_{k \in \mathbb{N}} \in \ell_2$. Then T is computable and $\|T\| = a$ is not right-computable.   ⊣

The following remark provides the topological counterpart of this observation.

REMARK 5.4. The mapping $\| \| :\subseteq C(\ell_2, \ell_2) \to \mathbb{R}, T \mapsto \|T\|$, defined for linear bounded and bijective operators $T : \ell_2 \to \ell_2$, is discontinuous with respect to the compact open topology on $C(\ell_2, \ell_2)$.

Now we want to study the *uniform bound map*

$$\mathcal{U} :\subseteq C(X, Y)^{\mathbb{N}} \longrightarrow \mathbb{R}, (T_i)_{i \in \mathbb{N}} \longmapsto \sup_{i \in \mathbb{N}} \|T_i\|,$$

where $\mathrm{dom}(\mathcal{U})$ is the set of all sequences $(T_i)_{i \in \mathbb{N}}$ of linear and bounded operators $T_i : X \to Y$ such that $\sup_{i \in \mathbb{N}} \|T_i\|$ exists. By the classical Uniform Boundedness Theorem, $\sup_{i \in \mathbb{N}} \|T_i\|$ in particular exists if X is a Banach space and $\{\|T_i x\| : i \in \mathbb{N}\}$ is bounded for each $x \in X$. In the following we will consider $\mathbb{R}_<$ and $\mathbb{R}_<^*$ instead of $\mathbb{R}$ in the target of $\mathcal{U}$ as well. Using Theorem 5.1 and the fact that $\sup :\subseteq \mathbb{R}_<^{\mathbb{N}} \to \mathbb{R}_<$ is computable, we obtain the following (not very surprising) computable version of the Uniform Boundedness Theorem which states that we can compute the uniform bound from below.

COROLLARY 5.5 (Semi-computable Uniform Boundedness Theorem). *Let X, Y be computable normed spaces. Then $\mathcal{U} :\subseteq C(X, Y)^{\mathbb{N}} \to \mathbb{R}_<$ is computable.*

We directly obtain the following corollary.

COROLLARY 5.6. *Let X be a computable Banach space and let Y be a computable normed space. If $(T_i)_{i \in \mathbb{N}}$ is a computable sequence of computable bounded operators $T_i : X \to Y$ such that $\{\|T_i x\| : i \in \mathbb{N}\}$ is bounded for each $x \in X$, then $\sup_{i \in \mathbb{N}} \|T_i\|$ is a left-computable real number.*

On the other hand, using the constant sequence with the computable operator $T : \ell_2 \to \ell_2$ from Example 5.3, one can see that the uniform bound is not computable in general. If we cannot compute the uniform bound $\sup_{i \in \mathbb{N}} \|T_i\|$ precisely, then it would be useful to compute at least *some* upper bound $n \geq \sup_{i \in \mathbb{N}} \|T_i\|$, as it was possible in case of the bound of a single operator in Theorem 5.1. However, a corresponding result does not hold in case of the uniform bound, not even in case of $X = Y = \mathbb{R}$.

REMARK 5.7. The map $\mathcal{U} :\subseteq C(\mathbb{R})^{\mathbb{N}} \to \mathbb{R}_<^*$ is discontinuous.

PROOF. Let us assume that $\mathcal{U} :\subseteq C(\mathbb{R})^{\mathbb{N}} \to \mathbb{R}^*_<$ is continuous. Given a real number $a \in \mathbb{R}_<$ with $a > 0$, we can effectively find an increasing sequence $(a_n)_{n\in\mathbb{N}}$ of rational numbers $a_n \in \mathbb{Q}$ such that $a_0 > 0$ and $a = \sup_{n\in\mathbb{N}} a_n$. We define operators $T_i : \mathbb{R} \to \mathbb{R}$ by $T_i(x) := a_i x$. Then $\|T_i\| = a_i$. Using evaluation and type conversion, we can actually determine the sequence $(T_i)_{i\in\mathbb{N}} \in C(\mathbb{R})^{\mathbb{N}}$ from a given $a \in (0,\infty)$. By applying $\mathcal{U}$ to this sequence we obtain some $n \geq \sup_{i\in\mathbb{N}} \|T_i\| = a$. But this implies that the identity id $: \mathbb{R}_< \to \mathbb{R}^*_<$ is continuous on $(0,\infty)$, which is a contradiction!   ⊣

Now the general question appears: which input information on a sequence $(T_i)_{i\in\mathbb{N}}$ of linear bounded and pointwise bounded operators T_i suffices to compute some upper bound on $\sup_{i\in\mathbb{N}} \|T_i\|$? As we have seen in the previous remark, the pure knowledge of pointwise boundedness does not suffice, not even in case of Euclidean space $X = Y = \mathbb{R}$. While in this case it would help to add the pointwise bound $\sup_{i\in\mathbb{N}} \|T_i 1\| \in \mathbb{R}$ as additional input information, the following result shows that in the general case it does not help if the pointwise bounds on some fundamental sequence of unit vectors are given as additional input information.

REMARK 5.8. The map $\mathcal{U}' :\subseteq C(\ell_2, \ell_2)^{\mathbb{N}} \times \mathbb{R}^{\mathbb{N}} \to \mathbb{R}^*_<$, $((T_i)_{i\in\mathbb{N}}, (s_j)_{j\in\mathbb{N}}) \mapsto \sup_{i\in\mathbb{N}} \|T_i\|$, defined for all sequences $((T_i)_{i\in\mathbb{N}}, (s_j)_{j\in\mathbb{N}})$ of linear bounded operators $T_i : \ell_2 \to \ell_2$ and bounds $s_j = \sup_{i\in\mathbb{N}} \|T_i e_j\|$, is discontinuous.

PROOF. Let us assume that $\mathcal{U}' :\subseteq C(\ell_2, \ell_2)^{\mathbb{N}} \times \mathbb{R}^{\mathbb{N}} \to \mathbb{R}^*_<$ is continuous. Given a real number $a \in \mathbb{R}_<$ with $a > 0$, we can effectively find an increasing sequence $(a_n)_{n\in\mathbb{N}}$ of rational numbers $a_n \in \mathbb{Q}$ such that $a_0 > 0$ and $a = \sup_{n\in\mathbb{N}} a_n$. Let

$$a_{ij} := \begin{cases} a_j & \text{for } j = 0, \ldots, i, \\ a_i & \text{for } j > i. \end{cases}$$

We define operators $T_i : \ell_2 \to \ell_2$ by $T_i(x_k)_{k\in\mathbb{N}} := (a_{ik} x_k)_{k\in\mathbb{N}}$ for all $(x_k)_{k\in\mathbb{N}} \in \ell_2$. Then $\|T_i\| = a_i$ and $\sup_{i\in\mathbb{N}} \|T_i e_j\|_2 = a_j$. Given some $x = (x_k)_{k\in\mathbb{N}}$, $i \in \mathbb{N}$ and a precision $m \in \mathbb{N}$ we can effectively find some $n \in \mathbb{N}$ and numbers $q_0, \ldots, q_n \in Q_{\mathbb{F}}$ such that $\|\sum_{j=0}^n q_j e_j - x\|_2 < \frac{1}{a_i} 2^{-m}$ and hence

$$\left\| T_i\left(\sum_{j=0}^n q_j e_j\right) - T_i(x) \right\|_2 \leq \|T_i\| \cdot \left\| \sum_{j=0}^n q_j e_j - x \right\|_2 < 2^{-m}.$$

By linearity of T_i we obtain $T_i(\sum_{j=0}^n q_j e_j) = \sum_{j=0}^n q_j T_i(e_j) = \sum_{j=0}^n q_j a_{ij}$ and thus we can evaluate each T_i effectively up to any given precision m. Using type conversion we can actually prove that there exists a computable operation $\sigma :\subseteq \mathbb{R}_< \rightrightarrows C(\ell_2, \ell_2)^{\mathbb{N}}$ that maps each $a > 0$ to a sequence $(T_i)_{i\in\mathbb{N}}$ of operators as described above. Now by assumption we can continuously find some $n \geq \sup_{i\in\mathbb{N}} \|T_i\| = \sup_{i\in\mathbb{N}} a_i = a$ Altogether, we have proved that the identity id $: \mathbb{R}_< \to \mathbb{R}^*_<$ is continuous on $(0,\infty)$, which is a contradiction!   ⊣

Although it does not help to know the pointwise bounds on some fundamental sequence of unit vectors, it is sufficient to have *all* pointwise bounds as additional input information, as the following result shows. Therefore, we will consider the input as operator $T : X \to \mathcal{B}(\mathbb{N}, Y)$ and for all such operators we define operators $T_i : X \to Y, x \mapsto T(x)(i)$. It is clear that an operator T is well-defined, if and only if $(T_i)_{i \in \mathbb{N}}$ is pointwise bounded. Moreover, T is linear, if and only if all T_i are linear. By the classical Uniform Boundedness Theorem a linear operator T is bounded, if and only if all T_i are bounded and the sequence $(T_i)_{i \in \mathbb{N}}$ is pointwise bounded. In this case we obtain

$$\|T\| = \sup_{x \in \overline{B}(0,1)} \|Tx\| = \sup_{x \in \overline{B}(0,1)} \sup_{i \in \mathbb{N}} \|T_i x\| = \sup_{i \in \mathbb{N}} \sup_{x \in \overline{B}(0,1)} \|T_i x\| = \sup_{i \in \mathbb{N}} \|T_i\|,$$

thus, the uniform bound of $(T_i)_{i \in \mathbb{N}}$ is nothing but the bound of T. The main technical obstacle in the proof of the following theorem is the fact that in general $\mathcal{B}(\mathbb{N}, Y)$ is a non-separable normed space and therefore we cannot derive the result directly from Theorem 5.1. Moreover, we recall that we have endowed $\mathcal{B}(\mathbb{N}, Y)$ with the weak$^=$ topology τ which does not coincide with the topology $\tau_{\| \ \|}$ induced by the supremum norm $\| \ \|$ on $\mathcal{B}(\mathbb{N}, Y)$. Therefore, we first have to compare these topologies and the corresponding notions of continuity.

LEMMA 5.9. *Let X, Y be normed spaces and let τ denote the weak$^=$ topology and $\tau_{\| \ \|}$ the norm topology on $\mathcal{B}(\mathbb{N}, Y)$. Then $\tau \subseteq \tau_{\| \ \|}$ and any bounded linear operator $T : X \to \mathcal{B}(\mathbb{N}, Y)$ is also continuous with respect to the weak$^=$ topology τ on $\mathcal{B}(\mathbb{N}, Y)$.*

PROOF. Let τ' be the subtopology on $\mathcal{B}(\mathbb{N}, Y)$ of the product topology on $Y^{\mathbb{N}}$ (where Y is endowed with the topology induced by its norm). Let τ'' be the weakest topology on $\mathcal{B}(\mathbb{N}, Y)$ which contains τ' and which makes the norm $\| \ \| : \mathcal{B}(\mathbb{N}, Y) \to \mathbb{R}$ continuous. The weak$^=$ topology τ is defined to be the sequentialization of τ''. Since the norm topology $\tau_{\| \ \|}$ is a topology which contains τ' and which makes the norm continuous, it follows that $\tau_{\| \ \|}$ also contains τ''. Thus, τ, i.e. the sequentialization of τ'', is contained in the sequentialization of $\tau_{\| \ \|}$. Since any topology induced by a norm is first countable and hence sequential, it coincides with its sequentialization. Hence τ is contained in $\tau_{\| \ \|}$. In particular, any operator $T : X \to \mathcal{B}(\mathbb{N}, Y)$ which is continuous with respect to the norm topology $\tau_{\| \ \|}$ is also continuous with respect to the weak$^=$ topology τ on $\mathcal{B}(\mathbb{N}, Y)$.   ⊣

In [9] we have proved a more general result in this direction. Since we have chosen the weak$^=$ topology as our standard topology on $\mathcal{B}(\mathbb{N}, Y)$, we consequently assume that $\mathcal{C}(X, \mathcal{B}(\mathbb{N}, Y))$ is the space of functions $T : X \to \mathcal{B}(\mathbb{N}, Y)$ which are continuous with respect to the weak$^=$ topology. By the previous lemma this space especially contains all linear bounded operators $T : X \to \mathcal{B}(\mathbb{N}, Y)$ and thus the mapping in the following theorem is well-defined.

THEOREM 5.10 (Computable Uniform Boundedness Theorem). *Let X, Y be computable normed spaces. Then $\| \; \| :\subseteq C(X, \mathcal{B}(\mathbb{N}, Y)) \to \mathbb{R}_<^*, T \mapsto \|T\|$, defined for linear bounded operators $T : X \to \mathcal{B}(\mathbb{N}, Y)$, is computable.*

PROOF. Since $\mathcal{B}(\mathbb{N}, Y) \to Y^{\mathbb{N}}$ is computable, it follows by evaluation and type conversion that $C(X, \mathcal{B}(\mathbb{N}, Y)) \to C(X, Y)^{\mathbb{N}}, T \mapsto (T_i)_{i \in \mathbb{N}}$ is computable. Hence, by Corollary 5.5 it follows that $\| \; \| :\subseteq C(X, \mathcal{B}(\mathbb{N}, Y)) \to \mathbb{R}_<$ is computable. It remains to prove the result with $\mathbb{R}_<^*$ in place of $\mathbb{R}_<$. Therefore, we have to show that we can also determine some upper bound of the operator bound. First of all, we prove that the operation

$$U : C(X, \mathcal{B}(\mathbb{N}, Y)) \longrightarrow \mathcal{O}(X), T \longmapsto T^{-1}B(0, 1)$$

is computable. Therefore, consider the function $f : \mathcal{B}(\mathbb{N}, Y) \to \mathbb{R}, (y_i)_{i \in \mathbb{N}} \mapsto \max\{1 - \|(y_i)_{i \in \mathbb{N}}\|, 0\}$. Since the norm $\| \; \| : \mathcal{B}(\mathbb{N}, Y) \to \mathbb{R}$ is computable, it follows that f is computable too. Thus, given $T \in C(X, \mathcal{B}(\mathbb{N}, Y))$ we can effectively find $g := f \circ T : X \to \mathbb{R}$, using evaluation and type conversion. Now $g^{-1}\{0\} = T^{-1}f^{-1}\{0\} = T^{-1}(\mathcal{B}(\mathbb{N}, Y) \setminus B(0, 1))$ and thus we can determine $U(T) = T^{-1}B(0, 1) = X \setminus g^{-1}\{0\} \in \mathcal{O}(X)$ effectively by Proposition 4.3. By linearity of T, we obtain $0 \in T^{-1}B(0, 1)$ and thus we can effectively find some rational number $r > 0$ such that $B(0, r) \subseteq U(T) = T^{-1}B(0, 1)$ by Lemma 4.2. Consequently, we obtain $\|T\| \leq \frac{1}{r}$ by linearity of T. Thus, any $n \geq \frac{1}{r}$ is an appropriate result.   ⊣

The reader might object that the Computable Uniform Boundedness Theorem, as stated here, does not deserve its name, since it does not imply the classical version of the theorem. However, it is straightforward how to combine this more general result with the classical Uniform Boundedness Theorem in order to get a version which actually implies the classical theorem. Our result shows how to compute uniform bounds if they exist while the classical result guarantees that they exist under certain additional assumptions.

§6. **The contraposition of the Uniform Boundedness Theorem.** Can we also effectivize the contraposition of the Uniform Boundedness Theorem? Thus, given a sequence of linear and bounded operators $T_i : X \to Y$ such that $\{\|T_i\| : i \in \mathbb{N}\}$ is unbounded, can we effectively find some witness $x \in X$ such that $\{\|T_i x\| : i \in \mathbb{N}\}$ is unbounded? The following theorem answers the question in the affirmative. The proof is a direct effectivization of the classical proof of the Uniform Boundedness Theorem [14] and it uses the computable Baire Category Theorem [6]. A corresponding version of the Uniform Boundedness Theorem is known in constructive analysis [4].

THEOREM 6.1 (Contra-computable Uniform Boundedness Theorem). *Let X be a computable Banach space and let Y be a computable normed space. There exists a computable multi-valued operation $\beta :\subseteq C(X, Y)^{\mathbb{N}} \rightrightarrows X$ with the following property: for all sequences $(T_i)_{i \in \mathbb{N}}$ of linear and bounded operators $T_i : X \to Y$ such that $\{\|T_i\| : i \in \mathbb{N}\}$ is unbounded, there exists an $x \in \beta(T_i)_{i \in \mathbb{N}}$ and $\{\|T_i x\| : i \in \mathbb{N}\}$ is unbounded for all such x.*

PROOF. Let us consider a sequence $(T_i)_{i \in \mathbb{N}}$ of linear and bounded operators $T_i : X \to Y$ such that $\{\|T_i\| : i \in \mathbb{N}\}$ is unbounded. Let $f_i : X \to \mathbb{R}$ be defined by $f_i(x) := \|T_i x\|$ and

$$A_n := \left\{ x \in X : (\forall i \in \mathbb{N})\, \|T_i x\| \leq n \right\} = \bigcap_{i=0}^{\infty} f_i^{-1}[0, n].$$

Thus, using evaluation and type conversion and the fact that the norm $\| \| : Y \to \mathbb{R}$ is computable, we obtain by Proposition 11.1 from [6] and Theorem 4.5 that the mapping $\alpha : C(X, Y)^{\mathbb{N}} \to \mathcal{A}_>(X)^{\mathbb{N}}, (T_i)_{i \in \mathbb{N}} \to (A_n)_{n \in \mathbb{N}}$ is computable, since $([0, n])_{n \in \mathbb{N}}$ is a computable sequence in $\mathcal{A}_>(\mathbb{R})$. Moreover, we obtain

$$\bigcup_{n=0}^{\infty} A_n = \left\{ x \in X : (\exists n \in \mathbb{N})(\forall i \in \mathbb{N})\, \|T_i x\| \leq n \right\}$$

$$= \left\{ x \in X : \{\|T_i x\| : i \in \mathbb{N}\} \text{ bounded} \right\}.$$

Now, let us assume that some set A_n is somewhere dense, i.e. there exists some $x \in X$ and some $\varepsilon > 0$ such that $B(x, \varepsilon) \subseteq A_n$. Then there is some $r \in \mathbb{N}$ such that

$$\overline{B}(0, 1) \subseteq r B(x, \varepsilon) \subseteq r A_n \subseteq A_{rn}.$$

Thus, $\|T_i\| = \sup_{x \in \overline{B}(0,1)} \|T_i x\| \leq rn$ for all $i \in \mathbb{N}$ which contradicts the assumption that $\{\|T_i\| : i \in \mathbb{N}\}$ is unbounded. Hence, $(A_n)_{n \in \mathbb{N}}$ is a sequence of nowhere dense sets. By the computable Baire Category Theorem, i.e. [6, Theorem 6], there exists a computable operation $\Delta :\subseteq \mathcal{A}_>(X)^{\mathbb{N}} \rightrightarrows X$ such that there exists some $x \in \Delta(A_n)_{n \in \mathbb{N}}$ whenever $(A_n)_{n \in \mathbb{N}}$ is a sequence of nowhere dense closed subsets $A_n \subseteq X$ and $x \in X \setminus \bigcup_{n=0}^{\infty} A_n$ for all such x. Thus, $\beta := \Delta \circ \alpha$ is a computable operation with the desired properties. ⊣

We obtain the following weaker non-uniform corollary.

COROLLARY 6.2. *Let X be a computable Banach space and let Y be a computable normed space. For any computable sequence $(T_i)_{i \in \mathbb{N}}$ of linear and computable operators $T_i : X \to Y$ such that $\{\|T_i\| : i \in \mathbb{N}\}$ is unbounded, there exists a computable $x \in X$ such that $\{\|T_i x\| : i \in \mathbb{N}\}$ is unbounded.*

§7. Divergent Fourier series.

One standard example of an application of the Uniform Boundedness Theorem is the construction of a continuous function $f : [0, 2\pi] \to \mathbb{R}$ whose Fourier series diverges at $t = 0$ (cf. [14]). We can directly transfer this to the computable setting and thus we can prove that there exists a computable function f whose Fourier series does not converge.

THEOREM 7.1. *There exists a computable function $f : [0, 2\pi] \to \mathbb{R}$ such that*

$$s_i := \frac{1}{2\pi} \int_0^{2\pi} f(t) \frac{\sin(i + \frac{1}{2})t}{\sin \frac{1}{2} t} \, dt$$

does not converge to $f(0)$ as $i \to \infty$.

PROOF. We consider the computable Banach space $C[0, 2\pi]$ and we define a sequence of operators $T_i : C[0, 2\pi] \to \mathbb{R}$ by

$$T_i(f) := \frac{1}{2\pi} \int_0^{2\pi} f(t) \frac{\sin(i + \frac{1}{2})t}{\sin \frac{1}{2} t} \, dt.$$

One can prove that $\|T_i\| = \frac{1}{2\pi} \int_0^{2\pi} |\frac{\sin(i+1/2)t}{\sin(t/2)}| dt$ and thus each T_i is bounded, whereas $\{\|T_i\| : i \in \mathbb{N}\}$ is unbounded [14]. Using evaluation and type conversion, one can prove that $(T_i)_{i \in \mathbb{N}}$ is a computable sequence of linear bounded and computable operators $T_i : C[0, 2\pi] \to \mathbb{R}$ in $C(C[0, 2\pi])$. This follows from the fact that integration is computable (cf. [31, Theorem 6.4.1.2]). Now Corollary 6.2 yields a computable function $f \in C[0, 2\pi]$ such that $\{\|T_i f\| : i \in \mathbb{N}\}$ is unbounded.   ⊣

Using the computable version of the Theorem on Condensation of Singularities [5] we could even prove that, given a sequence of computable numbers $(t_n)_{n \in \mathbb{N}}$ in $[0, 2\pi]$, we can effectively find a computable function $f : [0, 2\pi] \to \mathbb{R}$ such that the Fourier series of f does not converge to f at t_n for all $n \in \mathbb{N}$. We will not formulate this result here.

§8. **The Banach-Steinhaus Theorem.** In this section we want to discuss a computable version of the Banach-Steinhaus Theorem 1.2. The following simple example shows (with a similar construction as in Remark 5.7) that a computable sequence of computable and pointwise converging operators does not need to converge to a computable operator.

EXAMPLE 8.1. Let $(a_n)_{n \in \mathbb{N}}$ be an increasing computable sequence of real numbers such that $a := \sup_{n \in \mathbb{N}} a_n$ exists, but is not computable. Then the sequence $(T_i)_{i \in \mathbb{N}}$ of mappings $T_i : \mathbb{R} \to \mathbb{R}, x \mapsto a_i x$ is a computable sequence of computable linear operators such that $Tx := \lim_{n \to \infty} T_n x$ exists for all $x \in \mathbb{R}$, but the operator $T : \mathbb{R} \to \mathbb{R}$ defined in this way, i.e. $Tx = ax$, is not computable.

Thus, for a computable version of the Banach-Steinhaus Theorem we need more information on the sequence $(T_i)_{i \in \mathbb{N}}$. It turns out that it suffices to know the uniform bound $\sup_{i \in \mathbb{N}} \|T_i\|$ and the moduli of convergence of the sequences $(T_i x)_{i \in \mathbb{N}}$. Because of linearity, it even suffices to know these moduli on some fundamental sequence.

THEOREM 8.2 (Computable Banach-Steinhaus Theorem). *Let X be a computable Banach space with some computable fundamental sequence $(e_j)_{j \in \mathbb{N}}$ and let Y be a computable normed space. Then the function*

$$\mathcal{L} :\subseteq \mathcal{C}(X, Y)^{\mathbb{N}} \times \mathbb{N}^{\mathbb{N}} \times \mathbb{N} \longrightarrow \mathcal{C}(X, Y), \left((T_i)_{i \in \mathbb{N}}, m, s\right) \longmapsto \left(x \longmapsto \lim_{n \to \infty} T_n x\right),$$

defined for all tuples $((T_i)_{i \in \mathbb{N}}, m, s)$ such that the operators $T_i : X \to Y$ are linear bounded and pointwise convergent as $i \to \infty$, and such that $\sup_{i \in \mathbb{N}} \|T_i\| \leq s$ and $\|T_k e_j - \lim_{n \to \infty} T_n e_j\| \leq 2^{-i}$ for all $i, j \in \mathbb{N}$ and $k \geq m\langle i, j \rangle$, is computable.

PROOF. Given a sequence $(T_i)_{i \in \mathbb{N}}$ of linear bounded and pointwise convergent operators $T_i : X \to Y$ together with a bound $s \geq \sup_{i \in \mathbb{N}} \|T_i\|$ and a modulus of convergence $m : \mathbb{N} \to \mathbb{N}$ for the sequence $(e_j)_{j \in \mathbb{N}}$ such that $\|T_k e_j - \lim_{n \to \infty} T_n e_j\| \leq 2^{-i}$ for all $k \geq m\langle i, j \rangle$, the classical Banach-Steinhaus Theorem guarantees that the operator $T := (x \mapsto \lim_{n \to \infty} T_n x) \in \mathcal{C}(X, Y)$ is defined and $\|T\| \leq s$. Given some $x \in X$ and some $k \in \mathbb{N}$ we can effectively find some finite linear combination $x' := \sum_{j=0}^{l} a_j e_j$ with $a_j \in \mathbb{Q}_{\mathbb{F}}$ such that $\|x - x'\| < \frac{1}{s} 2^{-k-2}$. Let $a := \max\{|a_j| : j = 0, \ldots, l\}$ and $i \in \mathbb{N}$ such that $(l + 1) \cdot a \cdot 2^{-i} < 2^{-k-2}$, and let $M := \max\{m\langle i, j \rangle : j = 0, \ldots, l\}$ and $y := T_M x'$. Then we obtain

$$\|Tx - y\| \leq \|Tx - Tx'\| + \|Tx' - y\|$$
$$\leq s \frac{1}{s} 2^{-k-2} + \left\| \sum_{j=0}^{l} a_j \left(T e_j - T_M e_j\right) \right\|$$
$$\leq 2^{-k-2} + (l + 1) \cdot a \cdot 2^{-i}$$
$$< 2^{-k-1}.$$

Thus, by producing some output $y' \in Y$ with $\|y - y'\| < 2^{-k-1}$ one can actually compute T with precision 2^{-k}. Using evaluation and type conversion, one can show that a given tuple $((T_i)_{i \in \mathbb{N}}, m, s)$ can be effectively transformed into T.   ⊣

A combination of the computable Banach-Steinhaus Theorem 8.2 with the computable Uniform Boundedness Theorem 5.10 leads to the following corollary.

COROLLARY 8.3. *Let X be a computable Banach space with some computable fundamental sequence $(e_j)_{j \in \mathbb{N}}$ and let Y be a computable normed space. Then the function*

$$\mathcal{L}' :\subseteq \mathcal{C}(X, \mathcal{B}(\mathbb{N}, Y)) \times \mathbb{N}^{\mathbb{N}} \longrightarrow \mathcal{C}(X, Y), (T, m) \longmapsto \left(x \longmapsto \lim_{n \to \infty} T_n x\right),$$

defined for all linear bounded operators $T : X \to \mathcal{B}(\mathbb{N}, Y)$ *such that* $(T_i x)_{i \in \mathbb{N}}$ *converges for each* $x \in X$ *as* $i \to \infty$ *and* $\|T_k e_j - \lim_{n \to \infty} T_n e_j\| \leq 2^{-i}$ *for all* $i, j \in \mathbb{N}$ *and* $k \geq m\langle i, j \rangle$, *is computable.*

Since for a single sequence $(T_i)_{i \in \mathbb{N}}$ an upper bound s on the uniform bound is always available, we obtain the following less uniform version of the computable Banach-Steinhaus Theorem 8.2.

COROLLARY 8.4. *Let* X *be a computable Banach space with some computable fundamental sequence* $(e_j)_{j \in \mathbb{N}}$ *and let* Y *be a computable normed space. If* $(T_i)_{i \in \mathbb{N}}$ *is a computable sequence of linear and computable operators* $T_i : X \to Y$ *which converges pointwise with a computable modulus of convergence* $m :$ $\mathbb{N} \to \mathbb{N}$, *i.e.* $\|T_k e_j - \lim_{n \to \infty} T_n e_j\| \leq 2^{-i}$ *for all* $i, j \in \mathbb{N}$ *and* $k \geq m\langle i, j \rangle$, *then by*

$$Tx := \lim_{n \to \infty} T_n x$$

a linear and computable operator $T : X \to Y$ *is defined.*

§9. **Conclusion.** In this paper we have studied the computational content of the Uniform Boundedness Theorem from the point of view of computable analysis. It turned out that the essential question was: how much information on a sequence of linear bounded and pointwise bounded operators is sufficient to determine some upper bound of the uniform bound? This question is important since such upper bounds can serve as moduli of continuity. With Theorem 5.10 we have found a conclusive answer to this question. Moreover, with Theorem 6.1 we have proved a computable version of the contraposition of the Uniform Boundedness Theorem and as an easy application we could conclude that there exists a computable function whose Fourier series does not converge. Finally, with Theorem 8.2 we have provided a computable version of the Banach-Steinhaus Theorem.

Some related results have been obtained by Bishop's school of constructive analysis [4]. For instance, the contraposition of the Uniform Boundedness Theorem admits a similar constructive version and it might even be possible to transfer the constructive version into a computable one (see [28] for a partial transfer result). Moreover, Ishihara proved that in the constructive setting, the Banach-Steinhaus Theorem is equivalent to BD-$\mathbb{N}$ (a principle which states that each "pseudobounded" subset of $\mathbb{N}$ is bounded and which cannot be proved in the constructive setting, see [16, 17]). Although this principle should be satisfied in a suitable realizability model, it seems to be impossible to derive the computable versions of the Banach-Steinhaus Theorem, given in Theorem 8.2 and Corollary 8.3, from this result. Finally, no analog of the computable Uniform Boundedness Theorem 5.10 appears to be known in constructive analysis.

In reverse mathematics, as proposed by Friedman and Simpson [26], several theorems of functional analysis have been analysed according to which axioms are needed to prove the corresponding theorems in the language of second order arithmetic. In particular, it has been shown that the subsystem RCA_0 of second order arithmetic (i.e. second order arithmetic with a restricted recursive comprehension axiom) suffices to prove the Baire Category Theorem [13, 26]. However, the proofs in reverse mathematics are not necessarily effective such that they cannot be directly transfered to computable analysis (an example is the proof of the Banach-Steinhaus [26, Theorem II.10.8], where the non-effective contraposition of the Baire Category Theorem is applied). Nevertheless, computable counterexamples, as provided by computable analysis, could be a relevant source for reverse mathematics (cf. the discussion in Remark I.8.5 of [26]) as well as for constructive analysis.

Acknowledgement. The author would like to thank Matthias Schröder for helpful comments on his work [25] which led to Lemma 5.9.

REFERENCES

[1] STEFAN BANACH, *Sur les opérations dans les ensembles abstraits et leur application aux équations intégrales*, **Fundamenta Mathematicae**, vol. 3 (1922), pp. 133–181.

[2] STEFAN BANACH and STANISŁAW MAZUR, *Sur les fonctions calculables*, **Annales de la Société Polonaise de Mathématique**, vol. 16 (1937), p. 223.

[3] STEFAN BANACH and HUGO STEINHAUS, *Sur le principe de la condensation des singularités*, **Fundamenta Mathematicae**, vol. 9 (1927), pp. 50–61.

[4] ERRETT BISHOP and DOUGLAS BRIDGES, **Constructive Analysis**, Grundlehren der Mathematischen Wissenschaften, vol. 279, Springer-Verlag, Berlin, 1985.

[5] VASCO BRATTKA, **Computability and Complexity in Analysis**, Informatik Berichte, vol. 286, FernUniversität Hagen, Fachbereich Informatik, Hagen, 2001.

[6] ———, *Computable versions of Baire's category theorem*, **Mathematical Foundations of Computer Science, 2001 (Mariánské Lázně)** (Jiří Sgall, Aleš Pultr, and Petr Kolman, editors), Lecture Notes in Computer Science, vol. 2136, Springer, Berlin, 2001, pp. 224–235.

[7] ———, *Computing uniform bounds*, **CCA 2002 Computability and Complexity in Analysis** (Vasco Brattka, Matthias Schröder, and Klaus Weihrauch, editors), Electronic Notes in Theoretical Computer Science, vol. 66, Elsevier, Amsterdam, 2002, 5th International Workshop, CCA 2002, Málaga, Spain, July 12–13, 2002.

[8] ———, *Computability over topological structures*, **Computability and Models** (S. Barry Cooper and Sergey S. Goncharov, editors), Kluwer/Plenum, New York, 2003, pp. 93–136.

[9] ———, *Computability on non-separable Banach spaces and Landau's theorem*, **From Sets and Types to Topology and Analysis: Towards Practicable Foundations for Constructive Mathematics** (Laura Crosilla and Peter Schuster, editors), Oxford University Press, 2005, pp. 316–333.

[10] VASCO BRATTKA and GERO PRESSER, *Computability on subsets of metric spaces*, **Theoretical Computer Science**, vol. 305 (2003), no. 1-3, pp. 43–76.

[11] VASCO BRATTKA and MATTHIAS SCHRÖDER, *Computing with sequences, weak topologies and the axiom of choice*, **Computer Science Logic** (Luke Ong, editor), Lecture Notes in Computer Science, vol. 3634, Springer, 2005, pp. 462–476.

[12] VASCO BRATTKA and KLAUS WEIHRAUCH, *Computability on subsets of Euclidean space. I. Closed and compact subsets*, Theoretical Computer Science, vol. 219 (1999), no. 1-2, pp. 65–93.

[13] DOUGLAS K. BROWN and STEPHEN G. SIMPSON, *The Baire category theorem in weak subsystems of second-order arithmetic*, The Journal of Symbolic Logic, vol. 58 (1993), no. 2, pp. 557–578.

[14] CASPER GOFFMAN and GEORGE PEDRICK, **First Course in Functional Analysis**, Prentice-Hall Inc., Englewood Cliffs, N.J., 1965.

[15] ANDRZEJ GRZEGORCZYK, *On the definitions of computable real continuous functions*, **Polska Akademia Nauk. Fundamenta Mathematicae**, vol. 44 (1957), pp. 61–71.

[16] HAJIME ISHIHARA, *Sequential continuity of linear mappings in constructive mathematics*, **The Journal of Universal Computer Science**, vol. 3 (1997), no. 11, pp. 1250–1254.

[17] ———, *Sequentially continuity in constructive mathematics*, **Combinatorics, computability and logic (Constanța, 2001)** (C. S. Calude, M. J. Dinneen, and S. Sburlan, editors), Springer Ser. Discrete Math. Theor. Comput. Sci., Springer, London, 2001, pp. 5–12.

[18] KER-I KO, **Complexity Theory of Real Functions**, Progress in Theoretical Computer Science, Birkhäuser, Boston, MA, 1991.

[19] BORIS ABRAMOVICH KUŠNER, **Lectures on Constructive Mathematical Analysis**, Translations of Mathematical Monographs, vol. 60, American Mathematical Society, Providence, RI, 1984.

[20] DANIEL LACOMBE, *Extension de la notion de fonction récursive aux fonctions d'une ou plusieurs variables réelles. I, II, III*, **Comptes Rendus Mathématique. Académie des Sciences. Paris**, vol. 240, 241 (1955), pp. 2478–2480, 13–14, 151–153.

[21] ———, *Quelques procédés de définition en topologie recursive*, **Constructivity in Mathematics: Proceedings of the Colloquium Held at Amsterdam, 1957** (A. Heyting, editor), Studies in Logic and the Foundations of Mathematics, North-Holland, Amsterdam, 1959, pp. 129–158.

[22] GEORGE METAKIDES, ANIL NERODE, and R. A. SHORE, *Recursive limits on the Hahn-Banach theorem*, **Errett Bishop: Reflections on Him and His Research (San Diego, Calif., 1983)** (Murray Rosenblatt, editor), Contemporary Mathematics, vol. 39, AMS, Providence, RI, 1985, pp. 85–91.

[23] YIANNIS NICHOLAS MOSCHOVAKIS, *Recursive metric spaces*, **Polska Akademia Nauk. Fundamenta Mathematicae**, vol. 55 (1964), pp. 215–238.

[24] MARIAN B. POUR-EL and J. IAN RICHARDS, **Computability in Analysis and Physics**, Perspectives in Mathematical Logic, Springer-Verlag, Berlin, 1989.

[25] MATTHIAS SCHRÖDER, *Extended admissibility*, **Theoretical Computer Science**, vol. 284 (2002), no. 2, pp. 519–538.

[26] STEPHEN G. SIMPSON, **Subsystems of Second Order Arithmetic**, Perspectives in Mathematical Logic, Springer-Verlag, Berlin, 1999.

[27] DIETER SPREEN, *On effective topological spaces*, **The Journal of Symbolic Logic**, vol. 63 (1998), no. 1, pp. 185–221.

[28] A. S. TROELSTRA, *Comparing the theory of representations and constructive mathematics*, **Computer Science Logic (Berne, 1991)** (E. Börger et al., editors), Lecture Notes in Computer Science, vol. 626, Springer, Berlin, 1992, pp. 382–395.

[29] ALAN M. TURING, *On computable numbers, with an application to the "Entscheidungsproblem"*, **Proceedings of the London Mathematical Society, Vol. 42**, no. 2, 1936, pp. 230–265.

[30] KLAUS WEIHRAUCH, *Computability on computable metric spaces*, **Theoretical Computer Science**, vol. 113 (1993), no. 2, pp. 191–210.

[31] ———, **Computable Analysis**, Springer-Verlag, Berlin, 2000.

[32] KLAUS WEIHRAUCH and NING ZHONG, *Is wave propagation computable or can wave computers beat the Turing machine?*, **Proceedings of the London Mathematical Society. Third Series**, vol. 85 (2002), no. 2, pp. 312–332.

[33] MARIKO YASUGI, TAKAKAZU MORI, and YOSHIKI TSUJII, *Effective properties of sets and functions in metric spaces with computability structure*, **Theoretical Computer Science**, vol. 219 (1999), no. 1-2, pp. 467–486.

[34] NING ZHONG, *Computability structure of the Sobolev spaces and its applications*, **Theoretical Computer Science**, vol. 219 (1999), no. 1-2, pp. 487–510.

LABORATORY OF FOUNDATIONAL ASPECTS OF COMPUTER SCIENCE
DEPARTMENT OF MATHEMATICS & APPLIED MATHEMATICS
UNIVERSITY OF CAPE TOWN
RONDEBOSCH 7701, SOUTH AFRICA
E-mail: BrattkaV@maths.uct.ac.za

SYMMETRY OF THE UNIVERSAL COMPUTABLE FUNCTION: A STUDY OF ITS AUTOMORPHISMS, HOMOMORPHISMS AND ISOMORPHIC EMBEDDINGS

ELÍAS F. COMBARRO

Abstract. In this paper we study the symmetry of the universal computable function by means of its automorphisms, homomorphisms and isomorphic embeddings. We show that all these functions are computable and also study other related functions with similar properties.

We also reproduce this study for the predicate expressing membership of an element to a recursively enumerable set. The results found for this predicate are quite different from those found for the universal computable function, especially when concerning homomorphisms. This shows that there exists a deep difference between these two, for all the other, closely related objects.

§1. Introduction. One of the most important objects in the study of the recursive functions is the universal computable function, which makes it possible to compute the values of those functions in a uniform and algorithmic way. That is, if $\{\varphi_e\}_{e \geq 0}$ is an acceptable system of indices (see [3]) for the unary partial computable functions, then the function

$$\phi(e, x) = \varphi_e(x)$$

is also computable.

Hence, it is interesting to study its properties, including its symmetry. The usual way to accomplish this is to study the automorphisms, isomorphic embeddings and homomorphisms of the object (see [1] for the definitions). In this case we will study the corresponding functions for the graph of ϕ. We will denote this predicate by P, so

$$P(e, x, y) \Longleftrightarrow \phi(e, x) = y \Longleftrightarrow \varphi_e(x) = y,$$

for all natural numbers e, x and y.

A permutation f of the natural numbers will be an automorphism [1] of this predicate if

$$P(e, x, y) \Longleftrightarrow P(f(e), f(x), f(y)),$$

for all natural numbers e, x and y; that is, if

$$\varphi_e(x) = y \Longleftrightarrow \varphi_{f(e)}(f(x)) = f(y),$$

for all natural numbers e, x and y.

Logic Colloquium '02
Edited by Z. Chatzidakis, P. Koepke, and W. Pohlers
Lecture Notes in Logic, 27

Thus, these automorphisms are changes of coordinates with respect to which the computations of the recursive functions are performed in exactly the same way as in the original model. In other words, these are the symmetric transformations which take into account the way the computable functions are actually computed.

Another interesting predicate closely related to P is the one that expresses membership of an element to an r.e. set. Since r.e. sets are exactly the domains of the unary partial computable functions (see, for instance, [4]), if we define (as usual)

$$W_n = \mathrm{Dom}\, \varphi_n,$$

the new predicate, which we will denote by W, will have the form

$$W(x, y) \iff x \in W_y,$$

for x, y natural numbers.

Then, an automorphism of W will be a permutation f such that

$$W(x, y) \iff W(f(x), f(y)),$$

for all natural numbers x, y; that is

$$x \in W_y \iff f(x) \in W_{f(y)},$$

for all natural numbers x, y.

We have studied the properties of the automorphisms of these two predicates in a recent paper [2]. Here, we will study some natural weakenings of the concept of automorphism: isomorphic embeddings (Section 4) and homomorphisms (Section 5). We will also introduce a new class of functions which have almost the same properties as the automorphisms of P and W (Section 3). But first of all, we summarize the results found for the automorphisms of P and W (Section 2).

1.1. Notation. Throughout this paper, we will denote the set of natural numbers (including zero) by ω (the first infinite ordinal). We will fix an acceptable system of indices for the unary partial computable functions and denote by φ_n the n-th such function and by W_n its domain.

Clearly, the set of the automorphisms of P (W, respectively) as defined above together with the usual composition of functions forms a group which we will denote by *Aut P* (by *Aut W*, respectively).

If f is a computable function we will denote by I_f the set of its indices, i.e.

$$I_f = \{n \in \omega : f = \varphi_n\}.$$

In particular we will denote by I_{id} the set of indices of the identity function and by $I_\uparrow$ the set of indices of the nowhere defined function. If $g = \varphi_n$ then we will simply set $I_n = I_g$.

If A is a r.e. set we will set

$$H_A = \{x \in \omega : W_x = A\}.$$

If $A = W_n$ we will simply set $H_n = H_A$.

If A is a set we will denote its cardinality by $|A|$.

§2. **Automorphisms of P and W.** Some of the proofs of the results found for the homomorphisms and the isomorphic embeddings of P and W are just variations of the proofs of the corresponding results for automorphisms. Hence, for sake of completeness, we summarize in this section the main results found for the automorphisms of P and W (cf. [2]) and also some of their proofs.

First, we state some useful lemmas.

LEMMA 1. *Let f be an automorphism of W. Then for every natural number n*

$$f(W_n) = W_{f(n)}$$

and

$$f(H_n) = H_{f(n)}.$$

LEMMA 2. *Let f be an automorphism of P. Then for every natural number n*

$$f \cdot \varphi_n \cdot f^{-1} = \varphi_{f(n)}$$

and

$$f(I_n) = I_{f(n)}.$$

Now, we show that there exists only a countable number of automorphisms of W and of P (the proofs can be found in [2]).

THEOREM 3. *There exists e in ω such that for all $h \in \operatorname{Aut} W$*

$$W_{h(e)} = W_e \Longleftrightarrow h = id.$$

COROLLARY 4. *There exists e in ω such that for all $h \in \operatorname{Aut} P$*

$$I_{h(e)} = I_e \Longleftrightarrow h = id.$$

Now, we state and prove the main theorem of the section.

THEOREM 5. *If f is an automorphism of W then it is computable.*

PROOF. Let e_0 be an index of the empty set, e_1 an index of $\{e_0\}$ and e_2 an index of $\{e_0, e_1\}$. We consider the set

$$S = \{n_0, \ldots, n_k, \ldots\}$$

where

$$W_{n_k} = \{b_1^k, b_2^k\},$$
$$W_{b_2^k} = \{k\},$$
$$W_{b_1^k} = \{m_0^k, e_0\},$$
$$W_{m_0^k} = \{m_1^k\},$$
$$\vdots$$
$$W_{m_k^k} = \{e_2\}.$$

It is clear that S is r.e. since, given k, we can compute, in turn, $m_k^k, \ldots, m_0^k, b_1^k$, b_2^k and n_k. Then, we can fix an index a of S as r.e. set.

Consider the r.e. set

$$W_{f(a)} = \{f(n_0), \ldots, f(n_k), \ldots\}.$$

It is clear (see Lemma 1) that

$$W_{f(n_k)} = \{f(b_1^k), f(b_2^k)\},$$
$$W_{f(b_2^k)} = \{f(k)\},$$
$$W_{f(b_1^k)} = \{f(m_0^k), f(e_0)\},$$
$$W_{f(m_0^k)} = \{f(m_1^k)\},$$
$$\vdots$$
$$W_{f(m_k^k)} = \{f(e_2)\},$$

and that

$$W_{f(e_0)} = \emptyset,$$
$$W_{f(e_1)} = \{f(e_0)\},$$
$$W_{f(e_2)} = \{f(e_0), f(e_1)\}.$$

Note that $|W_{f(e_0)}| = 0$ but $|W_{f(e_1)}| = 1$. Thus we have $f(e_0) \neq f(e_1)$ and $|W_{f(e_2)}| = 2$. Note also that $f(n_i) \neq f(n_j)$ if $i \neq j$.

To enumerate the set $A = \{(x, y) : f(x) = y\}$ we repeat the following steps:

- Enumerate a new element m of the set $W_{f(a)}$.
- Enumerate W_m till two values r and s are obtained.
- Enumerate W_r and W_s. One of them will have two members and the other only one. Suppose that

$$W_r = \{i, j\},$$
$$W_s = \{y\}.$$

- Enumerate W_i and W_j. One of them will be empty and the other will have one member. Suppose that

$$W_i = \{c_1\},$$
$$W_j = \emptyset.$$

- Enumerate W_{c_1} till obtain a value c_2, say, is obtained; dovetail the enumeration of W_{c_1} with the enumeration of W_{c_2}; when it is found that $c_3 \in W_{c_2}$, dovetail the enumerations of W_{c_1}, W_{c_2} and W_{c_3}; ... and so on. Eventually in one of these sets, say W_{c_n}, there will appear two elements. From the structure of the set W_a (and of $W_{f(a)}$) it follows that

$$y = f(n-1).$$

Then we add to A the value $(n-1, y)$.

It is clear that the process enumerates the set A, so it is r.e. and f is computable. ⊣

COROLLARY 6. *Every automorphism of P is computable.*

PROOF. Clearly, W can be defined in terms of P, so every automorphism of P is also an automorphism of W and, hence, it is computable. ⊣

§3. **Functions commuting with P and W.** In the study of the automorphisms of the predicates P and W two properties play a key role: the way they act on r.e. sets and on partial computable functions, respectively. As we have already mentioned, if f is an automorphism of W then for all n we have

$$f(W_n) = W_{f(n)},$$

and if g is an automorphism of P then for all e we have

$$g \cdot \varphi_e = \varphi_{g(e)} \cdot g.$$

Thus, it is interesting studying these properties on their own. We will give a name to functions with these properties.

DEFINITION 7. We say that a function f commutes with W if

$$f(W_n) = W_{f(n)},$$

for all n in ω.

We say that a function g commutes with P if

$$g \cdot \varphi_e = \varphi_{g(e)} \cdot g,$$

for all e in ω.

The main result of this section is that, indeed, the above properties are enough to prove that a function is computable (and, thus, in the proof of Theorem 5 we do not need the function being a permutation).

3.1. Functions commuting with W. We begin studying the functions which commute with W.

THEOREM 8. *If a function commutes with W then it is computable.*

PROOF. Let f be a function commuting with W. Note that, for the proof of Theorem 5 to be valid here, we only need to show that

1. $|W_{f(n_i)}| = 2$ for each i in ω,
2. $|W_{f(b_1^k)}| = 2$,
3. $|W_{f(e_0)}| = 0$,
4. $|W_{f(e_2)}| = 2$,

since the cardinality of the other sets in the proof is, trivially, 1.

It is clear that

$$|W_{f(e_0)}| = |f(W_{e_0})| = |f(\emptyset)| = 0$$

and then since $|W_{f(e_1)}| = 1$, we will have $f(e_0) \neq f(e_1)$ and hence

$$|W_{f(e_2)}| = 2.$$

On the other hand, $|W_{f(m_0^k)}| = 1$ so $f(e_0) \neq f(m_0^k)$ and

$$|W_{f(b_1^k)}| = 2.$$

Analogously, $f(b_1^k) \neq f(b_2^k)$ and

$$|W_{f(n_i)}| = 2.$$ $\dashv$

Implicit in the proof of the previous theorem we find the following results.

THEOREM 9. *There exists an element $a \in \omega$ such that if f is a function commuting with W and $W_{f(a)} = W_a$, then f is the identity function.*

PROOF. It is enough to take a as in the proof of Theorem 5 $\dashv$

PROPOSITION 10. *If f is a function commuting with W then its range is infinite.*

PROOF. Let e_0 be an index of the empty set, e_1 an index of $\{e_0\}$, e_2 an index of $\{e_0, e_1\}$ ... In general, e_n will be an index of $\{e_0, \ldots, e_{n-1}\}$.

We have already seen in the proof of Theorem 8 that

1. $|W_{f(e_0)}| = 0$,
2. $|W_{f(e_1)}| = 1$,
3. $|W_{f(e_2)}| = 2$,

and it is easy to prove by induction that $|W_{f(e_i)}| = i$, so $f(e_i) \neq f(e_j)$ if $i \neq j$ and then the range of f must be infinite. $\dashv$

This last result is the best we can get in general in view of the following proposition.

PROPOSITION 11. *There exists a 1-1 function f which commutes with W and has co-infinite range.*

PROOF. The result can be proved by modifying the constructions used in [2] to show the existence of the automorphisms of W and P. Here, we present a sketch of the proof.

Let A be an infinite recursive set. Fix an effective enumeration without repetitions $a_0, a_1, \ldots$ of A (see [4]). We define a total and computable function h by:

Given a natural number z, $h(z)$ is an index of the function computed in the following way:

Step t: Let m be the least number not yet in the domain of $\varphi_{h(z)}$. We effectively enumerate indices of the set

$$\varphi_z(W_m)$$

until we obtain one, say c, satisfying:

1. The range of $\varphi_{h(z)}$ does not yet include c.
2. The index c is not in $\{a_0, a_1, \ldots, a_{\frac{(t+1)(t+2)}{2}-1}\}$.

Then, we set $\varphi_{h(z)}(m) = c$, and we have

$$\varphi_z(W_m) = W_c = W_{\varphi_{h(z)}(m)}.$$

Applying the Recursion Theorem (see [4]) we obtain z_0 such that $\varphi_{z_0} = \varphi_{h(z_0)}$ and we can consider $f = \varphi_{z_0}$. It can be checked that f is the desired function. ⊣

On the other hand, a function commuting with W can be onto and, at the same time, it can be very far away from being 1-1.

PROPOSITION 12. *There exists a function f commuting with W and such that for all x in ω the set $f^{-1}(x)$ is infinite.*

PROOF. The proof is similar to that of the previous proposition, but substituting step t by:

Step 2t: Let p be the least number not yet in the domain of $\varphi_{h(z)}$. Calculate an index c of the set

$$\varphi_z(W_p).$$

Set $\varphi_{h(z)}(p) = c$, to obtain

$$\varphi_z(W_p) = W_c = W_{\varphi_{h(z)}(p)}.$$

Step 2t + 1: Suppose that t is the codification of (m, n), i.e. $t = \langle m, n \rangle$ (see [4]). Define $\varphi_{\bar{z}}$ as in [2] and compute indices of the set

$$\varphi_{\bar{z}}(W_m)$$

until one, say c, not yet in the domain of $\varphi_{h(z)}$ is found. Define $\varphi_{h(z)}(c) = m$ to obtain

$$\varphi_{\bar{z}}(W_{\varphi_{h(z)}(c)}) = W_c.$$

⊣

We will also show the relationship of these functions with the ones which will be studied in the following sections.

PROPOSITION 13. *Let f be a function commuting with W. Then f is a homomorphism of W.*

PROOF. Let x, y be two natural numbers. Suppose that $x \in W_y$. Since f commutes with W we will have

$$f(x) \in f(W_y) = W_{f(y)}. \qquad \dashv$$

PROPOSITION 14. *Let f be a 1-1 function commuting with W. Then, for all $x, y \in \omega$ we have*

$$x \in W_y \Longleftrightarrow f(x) \in W_{f(y)}.$$

PROOF. From the previous proposition, it only remains to show the right-to-left implication. Let x, y be two natural numbers and suppose that $f(x) \in W_{f(y)}$. We then have $f(x) \in f(W_y)$ and, since f is 1-1, it holds that $x \in W_y$. $\qquad \dashv$

3.2. Functions commuting with P. In this paragraph we will show that the functions commuting with P are also computable. The following result shows that we cannot use Theorem 8.

PROPOSITION 15. *There exists a 1-1 function f commuting with P such that for all $n \in \omega$ we have*

$$f(W_n) \neq W_{f(n)}.$$

PROOF. We will use the Recursion Theorem again. We fix an element a and construct a function which will never attain that value.

Given z, let $h(z)$ be an index of the function that is computed by:
Step t: Let m be the least number not yet in the domain of $\varphi_{h(z)}$. Consider the function

$$g_m(x) = \begin{cases} \varphi_z(\varphi_m(\varphi_{\bar{z}}(x))) & \text{if } x \neq a, \\ a & \text{if } x = a \end{cases}$$

where $\varphi_{\bar{z}}$ is defined as in [2]. Clearly, g_m is computable. Enumerate indices of g until one, say c, different from a and not yet in the range of $\varphi_{h(z)}$ is found. Set $\varphi_{h(z)}(m) = c$.

Applying the Recursion Theorem there exists z_0 such that $\varphi_{z_0} = \varphi_{h(z_0)}$. It can be checked that $f = \varphi_{z_0}$ is the desired function. $\qquad \dashv$

Thus, we will adopt a more direct approach.

LEMMA 16. *Let n be an index of the computable function $h(x) = x + 1$. Let f be a function such that $f \cdot \varphi_n = \varphi_{f(n)} \cdot f$. Then, f is computable.*

PROOF. Clearly, we have

$$f(1) = f(\varphi_n(0)) = \varphi_{f(n)}(f(0)),$$
$$f(2) = f(\varphi_n(1)) = \varphi_{f(n)}(f(1)),$$
$$f(3) = f(\varphi_n(2)) = \varphi_{f(n)}(f(2)),$$
$$\vdots$$
$$f(k+1) = f(\varphi_n(k)) = \varphi_{f(n)}(f(k)),$$
$$\vdots$$

and, hence, we can compute f from $f(n)$ and $f(0)$.   ⊣

We immediately obtain the following result.

THEOREM 17. *If f commutes with P then it is computable.*

Finally, we will show the relationship between these functions and the ones which we will study below.

PROPOSITION 18. *If f commutes with P, then it is an isomorphic embedding of W and of P.*

PROOF. First, we will show that

$$x \in W_n \iff f(x) \in W_{f(n)},$$

for all x, n in ω.

Suppose that $x \in W_n$. Then, there exists z such that $\varphi_n(x) = z$. Since f commutes with P it holds that $\varphi_{f(n)}(f(x)) = f(z)$, and then $f(x) \in W_{f(n)}$.

On the other hand, if $f(x) \in W_{f(n)}$ there must exist z such that $\varphi_{f(n)}(f(x)) = z$. Then, since $f \cdot \varphi_n = \varphi_{f(n)} \cdot f$, it follows that $f(\varphi_n(x)) = z$ and consequently $\varphi_n(x)$ is defined. Then we have $x \in W_n$.

Now, we will show that f is 1-1.

Suppose that there exist x_1 and x_2 such that $f(x_1) = f(x_2)$. Consider $W_{y_1} = \{x_1\}$. Then

$$x_1 \in W_{y_1} \implies f(x_1) \in W_{f(y_1)}$$
$$\implies f(x_2) \in W_{f(y_1)}$$
$$\implies x_2 \in W_{y_1} = \{x_1\}$$
$$\implies x_1 = x_2$$

and, consequently, f must be 1-1.

Finally, we will prove that

$$\varphi_e(x) = y \iff \varphi_{f(e)}(f(x)) = f(y),$$

for all $e, x, y \in \omega$.

Suppose that $\varphi_e(x) = y$. We know that f commutes with P, so $\varphi_{f(e)}(f(x)) = f(\varphi_e(x)) = f(y)$.

Now suppose that $\varphi_{f(e)}(f(x)) = f(y)$. Then, since f commutes with P, we will have $\varphi_{f(e)}(f(x)) = f(\varphi_e(x))$, and since f is 1-1, it follows that $\varphi_e(x) = y$. ⊣

§4. **Isomorphic embeddings of W and P.** Another natural weakening of the concept of automorphism is that of isomorphic embedding. We drop the condition of surjectivity while keeping the way the functions act on the predicates. The resulting image of the structure is then isomorphic to the original one.

We begin by defining these concepts in the case of our predicates.

DEFINITION 19. A 1-1 function f is an isomorphic embedding of W if

$$x \in W_y \iff f(x) \in W_{f(y)},$$

for all x and y in ω.

A 1-1 function g is an isomorphic embedding of P if

$$\varphi_e(x) = y \iff \varphi_{g(e)}(g(x)) = g(y),$$

for all e, x and y in ω.

As in the previous section, we will focus first on the predicate W.

4.1. Isomorphic embeddings of W. We begin by showing that, in fact, the property of being 1-1 is implicit in the main property of the isomorphic embeddings of W.

PROPOSITION 20. *Let f be a function such that*

$$x \in W_y \iff f(x) \in W_{f(y)},$$

for all $x, y \in \omega$. Then f is 1-1.

PROOF. It is contained in the proof of Proposition 18. ⊣

We would like to reproduce here the results found for the functions commuting with W. However, the following result suggests that it might not be possible.

PROPOSITION 21. *There exists an isomorphic embedding f of W such that for all $n \in \omega$*

$$f(W_n) \neq W_{f(n)}.$$

PROOF. The proof is similar to that or proposition 15, but step t now is

Step t: Let m be the least number not yet in the domain of $\varphi_{h(z)}$. Compute indices of the set

$$\varphi_z(W_m) \cup \{a\}$$

until one, say c, different from a and not yet in the range of $\varphi_{h(z)}$ is found. Set $\varphi_{h(z)}(m) = c$ to obtain

$$W_{\varphi_{h(z)}(m)} = W_c = \varphi_z(W_m) \cup \{a\}. \qquad \dashv$$

Nonetheless, we can prove the following "weak commutation".

PROPOSITION 22. *If f is an isomorphic embedding of W then for all $n \in \omega$ we have*

$$f(W_n) = W_{f(n)} \cap Rg \ f.$$

PROOF. Suppose that $x \in W_n$. Then $f(x) \in W_{f(n)}$ and hence

$$f(x) \in W_{f(n)} \cap Rg \ f.$$

Consequently, we will have

$$f(W_n) \subseteq W_{f(n)} \cap Rg \ f.$$

Suppose now that $x \in W_{f(n)} \cap Rg \ f$. There must exist y such that $f(y) = x \in W_{f(n)}$. Then $y \in W_n$ and, hence, $x = f(y) \in f(W_n)$. We will have

$$f(W_n) \supseteq W_{f(n)} \cap Rg \ f,$$

and the result follows. $\qquad \dashv$

Now, we aim to prove an analogue of Theorem 5. First, we will prove a lemma which we will find particularly useful.

LEMMA 23. *There exists an index $e \in \omega$ such that if f is a function satisfying the conditions*

1. $f(W_n) = W_{f(n)} \cap Rg \ f$ *for all $n \in \omega$,*
2. $W_{f(e)} \subseteq W_e$,

then f is the identity function.

PROOF. Consider $n_0, n_1, \ldots$ as in the proof of Theorem 5.

Let f be a function in the conditions of the lemma. First, we will prove that if $f(n_i) = n_j$, then $i = j$ and $f(i) = j = i$.

Suppose that $f(n_i) = n_j$. Then, we will have $W_{f(n_i)} = W_{n_j}$. We know (see Proposition 22) that

$$
\begin{aligned}
\left|W_{f(e_0)} \cap Rg \ f\right| &= \left|f(W_{e_0})\right| &= \left|f(\emptyset)\right| &= 0, \\
\left|W_{f(e_1)} \cap Rg \ f\right| &= \left|f(\{e_0\})\right| &= \left|\{f(e_0)\}\right| &= 1, \\
\left|W_{f(e_2)} \cap Rg \ f\right| &= \left|f(\{e_0, e_1\})\right| &= \left|\{f(e_0), f(e_1)\}\right| &= 2.
\end{aligned}
$$

What's more

$$\left|W_{f(m_0^i)} \cap Rg \ f\right| = \left|f(W_{m_0^i})\right| = 1,$$

so $f(e_0) \neq f(m_0^i)$ and

$$\left|W_{f(b_1^i)} \cap Rg \ f\right| = \left|f(W_{b_1^i})\right| = \left|\{f(e_0), f(m_0^i)\}\right| = 2.$$

Analogously, $f(b_1^i) \neq f(b_2^i)$ since

$$\left| W_{f(b_2^i)} \cap Rg\, f \right| = \left| f\left(W_{b_2^i} \right) \right| = \left| \{ f(i) \} \right| = 1.$$

We know that

$$\left\{ f\left(b_1^i \right), f\left(b_2^i \right) \right\} = f\left(W_{n_i} \right) \subseteq W_{f(n_i)} = W_{n_j} = \left\{ b_1^j, b_2^j \right\}.$$

There exists two possibilities:
1. $f(b_1^i) = b_2^j$ and $f(b_2^i) = b_1^j$,
2. $f(b_1^i) = b_1^j$ and $f(b_2^i) = b_2^j$.

The first one is contradictory because if it holds, we would have

$$W_{f(b_1^i)} = W_{b_2^j},$$

and consequently

$$2 = \left| f\left(W_{b_1^i} \right) \right| \leq \left| W_{f(b_1^i)} \right| = \left| W_{b_2^j} \right| = 1.$$

Then, it must be the case that

$$\{ f(i) \} = f\left(W_{b_2^i} \right) \subseteq W_{f(b_2^i)} = W_{b_2^j} = \{ j \},$$

and so

$$f(i) = j.$$

Moreover, we will have

$$\left\{ f\left(m_0^i \right), f(e_0) \right\} \subseteq W_{f(b_1^i)} = W_{b_1^j} = \left\{ m_0^j, e_0 \right\},$$

so

$$\left\{ f\left(m_0^i \right), f(e_0) \right\} = \left\{ m_0^j, e_0 \right\}.$$

Again, there exist two different possibilities, namely:
1. $f(m_0^i) = e_0$,
2. $f(m_0^i) = m_0^j$.

The first one is, in fact, impossible, since we would have $W_{f(m_0^i)} = W_{e_0}$ and, consequently

$$1 \leq \left| f\left(W_{m_0^i} \right) \right| \leq \left| W_{f(m_0^i)} \right| = \left| W_{e_0} \right| = 0.$$

Hence, we have $f(m_0^i) = m_0^j$ and $f(e_0) = e_0$.

Now, we will prove by induction that if $k \leq \min(i, j)$ then $f(m_k^i) = m_k^j$:
$k = 0$: It's already proven
$k > 0$: We know that

$$W_{m_{k-1}^i} = \left\{ m_k^i \right\},$$

$$W_{m_{k-1}^j} = \left\{ m_k^j \right\},$$

and, then, by the induction hypothesis

$$f\left(m_{k-1}^i\right) = m_{k-1}^j.$$

Then

$$\left\{f\left(m_k^i\right)\right\} = f\left(W_{m_{k-1}^i}\right) \subseteq W_{f(m_{k-1}^i)} = W_{m_{k-1}^j} = \left\{m_k^j\right\}$$

and, consequently, $f(m_k^i) = m_k^j$.

We have the following three possibilities

1. $i < j$,
2. $j > i$,
3. $i = j$,

but the first and the second are contradictory. Indeed:

If $i < j$, then

$$\left\{f(e_1), f(e_0)\right\} = f\left(W_{m_i^i}\right) \subseteq W_{f(m_i^i)} = W_{m_i^j} = \left\{m_{i+1}^j\right\},$$

but we had already proven that $f(e_0) \neq f(e_1)$.

If $i > j$, we have

$$\{e_0, e_1\} = W_{m_j^j} = W_{f(m_j^j)} \supseteq f\left(W_{m_j^j}\right) = \left\{f\left(m_{j+1}^i\right)\right\},$$

and, then, another two possibilities:

1. $f(m_{j+1}^i) = e_0$, which leads to the following contradiction

$$1 \leq \left|f\left(W_{m_{j+1}^i}\right)\right| \leq \left|W_{f(m_{j+1}^i)}\right| = \left|W_{e_0}\right| = 0.$$

2. $f(m_{j+1}^i) = e_1$. In this case, by the construction of m_{j+1}^i it is clear that there exists $x \in W_{m_{j+1}^i}$ such that $|W_x| = 1$. Then

$$\{f(x)\} \subseteq f\left(W_{m_{j+1}^i}\right) \subseteq W_{f(m_{j+1}^i)} = W_{e_1} = \{e_0\},$$

from which $f(x) = e_0$. But, again by the choice of m_{j+1}^i it would hold

$$1 \leq |f(W_x)| \leq |W_{f(x)}| = |W_{e_0}| = 0,$$

which is also a contradiction.

Thus only the case $i = j$ remains.

Now, consider $W_e = \{n_0, n_1, \dots\}$. Since $W_{f(e)} \subseteq W_e$, we will have

$$f(W_e) = W_{f(e)} \cap Rg \ f \subseteq W_e \cap Rg \ f,$$

and, particularly,

$$f(W_e) \subseteq W_e.$$

That is, we have that for each i there exists j such that $f(n_i) = n_j$. Then, we have just proven that $f(i) = i$ holds for each i.   ⊣

REMARK 24. Notice that this last result gives us a different way of proving Theorem 9.

We have the following consequence.

PROPOSITION 25. *There exists e such that if f and g are isomorphic embeddings W satisfying*

$$Rg \ f \subseteq Rg \ g,$$
$$W_{f(e)} \subseteq W_{g(e)},$$

then we have f = g.

PROOF. Let f and g be functions as in the hypothesis of the proposition. We know that f and g are 1-1. Thus, we can consider the function $h = g^{-1} \cdot f$. Since $Rg \ f \subseteq Rg \ g$, we will have

$$x \in W_y \iff f(x) \in W_{f(y)}$$
$$\iff g\left(g^{-1}(f(x))\right) \in W_{g(g^{-1}(f(x)))}$$
$$\iff g^{-1}(f(x)) \in W_{g^{-1}(f(x))}$$
$$\iff h(x) \in W_{h(y)},$$

that is, h is also an isomorphic embedding of W.

Moreover, there exists a such that $f(e) = g(a)$ (since $Rg \ f \subseteq Rg \ g$) so

$$W_{g(a)} = W_{f(e)} \subseteq W_{g(e)},$$

and hence

$$g(W_a) = W_{g(a)} \cap Rg \ g \subseteq W_{g(e)} \cap Rg \ g = g(W_e).$$

Since g is 1-1, we will have $W_a \subseteq W_e$. But

$$a = g^{-1}(g(a)) = g^{-1}(f(e)) = h(e)$$

so $W_{h(e)} \subseteq W_e$, and applying Proposition 22 and the previous lemma, we conclude that $h = id$ and, consequently, $f = g$.   ⊣

To conclude this section, we will prove a weak version of Theorem 5 for isomorphic embeddings of W.

THEOREM 26. *Let f be an isomorphic embedding of W. Then $f \leq_T Rg \ f$.*

PROOF. We can use the argument in the proof of Theorem 5 if instead of the sets

$$\begin{array}{lll}
W_{f(a)} & W_{f(n_k)} & \\
W_{f(b_1^k)} & W_{f(b_2^k)} & \\
W_{f(m_0^k)} & \cdots & W_{f(m_k^k)} \\
W_{f(e_0)} & W_{f(e_1)} & W_{f(e_2)}
\end{array}$$

we consider their intersections with the range of f. Since we have

$$f(W_n) = W_{f(n)} \cap Rg \ f$$

and f is 1-1, we can conclude that, if we have an algorithm to enumerate the range of f, we can also enumerate the sets

$$\begin{array}{ccc}
f(W_a) & f(W_{n_k}) & \\
f(W_{b_1^k}) & f(W_{b_2^k}) & \\
f(W_{m_0^k}) & \cdots & f(W_{m_k^k}) \\
f(W_{e_0}) & f(W_{e_1}) & f(W_{e_2})
\end{array}$$

and hence compute f. $\qquad\qquad\qquad\qquad\qquad\qquad\qquad\qquad\qquad\dashv$

4.2. Isomorphic embeddings of P. As in the case of W we begin by showing that all functions acting like isomorphic embeddings of P are, in fact, 1-1.

PROPOSITION 27. *Let f be a function such that*

$$\varphi_e(x) = y \Longleftrightarrow \varphi_{f(e)}(f(x)) = f(y),$$

for all $e, x, y \in \omega$. Then, f is 1-1.

PROOF. Suppose there exist $x_1, x_2 \in \omega$ such that $f(x_1) = f(x_2)$. Consider the function

$$\varphi_e(z) \begin{cases} 0 & \text{if } z = x_1, \\ \uparrow & \text{otherwise.} \end{cases}$$

Then

$$\begin{aligned}
\varphi_e(x_1) = 0 &\Longrightarrow \varphi_{f(e)}(f(x_1)) = f(0) \\
&\Longrightarrow \varphi_{f(e)}(f(x_2)) = f(0) \\
&\Longrightarrow \varphi_e(x_2) = 0 \\
&\Longrightarrow x_2 = x_1
\end{aligned}$$

and f must be 1-1. $\qquad\qquad\qquad\qquad\qquad\qquad\qquad\qquad\qquad\qquad\dashv$

Next, we prove that not all isomorphic embeddings of P do commute with P.

PROPOSITION 28. *There exist an isomorphic embedding f of P and an index n such that*

$$f \cdot \varphi_n \neq \varphi_{f(n)} \cdot f.$$

PROOF. In the proof of Proposition 15 rename step t as step $t + 2$. Let n_0, n_1, n_2 be three different indices of the empty set each one also different from a. Add the following preliminary steps:

Step 0: Set $\varphi_{h(z)}(n_1) = n_2$.

Step 1: Consider the function

$$g(x) = \begin{cases} a & \text{if } x = n_2 \\ \uparrow & \text{otherwise.} \end{cases}$$

Let c be an index of the function g (which is clearly computable) different from a. Set $\varphi_{h(z)}(n_0) = c$. $\qquad\qquad\qquad\qquad\qquad\qquad\qquad\dashv$

REMARK 29. In fact, we could also prove that there exist isomorphic embeddings of P not commuting with W.

However, it is easy to prove that all isomorphic embeddings of P are computable.

LEMMA 30. *Let n be an index of $h(x) = x + 1$. Let f be an isomorphic embedding of P. Then $f \cdot \varphi_n = \varphi_{f(n)} \cdot f$.*

PROOF. It is clear that $W_n = \omega$. Hence, for all x there exists y such that $\varphi_n(x) = y$ and, since f is an isomorphic embedding of P, we have

$$\varphi_{f(n)}(f(x)) = f(y) = f(\varphi_n(x))$$

and then

$$\varphi_{f(n)} \cdot f = f \cdot \varphi_n. \hspace{2cm} \dashv$$

Thus, Lemma 16 leads us to the result we were seeking.

THEOREM 31. *All the isomorphic embeddings of P are computable.*

§5. **Homomorphisms of W and P.** Finally, we will study the next natural weakening of the concept of automorphism: homomorphisms.

DEFINITION 32. A function f is a homomorphism of W if

$$x \in W_y \implies f(x) \in W_{f(y)},$$

for all x and y in ω.

A function g is a homomorphism of P if

$$\varphi_e(x) = y \implies \varphi_{g(e)}(g(x)) = g(y),$$

for all e, x and y in ω.

We will study the homomorphisms of both predicates, first those of W and then those of P, showing that their algorithmic properties are, in this case, completely different.

5.1. Homomorphisms of W. To begin with, we will prove a result about the cardinality of the range of the homomorphisms of W.

PROPOSITION 33. *For all $n \in \omega^+ = \omega \cup \{\omega\}$ greater than 0 there exists a homomorphism of W whose range has cardinality equal to n.*

PROOF. Take $A \subseteq \{e \in \omega : W_e = \omega\}$ with n elements and a function f such that $Rg\, f = A$. Then $x \in W_{f(y)} = \omega$ for all $x, y \in \omega$ and hence we have

$$x \in W_y \implies f(x) \in W_{f(y)},$$

for all $x, y \in \omega$.

Clearly, f is the desired function. $\hspace{2cm} \dashv$

The idea in the proof of the proposition above is fully exploited in the next theorem, which is the main result of this paragraph.

THEOREM 34. *For every subset A of ω there exists a 1-1 homomorphism h of W such that $h \equiv_T A$.*

PROOF. Let A be a subset of ω. Take $B, C \subseteq H_\omega$ such that

- B and C are both recursive,
- $|A| = |B|$,
- $|\overline{A}| = |C|$,
- $B \cap C = \emptyset$,

where $\overline{A}$ is the complement of A in ω. Obviously, such sets exit for whatever A.

Now suppose that

$$A = \{a_0 < a_1 < \ldots\},$$

$$\overline{A} = \{\overline{a}_0 < \overline{a}_1 < \ldots\},$$

$$B = \{b_0 < b_1 < \ldots\},$$

$$C = \{c_0 < c_1 < \ldots\},$$

where the sequences are finite if the corresponding sets are, and infinite otherwise.

Consider the functions

$$f : \quad A \quad \rightarrow \quad B$$
$$a_i \quad \rightsquigarrow \quad b_i,$$

$$g : \quad \overline{A} \quad \rightarrow \quad C$$
$$\overline{a}_i \quad \rightsquigarrow \quad c_i.$$

Clearly, we have $f, g \leq_T A$, since B and C are recursive.

Define

$$h(x) = \begin{cases} f(x) & \text{if } x \in A, \\ g(x) & \text{otherwise,} \end{cases}$$

which is a 1-1 homomorphism of W since for all x, y we have $x \in W_{h(y)} = \omega$ and, consequently, for all x, y we have

$$x \in W_y \implies h(x) \in W_{h(y)}.$$

From the definition of h it easily follows that $h \leq_T A$ and, since the sets B and C are recursive, we also have $A \leq_T h$. ⊣

5.2. Homomorphisms of P. As we did in the previous paragraph we begin studying the cardinality of the ranges of the homomorphisms of P.

PROPOSITION 35. *There exist constant homomorphisms of P.*

PROOF. Consider the total and computable function defined by

$$\varphi_{h(n)}(x) = n,$$

for all $x, n \in \omega$. That is, $h(n)$ is an index of the constant function which always returns n.

Applying the Recursion Theorem (consult [4]), we obtain n_0 such that $\varphi_{h(n_0)} = \varphi_{n_0}$, and then $\varphi_{n_0}(x) = n_0$ for all $x \in \omega$.

Consider the constant function $f = \varphi_{n_0}$. For all e, x, y we have

$$\varphi_{f(e)}(f(x)) = \varphi_{n_0}(n_0) = f(n_0) = n_0 = f(y),$$

so

$$\varphi_e(x) = y \Longrightarrow \varphi_{f(e)}(f(x)) = f(y),$$

and f is a homomorphism of P.   ⊣

However, this is the only case in which a homomorphism of P has finite range, as the following result shows (cf. Proposition 33)

PROPOSITION 36. *Let f be a homomorphism of P. Then*

$$|Rg\ f| \in \{1, \omega\}.$$

In fact, if f is not constant, then

$$\left|\{I_m : m \in Rg\ f\}\right| = \omega,$$

where I_m is the set of indices of φ_m.

PROOF. Let f be a nonconstant homomorphism of f and suppose that

$$\left|\{I_m : m \in Rg\ f\}\right| = n,$$

with n finite. We consider two possibilities

$n = 1$: In this case $\varphi_{f(x)} = \varphi_{f(y)}$ for all $x, y \in \omega$. Let a_1, a_2 be two different elements in the range of f and let x_1, x_2 be natural number such that

$$f(x_1) = a_1,$$
$$f(x_2) = a_2.$$

Let m_1, m_2 be such that

$$\varphi_{m_1}(x_1) = x_1,$$
$$\varphi_{m_2}(x_1) = x_2.$$

Then, since f is a homomorphism of P

$$a_1 = f(x_1) = f\big(\varphi_{m_1}(x_1)\big)$$
$$= \varphi_{f(m_1)}(f(x_1)) = \varphi_{f(m_2)}(f(x_1))$$
$$= f\big(\varphi_{m_2}(x_1)\big) = f(x_2)$$
$$= a_2,$$

which is a contradiction, because a_1 and a_2 were chosen to be different.

$n > 1$: Suppose that

$$\{I_m : m \in Rg\ f\} = \{I_{a_1}, \ldots, I_{a_n}\}.$$

Let $x_1, \ldots, x_n$ be elements such that $\varphi_{f(x_i)} = \varphi_{a_i}$ for all $i = 1, \ldots, n$. Define $X = \{x_1, \ldots, x_n\}$. Consider $m_1, \ldots, m_{n^n}$ such that

$$\{\varphi_{m_i|X} : i = 1, \ldots, n^n\} = X^X,$$

where X^X is the set of all total function from X into itself, and $\varphi_{m_i|X}$ is the restriction of φ_{m_i} to X.

Since $n > 1$ we have $|X^X| = n^n > n$, and then there exist $i \neq j$ such that

$$\varphi_{f(m_i)} = \varphi_{f(m_j)}.$$

Since $m_i \neq m_j$ there exist x_k such that

$$x_{l_1} = \varphi_{m_i}(x_k) \neq \varphi_{m_j}(x_k) = x_{l_2},$$

but f is a homomorphism of P and consequently

$$\begin{aligned}
f(x_{l_1}) &= f\left(\varphi_{m_i}(x_k)\right) \\
&= \varphi_{f(m_i)}(f(x_k)) \\
&= \varphi_{f(m_j)}(f(x_k)) \\
&= f\left(\varphi_{m_j}(x_k)\right) \\
&= f\left(x_{l_2}\right).
\end{aligned}$$

Then

$$\varphi_{a_{l_1}} = \varphi_{f(x_{l_1})} = \varphi_{f(x_{l_2})} = \varphi_{a_{l_2}},$$

from which $l_1 = l_2$, which is a contradiction.   ⊣

Finally, we prove that all the homomorphisms of P are computable.

LEMMA 37. *Let n be an index of the function $h(x) = x + 1$. Let f be a homomorphism of P. Then $f \cdot \varphi_n = \varphi_{f(n)} \cdot f$.*

PROOF. The proof of Lemma 30 works here with no change at all.   ⊣

THEOREM 38. *All the homomorphisms of P are computable.*

PROOF. Just apply the previous lemma and Lemma 16.   ⊣

REFERENCES

[1] C. C. CHANG and H. J. KEISLER, *Model Theory*, Elsevier, 1998.
[2] E. F. COMBARRO, *Automorphism groups of computably enumerable predicates*, **Algebra and Logic**, vol. 42 (2002), no. 5, pp. 285–294.
[3] P. ODIFREDDI, *Classical Recursion Theory*, vol. 1, North-Holland, 1989.
[4] R. ROGERS, *Theory of Recursive Functions and Effective Computability*, McGraw-Hill, 1967.

DPTO. INFORMÁTICA, UNIVERSIDAD DE OVIEDO
  EDIFICIOS DEPARTAMENTALES
    DESPACHO 1.1.36
      GIJÓN 33271, SPAIN
E-mail: elias@aic.uniovi.es

PCF THEORY AND WOODIN CARDINALS

MOTI GITIK[†], RALF SCHINDLER, AND SAHARON SHELAH[†]

Abstract. THEOREM 1.1. Let α be a limit ordinal. Suppose that $2^{|\alpha|} < \aleph_\alpha$ and $2^{|\alpha|^+} < \aleph_{|\alpha|^+}$, whereas $\aleph_\alpha^{|\alpha|} > \aleph_{|\alpha|^+}$. Then for all $n < \omega$ and for all bounded $X \subset \aleph_{|\alpha|^+}$, $M_n^\#(X)$ exists.

THEOREM 1.4. Let κ be a singular cardinal of uncountable cofinality. If $\{\alpha < \kappa \mid 2^\alpha = \alpha^+\}$ is stationary as well as co-stationary then for all $n < \omega$ and for all bounded $X \subset \kappa$, $M_n^\#(X)$ exists.

Theorem 1.1 answers a question of Gitik and Mitchell (cf. [11, Question 5, p. 315]), and Theorem 1.4 yields a lower bound for an assertion discussed in [10] (cf. [10, Problem 4]).

The proofs of these theorems combine pcf theory with core model theory. Along the way we establish some ZFC results in cardinal arithmetic, motivated by Silver's theorem [29], and we obtain results of core model theory, motivated by the task of building a "stable core model." Both sets of results are of independent interest.

§1. Introduction and statements of results.

In this paper we prove results which were announced in the first two authors' talks at the Logic Colloquium 2002 in Münster. Specifically, we shall obtain lower bounds for the consistency strength of statements of cardinal arithmetic.

Cardinal arithmetic deals with possible behaviours of the function $(\kappa, \lambda) \mapsto \kappa^\lambda$ for infinite cardinals κ, λ. Easton, inventing a class version of Cohen's set forcing (cf. [4]) had shown that if $V \models$ GCH and $\Phi : \text{Reg} \to \text{Card}$ is monotone and such that $\text{cf}(\Phi(\kappa)) > \kappa$ for all $\kappa \in \text{Reg}$ then there is a forcing extension of V in which $\Phi(\kappa) = 2^\kappa$ for all $\kappa \in \text{Reg}$. (Here, Card denotes the class of all infinite cardinals, and Reg denotes the class of all infinite regular cardinals.) However, in any of Easton's models, the so-called Singular Cardinal Hypothesis (abbreviated by SCH) holds true (cf. [14, Exercise 20.7]), i.e., $\kappa^{\text{cf}(\kappa)} = \kappa^+ \cdot 2^{\text{cf}(\kappa)}$ for all infinite cardinals κ. If SCH holds then cardinal arithmetic is in some sense simple, cf. [14, Lemma 8.1].

On the other hand, the study of situations in which SCH fails turned out to be an exciting subject. Work of Silver and Prikry showed that SCH may

2000 *Mathematics Subject Classification.* Primary 03E04, 03E45. Secondary 03E35, 03E55.

Key words and phrases. cardinal arithmetic/pcf theory/core models/large cardinals.

[†]The first and the third author's research was supported by The Israel Science Foundation. This is publication #805 in the third author's list of publications.

indeed fail (cf. [21] and [29]), and Magidor showed that GCH below $\aleph_\omega$ does not imply $2^{\aleph_\omega} = \aleph_{\omega+1}$ (cf. [16] and [17]). Both results had to assume the consistency of a supercompact cardinal. Jensen showed that large cardinals are indeed necessary: if SCH fails then $0^\#$ exists (cf. [3]). We refer the reader to [15] for an excellently written account of the history of the investigation of ¬SCH.

The study of ¬SCH in fact inspired pcf theory as well as core model theory. We know today that ¬SCH is equiconsistent with the existence of a cardinal κ with $o(\kappa) = \kappa^{++}$ (cf. [6] and [7]). By now we actually have a fairly complete picture of the possible behaviours of $(\kappa, \lambda) \mapsto \kappa^\lambda$ under the assumption that $0^\P$ does not exist (cf. for instance [8] and [10]).

In contrast, very little is known if we allow $0^\P$ (or more) to exist. (The existence of $0^\P$ is equivalent with the existence of indiscernibles for an inner model with a strong cardinal.) This paper shall be concerned with strong violations of SCH, where we take "strong" to mean that they imply the existence of $0^\P$ and much more.

It is consistent with the non-existence of $0^\P$ that $\aleph_\omega$ is a strong limit cardinal (in fact that GCH holds below $\aleph_\omega$) whereas $2^{\aleph_\omega} = \aleph_\alpha$, where α is a countable ordinal at least as big as some arbitrary countable ordinal fixed in advance (cf. [16]). As of today, it is not known, though, if $\aleph_\omega$ can be a strong limit and $2^{\aleph_\omega} > \aleph_{\omega_1}$. The only limitation known to exists is the third author's thorem according to which $\aleph_\omega^{\aleph_0} \leq 2^{\aleph_0} + \aleph_{\omega_4}$ (cf. [27]).

Mitchell and the first author have shown that if $2^{\aleph_0} < \aleph_\omega$ and $\aleph_\omega^{\aleph_0} > \aleph_{\omega_1}$ then $0^\P$ exists (cf. [11, Theorem 5.1]). Our first main theorem strengthens this result. The objects $M_n^\#(X)$, where $n < \omega$, are defined in [30, p. 81] or [5, p. 1841].

THEOREM 1.1. *Let α be a limit ordinal. Suppose that $2^{|\alpha|} < \aleph_\alpha$ and $2^{|\alpha|^+} < \aleph_{|\alpha|^+}$, whereas $\aleph_\alpha^{|\alpha|} > \aleph_{|\alpha|^+}$. Then for all $n < \omega$ and for all bounded $X \subset \aleph_{|\alpha|^+}$, $M_n^\#(X)$ exists.*

Theorem 1.1 gives an affirmative answer to [11, Question 5, p. 315]. One of the key ingredients of its proof is a new technique for building a "stable core model" of height κ, where κ will be the $\aleph_{|\alpha|^+}$ of the statement of Theorem 1.1 and will therefore be a cardinal which is *not* countably closed (cf. Theorems 3.7, 3.9, and 3.11 below).

Let κ be a singular cardinal of uncountable cofinality. Silver's celebrated theorem [29] says that if $2^\kappa > \kappa^+$ then the set $\{\alpha < \kappa \mid 2^\alpha > \alpha^+\}$ contains a club. But what if $2^\kappa = \kappa^+$, should then either $\{\alpha < \kappa \mid 2^\alpha > \alpha^+\}$ or else $\{\alpha < \kappa \mid 2^\alpha = \alpha^+\}$ contain a club? We formulate a natural (from both forcing and pcf points of view) principle which implies an affirmative answer. We let $(*)_\kappa$ denote the assertion that there is a strictly increasing and continuous sequence $(\kappa_i \mid i < \mathrm{cf}(\kappa))$ of singular cardinals which is cofinal in κ and such that for every limit ordinal $i < \mathrm{cf}(\kappa)$, $\kappa_i^+ = \max(\mathrm{pcf}(\{\kappa_j^+ \mid j < i\}))$. Note that if $cf(i) > \omega$ then this is always the case on a club by [27, Claim 2.1, p. 55]. We show:

THEOREM 1.2. *Let κ be a singular cardinal of uncountable cofinality. If $(*)_\kappa$ holds then either $\{\alpha < \kappa \mid 2^\alpha = \alpha^+\}$ contains a club, or else $\{\alpha < \kappa \mid 2^\alpha > \alpha^+\}$ contains a club.*

The next theorem shows that $\neg(*)_\kappa$, for κ a singular cardinal of uncountable cofinality, is pretty strong.

THEOREM 1.3. *Let κ be a singular cardinal of uncountable cofinality. If $(*)_\kappa$ fails then for all $n < \omega$ and for all bounded $X \subset \kappa$, $M_n^\#(X)$ exists.*

The second main theorem is an immediate consequence of Theorems 1.2 and 1.3:

THEOREM 1.4. *Let κ be a singular cardinal of uncountable cofinality. If $\{\alpha < \kappa \mid 2^\alpha = \alpha^+\}$ is stationary as well as co-stationary then for all $n < \omega$ and for all bounded $X \subset \kappa$, $M_n^\#(X)$ exists.*

After this paper had been written, the first author showed that the hypothesis of Theorem 1.4 is in fact consistent relative to the consistency of a supercompact cardinal (cf. [9]).

The proofs of Theorems 1.1 and 1.3 will use the first ω many steps of Woodin's core model induction. The reader may find a published version of this part of Woodin's induction in [5]. By work of Martin, Steel, and Woodin, the conclusions of Theorems 1.1 and 1.4 both imply that PD (Projective Determinacy) holds. The respective hypotheses of Theorems 1.1 and 1.4 are thereby the first statements in cardinal arithmetic which provably yield PD and which are (in the case of Theorem 1.1) not known to be inconsistent or even (in the case of Theorem 1.4) known to be consistent.

It is straightforward to verify that both hypotheses of Theorems 1.1 and 1.4 imply that SCH fails. The hypothesis of Theorem 1.1 implies that, setting $a = \{\kappa < \aleph_\alpha \mid |\alpha|^+ \leq \kappa \wedge \kappa \in \text{Reg}\}$, we have $\text{Card}(\text{pcf}(a)) > \text{Card}(a)$. The question if some such a can exist is one of the key open problems in pcf theory. At this point neither of the hypotheses of our main theorems is known to be consistent. We expect future research to uncover the status of the hypotheses of our main theorems.

Theorems 2.1, 1.2, and 2.5 were originally proven by the first author; subsequently, the third author found much simpler proofs for them. Theorems 2.4 and 2.7 are due to the third author, and theorems 2.6 and 2.8 are due to the first author. The results contained in the section on core model theory are due to the second author.

We wish to thank the members of the logic groups of Bonn and Münster, in particular Professors P. Koepke and W. Pohlers, for their warm hospitality during the Münster meeting.

The first author thanks Andreas Liu for his comments on the section on pcf theory of an earlier version of this paper.

The second author thanks John Steel for fixing a gap in an earlier version of the proof of Lemma 3.5 and for a discussion that led to a proof of Lemma 3.10. He also thanks R. Jensen, B. Mitchell, E. Schimmerling, J. Steel, and M. Zeman for the many pivotal discussions held at Luminy in Sept. 02.

§2. Some pcf theory.

We refer the reader to [27], [1], [2], and to [13] for introductions to the third author's pcf theory.

Let κ be a singular strong limit cardinal of uncountable cofinality. Set $S_1 = \{\alpha < \kappa \mid 2^\alpha = \alpha^+\}$ and $S_2 = \{\alpha < \kappa \mid 2^\alpha > \alpha^+\}$. Silver's famous theorem states that if $2^\kappa > \kappa^+$ then S_2 contains a club (cf. for instance [13, Corollary 2.3.12]). But what if $2^\kappa = \kappa^+$? We would like to show that unless certain large cardinals are consistent either S_1 or S_2 contains a club.

The third author showed that it is possible to replace the power set operation by pp in Silver's theorem (cf. for instance [13, Theorem 9.1.6]), providing nontrivial information in the case where κ is not a strong limit cardinal, for example if $\kappa < 2^{\aleph_0}$. Thus, if κ is a singular cardinal of uncountable cofinality, and if $S_1 = \{\alpha < \kappa \mid \mathrm{pp}(\alpha) = \alpha^+\}$, $S_2 = \{\alpha < \kappa \mid \mathrm{pp}(\alpha) > \alpha^+\}$, and $\mathrm{pp}(\kappa) > \kappa^+$ then S_2 contains a club.

The following result, or rather its corollary, will be needed for the proof of Theorem 1.4. The statement $(*)_\kappa$ was already introduced in the introduction.

THEOREM 2.1. *Let κ be a singular cardinal of uncountable cofinality. Suppose that*

$(*)_\kappa$ *there is a strictly increasing and continuous sequence $(\kappa_i \mid i < \mathrm{cf}(\kappa))$ of singular cardinals which is cofinal in κ and such that for every limit ordinal $i < \mathrm{cf}(\kappa)$, $\kappa_i^+ = \max(\mathrm{pcf}(\{\kappa_j^+ \mid j < i\}))$.*

Then either $\{\alpha < \kappa \mid \mathrm{pp}(\alpha) = \alpha^+\}$ contains a club, or else $\{\alpha < \kappa \mid \mathrm{pp}(\alpha) > \alpha^+\}$ contains a club.

PROOF. Let $(\kappa_i \mid i < \mathrm{cf}(\kappa))$ be a sequence witnessing $(*)_\kappa$. Assume that both S_1 and S_2 are stationary, where $S_1 \subset \{\alpha < \kappa \mid \mathrm{pp}(\alpha) = \alpha^+\}$ and $S_2 \subset \{\alpha < \kappa \mid \mathrm{pp}(\alpha) > \alpha^+\}$. We may and shall assume that $S_1 \cup S_2 \subset \{\kappa_i \mid i < \mathrm{cf}(\kappa)\}$ and $\kappa_0 > \mathrm{cf}(\kappa)$.

Let $\chi > \kappa$ be a regular cardinal, and let $M \prec H_\chi$ be such that $\mathrm{Card}(M) = \mathrm{cf}(\kappa)$, $M \supset \mathrm{cf}(\kappa)$, and $(\kappa_i \mid i < \mathrm{cf}(\kappa))$, S_1, $S_2 \in M$. Set $\mathsf{a} = (M \cap \mathrm{Reg}) \setminus (\mathrm{cf}(\kappa) + 1)$.

We may pick a smooth sequence $(b_\theta \mid \theta \in \mathsf{a})$ of generators for a (cf. [28, Claim 6.7], [1, Theorem 6.3]). I.e., if $\theta \in \mathsf{a}$ and $\bar{\theta} \in b_\theta$ then $b_{\bar{\theta}} \subset b_\theta$ (smooth), and if $\theta \in \mathsf{a}$ then $\mathcal{J}_{\leq\theta}(\mathsf{a}) = \mathcal{J}_{<\theta}(\mathsf{a}) + b_\theta$ (generating).

Let $\kappa_j \in S_1$. As $\mathrm{pp}(\kappa_j) = \kappa_j^+$, we have that $\mathsf{a} \cap \kappa_j \in \mathcal{J}_{<\kappa_j^+}(\mathsf{a})$, since by [13, Lemma 9.1.5], $\mathrm{pp}(\kappa_j) = \mathrm{pp}_\delta(\kappa_j)$ for all δ with $\mathrm{cf}(\kappa_j) \leq \delta < \kappa_j$. Thus

$(a \cap \kappa_j) \setminus b_{\kappa_j^+} \in \mathcal{J}_{<\kappa_j^+}(a)$, as $b_{\kappa_j^+}$ generates $\mathcal{J}_{\leq\kappa_j^+}(a)$ over $\mathcal{J}_{<\kappa_j^+}(a)$. Hence $(a \cap \kappa_j) \setminus b_{\kappa_j^+}$ must be bounded below κ_j, as κ_j is singular and an unbounded subset of $(a \cap \kappa_j)$ can thus not force $\prod(a \cap \kappa_j)$ to have cofinality $\leq \kappa_j$. We may therefore pick some $v_j < \kappa_j$ such that $b_{\kappa_j^+} \supset a \cap [v_j, \kappa_j)$. By Fodor's Lemma, there is now some $v^* < \kappa$ and some stationary $S_1^* \subset S_1$ such that for each $\kappa_j \in S_1^*$, $b_{\kappa_j^+} \supset a \cap [v^*, \kappa_j)$.

Let us fix $\kappa_i \in S_2$, a limit of elements of S_1^*. By $(*)_\kappa$, $\max(\text{pcf}(\{\kappa_j^+ \mid j < i\})) = \kappa_i^+$, i.e., $\{\kappa_j^+ \mid j < i\} \in \mathcal{J}_{\leq\kappa_i^+}(a)$. Therefore, by arguing as in the preceeding paragraph, there is some $i^* < i$ such that $\kappa_j^+ \in b_{\kappa_i^+}$ whenever $i^* < j < i$. If $\kappa_j \in S_1^*$, where $i^* < j < i$, then by the smoothness of $(b_\theta \mid \theta \in a)$, $b_{\kappa_j^+} \subset b_{\kappa_i^+}$, and so $b_{\kappa_i^+} \supset a \cap [v^*, \kappa_j)$. As the set of j with $i^* < j < i$ and $\kappa_j \in S_1^*$ is unbounded in i, we therefore get that $b_{\kappa_i^+} \supset a \cap [v^*, \kappa_i)$. This means that $a \cap [v^*, \kappa_i) \in \mathcal{J}_{\leq\kappa_i^+}(a)$, which clearly implies that $\text{pp}(\kappa_i) = \kappa_i^+$ by the choice of a. However, $\text{pp}(\kappa_i) > \kappa_i^+$, since $\kappa_i \in S_2$. Contradiction!   ⊣

PROOF OF THEOREM 1.2. If κ is not a strong limit then, obviously, $\{\alpha < \kappa \mid 2^\alpha > \alpha^+\}$ contains a club. So assume that κ is a strong limit. Then the set $C = \{\alpha < \kappa \mid \alpha$ is a strong limit$\}$ is closed unbounded. If $\alpha \in C$ has uncountable cofinality, then $\text{pp}(\alpha) = 2^\alpha$, by [13, Theorem 9.1.3]. For countable cofinality this equality is an open problem. But by [27, Chapter IX, Conclusion 5.9], for $\alpha \in C$ of countable cofinality, $\text{pp}(\alpha) < 2^\alpha$ implies that the set $\{\mu \mid \alpha < \mu = \aleph_\mu < \text{pp}(\alpha)\}$ is uncountable. Certainly, in this case $\text{pp}(\alpha) > \alpha^+$. Hence, for every $\alpha \in C$, $\text{pp}(\alpha) = \alpha^+$ if and only if $2^\alpha = \alpha^+$. So Theorem 2.1 applies and gives the desired conclusion.   ⊣

Before proving a generalization of Theorem 2.1 let us formulate a simple "combinatorial" fact, Lemma 2.2, which shall be used in the proofs of Theorems 1.1 and 1.4. We shall also state a consequence of Lemma 2.2, namely Lemma 2.3, which we shall need in the proof of Theorem 1.4.

Let $\lambda \leq \theta$ be infinite cardinals. Then H_θ is the set of all sets which are hereditarily smaller than θ, and $[H_\theta]^\lambda$ is the set of all subsets of H_θ of size λ. If H is any set of size at least λ then a set $S \subset [H]^\lambda$ is stationary in $[H]^\lambda$ if for every model $\mathfrak{M} = (H; \dots)$ with universe H and whose type has cardinality at most λ there is some $(X; \dots) \prec \mathfrak{M}$ with $X \in S$. Let $\vec{\kappa} = (\kappa_i \mid i \in A) \subset H_\theta$ with $\lambda \leq \kappa_i$ for all $i \in A$, and let $X \in [H_\theta]^\lambda$. Then we write $\text{char}_{\vec{\kappa}}^X$ for the function $f \in \prod_{i \in A} \kappa_i^+$ which is defined by $f(\kappa_i^+) = \sup(X \cap \kappa_i^+)$. If f, $g \in \prod_{i \in A} \kappa_i^+$ then we write $f < g$ just in case that $f(\kappa_i^+) < g(\kappa_i^+)$ for all $i \in A$. Recall that by [2, Corollary 7.10] if $|A|^+ \leq \kappa_i$ for all $i \in A$ then

$$\max(\text{pcf}(\{\kappa_i^+ \mid i \in A\})) = \text{cf}\left(\prod(\{\kappa_i^+ \mid i \in A\}), <\right).$$

If δ is a regular uncountable cardinal then NS_δ is the non-stationary ideal on δ.

LEMMA 2.2. *Let κ be a singular cardinal with $\mathrm{cf}(\kappa) = \delta \geq \aleph_1$. Let $\vec{\kappa} = (\kappa_i \mid i < \delta)$ be a strictly increasing and continuous sequence of cardinals which is cofinal in κ and such that $2^\delta \leq \kappa_0 < \kappa$. Let $\delta \leq \lambda < \kappa$ Let $\theta > \kappa$ be regular, and let $\Phi : [H_\theta]^\lambda \to NS_\delta$. Let $S \subset [H_\theta]^\lambda$ be stationary in $[H_\theta]^\lambda$.*

There is then some club $C \subset \delta$ such that for all $f \in \prod_{i \in C} \kappa_i^+$ there is some $Y \prec H_\theta$ such that $Y \in S$, $C \cap \Phi(Y) = \emptyset$, and $f < \mathrm{char}_{\vec{\kappa}}^Y$.

PROOF. Suppose not. Then for every club $C \subset \delta$ we may pick some $f_C \in \prod_{i \in C} \kappa_i^+$ such that if $Y \prec H_\theta$ is such that $Y \in S$ and $f_C < \mathrm{char}_{\vec{\kappa}}^Y$, then $C \cap \Phi(Y) \neq \emptyset$. Define $\tilde{f} \in \prod_{i \in C} \kappa_i^+$ by $\tilde{f}(\kappa_i^+) = \sup_C f_C(\kappa_i^+)$, and pick some $Y \prec H_\theta$ which is such that $Y \in S$ and $\tilde{f}(\kappa_i^+) < \mathrm{char}_{\vec{\kappa}}^Y$. We must then have that $C \cap \Phi(Y) \neq \emptyset$ for every club $C \subset \delta$, which means that $\Phi(Y)$ is stationary. Contradiction! ⊣

The function Φ to which we shall apply Lemma 2.2 will be chosen by inner model theory. Lemma 2.2 readily implies the following.

LEMMA 2.3. *Let κ be a singular cardinal with $\mathrm{cf}(\kappa) = \delta \geq \aleph_1$. Let $\vec{\kappa} = (\kappa_i \mid i < \delta)$ be a strictly increasing and continuous sequence of cardinals which is cofinal in κ and such that $2^\delta \leq \kappa_0 < \kappa$. Suppose that $(*)_\kappa$ fails. Let $\theta > \kappa$ be regular, and let $\Phi : [H_\theta]^{2^\delta} \to NS_\delta$.*

There is then a club $C \subset \delta$ and a limit point ξ of C with $\mathrm{cf}(\prod(\{\kappa_i^+ \mid i < \xi\}), <) > \kappa_\xi^+$ and such that for all $f \in \prod_{i \in C} \kappa_i^+$ there is some $Y \prec H_\theta$ such that $\mathrm{Card}(Y) = 2^\delta$, $^\omega Y \subset Y$, $C \cap \Phi(Y) = \emptyset$, and $f < \mathrm{char}_{\vec{\kappa}}^Y$.

Let us now turn towards our generalization of Theorem 2.1. This will not be needed for the proofs of our main theorems.

THEOREM 2.4. *Suppose that the following hold true.*

(a) *κ is a singular cardinal of uncountable cofinality δ, and $(\kappa_i \mid i < \delta)$ is an increasing continuous sequence of singular cardinals cofinal in κ with $\kappa_0 > \delta$,*

(b) *$S \subset \delta$ is stationary, and $(\gamma_i^* \mid i \in S)$ and $(\gamma_i^{**} \mid i \in S)$ are two sequences of ordinals such that $1 \leq \gamma_i^* \leq \gamma_i^{**} < \delta$ and $\kappa_i^{+\gamma_i^{**}} < \kappa_{i+1}$ for $i < \delta$,*

(c) *for any $\xi \in S$ which is a limit point of S, for any $A \subset S \cap \xi$ with $\sup(A) = \xi$ and for any sequence $(\beta_i \mid i \in A)$ with $\beta_i < \gamma_i^*$ for all $i \in A$ we have that*

$$\mathrm{pcf}(\{\kappa_i^{+\beta_i+1} \mid i \in A\}) \cap (\kappa_\xi, \kappa_\xi^{+\gamma_\xi^*}] \neq \emptyset,$$

(d) *$\mathrm{pcf}(\{\kappa_i^{+\beta+1} \mid \beta < \gamma_i^*\}) = \{\kappa_i^{+\beta+1} \mid \beta < \gamma_i^*\}$ for every $i \in S$,*

(e) *$\mathrm{pp}(\kappa_i) = \kappa_i^{+\gamma_i^{**}}$ for every $i \in S$, and*

(f) *S^* is the set of all $\xi \in S$ such that either*

    (α) *$\xi > \sup(S \cap \xi)$, or*

    (β) *$\mathrm{cf}(\xi) > \aleph_0$ and $\{i \in S \cap \xi \mid \gamma_i^* = \gamma_i^{**}\}$ is a stationary subset of ξ, or*

    (γ) *$\mathrm{pcf}(\{\kappa_j^{+\beta+1} \mid \beta < \gamma_j^{**}, j < \xi\}) \supset \{\kappa_\xi^{+\beta+1} \mid \gamma_\xi^* \leq \beta < \gamma_\xi^{**}\}$.*

Then there is a club $C \subset \delta$ such that one of the two sets $S_1 = \{i \in S^* \mid \gamma_i^* = \gamma_i^{**}\}$ and $S_2 = \{i \in S^* \mid \gamma_i^* < \gamma_i^{**}\}$ contains $C \cap S^*$.

It is easy to see that Theorem 2.1 (with some limitations on the size of $pp(\kappa_i)$ as in (b) and (e) above) can be deduced from Theorem 2.4 by taking $S = \{i < cf(\kappa) \mid pp(\kappa_i) > \kappa_i^+\}$, $\gamma_i^* = 1$, and $\gamma_i^{**} = 2$. Condition (c) in the statement of Theorem 2.4 plays the role of the assumption $(*)_\kappa$ in the statement of Theorem 2.1.

PROOF OF THEOREM 2.4. Let us suppose that the conclusion of the statement of Theorem 2.4 fails.

Let, for $i \in S$, $a_i = \{\kappa_i^{+\beta+1} \mid \beta < \gamma_i^{**}\}$, and set $a = \bigcup\{a_i \mid i \in S\}$. We may fix a smooth and closed sequence $(b_\theta \mid \theta \in a)$ of generators for a (cf. [28, Claim 6.7]). I.e., $(b_\theta \mid \theta \in a)$ is smooth and generating, and if $\theta \in a$ then $b_\theta = a \cap pcf(b_\theta)$ (closed).

For each $\xi \in S^* = S_1 \cup S_2$, by [27, I Fact 3.2] and hypothesis (d) in the statement of Theorem 2.4 we may pick a finite $d_\xi \subset \{\kappa_\xi^{+\beta+1} \mid \beta < \gamma_\xi^*\}$ such that $\bigcup\{b_\theta \mid \theta \in d_\xi\} \supset \{\kappa_\xi^{+\beta+1} \mid \beta < \gamma_\xi^*\}$.

If $\xi \in S_1$ then $\gamma_\xi^* = \gamma_\xi^{**}$ and so $pp(\kappa_\xi) = \kappa_\xi^{+\gamma_\xi^*}$ by (e) in the statement of Theorem 2.4. By [13, Corollary 5.3.4] we may and shall assume that $\bigcup\{b_\theta \mid \theta \in d_\xi\}$ contains a final segment of $\bigcup\{a_j \mid j < \xi\}$, and we may therefore choose some $\varepsilon(\xi) < \xi$ such that

$$\bigcup\{b_\theta \mid \theta \in d_\xi\} \supset \bigcup\{a_j \mid \varepsilon(\xi) \leq j < \xi\}.$$

There is then some ε^* and some stationary $S_1^* \subset S_1$ such that $\varepsilon(\xi) = \varepsilon^*$ for every $\xi \in S_1^*$. Let C be the club set $\{\xi < \delta \mid \xi = \sup(\xi \cap S_1^*)\}$.

If $\xi \in S$ then condition (c) in the statement of Theorem 2.4 implies that we may assume that $\bigcup\{b_\theta \mid \theta \in d_\xi\}$ contains a final segment of the set $\{\kappa_i^{+\beta+1} \mid \beta < \gamma_i^* \wedge i < \xi\}$.

Now let $\xi \in S_2 \cap C$. Trivially, by the choice of ξ, (α) of the condition (f) in the statement of Theorem 2.4 cannot hold. If (β) of the condition (f) holds then by [27, Chapter II, Claim 2.4 (2)] we would have that $\gamma_\xi^* = \gamma_\xi^{**}$, so that $\xi \notin S_2$. Let us finally suppose that (γ) of the condition (f) holds. Because $pp(\kappa_\xi) = \kappa_\xi^{+\gamma_\xi^{**}}$, $\gamma_\xi^* < \gamma_\xi^{**}$, and (γ) of (f) holds, there must be some $\theta \in (\gamma_\xi^*, \gamma_\xi^{**}]$ such that b_θ contains a cofinal subset of $\bigcup\{a_j \mid j < \xi\}$. On the other hand, this is impossible, as

$$\bigcup\{b_\theta \mid \theta \in d_i \wedge i < \xi\} \supset \bigcup\{a_i \mid \varepsilon^* \leq i < \xi\}.$$

We have reached a contradiction!   ⊣

The next two theorems put serious limitations on constructions of models of $\neg(*)_\kappa$, where $(*)_\kappa$ is as in Theorem 2.1. Thus, for example, the "obvious candidate" iteration of short extenders forcing of [8] does not work (but see [9]). The reason is that powers of singular cardinals δ are blown up and this leaves no room for indiscernibles between δ and its power.

THEOREM 2.5. *Let κ be a singular cardinal of uncountable cofinality, and let $(\kappa_i \mid i < \mathrm{cf}(\kappa))$ be a strictly increasing and continuous sequence which is cofinal in κ and such that $\kappa_0 > \mathrm{cf}(\kappa)$. Suppose that $S \subset \mathrm{cf}(\kappa)$ is such that*

(1) *there is a sequence $(\tau_{i\alpha} \mid i \in S \wedge \alpha < \mathrm{cf}(i))$ such that $\mathrm{cf}(\prod_{\alpha < \mathrm{cf}(i)} \tau_{i\alpha}/D_i)$ $= \kappa_i^{++}$ for some ultrafilter D_i extending the Fréchet filter on $\mathrm{cf}(i)$ and*

$$\forall j < \mathrm{cf}(\kappa) \, |\kappa_j \cap \{\tau_{i\alpha} \mid i \in S \wedge \alpha < \mathrm{cf}(i)\}| < \mathrm{cf}(\kappa),$$

and
(2) *if $i \in S$ is a limit point of S then $\max(\mathrm{pcf}(\{\kappa_j^{++} \mid j \in i \cap S\})) = \kappa_i^+$.*

Then S is not stationary.

PROOF. Let us suppose that S is stationary. Set $a = \{\kappa_i^+ \mid i \in S\} \cup \{\kappa_i^{++} \mid i \in S\} \cup \{\tau_{i\alpha} \mid i \in S \wedge \alpha \in \mathrm{cf}(i)\}$. Let $(b_\theta \mid \theta \in a)$ be a smooth sequence of generators for a.

Let C be the set of all limit ordinals $\delta < \mathrm{cf}(\kappa)$ such that for every i with $0 < i < \delta$, if $j \geq \delta$ with $j \in S$ and if $\theta \in a \cap \kappa_{i+1} \cap b_{\kappa_j^{++}}$ then $\delta = \sup(\{j \in S \cap \delta \mid \theta \in b_{\kappa_j^{++}}\})$. Clearly, C is club.

By (2) in the statement of Theorem 2.5, we may find $\delta^* \in C \cap S$ and some $i^* < \delta^*$ such that for every j with $i^* < j < \delta^*$, if $j \in S$ then $\kappa_j^{++} \in b_{\kappa_{\delta^*}^+}$ (cf. the proof of Theorem 2.1).

Let $\theta \in b_{\kappa_{\delta^*}^{++}}$. Then $\theta \in a \cap \kappa_{i+1}$ for some i with $0 < i < \delta^*$. By the choice of C there is then some $j \in S$ with $i^* < j < \delta^*$ and such that $\theta \in b_{\kappa_j^{++}}$. By the smoothness of the sequence of generators we'll have $b_{\kappa_j^{++}} \subset b_{\kappa_{\delta^*}^+}$, and hence $\theta \in b_{\kappa_{\delta^*}^+}$.

We have shown that $b_{\kappa_{\delta^*}^{++}} \subset b_{\kappa_{\delta^*}^+}$, which is absurd because $\mathrm{pp}(\kappa_{\delta^*}) \geq \kappa_{\delta^*}^{++}$ by (1) in the statement of Theorem 2.5. $\dashv$

If δ is a cardinal and $\kappa = \aleph_\delta > \delta$ then condition (1) in the statement of Theorem 2.5 can be replaced by "$i \in S \Rightarrow \mathrm{pp}(\kappa_i) \geq \kappa_i^{++}$," giving the same conclusion.

THEOREM 2.6. *Let κ be a singular cardinal of uncountable cofinality, and let $(\kappa_i \mid i < \mathrm{cf}(\kappa))$ be strictly increasing and continuous sequence which is cofinal in κ and such that $\kappa_0 > \mathrm{cf}(\kappa)$. Suppose that there is $\mu_0 < \kappa$ such that for every μ with $\mu_0 < \mu < \kappa$, $\mathrm{pp}(\mu) < \kappa$. Let $S \subset \mathrm{cf}(\kappa)$ be such that*

(1) $i \in S \Rightarrow \mathrm{pp}(\kappa_i) > \kappa_i^+$, *and*
(2) *if $i \in S$ is a limit point of S and $X = \{\lambda_j \mid j \in i \cap S\}$ with $\kappa_j < \lambda_j \leq \mathrm{pp}(\kappa_j)$ regular then $\max(\mathrm{pcf}(X)) = \kappa_i^+$.*

Then S is not stationary.

PROOF. Let us suppose that S is stationary. Assume that $\mu_0 = 0$ otherwise just work above it. We can assume that for every $i < \text{cf}(\kappa)$ if $\mu < \kappa_i$ then also $\text{pp}(\mu) < \kappa_i$. Let $\chi > \kappa$ be a regular cardinal, and let $M \prec H\chi$ be such that $\text{Card}(M) = \text{cf}(\kappa)$, $M \supset \text{cf}(\kappa)$, and $(\kappa_i \mid i < \text{cf}(\kappa))$, $S \in M$. Set $a = (M \cap \text{Reg}) \setminus (\text{cf}(\kappa) + 1)$. If there is $\mu < \kappa$ such that for each $i \in S$ $|(a \setminus \mu) \cap \kappa_i| < \text{cf}(\kappa)$ then the previous theorem applies. Suppose otherwise. Without loss of generality we can assume that for every $i \in S$ and $\mu < \kappa_i$ $|(a \setminus \mu) \cap \kappa_i| = \text{cf}(\kappa)$.

Let $(b_\theta \mid \theta \in a)$ be a smooth and closed (i.e., $\text{pcf}(b_\theta) = b_\theta$) sequence of generators for a.

CLAIM. For every limit point $i \in S$ $\max(\text{pcf}(a \cap \kappa_i)) \leq \text{pp}(\kappa_i)$.

PROOF. Fix an increasing sequence $(\mu_j \mid j < \text{cf}(i))$ of cardinals of cofinality $\text{cf}(\kappa)$ with limit κ_i and so that $\bigcup(a \cap \mu_j) = \mu_j$. Now $\max(\text{pcf}(a \cap \mu_j)) \leq \text{pp}(\mu_j)$ for every $j < i$, since $|a \cap \mu_j| = \text{cf}(\kappa) = \text{cf}(\mu_j)$. There is a finite $f_j \subset \text{pcf}(a \cap \mu_j)$ such that $a \cap \mu_j \subset \bigcup\{b_\theta \mid \theta \in f_j\}$. Assume that $|i| = \text{cf}(i)$. Otherwise just run the same argument replacing i by a cofinal sequence of the type $\text{cf}(i)$. Consider $v = \max(\text{pcf}(\bigcup\{f_j \mid j < i\}))$. Then $v \leq \text{pp}(\kappa_i)$, since $|\bigcup\{f_j \mid j < i\}| \leq \text{cf}(i)$. So, there is a finite $g \subset \text{pcf}(\bigcup\{f_j \mid j < i\}) \subset v + 1 \subset \text{pp}(\kappa_i) + 1$ such that $\bigcup\{f_j \mid j < i\} \subset \bigcup\{b_\theta \mid \theta \in g\}$. By smoothness, then $a \cap \kappa_i \subset \bigcup\{b_\theta \mid \theta \in g\}$. Since the generators are closed and g is finite, also $\text{pcf}(a \cap \kappa_i) \subset \bigcup\{b_\theta \mid \theta \in g\}$. Hence, $\max(\text{pcf}(a \cap \kappa_i)) \leq \max(g)$ which is at most $\text{pp}(\kappa_i)$.   ⊣

For every limit point i of S find a finite set $c_i \subset \text{pcf}(a \cap \kappa_i)$ such that $a \cap \kappa_i \subset \bigcup\{b_\theta \mid \theta \in c_i\}$ By the claim, $\max(\text{pcf}(a \cap \kappa_i)) \leq \text{pp}(\kappa_i)$. So, $c_i \subset \text{pp}(\kappa_i) + 1$. Set $d_i = c_i \setminus \kappa_i$. Then the set $a \cap \kappa_i \setminus \bigcup\{b_\theta \mid \theta \in d_i\}$ is bounded in κ_i, since we just removed a finite number of b_θ's for $\theta < \kappa_i$. So, there is $\alpha(i) < i$ such that $\kappa_{\alpha(i)} \supset a \cap \kappa_i \setminus \bigcup\{b_\theta \mid \theta \in d_i\}$. Find a stationary $S^* \subset S$ and α^* such that for each $i \in S^*$ $\alpha(i) = \alpha^*$. Let now $i \in S$ be a limit of elements of S^*. Then there is $\tau < \kappa_i$ such that $\bigcup\{d_j \mid j < i\} \setminus \tau \subset b_{\kappa_i^+}$. Since otherwise it is easy to construct $X = \{\lambda_j \mid j \in i \cap S\}$ with $\kappa_j < \lambda_j \leq \text{pp}(\kappa_j)$ regular and $\max(\text{pcf}(X)) > \kappa_i^+$. Now, by smoothness of the generators, $b_{\kappa_i^+}$ should contain a final segment of $a \cap \kappa_i$. Which is impossible, since $\text{pp}(\kappa_i) > \kappa_i^+$. Contradiction.   ⊣

The same argument works if we require only $\text{pp}(\mu) < \kappa$ for μ's of cofinality $\text{cf}(\kappa)$. The consistency of the negation of this (i.e., of: there are unbounded in κ many μ's with $\text{pp}(\mu) \geq \kappa$) is unknown. Shelah's Weak Hypothesis states that this is impossible.

The claim in the proof of Theorem 2.6 can be deduced from general results like [27, Chapter 8, 1.6].

Let again κ be a singular cardinal of uncountable cofinality, and let $(\kappa_i \mid i < \text{cf}(\kappa))$ be a strictly increasing and continuous sequence which is cofinal in κ and such that $\kappa_0 > \mathrm{cf}(\kappa)$. Let, for $n \leq \omega + 1$, S_n denote the set $\{\kappa_i^{+n} \mid i < \mathrm{cf}(\kappa)\}$. By [11], if there is no inner model with a strong cardinal and $\kappa_i^{+\omega} = (\kappa_i^{+\omega})^K$ for every $i < \mathrm{cf}(\kappa)$ then for every $n < \omega$, if for each $i < \mathrm{cf}(\kappa)$ we have $2^{\kappa_i} \geq \kappa_i^{+n}$ then there is a club $C \subset \mathrm{cf}(\kappa)$ such that $\kappa \cap \mathrm{pcf}(\{\kappa_i^{+n} \mid i \in C\}) \subset S_n$. Notice that $(*)$ from the statement of 2.1 just says that $\kappa \cap \mathrm{pcf}(\{\kappa_i^{+} \mid i < \mathrm{cf}(\kappa)\}) \subset S_1$, or equivalently $\kappa \cap \mathrm{pcf}(S_1) = S_1$.

The following says that the connection between the κ_i^{+n}'s and the S_n's cannot be broken for the first time at $\omega + 1$.

THEOREM 2.7. *Let κ be a singular cardinal of uncountable cofinality, and let $(\kappa_i \mid i < \mathrm{cf}(\kappa))$ be a strictly increasing and continuous sequence which is cofinal in κ and such that $\kappa_0 > \mathrm{cf}(\kappa)$. Let, for $n \leq \omega + 1$, S_n denote the set $\{\kappa_i^{+n} \mid i < \mathrm{cf}(\kappa)\}$. Suppose that for every $i < \mathrm{cf}(\kappa)$, $\mathrm{pp}(\kappa_i^{+\omega}) = \kappa_i^{+\omega+1}$.*

If for every $n < \omega$ there is a club $C_n \subset \mathrm{cf}(\kappa)$ such that $\mathrm{pcf}(\{\kappa_i^{+n} \mid i \in C_n\}) \cap \kappa \subset S_n$ then there is a club $C_{\omega+1} \subset \mathrm{cf}(\kappa)$ such that $\mathrm{pcf}(\{\kappa_i^{+\omega+1} \mid i \in C_{\omega+1}\}) \cap \kappa \subset S_{\omega+1}$.

PROOF. set $C = \bigcap_{n<\omega} C_n$. Let $\chi > \kappa$ be a regular cardinal, and let $M \prec H_\chi$ be such that $\mathrm{Card}(M) = \mathrm{cf}(\kappa)$, $M \supset \mathrm{cf}(\kappa)$, and $(\kappa_i \mid i < \mathrm{cf}(\kappa)) \cup \{C_n \mid n < \omega\} \cup \{S_n \mid n \leq \omega+1\} \in M$. Set $a = (M \cap \mathrm{Reg}) \setminus (\mathrm{cf}(\kappa)+1)$. Let $(b_\theta \mid \theta \in a)$ be a smooth and closed sequence of generators for a.

For every $n < \omega$ we find a stationary $E_n \subset C$ and some $\varepsilon_n < \mathrm{cf}(\kappa)$ such that for every $i \in E_n$,

$$\{\kappa_j^{+n} \mid \varepsilon_n < j < i \wedge j \in C_n\} \subset b_{\kappa_i^{+n}}.$$

This is possible since our assumption implies that

$$\mathrm{tcf}\left(\prod_{j \in i \cap C_n} \kappa_j^{+n}/\mathrm{Frechet}\right) = \kappa_i^{+n}$$

for each limit point i of C_n.

Set $\varepsilon = \bigcup_{n<\omega} \varepsilon_n$. Let $C'_{\omega+1}$ be the set of all $i < \mathrm{cf}(\kappa)$ such that for every $n < \omega$, i is a limit of points in E_n. Then, for every $\alpha \in C'_{\omega+1}$ and for every $n < \omega$,

$$b_{\kappa_\alpha^{+n}} \supset \{\kappa_j^{+n} \mid \varepsilon < j < \alpha \wedge j \in C\},$$

since $b_{\kappa_\alpha^{+n}}$ contains a final segment of $\{\kappa_i^{+n} \mid i \in E_n \cap \alpha\}$, and so, by the smoothness of $(b_\theta \mid \theta \in a)$, $b_{\kappa_\alpha^{+n}} \supset b_{\kappa_i^{+n}}$. Moreover, $b_{\kappa_i^{+n}}$ in turn contains $\{\kappa_j^{+n} \mid \varepsilon_n < j < i \wedge j \in C_n\}$.

Let $\alpha \in C'_{\omega+1}$. As $\mathrm{pp}(\kappa_\alpha^{+\omega}) = \kappa_\alpha^{+\omega+1}$, there is some $n(\alpha) < \omega$ such that for every n with $n(\alpha) \leq n < \omega$, $\kappa_\alpha^{+n} \in b_{\kappa_\alpha^{+\omega+1}}$. Again by the smoothness of $(b_\theta \mid \theta \in a)$, $b_{\kappa_\alpha^{+\omega+1}} \supset \bigcup_{n(\alpha) \leq n<\omega} b_{\kappa_\alpha^{+n}}$. Therefore $\kappa_j^{+n} \in b_{\kappa_\alpha^{+\omega+1}}$ for every $j \in C, \varepsilon < j < \alpha$, and $n(\alpha) \leq n < \omega$.

Fix some $j \in C$ with $\varepsilon < j < \alpha$. The fact that $\mathrm{pp}(\kappa_j^{+\omega}) = \kappa_j^{+\omega+1}$ implies that $\kappa_j^{+\omega+1} \in \mathrm{pcf}(\{\kappa_j^{+n} \mid n(\alpha) \leq n < \omega\})$, and hence $\kappa_j^{+\omega+1} \in \mathrm{pcf}(b_{\kappa_\alpha^{+\omega+1}})$. By the closedness of $(b_\theta \mid \theta \in \mathsf{a})$, $\mathsf{a} \cap \mathrm{pcf}(b_{\kappa_\alpha^{+\omega+1}}) = b_{\kappa_\alpha^{+\omega+1}}$. Thus $\kappa_j^{+\omega+1} \in b_{\kappa_\alpha^{+\omega+1}}$.

We may now pick a stationary set $E \subset C'_{\omega+1}$ and some $n^* < \omega$ such that $n(\alpha) = n^*$ for every $\alpha \in E$. Let $C_{\omega+1}$ be the intersection of the limit points of E with $C \setminus (\varepsilon + 1)$.

CLAIM 1. For every $\alpha \in C'_{\omega+1}$, $\mathrm{pcf}(\{\kappa_i^{+\omega+1} \mid i \in (C \cap \alpha) \setminus (\varepsilon + 1)\}) \setminus \kappa_\alpha \subset \{\kappa_\alpha^{+n} \mid 0 < n \leq \omega + 1\}$.

PROOF. Suppose otherwise. By elementarity, we may then find some $\lambda \in \mathsf{a} \cap \mathrm{pcf}(\{\kappa_i^{+\omega+1} \mid i \in (C \cap \alpha) \setminus (\varepsilon + 1)\}) \setminus \kappa_\alpha$ which is above $\kappa_\alpha^{+\omega+1}$. For every $i \in C \cap \alpha$ and $m < \omega$, $\kappa_i^{+\omega+1} \in \mathrm{pcf}(\{\kappa_i^{+n} \mid m < n < \omega\})$, since $\mathrm{pp}(\kappa_i^{+\omega}) = \kappa_i^{+\omega+1}$. Hence $\lambda \in \mathrm{pcf}(\{\kappa_i^{+n} \mid i \in (C \cap \alpha) \setminus (\varepsilon + 1) \wedge m < n < \omega\})$ for each $m < \omega$. But $\alpha \in C'_{\omega+1}$, so for every $n < \omega$, $b_{\kappa_\alpha^{+n}} \supset \{\kappa_i^{+n} \mid i \in (C \cap \alpha) \setminus (\varepsilon + 1)\}$. The fact that $\mathrm{pp}(\kappa_\alpha^{+\omega}) = \kappa_\alpha^{+\omega+1}$ implies that there is some $m < \omega$ such that for every n with $m < n < \omega$, $\kappa_\alpha^{+n} \in b_{\kappa_\alpha^{+\omega+1}}$. The smoothness of $(b_\theta \mid \theta \in \mathsf{a})$ then yields

$$b_{\kappa_\alpha^{+\omega+1}} \supset \{\kappa_i^{+n} \mid m < n < \omega \wedge i \in (C \cap \alpha) \setminus (\varepsilon + 1)\}.$$

Finally, the closedness of $(b_\theta \mid \theta \in \mathsf{a})$ implies that $\mathrm{pcf}(b_{\kappa_\alpha^{+\omega+1}}) \cap \mathsf{a} = b_{\kappa_\alpha^{+\omega+1}}$, and so $\lambda \in b_{\kappa_\alpha^{+\omega+1}}$. Hence $b_\lambda \subset b_{\kappa_\alpha^{+\omega+1}}$, which is possible only when $\lambda \leq \kappa_\alpha^{+\omega+1}$. Contradiction!   ⊣

Now let α be a limit point of $C_{\omega+1}$ and let $\lambda \in \mathrm{pcf}(\{\kappa_i^{+\omega+1} \mid i \in C_{\omega+1} \cap \alpha\}) \setminus \kappa_\alpha$. Then by Claim 1, $\lambda \in \{\kappa_\alpha^{+n} \mid n \leq \omega + 1\}$. We need to show that $\lambda = \kappa_\alpha^{+\omega+1}$.

CLAIM 2. $\mathrm{pcf}(\{\kappa_i^{+\omega+1} \mid i \in E\}) \cap \kappa \subset S_{\omega+1}$.

PROOF. Let $\beta < \mathrm{cf}(\kappa)$ be a limit of ordinals from E. We need to show that $\mathrm{pcf}(\{\kappa_i^{+\omega+1} \mid i \in E \cap \beta\}) \setminus \kappa_\beta = \{\kappa_\beta^{+\omega+1}\}$.

Suppose otherwise. By Claim 1, there is then some $m < \omega$ such that $\kappa_\beta^{+m} \in \mathrm{pcf}(\{\kappa_i^{+\omega+1} \mid i \in E \cap \beta\})$. Then for some unbounded $A \subset E \cap \beta$ we'll have that for every $i \in A$, $\kappa_i^{+\omega+1} \in b_{\kappa_\beta^{+m}}$. By the choice of E, $\kappa_j^{+n} \in b_{\kappa_i^{+\omega+1}}$ for every $j \in C$, $\varepsilon < j < i$, and $n^* \leq n < \omega$.

Fix some $\tilde{n} > \max(m, n^*)$. By the smoothness of $(b_\theta \mid \theta \in \mathsf{a})$, $\kappa_j^{+\tilde{n}} \in b_{\kappa_\beta^{+m}}$ for every $j \in (C \cap \beta) \setminus (\varepsilon + 1)$. But

$$\mathrm{pcf}\left(\{\kappa_j^{+\tilde{n}} \mid j \in (C \cap \beta) \setminus (\varepsilon + 1)\}\right) \setminus \kappa_\beta = \{\kappa_\beta^{+\tilde{n}}\}.$$

So $\kappa_\beta^{+\tilde{n}} \in b_{\kappa_\beta^{+m}}$ and hence $b_{\kappa_\beta^{+\tilde{n}}} \subset b_{\kappa_\beta^{+m}}$. This, however, is impossible, since $\tilde{n} > m$. Contradiction!   ⊣

We now have that $\kappa_i^{+\omega+1} \in b_\lambda$ for unboundedly many $i \in C_{\omega+1} \cap \alpha$. By Claim 2, by the smoothness of $(b_\theta \mid \theta \in \mathsf{a})$, and by the choice of $C_{\omega+1}$, we therefore get that $\kappa_j^{+\omega+1} \in b_\lambda$ for unboundedly many $j \in E \cap \alpha$. Hence again by Claim 2 and by the closedness of $(b_\theta \mid \theta \in a)$, $\kappa_\alpha^{+\omega+1} \in b_\lambda$. Bo by the smoothness of $(b_\theta \mid \theta \in a)$, $b_{\kappa_\alpha^{+\omega+1}} \subset b_\lambda$. This implies that $\lambda = \kappa_\alpha^{+\omega+1}$, and we are done. $\dashv$

M. Magidor asked the following question. Let κ be a singular cardinal of uncountable cofinality, and let $(\kappa_i \mid i < \mathrm{cf}(\kappa))$ be a strictly increasing and continuous sequence which is cofinal in κ. Is it possible to have a stationary and co-stationary set $S \subset \mathrm{cf}(\kappa)$ such that

$$\mathrm{tcf}\left(\prod_{i<\mathrm{cf}(\kappa)} \kappa_i^{++}/(\mathrm{Club}_{\mathrm{cf}(\kappa)} + S) \right) = \kappa^{++}$$

and

$$\mathrm{tcf}\left(\prod_{i<\mathrm{cf}(\kappa)} \kappa_i^{++}/(\mathrm{Club}_{\mathrm{cf}(\kappa)} + (\mathrm{cf}(\kappa) \setminus S)) \right) = \kappa^{+} ?$$

The full answer to this question is unknown. By methods of [11] it is possible to show that at least an inner model with a strong cardinal is needed, provided that $\mathrm{cf}(\kappa) \geq \aleph_2$.

We shall now give a partial negative answer to Magidor's question. A variant of this result was also proved by T. Jech.

THEOREM 2.8. *Let κ be a singular cardinal of uncountable cofinality, and let $(\kappa_i \mid i < \mathrm{cf}(\kappa))$ be a strictly increasing and continuous sequence which is cofinal in κ. Suppose that for some n, $1 \leq n < \omega$, $\mathrm{pp}(\kappa) = \kappa^{+n}$ and $\mathrm{pp}(\kappa_i) = \kappa_i^{+n}$ for each $i < \mathrm{cf}(\kappa)$.*

Then there is a club $C^* \subset \mathrm{cf}(\kappa)$ so that $\mathrm{pcf}(\{\kappa_i^{+k} \mid i \in C^*\}) \setminus \kappa = \{\kappa^{+k}\}$ for every k with $1 \leq k \leq n$.

PROOF. Let $a = \{\kappa_i^{+k} \mid 1 \leq k \leq n \wedge i < \mathrm{cf}(\kappa)\} \cup \{\kappa^{+k} \mid 1 \leq k \leq n\}$. Then $\mathrm{pcf}(a) = a$ by the assumptions of theorem. Without loss of generality, $\min(a) = \kappa_0^+ > \mathrm{cf}(\kappa) = \mathrm{Card}(a)$. Fix a smooth and closed set $(b_\theta \mid \theta \in a)$ of generators for a.

By [2, Lemma 6.3] there is a club $C \subset \mathrm{cf}(\kappa)$ such that for every k with $1 \leq k \leq n$,

$$\{\kappa_i^{+k} \mid i \in C\} \subset \bigcup \{b_{\kappa^{+k'}} \mid 1 \leq k' \leq k\}.$$

Let C^* be the set of all $i \in C$ such that for every j with $1 \leq j \leq i$, κ_i is a limit point of $b_{\kappa^{+j}} \setminus \bigcup\{b_{\kappa^{-j'}} \mid j' < j\}$. Clearly, C^* is club.

Let us show that C^* is as desired. It is enough to prove that for every k with $1 \leq k \leq n$ and $i \in C^*$,

$$\kappa_i^{+k} \in b_{\kappa^{+k}} \setminus \bigcup \{b_{\kappa^{+l}} \mid 1 \leq l < k\}.$$

Suppose otherwise. Then for some $i \in C^*$ and some k with $1 < k \leq n$, $\kappa_i^{+k} \in \bigcup\{b_{\kappa^{+l}} \mid 1 \leq l < k\}$. Define δ_j to be

$$\max \text{pcf}\left(\left(b_{\kappa^{+j}} \setminus \bigcup\{b_{\kappa^{+j'}} \mid 1 \leq j' < j\}\right) \cap \kappa_i\right)$$

for every j, $1 \leq j \leq n$. Then $\delta_j \in b_{\kappa^{+j}}$ by the closedness of $(b_\theta \mid \theta \in a)$. Also, $\delta_j \in \{\kappa_i^{+s} \mid 1 \leq s \leq n\}$, since $\text{pp}(\kappa_i) = \kappa_i^{+n}$.

CLAIM. For every j with $1 \leq j \leq n$, $\delta_j \geq \kappa_i^{+j}$.

PROOF. As $i \in C$, $\kappa_i^{+j'} \in \bigcup\{b_{\kappa^{+j''}} \mid 1 \leq j'' \leq j'\}$ for any j' with $1 \leq j' \leq n$. By the smoothness of $(b_\theta \mid \theta \in a)$, $\bigcup\{b_{\kappa^{+j'}} \mid 1 \leq j' \leq j\} \subset \bigcup\{b_{\kappa^{+j'}} \mid 1 \leq j' \leq j\}$. Recall that $b_{\kappa^{+j}} \setminus \bigcup\{b_{\kappa^{+j'}} \mid 1 \leq j' \leq j\}$ is unbounded in κ_i. Hence

$$\delta_j = \max \text{pcf}\left(\left(b_{\kappa^{+j}} \setminus \bigcup\{b_{\kappa^{+j'}} \mid 1 \leq j' < j\}\right) \cap \kappa_i\right)$$

should be at least κ_i^{+j}.   ⊣

Let us return to κ_i^{+k}. By the claim, $\delta_k \geq \kappa_i^{+k}$. But $\kappa_i^{+k} \in \bigcup\{b_{\kappa^{+l}} \mid 1 \leq l < k\}$. So $b_{\kappa_i^{+k}} \subset \bigcup\{b_{\kappa^{+l}} \mid 1 \leq l < k\}$. Let $l^* \leq k - 1$ be least such that $b_{\kappa^{+l^*}} \cap b_{\kappa_i^{+k}}$ is unbounded in κ_i. Then $\delta_{l^*} \geq \kappa_i^{+k}$. Hence for some $j_1 < j_2 \leq n$, $\delta_{j_1} = \delta_{j_2}$.

Let $\kappa_i^{+l} = \delta_{j_1} = \delta_{j_2}$, where $l \leq n$. Then $b_{\kappa_i^{+l}} \subset b_{\kappa^{+j_1}} \cap b_{\kappa^{+j_2}}$ by the smoothness of $(b_\theta \mid \theta \in a)$, since $\delta_{j_1} \in b_{\kappa^{+j_1}}$ and $\delta_{j_2} \in b_{\kappa^{+j_2}}$. But now $b_{\kappa^{+j_2}} \setminus \bigcup_{1 \leq j < j_2} b_{\kappa^{+j}}$ and $b_{\kappa_i^{+l}}$ should be disjoint. This, however, is impossible, as

$$\kappa_i^{+l} = \delta_{j_2} = \max \text{pcf}\left(\left(b_{\kappa^{+j_2}} \setminus \bigcup_{1 \leq j < j_2} b_{\kappa^{+j}}\right) \cap \kappa_i\right).$$

Contradiction!   ⊣

The previous theorem may break down if we replace n by ω. I.e., it is possible to have a model satisfying $\text{pp}(\kappa) = \kappa^{+\omega+1}$, $\text{pp}(\kappa_i) = \kappa_i^{+\omega+1}$ for $i < \text{cf}(\kappa) = \omega_1$, but

$$\max \text{pcf}\left(\{\kappa_i^{++} \mid i < \omega_1\}\right) = \kappa^+.$$

The construction is as follows. Start from a coherent sequence $\vec{E} = (E_{(\alpha,\beta)} \mid \alpha \leq \kappa \wedge \beta < \omega_1)$ of $(\alpha, \alpha + \omega + 1)$-extenders. Collapse κ^{++} to κ^+. Then force with the extender based Magidor forcing with $\vec{E}$ to change the cofinality of κ to ω_1 and to blow up 2^κ to $\kappa^{+\omega+1}$. The facts that κ^{++V} will have cofinality κ^+ in the extension and no cardinal below κ will be collapsed ensure that $\max \text{pcf}(\{\kappa_i^{++} \mid i < \omega_1\}) = \kappa^+$.

§3. Some core model theory. This paper will exploit the core model theory of [31] and its generalization [32]. We shall also have to take another look at the argument of [19] and [18] which we refer to as the "covering argument."

Our Theorems 1.1 and 1.4 will be shown by running the first ω many steps of Woodin's core model induction. The proof of Theorem 1.1 in [5] uses the very same method, and we urge the reader to at least gain some acquaintance with the inner model theoretic part of [5, §2].

The proof of Theorem 1.1 needs a refinement of the technique of "stabilizing the core model" which is introduced by [23, Lemma 3.1.1]. This is what we shall deal with first in this section.

LEMMA 3.1. *Let* $\mathcal{M}$ *be an iterable premouse, and let* $\delta \in \mathcal{M}$. *Let* T *be a normal iteration tree on* $\mathcal{M}$ *of length* $\theta + 1$ *such that* $\mathrm{lh}(E_\xi^T) \geq \delta$ *whenever* $\xi < \theta$ *and* δ *is a cardinal of* $\mathcal{M}_\theta^T$. *Then the phalanx* $((\mathcal{M}_\theta^T, \mathcal{M}), \delta)$ *is iterable.*

PROOF. Let $\mathcal{U}$ be an iteration tree on $((\mathcal{M}_\theta^T, \mathcal{M}), \delta)$. We want to "absorb" $\mathcal{U}$ by an iteration tree $\mathcal{U}^*$ on $\mathcal{M}$. The bookkeeping is simplified if we assume that whenever an extender $E_\xi^{\mathcal{U}}$ is applied to $\mathcal{M}_\theta^T$ to yield $\pi_{0\xi+1}^{\mathcal{U}}$ then right before that there are θ many steps of "padding." I.e., letting P denote the set of all $\eta + 1 \leq \mathrm{lh}(\mathcal{U})$ with $E_\eta^{\mathcal{U}} = \emptyset$, we want to assume that if $\mathrm{crit}(E_\xi^{\mathcal{U}}) < \delta$ then $\xi + 1 = \bar{\xi} + 1 + \theta$ for some $\bar{\xi}$ such that $\eta + 1 \in P$ for all $\eta \in [\bar{\xi}, \xi)$.

Let us now construct $\mathcal{U}^*$. We shall simultaneously construct embeddings

$$\pi_\alpha : \mathcal{M}_\alpha^{\mathcal{U}} \longrightarrow \mathcal{M}_\alpha^{\mathcal{U}^*},$$

where $\alpha \in \mathrm{lh}(\mathcal{U}) \setminus P$, such that $\pi_\alpha \upharpoonright \mathrm{lh}(E_\xi^{\mathcal{U}}) = \pi_\beta \upharpoonright \mathrm{lh}(E_\xi^{\mathcal{U}})$ whenever $\alpha < \beta \in \mathrm{lh}(\mathcal{U}) \setminus P$ and $\xi \leq \alpha$ or else $(\alpha, \xi] \subset P$. The construction of $\mathcal{U}^*$ and of the maps π_α is a standard recursive copying construction as in the proof of [20, Lemma p. 54f.], say, except for how to deal with the situation when an extender is applied to $\mathcal{M}_\theta^T$.

Suppose that we have constructed $\mathcal{U}^* \upharpoonright \bar{\xi} + 1$ and $(\pi_\alpha \mid \alpha \in (\bar{\xi} + 1) \setminus P)$, that $\eta + 1 \in P$ for all $\eta \in [\bar{\xi}, \bar{\xi} + 1 + \theta - 1)$, and that $\mathrm{crit}(E_{\bar{\xi}+\theta}^{\mathcal{U}}) < \delta$. We then proceed as follows. Let $\xi + 1 = \bar{\xi} + 1 + \theta$. We first let

$$\sigma : \mathcal{M} \longrightarrow_{\pi_{\bar{\xi}}(E_\xi^{\mathcal{U}})} \mathcal{M}_{\bar{\xi}+1}^{\mathcal{U}^*},$$

and we let

$$\tau : \mathcal{M} \longrightarrow_{E_\xi^{\mathcal{U}}} \mathrm{ult}\left(\mathcal{M}, E_\xi^{\mathcal{U}}\right).$$

We may define

$$k : \mathrm{ult}\left(\mathcal{M}, E_\xi^{\mathcal{U}}\right) \longrightarrow \mathcal{M}_{\bar{\xi}+1}^{\mathcal{U}^*}$$

by setting

$$k\left([a, f]_{E_\xi^{\mathcal{U}}}^{\mathcal{M}}\right) = [\pi_{\bar{\xi}}(a), f]_{\pi_{\bar{\xi}}(E_\xi^{\mathcal{U}})}^{\mathcal{M}} = \sigma(f)(\pi_{\bar{\xi}}(a))$$

for appropriate a and f. This works, because $\pi_{\bar{\xi}} \upharpoonright \mathcal{P}(\mathrm{crit}(E_\xi^{\mathcal{U}})) \cap \mathcal{M} = \mathrm{id}$. Notice that $k \upharpoonright \mathrm{lh}(E_\xi^{\mathcal{U}}) = \pi_{\bar{\xi}} \upharpoonright \mathrm{lh}(E_\xi^{\mathcal{U}})$.

We now let the models $\mathcal{M}^{\mathcal{U}^*}_{\xi+1+\eta}$ and maps

$$\sigma_\eta : \mathcal{M}^{\mathcal{T}}_\eta \longrightarrow \mathcal{M}^{\mathcal{U}^*}_{\xi+1+\eta},$$

for $\eta \leq \theta$, arise by copying the tree $\mathcal{T}$ onto $\mathcal{M}^{\mathcal{U}^*}_{\xi+1}$, using σ. We shall also have models $\mathcal{M}^{\tau\mathcal{T}}_\eta$ and maps

$$\tau_\eta : \mathcal{M}^{\mathcal{T}}_\eta \longrightarrow \mathcal{M}^{\tau\mathcal{T}}_\eta,$$

for $\eta \leq \theta$, which arise by copying the tree $\mathcal{T}$ onto $\mathrm{ult}(\mathcal{M}, E^{\mathcal{U}}_\xi)$, using τ. Notice that for $\eta \leq \theta$ there are also copy maps

$$k_\eta : \mathcal{M}^{\tau\mathcal{T}}_\eta \longrightarrow \mathcal{M}^{\mathcal{U}^*}_{\xi+1+\eta}$$

with $k_0 = k$. Because $\mathrm{lh}(E^{\mathcal{T}}_\xi) \geq \delta$ whenever $\xi < \theta$, $\mathrm{lh}(E^{\tau\mathcal{T}}_\xi) \geq \tau(\delta)$ whenever $\xi < \theta$, so that in particular $k_\theta \restriction \tau(\delta) = k \restriction \tau(\delta)$, and thus $k_\theta \restriction \mathrm{lh}(E^{\mathcal{U}}_\xi) = k \restriction \mathrm{lh}(E^{\mathcal{U}}_\xi)$.

We shall also have that $\tau_\theta \restriction \delta = \tau \restriction \delta$, so that we may define

$$k' : \mathcal{M}^{\mathcal{U}}_{\xi+1} = \mathrm{ult}\left(\mathcal{M}, E^{\mathcal{U}}_\xi\right) \longrightarrow \mathcal{M}^{\tau\mathcal{T}}_\theta$$

by setting

$$k'([a, f]^{\mathcal{M}}_{E^{\mathcal{U}}_\xi}) = \tau_\theta(f)(a)$$

for appropriate a and f. Let us now define

$$\pi_{\xi+1} : \mathcal{M}^{\mathcal{U}}_{\xi+1} = \mathrm{ult}\left(\mathcal{M}, E^{\mathcal{U}}_\xi\right) \longrightarrow \mathcal{M}^{\mathcal{U}^*}_{\xi+1}$$

by $\pi_{\xi+1} = k_\theta \circ k'$. We then get that $\pi_{\xi+1} \restriction \mathrm{lh}(E^{\mathcal{U}}_\xi) = k_\theta \restriction \mathrm{lh}(E^{\mathcal{U}}_\xi) = k \restriction \mathrm{lh}(E^{\mathcal{U}}_\xi) = \pi_\xi \restriction \mathrm{lh}(E^{\mathcal{U}}_\xi)$. $\dashv$

Let Ω be an inaccessible cardinal. We say that V_Ω is *n-suitable* if $n < \omega$ and V_Ω is closed under $M^\#_n$, but $M^\#_{n+1}$ does not exist (cf. [30, p. 81] or [5, p. 1841]). We say that V_Ω is *suitable* if there is some n such that V_Ω is n-suitable. If Ω is measurable and V_Ω is n-suitable then the core model K "below $n + 1$ Woodin cardinals" of height Ω exists (cf. [32]).

The following lemma is a version of Lemma 3.1 for $\mathcal{M} = K$. It is related to [19, Fact 3.19.1].

LEMMA 3.2 (Steel). *Let Ω be a measurable cardinal, and suppose that V_Ω is suitable. Let K denote the core model of height Ω. Let $\mathcal{T}$ be a normal iteration tree on K of length $\theta + 1 < \Omega$. Let $\theta' \leq \theta$, and let δ be a cardinal of $\mathcal{M}^{\mathcal{T}}_\theta$ such that $\nu(E^{\mathcal{T}}_\xi) > \delta$ whenever $\xi \in [\theta', \theta)$. Then the phalanx $((\mathcal{M}^{\mathcal{T}}_{\theta'}, \mathcal{M}^{\mathcal{T}}_\theta), \delta)$ is iterable.*

PROOF SKETCH. As K^c is a normal iterate of K (cf. [23, Theorem 2.3]), it suffices to prove Lemma 3.2 for K^c rather than for K.

We argue by contradiction. Let $\mathcal{T}$ be a normal iteration tree on K^c of length $\theta + 1 < \Omega$, let $\theta' \leq \theta$, and let δ be a cardinal of $\mathcal{M}^{\mathcal{T}}_\theta$ such that $\nu(E^{\mathcal{T}}_\xi) > \delta$

whenever $\xi \in [\theta', \theta)$. Suppose that $\mathcal{U}$ is an "ill behaved" putative normal iteration tree on the phalanx $((\mathcal{M}_{\theta'}^{\mathcal{T}}, \mathcal{M}_{\theta}^{\mathcal{T}}), \delta)$. Let $\pi : \bar{V} \to V_{\Omega+2}$ be such that $\bar{V}$ is countable and transitive and $\{K^c, \mathcal{T}, \theta', \delta, \mathcal{U}\} \in \operatorname{ran}(\pi)$. Set $\bar{\mathcal{T}} = \pi^{-1}(\mathcal{T})$, $\bar{\theta} = \pi^{-1}(\theta)$, $\bar{\theta}' = \pi^{-1}(\theta')$, $\bar{\delta} = \pi^{-1}(\delta)$, and $\bar{\mathcal{U}} = \pi^{-1}(\mathcal{U})$.

By [31, §9] there are ξ' and ξ and maps $\sigma' : \mathcal{M}_{\bar{\theta}'}^{\bar{\mathcal{T}}} \to \mathcal{N}_{\xi'}$ and $\sigma : \mathcal{M}_{\bar{\theta}}^{\bar{\mathcal{T}}} \to \mathcal{N}_{\xi}$ such that $\mathcal{N}_{\xi'}$ and $\mathcal{N}_{\xi}$ agree below $\sigma'(\bar{\delta})$, and $\sigma' \upharpoonright \sigma'(\bar{\delta}) = \sigma \upharpoonright \sigma'(\bar{\delta})$. (Here $\mathcal{N}_{\xi'}$ and $\mathcal{N}_{\xi}$ denote models from the K^c construction.) We may now run the argument of [31, §9] once more to get that in fact $\bar{\mathcal{U}}$ is "well behaved." But then also $\mathcal{U}$ is "well behaved" after all. ⊣

In the proofs to follow we shall sometimes tacitly use the letter K to denote not K but rather a canonical very soundness witness for a segment of K which is long enough. If $\mathcal{M}$ is a premouse then we shall denote by $\mathcal{M}|\alpha$ the premouse $\mathcal{M}$ as being cut off at α *without* a top extender (even if $E_\alpha^{\mathcal{M}} \neq \emptyset$), and we shall denote by $\mathcal{M}\|\alpha$ the premouse $\mathcal{M}$ as being cut off at α *with* $E_\alpha^{\mathcal{M}}$ as a top extender (if $E_\alpha^{\mathcal{M}} \neq \emptyset$, otherwise $\mathcal{M}\|\alpha = \mathcal{M}|\alpha$). If $\beta \in \mathcal{M}$ then $\beta^{+\mathcal{M}}$ will either denote the cardinal successor of β in $\mathcal{M}$ (if there is one) or else $\beta^{+\mathcal{M}} = \mathcal{M} \cap \operatorname{OR}$.

LEMMA 3.3. *Let Ω be a measurable cardinal, and suppose that V_Ω is suitable. Let K denote the core model of height Ω. Let $\kappa \geq \aleph_2$ be a regular cardinal, and let $\mathcal{M} \trianglerighteq K\|\kappa$ be an iterable premouse. Then the phalanx $((K, \mathcal{M}), \kappa)$ is iterable.*

PROOF. We shall exploit the covering argument. Let

$$\pi : N \cong X \prec V_{\Omega+2}$$

such that N is transitive, $\operatorname{Card}(N) < \kappa$, $\{K, \mathcal{M}, \kappa\} \subset X$, $X \cap \kappa \in \kappa$, and $\bar{K} = \pi^{-1}(K)$ is a normal iterate of K, hence of $K\|\kappa$, and hence of $\mathcal{M}$. Such a map π exists by [18]. Set $\bar{\mathcal{M}} = \pi^{-1}(\mathcal{M})$ and $\bar{\kappa} = \pi^{-1}(\kappa)$. By the relevant version of [31, Lemma 2.4] it suffices to verify that $((\bar{K}, \bar{\mathcal{M}}), \bar{\kappa})$ is iterable. However, the iterability of $((\bar{K}, \mathcal{M}), \bar{\kappa})$ readily follows from Lemma 3.1. Using the map π, we may thus infer that $((\bar{K}, \bar{\mathcal{M}}), \bar{\kappa})$ is iterable as well. ⊣

LEMMA 3.4. *Let Ω be a measurable cardinal, and suppose that V_Ω is suitable. Let K denote the core model of height Ω. Let κ be a cardinal of K, and let $\mathcal{M}$ be a premouse such that $\mathcal{M}|\kappa^{+\mathcal{M}} = K|\kappa^{+\mathcal{M}}$, $\rho_\omega(\mathcal{M}) \leq \kappa$, and $\mathcal{M}$ is sound above κ. Suppose further that the phalanx $((K, \mathcal{M}), \kappa)$ is iterable. Then $\mathcal{M} \triangleleft K$.*

PROOF. This follows from the proof of [19, Lemma 3.10]. This proof shows that $((K, \mathcal{M}), \kappa)$ cannot move in the comparison with K, and that either $\mathcal{M} \triangleleft K$ or else, setting $\nu = \kappa^{+\mathcal{M}}$, $E_\nu^K \neq \emptyset$ and $\mathcal{M}$ is the ultrapower of an initial segment of K by E_ν^K. However, the latter case never occurs, as we'd have that $\mu = \operatorname{crit}(E_\nu^K) < \kappa$ so that $\mu^{+K\|\nu} = \mu^{+K}$ and hence E_ν^K would be a total extender on K. ⊣

Let W be a weasel. We shall write $\kappa(W)$ for the class projectum of W, and $c(W)$ for the class parameter of W (cf. [19, §2.2]). Let E be an extender or an extender fragment. We shall then write $\tau(E)$ for the Dodd projectum of E, and $s(E)$ for the Dodd parameter of E (cf. [19, §2.1]).

The following lemma generalizes [25, Lemma 2.1].

LEMMA 3.5. *Let Ω be a measurable cardinal, and suppose that V_Ω is suitable. Let K denote the core model of height Ω. Let $\kappa \geq \aleph_2$ be a cardinal of K, and let $\mathcal{M} \unrhd K\|\kappa$ be an iterable premouse such that $\rho_\omega(\mathcal{M}) \leq \kappa$, and $\mathcal{M}$ is sound above κ. Then $\mathcal{M} \lhd K$.*

PROOF. The proof is by "induction on $\mathcal{M}$." Let us fix $\kappa \geq \aleph_2$, a cardinal of K. Let $\mathcal{M} \unrhd K\|\kappa$ be an iterable premouse such that $\rho_\omega(\mathcal{M}) \leq \kappa$ and $\mathcal{M}$ is sound above κ. Let us further assume that for all $\mathcal{N} \lhd \mathcal{M}$ with $\rho_\omega(\mathcal{N}) \leq \kappa$ we have that $\mathcal{N} \lhd K$. We aim to show that $\mathcal{M} \lhd K$.

By Lemma 3.4 it suffices to prove that the phalanx $((K, \mathcal{M}), \kappa)$ is iterable. Let us suppose that this is not the case.

We shall again make use of the covering argument. Let

$$\pi : N \cong X \prec V_{\Omega+2}$$

be such that N is transitive, $\mathrm{Card}(N) = \aleph_1$, $\{K, \mathcal{M}, \kappa\} \subset X$, $X \cap \aleph_2 \in \aleph_2$, and $\bar{K} = \pi^{-1}(K)$ is a normal iterate of K. Such a map π exists by [18]. Set $\bar{\mathcal{M}} = \pi^{-1}(\mathcal{M})$, $\bar{\kappa} = \pi^{-1}(\kappa)$, and $\delta = \pi^{-1}(\aleph_2)$. By the relevant version of [31, Lemma 2.4], we may and shall assume to have chosen π so that the phalanx $((\bar{K}, \bar{\mathcal{M}}), \bar{\kappa})$ is not iterable.

We may and shall moreover assume that all objects occuring in the proof of [18] are iterable. Let $\mathcal{T}$ be the normal iteration tree on K arising from the coiteration with $\bar{K}$. Set $\theta + 1 = \mathrm{lh}(\mathcal{T})$. Let $(\kappa_i \mid i \leq \varphi)$ be the strictly monotone enumeration of $\mathrm{Card}^{\bar{K}} \cap (\bar{\kappa} + 1)$, and set $\lambda_i = \kappa_i^{+\bar{K}}$ for $i \leq \varphi$. Let, for $i \leq \varphi$, $\alpha_i < \theta$ be the least α such that $\kappa_i < \nu(E_\alpha^{\mathcal{T}})$, if there is some such α; otherwise let $\alpha_i = \theta$. Notice that $\mathcal{M}_{\alpha_i}^{\mathcal{T}}|\lambda_i = \bar{K}|\lambda_i$ for all $i \leq \varphi$. Let, for $i \leq \varphi$, $\mathcal{P}_i$ be the longest initial segment of $\mathcal{M}_{\alpha_i}^{\mathcal{T}}$ such that $\mathcal{P}(\kappa_i) \cap \mathcal{P}_i \subset \bar{K}$. Let

$$\mathcal{R}_i = \mathrm{ult}(\mathcal{P}_i, E_\pi \upharpoonright \pi(\kappa_i)),$$

where $i \leq \varphi$. Some of the objects $\mathcal{R}_i$ might be proto-mice rather than premice. We recursively define $(\mathcal{S}_i \mid i \leq \varphi)$ as follows. If $\mathcal{R}_i$ is a premouse then we set $\mathcal{S}_i = \mathcal{R}_i$. If $\mathcal{R}_i$ is not a premouse then we set

$$\mathcal{S}_i = \mathrm{ult}\left(\mathcal{S}_j, \dot{F}^{\mathcal{R}_i}\right),$$

where $\kappa_j = \mathrm{crit}(\dot{F}^{\mathcal{P}_i})$ (we have $j < i$). Set $\Lambda_i = \sup(\pi''\lambda_i)$ for $i \leq \varphi$.

The proof of [18] now shows that we may and shall assume that the following hold true, for every $i \leq \varphi$.

CLAIM 1. *If $\mathcal{S}_i$ is a set premouse then $\mathcal{S}_i \lhd K\|\pi(\lambda_i)$.*

PROOF. This readily follows from the proof of [19, Lemma 3.10]. Cf. the proof of Lemma 3.4 above. ⊣

CLAIM 2. If S_i is a weasel then either $S_i = K$ or else $S_i = \mathrm{ult}(K, E_\nu^K)$ where $\nu \geq \Lambda_i$ is such that $\mathrm{crit}(E_\nu^K) < \delta$ and $\tau(E_\nu^K) \leq \pi(\kappa_i)$.

PROOF. This follows from the proof of [19, Lemma 3.11].

Fix i, and suppose that S_i is a weasel with $S_i \neq K$. Let

$$\mathcal{R}_{i_k} = S_{i_k} \longrightarrow S_{i_{k-1}} \longrightarrow \cdots \longrightarrow S_{i_0} = S_i$$

be the decomposition of S_i, and let $\sigma_j : S_{i_j} \to S_i$ for $j \leq k$ (cf. [19, Lemma 3.6]). We also have

$$\pi_{0\alpha_{i_k}}^{\mathcal{T}} : K \longrightarrow \mathcal{M}_{\alpha_{i_k}}^{\mathcal{T}} = \mathcal{P}_{i_k}.$$

Notice that we must have

$$\mu = \mathrm{crit}\left(\pi_{0\alpha_{i_k}}^{\mathcal{T}}\right) < \delta = \mathrm{crit}(\pi),$$

as otherwise $\mathcal{M}_{\alpha_{i_k}}^{\mathcal{T}}$ couldn't be a weasel. Let us write

$$\rho : \mathcal{P}_{i_k} \longrightarrow \mathrm{ult}(\mathcal{P}_{i_k}, E_\pi \restriction \pi(\kappa_{i_k})) = \mathcal{R}_{i_k}.$$

Notice that we have $\kappa(\mathcal{R}_{i_k}) \leq \pi(\kappa_{i_j})$ and $c(\mathcal{R}_{i_k}) = \emptyset$ (cf. [19, Lemma 3.6]. It is fairly easy to see that the proof of [19, Lemma 3.11] shows that we must indeed have $E_{\Lambda_{i_k}}^K \neq \emptyset$, $\mathrm{crit}(E_{\Lambda_{i_k}}^K) = \mu$, $\tau(E_{\Lambda_{i_k}}^K) \leq \pi(\kappa_{i_k})$, $s(E_{\Lambda_{i_k}}^K) = \emptyset$, and

$$\mathcal{R}_{i_k} = \mathrm{ult}\left(K, E_{\Lambda_{i_k}}^K\right).$$

The argument which gives this very conclusion is actually a simplified version of the argument which is to come.

We are hence already done if $k = 0$. Let us assume that $k > 0$ from now on. We now let F be the (μ, λ)-extender derived from $\sigma_{i_k} \circ \rho \circ \pi_{0\alpha_{i_k}}^{\mathcal{T}}$, where

$$\lambda = \max(\{\pi(\kappa_i)\} \cup c(S_i))^{+S_i}.$$

We shall have that $\tau(F) \leq \pi(\kappa_i)$ and $s(F) \setminus \pi(\kappa_i) = c(S_i) \setminus \pi(\kappa_i)$. Let us write $t = s(F) \setminus \pi(\kappa_i)$. We in fact have that

$$t = \bigcup_{j<k} \sigma_j\left(s\left(\dot{F}^{\mathcal{R}_{i_j}}\right)\right) \setminus \pi(\kappa_i)$$

(cf. [19, Lemma 3.6]). Using the facts that $E_{\Lambda_{i_k}}^K \in S_i$ and that every $\mathcal{R}_{i_j}$ is Dodd-solid above $\pi(\kappa_{i_j})$ for every $j < k$, it is easy to verify that we shall have that

$$F \restriction (t(l) \cup t \restriction l) \in S_i$$

for every $l < \mathrm{lh}(t)$.

Now let $\mathcal{U}, \mathcal{V}$ denote the iteration trees arising from the coiteration of K with $((K, S_i), \pi(\kappa_i))$. The proof of [19, Lemma 3.11] shows that $1 \in (0, \infty]_{\mathcal{U}}$, and that $\mathrm{crit}(E_0^{\mathcal{U}}) = \mu$ and $\tau(E_0^{\mathcal{U}}) \leq \pi(\kappa_i)$. Let us write $s = s(E_0^{\mathcal{U}})$. If $\bar{s} < s$

then $E_0^{\mathcal{U}} \upharpoonright (\pi(\kappa_i) \cup \bar{s}) \in \mathcal{M}_1^{\mathcal{U}}$, which implies that $\mathcal{M}_1^{\mathcal{U}}$ does not have the $\bar{s}$-hull property at $\pi(\kappa_i)$. Thus, s is the least $\bar{s}$ such that $\mathrm{ult}(K, E_0^{\mathcal{U}}) = \mathcal{M}_1^{\mathcal{U}}$ has the $\bar{s}$-hull property at $\pi(\kappa_i)$.

Let $\mathcal{N} = \mathcal{M}_\infty^{\mathcal{U}} = \mathcal{M}_\infty^{\mathcal{V}}$. We know that s is the least $\bar{s}$ such that $\mathcal{N} = \mathcal{M}_\infty^{\mathcal{U}}$ has the $\bar{s}$-hull property at $\pi(\kappa_i)$.

The proof of [19, Lemma 3.11] also gives that $1 = \mathrm{root}^{\mathcal{V}}$, i.e., that $\mathcal{N}$ sits above $\mathcal{S}_i$ rather than K. We have $\pi_{1\infty}^{\mathcal{V}} : \mathcal{R}_i \to \mathcal{N}$. As $\mathcal{R}_i$ has the t-hull property at $\pi(\kappa_i)$, $\mathcal{N}$ has the $\pi_{1\infty}^{\mathcal{V}}(t)$-hull property at $\pi(\kappa_i)$. Therefore, we must have that $s \leq \pi_{1\infty}^{\mathcal{V}}(t)$.

SUBCLAIM. $s = \pi_{1\infty}^{\mathcal{V}}(t)$.

PROOF. Suppose that $s < \pi_{1\infty}^{\mathcal{V}}(t)$. Let l be largest such that $s \upharpoonright l = \pi_{1\infty}^{\mathcal{V}}(t) \upharpoonright l$. Set $\bar{F} = F \upharpoonright (t(l) \cup t \upharpoonright l)$. We know that $\bar{F} \in \mathcal{S}_i$, which implies that $\pi_{1\infty}^{\mathcal{V}}(\bar{F}) \in \mathcal{N}$. In particular,

$$G = \pi_{1\infty}^{\mathcal{V}}(\bar{F}) \upharpoonright (\pi(\kappa_i) \cup s) \in \mathcal{N}.$$

Let us verify that $G = E_0^{\mathcal{U}}$.

Let us write $\tilde{\pi} = \pi_{1\infty}^{\mathcal{V}}$. Pick $a \in [\pi(\kappa_i) \cup s]^{<\omega}$, and let $X \in \mathcal{P}([\mu]^{\mathrm{Card}(a)}) \cap K$. We have that $X \in G_a$ if and only if $\tilde{\pi}(X) \in G_a$ (because $\mathrm{crit}(\tilde{\pi}) \geq \pi(\kappa_i) > \mu$) if and only if $\tilde{\pi}(X) \in \tilde{\pi}(F \upharpoonright (t(l) \cup t \upharpoonright l))_a$ if and only if

$$a \in \tilde{\pi}(\{u \mid X \in F \upharpoonright (t(l) \cup t \upharpoonright l)_u\}),$$

which is the case if and only if

$$a \in \tilde{\pi}(\{u \mid u \in \sigma_{i_k} \circ \rho \circ \pi_{0\alpha_{i_k}}^{\mathcal{T}}(X)\}) = \{u \mid u \in \tilde{\pi} \circ \sigma_{i_k} \circ \rho \circ \pi_{0\alpha_{i_k}}^{\mathcal{T}}(X)\}.$$

However, this holds if and only if $a \in \pi_{01}^{\mathcal{U}}(X)$, i.e., if and only if $X \in (E_0^{\mathcal{U}})_a$, because, using the hull- and definability properties of K, $\tilde{\pi} \circ \sigma_{i_k} \circ \rho \circ \pi_{0\alpha_{i_k}}^{\mathcal{T}}(X) = \pi_{0\infty}^{\mathcal{U}}(X) = \pi_{01}^{\mathcal{U}}(X)$.

We have indeed shown that $G = E_0^{\mathcal{U}}$. But we have that $G \in \mathcal{N} = \mathcal{M}_\infty^{\mathcal{U}}$. This is a contradiction! $\dashv$

By the subclaim, $s \in \mathrm{ran}(\pi_{1\infty}^{\mathcal{V}})$, and we may define an elementary embedding

$$\Phi : \mathcal{M}_1^{\mathcal{U}} \longrightarrow \mathcal{S}_i$$

by setting

$$\tau^{\mathcal{M}_1^{\mathcal{U}}}[\vec{\xi}_1, \vec{\xi}_2, \vec{\xi}_3] \longmapsto \tau^{\mathcal{S}_i}[\vec{\xi}_1, (\pi_{1\infty}^{\mathcal{V}})^{-1}(\vec{\xi}_2), \vec{\xi}_3],$$

where τ is a Skolem term, $\vec{\xi}_1 < \pi(\kappa_i)$, $\vec{\xi}_2 \in s$, and $\vec{\xi}_3 \in \Gamma$ for some appropriate thick class Γ. However, $t = (\pi_{1\infty}^{\mathcal{V}})^{-1}(s)$, and $\mathcal{S}_i = H_\omega^{\mathcal{S}_i}(\pi(\kappa_i) \cup t \cup \Gamma)$. Hence Φ is onto, and thus $\mathcal{S}_i = \mathcal{M}_1^{\mathcal{U}} = \mathrm{ult}(K, E_0^{\mathcal{U}})$.

If we now let v be such that $E_v^K = E_0^{\mathcal{U}}$ then v is as in the statement of Claim 2. $\dashv$

Let us abbreviate by $\vec{S}_{\mathcal{M}}$ the phalanx

$$((S_i \mid i < \varphi)^\frown \mathcal{M}, (\Lambda_i \mid i < \varphi)).$$

CLAIM 3. $\vec{S}_{\mathcal{M}}$ is a special phalanx which is iterable with respect to special iteration trees.

PROOF. Let $\mathcal{V}$ be a putative special iteration tree on the phalanx $\vec{S}_{\mathcal{M}}$. By Claims 1 and 2, we may construe $\mathcal{V}$ as an iteration of the phalanx

$$((K, \mathcal{M}), \delta).$$

The only wrinkle here is that if $\operatorname{crit}(E_\xi^{\mathcal{V}}) = \pi(\kappa_i)$ for some $i < \varphi$, where $S_i = \operatorname{ult}(K, E_\nu^K)$, then we have to observe that

$$\operatorname{ult}\left(K, \dot{F}^{\operatorname{ult}(K\|\nu, E_\xi^{\mathcal{V}})}\right) = \operatorname{ult}\left(\operatorname{ult}(K, E_\nu^K), E_\xi^{\mathcal{V}}\right),$$

and the resulting ultrapower maps are the same.

Lemma 3.3 now tells us that the phalanx $((K, \mathcal{M}), \delta)$ is iterable, so that $\mathcal{V}$ turns out to be "well behaved." $\dashv$

By [19, Lemma 3.18], Claim 3 gives that

$$((\mathcal{R}_i \mid i < \varphi)^\frown \mathcal{M}, (\Lambda_i \mid i < \varphi))$$

is a very special phalanx which is iterable with respect to special iteration trees. By [19, Lemma 3.17], the phalanx

$$((\mathcal{P}_i \mid i < \varphi)^\frown \bar{\mathcal{M}}, (\lambda_i \mid i < \varphi)),$$

call it $\vec{\mathcal{P}}_{\bar{\mathcal{M}}}$, is finally iterable as well.

CLAIM 4. Either $\bar{\mathcal{M}}$ is an iterate of K, or else $\bar{\mathcal{M}} \rhd \mathcal{P}_\varphi$.

PROOF. Because $\vec{\mathcal{P}}_{\bar{\mathcal{M}}}$ is iterable, we may coiterate $\vec{\mathcal{P}}_{\bar{\mathcal{M}}}$ with the phalanx

$$\vec{\mathcal{P}} = ((\mathcal{P}_i \mid i \le \varphi), (\lambda_i \mid i < \varphi)),$$

giving iteration trees $\mathcal{V}$ on $\vec{\mathcal{P}}_{\bar{\mathcal{M}}}$ and $\mathcal{V}'$ on $\vec{\mathcal{P}}$. An argument exactly as for (b) $\Rightarrow$ (a) in the proof of [31, Theorem 8.6] shows that the last model $\mathcal{M}_\infty^{\mathcal{V}}$ of $\mathcal{V}$ must sit above $\bar{\mathcal{M}}$, and that in fact $\mathcal{M}_\infty^{\mathcal{V}} = \bar{\mathcal{M}}$, i.e., $\mathcal{V}$ is trivial. But as $\rho_\omega(\bar{\mathcal{M}}) \le \bar{\kappa}$ and $\bar{\mathcal{M}}$ is sound above $\bar{\kappa}$, the fact that $\mathcal{V}$ is trivial readily implies that either $\mathcal{V}'$ is trivial as well, or else $\operatorname{lh}(\mathcal{V}') = 2$, and $\mathcal{M}_\infty^{\mathcal{V}'} = \mathcal{M}_1^{\mathcal{V}'} = \operatorname{ult}(\mathcal{P}_i, E_0^{\mathcal{V}'})$ where $\operatorname{crit}(E_0^{\mathcal{V}'}) = \kappa_i < \bar{\kappa}$, $\rho_\omega(\mathcal{M}_1^{\mathcal{V}'}) = \rho_\omega(\mathcal{P}_i) \le \kappa_i$, $\tau(E_0^{\mathcal{V}'}) \le \bar{\kappa}$, and $s(E_0^{\mathcal{V}'}) = \emptyset$.

We now have that $\bar{\mathcal{M}}$ is an iterate of K if either $\mathcal{V}'$ is trivial and $\bar{\mathcal{M}} \unlhd \mathcal{P}_\varphi$ or else if $\mathcal{V}'$ is non-trivial. On the other hand, if $\bar{\mathcal{M}}$ is not an iterate of K then we must have that $\bar{\mathcal{M}} \rhd \mathcal{P}_\varphi$. $\dashv$

Let us verify that $\bar{\mathcal{M}} \rhd \mathcal{P}_\varphi$ is impossible. Otherwise $\mathcal{P}_\varphi$ is a set premouse with $\rho_\omega(\mathcal{P}_\varphi) \le \bar{\kappa}$, and we may pick some $a \in \mathcal{P}(\bar{\kappa}) \cap (\Sigma_\omega(\mathcal{P}_\varphi) \setminus \mathcal{P}_\varphi)$. As $\bar{\mathcal{M}} \rhd \mathcal{P}_\varphi$, $a \in \bar{\mathcal{M}}$. However, by our inductive assumption on $\mathcal{M}$ (and by elementarity of π) we must have that $\mathcal{P}(\bar{\kappa}) \cap \bar{\mathcal{M}} \subset \mathcal{P}(\bar{\kappa}) \cap \bar{K} \subset \mathcal{P}_{\varphi}$. Therefore we'd get that $a \in \mathcal{P}_{\varphi}$ after all. Contradiction!

By Claim 4 we therefore must have that $\bar{\mathcal{M}}$ is an iterate of K. I.e., $\bar{K}$ and $\bar{\mathcal{M}}$ are hence both iterates of K, and we may apply Lemma 3.2 and deduce that the phalanx $((\bar{K}, \bar{\mathcal{M}}), \bar{\kappa})$ is iterable. This, however, is a contradiction as we chose π so that $((\bar{K}, \bar{\mathcal{M}}), \bar{\kappa})$ is not iterable. $\dashv$

Jensen has shown that Lemma 3.5 is false if in its statement we remove the assumption that $\kappa \geq \aleph_2$. He showed that if K has a measurable cardinal (but $0^\dagger$ may not exist) then there can be arbitrary large K-cardinals $\kappa < \aleph_2$ such that there is an iterable premouse $\mathcal{M} \rhd K \| \kappa$ with $\rho_\omega(\mathcal{M}) \leq \kappa$, $\mathcal{M}$ is sound above κ, but $\mathcal{M}$ is not an initial segment of K. In fact, the forcing presented in [22] can be used for constructing such examples.

To see that there can be arbitrary large K-cardinals $\kappa < \aleph_1$ such that there is an iterable premouse $\mathcal{M} \rhd K \| \kappa$ with $\rho_\omega(\mathcal{M}) \leq \kappa$, $\mathcal{M}$ is sound above κ, but $\mathcal{M}$ is not an initial segment of K, one can also argue as follows. $K \cap \mathrm{HC}$ need not projective (cf. [12]). If there is some $\eta < \aleph_1$ such that Lemma 3.5 holds for all K-cardinals in $[\eta, \aleph_1)$ then $K \cap \mathrm{HC}$ is certainly projective (in fact Σ_4^1).

By a *coarse premouse* we mean an amenable model of the form $\mathcal{P} = (P; \in, U)$ where P is transitive, $(P; \in) \models \mathrm{ZFC}^-$ (i.e., ZFC without the power set axiom), P has a largest cardinal, $\Omega = \Omega^\mathcal{P}$, and $\mathcal{P} \models$ "U is a normal measure on Ω." We shall say that the coarse premouse $\mathcal{P} = (P; \in, U)$ is *n-suitable* if $(P; \in) \models$ "$V_\Omega^\mathcal{P}$ is n-suitable," and $\mathcal{P}$ is *suitable* if $\mathcal{P}$ is n-suitable for some n. If $\mathcal{P}$ is n-suitable then $K^\mathcal{P}$, the core model "below $n + 1$ Woodin cardinals" inside $\mathcal{P}$ exists (cf. [32]).

DEFINITION 3.6. Let κ be an infinite cardinal. Suppose that for each $x \in H = \bigcup_{\theta < \kappa} H_{\theta^+}$ there is a suitable coarse premouse $\mathcal{P}$ with $x \in \mathcal{P} \in H$. Let $\alpha < \kappa$. We say that $K \| \alpha$ *stabilizes on a cone of elements of* H if there is some $x \in H$ such that for all suitable coarse premice $\mathcal{P}, \mathcal{Q} \in H$ with $x \in \mathcal{P} \cap \mathcal{Q}$ we have that $K^\mathcal{Q} \| \alpha = K^\mathcal{P} \| \alpha$. We say that K *stabilizes in* H if for all $\alpha < \kappa$, $K \| \alpha$ stabilizes on a cone of elements of H.

Notice that we might have $\alpha < \kappa < \lambda$ such that $K \| \alpha$ does not stabilize on a cone of elements of $\bigcup_{\theta < \kappa} H_{\theta^+}$, whereas $K \| \alpha$ does stabilize on a cone of elements of $\bigcup_{\theta < \lambda} H_{\theta^+}$. However, if we still have $\alpha < \kappa < \lambda$ and $K \| \alpha$ stabilizes on a cone of elements of $\bigcup_{\theta < \kappa} H_{\theta^+}$ then it also stabilizes on a cone of elements of $\bigcup_{\theta < \lambda} H_{\theta^+}$. The paper [23] shows that $K \| \alpha$ stabilizes on a cone of elements of $H_{(|\alpha|^{\aleph_0})^+}$ (cf. [23, Lemma 3.1.1]). What we shall need is that [23, Lemma 3.1.1] shows that $K \| \aleph_2$ stabilizes on a cone of elements of $H_{\aleph_3 \cdot (2^{\aleph_0})^+}$.

In the discussion of the previous paragraph we were assuming that enough suitable coarse premice exist.

THEOREM 3.7. *Let* $\kappa \geq \aleph_3 \cdot (2^{\aleph_0})^+$ *be a cardinal, and set* $H = \bigcup_{\theta < \kappa} H_{\theta^+}$. *Suppose that for each* $x \in H$ *there is a suitable coarse premouse* $\mathcal{P}$ *with* $x \in \mathcal{P}$. *Then* K *stabilizes in* H.

PROOF. By [23, Lemma 3.1.1], $K \| \aleph_2$ stabilizes on a cone of elements of H, because $\kappa \geq \aleph_3 \cdot (2^{\aleph_0})^+$. By Lemma 3.5 we may then work our way up to κ by just "stacking collapsing mice." $\dashv$

Theorem 3.7 gives a partial affirmative answer to [26, Question 5]. It can be used in a straighforward way to show that if $\square_\kappa$ fails, where $\kappa > 2^{\aleph_0}$ is a singular cardinal, then there is an inner model with a Woodin cardinal (cf. [23, Theorem 4.2]). One may use Theorems 3.9 and 3.11 below to show that if $\square_\kappa$ fails, where $\kappa > 2^{\aleph_0}$ is a singular cardinal, then for each $n < \omega$ there is an inner model with n Woodin cardinals.

Let Ω be an inaccessible cardinal, and let $X \in V_\Omega$. We say that V_Ω is (n, X)-*suitable* if $n < \omega$ and V_Ω is closed under $M_n^\#$, but $M_{n+1}^\#(X)$ does not exist (cf. [30, p. 81] or [5, p. 1841]). We say that V_Ω is X-*suitable* if there is some n such that V_Ω is (n, X)-suitable. If Ω is measurable and V_Ω is (n, X)-suitable then the core model $K(X)$ over X "below $n + 1$ Woodin cardinals" of height Ω exists (cf. [32]).

We shall say that the coarse premouse $\mathcal{P} = (P; \in, U)$ is (n, X)-*suitable* if $(P; \in) \models$ "$V_\Omega^{\mathcal{P}}$ is (n, X)-suitable," and $\mathcal{P}$ is X-*suitable* if $\mathcal{P}$ is (n, X)-suitable for some n. If $\mathcal{P}$ is (n, X)-suitable then $K(X)^{\mathcal{P}}$, the core model over X "below $n + 1$ Woodin cardinals" inside $\mathcal{P}$ exists (cf. [32]).

DEFINITION 3.8. Let κ be an infinite cardinal, and let $X \in H = \bigcup_{\theta < \kappa} H_{\theta^+}$. Suppose that for each $x \in H$ there is an X-suitable coarse premouse $\mathcal{P}$ with $x \in \mathcal{P} \in H$. Let $\alpha < \kappa$. We say that $K(X) \| \alpha$ *stabilizes on a cone of elements of* H if there is some $x \in H$ such that for all suitable coarse premice $\mathcal{P}, \mathcal{Q} \in H$ with $x \in \mathcal{P} \cap \mathcal{Q}$ we have that $K(X)^{\mathcal{Q}} \| \alpha = K(X)^{\mathcal{P}} \| \alpha$. We say that $K(X)$ *stabilizes in* H if for all $\alpha < \kappa$, $K(X) \| \alpha$ stabilizes on a cone of elements of H.

THEOREM 3.9. *Let* $\kappa \geq \aleph_3 \cdot (2^{\aleph_0})^+$ *be a cardinal, and let* $X \in H = \bigcup_{\theta < \kappa} H_{\theta^+}$. *Assume that, setting* $\xi = \mathrm{Card}(\mathrm{TC}(X))$, $\xi^{\aleph_0} < \kappa$. *Suppose that for each* $x \in H$ *there is an* X-*suitable coarse premouse* $\mathcal{P}$ *with* $x \in \mathcal{P}$. *Then* $K(X)$ *stabilizes in* H.

PROOF. Set $\alpha = \xi^+ \cdot \aleph_2$. By the appropriate version of [23, Lemma 3.1.1] for $K(X)$, $K(X) \| \alpha$ stabilizes on a cone of elements of H_λ, where $\lambda = (\xi^{\aleph_0})^+ \cdot \aleph_3 \cdot (2^{\aleph_0})^+$. Hence $K(X) \| \alpha$ stabilizes on a cone of elements H. But then $K(X)$ stabilizes in H by an appropriate version of Lemma 3.5. $\dashv$

We do not know how to remove the assumption that $\xi^{\aleph_0} < \kappa$ from Theorem 3.9. For our application we shall therefore need a different method for working ourselves up to a given cardinal.

LEMMA 3.10. *Let* $\aleph_2 \leq \kappa \leq \lambda < \Omega$ *be such that* κ *and* λ *are cardinals and* Ω *is a measurable cardinal. Let* $n < \omega$ *be such that for every bounded* $X \subset \kappa$, $M_{n+1}^{\#}(X)$ *exists. Let* $X \subset \kappa$ *be such that* V_Ω *is* (n, X)-*suitable.*

Let $\mathcal{M} \trianglerighteq K(X)\|\lambda$ *be an iterable* X-*premouse such that* $\rho_\omega(\mathcal{M}) \leq \lambda$ *and* $\mathcal{M}$ *is sound above* λ. *Then* $\mathcal{M} \triangleleft K(X)$.

PROOF. The proof is by "induction on $\mathcal{M}$." Let us fix κ, Ω, n, and X. Let us suppose that λ is least such that there is an X-premouse $\mathcal{M} \trianglerighteq K(X)\|\lambda$ such that $\rho_\omega(\mathcal{M}) \leq \lambda$, $\mathcal{M}$ is sound above λ, but $\mathcal{M}$ is not an initial segment of $K(X)$. Let $\mathcal{M} \trianglerighteq K(X)\|\lambda$ be such that $\rho_\omega(\mathcal{M}) \leq \lambda$, $\mathcal{M}$ is sound above λ, $\mathcal{M}$ is not an initial segment of $K(X)$, but if $K(X)\|\lambda \trianglelefteq \mathcal{N} \triangleleft \mathcal{M}$ is such that $\rho_\omega(\mathcal{N}) \leq \lambda$ and $\mathcal{N}$ is sound above λ then $\mathcal{N}$ is an initial segment of $K(X)$. In order to derive a contradiction it suffices to prove that the phalanx $((K(X), \mathcal{M}), \lambda)$ is not iterable.

Let us now imitate the proof of Lemma 3.5. Let

$$\pi : N \longrightarrow V_{\Omega+2}$$

be such that N is transitive, $\mathrm{Card}(N) = \aleph_1$, $\{K(X), \mathcal{M}, \kappa, \lambda\} \subset \mathrm{ran}(\pi)$, and $\pi''N \cap \aleph_2 \in \aleph_2$. Let $\bar{X} = \pi^{-1}(X)$, $\bar{\Omega} = \pi^{-1}(\Omega)$, $\bar{K}(\bar{X}) = \pi^{-1}(K(X))$, $\bar{\mathcal{M}} = \pi^{-1}(\bar{M})$, $\bar{\kappa} = \pi^{-1}(\kappa)$, $\bar{\lambda} = \pi^{-1}(\lambda)$, and $\delta = \pi^{-1}(\aleph_2) = \mathrm{crit}(\pi)$. We may and shall assume that $((\bar{K}(\bar{X}), \bar{\mathcal{M}}), \bar{\lambda})$ is not iterable. Furthermore, by the method of [18], we may and shall assume that the phalanxes occuring in the proof to follow are all iterable.

Let $\bar{\lambda}' \leq \bar{\Omega}$ be largest such that $\bar{K}(\bar{X})|\bar{\lambda}'$ does not move in the coiteration with $M_{n+1}^{\#}(\bar{X})$. Let $\mathcal{T}$ be the canonical normal iteration tree on $M_{n+1}^{\#}(\bar{X})$ of length $\theta + 1$ such that $\bar{K}(\bar{X})|\bar{\lambda}' \trianglelefteq \mathcal{M}_\theta^{\mathcal{T}}$. Let $(\kappa_i \mid i \leq \varphi)$ be the strictly monotone enumeration of the set of cardinals of $\bar{K}(\bar{X})|\bar{\lambda}'$, including $\bar{\lambda}'$, which are $\geq \delta$. For each $i \leq \varphi$, let the objects $\mathcal{P}_i$, $\mathcal{R}_i$, and $\mathcal{S}_i$ be defined exactly as in the proof of Lemma 3.5. For $i < \varphi$, let $\lambda_i = \kappa_{i+1}$.

Because $\rho_\omega(M_{n+1}^{\#}(\bar{X})) \leq \delta$, and as $\bar{K}(\bar{X})$ is n-small, whereas $M_{n+1}^{\#}(\bar{X})$ is not, we have that for each $i \leq \varphi$, $\mathcal{P}_i$ is a set-sized premouse with $\rho_\omega(\mathcal{P}_i) \leq \kappa_i$ such that $\mathcal{P}_i$ is sound above κ_i. Therefore, for each $i \leq \varphi$, $\mathcal{S}_i$ is a set-sized premouse with $\rho_\omega(\mathcal{S}_i) \leq \pi(\kappa_i)$ such that $\mathcal{S}_i$ is sound above $\pi(\kappa_i)$.

Let us verify that the phalanx

$$\vec{\mathcal{P}}_{\bar{\mathcal{M}}} = ((\mathcal{P}_i \mid i < \varphi)^\frown \bar{\mathcal{M}}, (\kappa_i \mid i < \varphi))$$

is coiterable with the phalanx

$$\vec{\mathcal{P}} = ((\mathcal{P}_i \mid i \leq \varphi), (\kappa_i \mid i < \varphi)).$$

In fact, by our inductive hypothesis, we shall now have that for each $i < \varphi$, $\mathcal{S}_i \triangleleft K(X)$ and hence $\mathcal{S}_i \triangleleft \mathcal{M}$. Setting $\Lambda_i = \sup(\pi''\lambda_i)$ for $i < \varphi$, we thus have that

$$((\mathcal{S}_i \mid i < \varphi)^\frown \mathcal{M}, (\Lambda_i \mid i < \varphi))$$

is a special phalanx which is iterable with respect to special iteration trees. As in the proof of [19], we therefore first get that

$$((\mathcal{R}_i \mid i < \varphi)^\frown \mathcal{M}, (\Lambda_i \mid i < \varphi))$$

is a very special phalanx which is iterable with respect to special iteration trees, and then that the phalanx

$$((\mathcal{P}_i \mid i < \varphi)^\frown \bar{\mathcal{M}}, (\Lambda_i \mid i < \varphi))$$

is iterable.

We may therefore coiterate $\vec{\mathcal{P}}_{\bar{\mathcal{M}}}$ with $\vec{\mathcal{P}}$. Standard arguments then show that this implies that $\bar{\lambda}'$ cannot be the index of an extender which is used in the comparison of $\mathcal{M}^\#_{n+1}(\bar{X})$ with $\bar{K}(\bar{X})$. We may conclude that $\bar{\lambda}' = \bar{\Omega}$, i.e., $\bar{K}(\bar{X})$ doesn't move in the comparison with $\mathcal{M}^\#_{n+1}(\bar{X})$. In other words, $\bar{K}(\bar{X}) = \mathcal{M}^T_\theta$.

However, we may now finish the argument exactly as in the proof of Lemma 3.5. The coiteration of $\vec{\mathcal{P}}_{\bar{\mathcal{M}}}$ with $\vec{\mathcal{P}}$ gives that either $\bar{\mathcal{M}}$ is an iterate of $\mathcal{M}^\#_{n+1}(\bar{X})$, or else that $\bar{\mathcal{M}} \rhd \mathcal{P}_\varphi$. By our assumptions on $\mathcal{M}$, we cannot have that $\mathcal{P}_\varphi \lhd \bar{\mathcal{M}}$. Therefore, $\bar{\mathcal{M}}$ is an iterate of $\mathcal{M}^\#_{n+1}(\bar{X})$. However, the proof of Lemma 3.2 implies that the phalanx $((\mathcal{M}^T_\theta, \bar{\mathcal{M}}), \bar{\lambda})$ is iterable. This is because the existence of $\mathcal{M}^\#_{n+1}(\bar{X})$ means that the K^c construction, when relativized to $\bar{X}$, is not n-small and reaches $\mathcal{M}^\#_{n+1}(\bar{X})$. But now $((\bar{K}(\bar{X}), \bar{\mathcal{M}}), \bar{\lambda})$ is iterable, which is a contradiction!   ⊣

THEOREM 3.11. *Let κ be a cardinal, and let $X \in H = \bigcup_{\theta<\kappa} H_{\theta^+}$. Let $n < \omega$. Assume that, setting $\xi = \mathrm{Card}(\mathrm{TC}(X))$, $\xi \geq \aleph_2$ and $M_{n+1}(\bar{X})$ exists for all bounded $\bar{X} \subset \xi$. Suppose further that for each $x \in H$ there is an (n, X)-suitable coarse premouse $\mathcal{P}$ with $x \in \mathcal{P}$. Then $K(X)$ stabilizes in H.*

PROOF. This immediately follows from 3.10.   ⊣

We now have to turn towards the task of majorizing functions in $\prod(\{\kappa_i^+ \mid i \in A\})$ by functions from the core model.

LEMMA 3.12. *Let Ω be a measurable cardinal, and suppose that V_Ω is suitable. Let K denote the core model of height Ω. Let $\kappa < \Omega$ be a limit cardinal with $\aleph_0 < \delta = \mathrm{cf}(\kappa) < \kappa$. Let $\vec{\kappa} = (\kappa_i \mid i < \delta)$ be a strictly increasing continuous sequence of singular cardinals below κ which is cofinal in κ and such that $\delta \leq \kappa_0$. Let $\mathfrak{M} = (V_{\Omega+2}; \ldots)$ be a model whose type has cardinality at most δ.*

There is then a pair (Y, f) such that $(Y; \ldots) \prec \mathfrak{M}$, $(\kappa_i \mid i < \delta) \subset Y$, $f : \kappa \to \kappa$, $f \in K$, and for all but nonstationarily many $i < \delta$, $f(\kappa_i) = \mathrm{char}^Y_\kappa(\kappa_i^+)$.

PROOF. Once more we shall make heavy use of the covering argument. Let

$$\pi : N \cong Y \prec V_{\Omega+2}$$

be such that $(Y; \dots) \prec \mathfrak{M}$, $(\kappa_i \mid i < \delta) \in Y$, and that all the objects occuring in the proof of [18] are iterable. Let $\bar{K} = \pi^{-1}(K)$, $\bar{\kappa} = \pi^{-1}(\kappa)$, and $\bar{\kappa}_i = \pi^{-1}(\kappa_i)$ for $i < \delta$.

We define $\mathcal{P}_i'$, $\mathcal{R}_i'$, and $\mathcal{S}_i'$ in exactly the same way as $\mathcal{P}_i$, $\mathcal{R}_i$, and $\mathcal{S}_i$ were defined in the proof of Lemma 3.5. Let $\mathcal{T}$ be the normal iteration tree on K arising from the coiteration with $\bar{K}$. Set $\theta + 1 = \mathrm{lh}(\mathcal{T})$. Let $(\kappa_i' \mid i \le \varphi)$ be the strictly monotone enumeration of $\mathrm{Card}^{\bar{K}} \cap (\bar{\kappa} + 1)$, and set $\lambda_i' = (\kappa_i')^{+\bar{K}}$ for $i \le \varphi$. Let, for $i \le \varphi$, $\alpha_i < \theta$ be the least α such that $\kappa_i' < \nu(E_\alpha^{\mathcal{T}})$, if there is some such α; otherwise let $\alpha_i = \theta$. Let, for $i \le \varphi$, $\mathcal{P}_i'$ be the longest initial segment of $\mathcal{M}_{\alpha_i}^{\mathcal{T}}$ such that $\mathcal{P}(\kappa_i) \cap \mathcal{P}_i' \subset \bar{K}$. Let

$$\mathcal{R}_i' = \mathrm{ult}(\mathcal{P}_i', E_\pi \restriction \pi(\kappa_i')),$$

where $i \le \varphi$. We recursively define $(\mathcal{S}_i' \mid i \le \varphi)$ as follows. If $\mathcal{R}_i'$ is a premouse then we set $\mathcal{S}_i' = \mathcal{R}_i'$. If $\mathcal{R}_i'$ is not a premouse then we set

$$\mathcal{S}_i' = \mathrm{ult}\left(\mathcal{S}_j', \dot{F}^{\mathcal{R}_i'}\right),$$

where $\kappa_j' = \mathrm{crit}(\dot{F}^{\mathcal{P}_i'})$ (we have $j < i$). Set $\Lambda_i' = \sup(\pi'' \lambda_i')$ for $i \le \varphi$.

We also want to define $\mathcal{P}_i$, $\mathcal{R}_i$, and $\mathcal{S}_i$. For $i < \delta$, we simply pick $i' < \varphi$ such that $\kappa_i = \kappa_{i'}'$, and then set $\mathcal{P}_i = \mathcal{P}_{i'}'$, $\mathcal{R}_i = \mathcal{R}_{i'}'$, and $\mathcal{S}_i = \mathcal{S}_{i'}'$; we also set $\theta_i = \alpha_{i'}$. Notice that we'll have that

$$\kappa_i^{+\mathcal{R}_i} = \kappa_i^{+\mathcal{S}_i} = \sup\left(N \cap \kappa_i^{+V}\right),$$

because $\bar{\kappa}_i^{+\mathcal{P}_i} = \bar{\kappa}_i^{+\bar{K}} = \bar{\kappa}_i^{+N}$ (the latter equality holds by [18]).

Let (A) denote the assertion (which might be true or false) that

$$\left\{\nu(E_\alpha^{\mathcal{T}}) \mid \alpha + 1 \le \theta\right\} \cap \bar{\kappa}$$

is unbounded in $\bar{\kappa}$. Let us define some $C \subset \delta$.

If (A) fails then $\mathcal{M}_{\theta_i}^{\mathcal{T}} = \mathcal{M}_\theta^{\mathcal{T}}$ for all but boundedly many $i < \delta$, which readily implies that there is some $\eta < \delta$ such that $\mathcal{P}_i = \mathcal{P}_j$ whenever $i, j \in \delta \setminus \eta$. In this case, we simply set $C = \delta \setminus \eta$.

Suppose now that (A) holds. Then $\mathrm{cf}(\theta) = \mathrm{cf}(\bar{\kappa}) = \delta > \aleph_0$, and both $[0, \theta)_T$ as well as $\{\theta_i \mid i < \delta\}$ are closed unbounded subsets of θ. Moreover, the set of all $i < \delta$ such that

$$\forall \alpha + 1 \in (0, \theta]_T \left(\mathrm{crit}\left(E_\alpha^{\mathcal{T}}\right) < \bar{\kappa}_i \Rightarrow \nu(E_\alpha^{\mathcal{T}}) < \bar{\kappa}_i\right)$$

is club in δ. There is hence some club $C \subset \delta$ such that whenever $i \in C$ then $\theta_i \in [0, \theta)_T$, $\forall \alpha + 1 \in (0, \theta]_T (\mathrm{crit}(E_\alpha^{\mathcal{T}}) < \bar{\kappa}_i \Rightarrow \nu(E_\alpha^{\mathcal{T}}) < \bar{\kappa}_i)$, and $[\theta_i, \theta]_T$ does not contain drops of any kind. By (A), $\bar{\kappa}$ is a cardinal in $\mathcal{M}_\theta^{\mathcal{T}}$, and it is thus easy to see that in fact for $i \in C$, $\mathcal{P}_i = \mathcal{M}_{\theta_i}^{\mathcal{T}}$. Moreover, if $i \le j \in C$ then $\pi_{\theta_i \theta_j}^{\mathcal{T}} : \mathcal{P}_i \to \mathcal{P}_j$ is such that $\pi_{\theta_i \theta_j}^{\mathcal{T}} \restriction \bar{\kappa}_i = \mathrm{id}$.

Let us now continue or discussion regardless of whether (A) holds true or not.

If $i \leq j \in C$ then we may define a map $\varphi_{ij} : \mathcal{R}_i \to \mathcal{R}_j$ by setting

$$[a, f]_{E_\pi \restriction \kappa_i}^{\mathcal{P}_i} \longmapsto [a, \pi_{\theta_i \theta_j}^{\mathcal{T}}(f)]_{E_\pi \restriction \kappa_j}^{\mathcal{P}_j},$$

where $a \in [\kappa_i]^{<\omega}$, and f ranges over those functions $f : [\bar{\kappa}_i]^k \to \mathcal{M}_{\theta_i}^{\mathcal{T}}$, some $k < \omega$, which are used for defining the long ultrapower of $\mathcal{P}_i$.

We now have to split the remaining argument into cases. We may and shall without loss of generality assume that C was chosen such that exactly one of the four following clauses holds true.

CLAUSE 1. For all $i \in C$, $\mathcal{P}_i$ is a set premouse, and $\mathcal{S}_i = \mathcal{R}_i$.

CLAUSE 2. For all $i \in C$, $\mathcal{P}_i$ is a weasel, and hence $\mathcal{S}_i = \mathcal{R}_i$.

CLAUSE 3. For all $i \in C$, $\mathcal{R}_i$ is a protomouse, $\mathcal{S}_i \neq \mathcal{R}_i$, and $\mathcal{S}_i$ is a set premouse.

CLAUSE 4. For all $i \in C$, $\mathcal{R}_i$ is a protomouse, $\mathcal{S}_i \neq \mathcal{R}_i$, and $\mathcal{S}_i$ is a weasel.

CASE 1. Clause 1, 3, or 4 holds true.

In this case we'll have that for all $i \in C$, $\mathcal{P}_i$ is a premouse with $\rho_\omega(\mathcal{P}_i) \leq \bar{\kappa}_i$. In fact, if i_0 is least in C then we shall have that $\rho_\omega(\mathcal{P}_i) \leq \bar{\kappa}_{i_0}$ for all $i \in C$. Moreover, $\mathcal{P}_i$ is sound above κ_i.

Let $n < \omega$ be such that $\rho_{n+1}(\mathcal{P}_i) \leq \bar{\kappa}_{i_0} < \rho_n(\mathcal{P}_i)$ for $i \in C$. Notice that for $i \leq j \in C$, $\mathcal{P}_i$ is the transitive collapse of

$$H_{n+1}^{\mathcal{P}_j}(\bar{\kappa}_i \cup \{p_{\mathcal{P}_j, n+1}\}),$$

where the inverse of the collapsing map is either $\pi_{\theta_i \theta_j}^{\mathcal{T}}$ (if (A) holds) or else is the identity (if (A) fails). Moreover, $\mathcal{R}_i$ is easily seen to be the transitive collapse of

$$H_{n+1}^{\mathcal{R}_j}(\kappa_i \cup \{p_{\mathcal{R}_j, n+1}\}),$$

where the inverse of the collapsing map is exactly φ_{ij}.

CASE 1.1. Clause 1 holds true.

In this case, $\mathcal{R}_i \in K$ for all $i \in C$, by [19, Lemma 3.10]. Let us define $f : \kappa \to \kappa$ as follows. We set $f(\xi) = \xi^+$ in the sense of the transitive collapse of

$$H_{n+1}^{\mathcal{R}_i}(\xi \cup \{p_{\mathcal{R}_i, n+1}\}),$$

where $i \in C$ is large enough so that $\xi \leq \kappa_i$. Due to the existence of the maps φ_{ij}, $f(\xi)$ is independent from the particular choice of i, and thus f is well-defined. Obviously, $f \restriction \gamma \in K$ for all $\gamma < \kappa$. Moreover, $f(\kappa_i) = \kappa_i^{+\mathcal{R}_i} = \sup(Y \cap \kappa_i^+)$ for all $i \in C$, as desired.

It is easy to verify that in fact $f \in K$. Let $\tilde{\mathcal{R}}$ be the premouse given by the direct limit of the system

$$(\mathcal{R}_i, \varphi_{ij} \mid i \leq j \in C).$$

As $\delta > \aleph_0$ this system does indeed have a well-founded direct limit which we can then take to be transitive; for the same reason, $\tilde{\mathcal{R}}$ will be iterable. We may then use Lemma 3.5 to deduce that actually $\tilde{\mathcal{R}} \in K$. However, we shall have that, for $\xi < \kappa$, $f(\xi) = \xi^+$ in the sense of the transitive collapse of

$$H_{n+1}^{\tilde{\mathcal{R}}}(\xi \cup \{p_{\tilde{\mathcal{R}},n+1}\}).$$

CASE 1.2. Clause 3 or 4 holds.

In this case, [19, Lemma 2.5.2] gives information on how $\mathcal{P}_i$ has to look like, for $i \in C$. In particular, $\mathcal{P}_i$ will have a top extender, $\dot{F}^{\mathcal{P}_i}$. By [19, Corollary 3.4], we'll have that $\tau(\dot{F}^{\mathcal{P}_i}) \leq \bar{\kappa}_i$.

Let $\mu = \mathrm{crit}(\dot{F}^{\mathcal{P}_i}) = \mathrm{crit}(\dot{F}^{\mathcal{P}_j})$ for $i, j \in C$. Of course, $\pi(\mu) = \mathrm{crit}(\dot{F}^{\mathcal{R}_i}) = \mathrm{crit}(\dot{F}^{\mathcal{R}_j})$ for $i, j \in C$. Let $\mu = \kappa'_k$, where $k < \varphi$. Setting $\mathcal{S} = \mathcal{S}'_k$, we have that

$$\mathcal{S}_i = \mathrm{ult}\left(\mathcal{S}, \dot{F}^{\mathcal{R}_i}\right)$$

for all $i \in C$.

By [19, Corollary 3.4], $\mathcal{P}_i$ is also Dodd-solid above $\bar{\kappa}_i$, for $i \in C$. By [19, Lemma 2.1.4], $\pi_{\theta_i \theta_j}^{\mathcal{T}}(s(\dot{F}^{\mathcal{P}_i})) = s(\dot{F}^{\mathcal{P}_j})$ for $i \leq j \in C$. Due to the existence of the maps φ_{ij}, it is then straightforward to verify that

$$\dot{F}^{\mathcal{R}_j} \upharpoonright \left(\kappa_i \cup s(\dot{F}^{\mathcal{R}_j})\right) = \dot{F}^{\mathcal{R}_i}$$

whenever $i \leq j \in C$.

Let us now define $f : \kappa \to \kappa$ as follows. We set $f(\xi) = \xi^+$ in the sense of

$$\mathrm{ult}\left(\mathcal{S}, \dot{F}^{\mathcal{R}_i} \upharpoonright \left(\xi \cup s(\dot{F}^{\mathcal{R}_i})\right)\right),$$

where $i \in C$ is large enough so that $\xi \leq \kappa_i$. $f(\xi)$ is then independent from the particular choice of i, and therefore f is well-defined. Moreover, $f(\kappa_i) = \kappa_i^{+S_i} = \sup(Y \cap \kappa_i^+)$.

CASE 1.2.1. Clause 3 holds.

By [19, Lemma 3.10], $\mathcal{S}_i \in K$ for all $i \in C$. Also, $\mathcal{S} \in K$.

In order to see that $f \upharpoonright \gamma \in K$ for all $\gamma < \kappa$ it suffices to verify that $\dot{F}^{\mathcal{R}_i} \in K$ for all $i \in C$. Fix $i \in C$. Let $m < \omega$ be such that $\rho_{m+1}(\mathcal{S}) \leq \pi(\mu) < \rho_m(\mathcal{S})$, and let

$$\sigma : \mathcal{S} \longrightarrow_{\dot{F}^{\mathcal{R}_i}} \mathcal{S}_i = K \| \beta_i,$$

some β_i. It is then straightforward to verify that

$$\mathrm{ran}(\sigma) = H_{m+1}^{K\|\beta_i}\left(\pi(\mu) \cup \sigma(p_{\mathcal{S},m+1}) \cup s(\dot{F}_i^{\mathcal{R}})\right).$$

This implies that $\sigma \in K$. But then $\dot{F}^{\mathcal{R}_i} \in K$ as well.

But now letting $\tilde{\mathcal{R}}$ be as in Case 1.1 we may actually conclude that $f \in K$.

CASE 1.2.2. Clause 4 holds.

We know that $S \neq K$, as π is discontinuous at $\mu^{+\tilde{K}}$. We also know that for all $i \in C$, $S_i \neq K$, as π is discontinuous at $\pi^{-1}(\kappa_i^+)$. We now have Claim 2 from the proof of Lemma 3.5 at our disposal, which gives the following. There is some v such that $S = \mathrm{ult}(K, E_v^K)$, where $\mathrm{crit}(E_v^K) < \mathrm{crit}(\pi)$, $v \geq \sup(\kappa_i^+ \cap Y)$, and $\tau(E_v^K) \leq \pi(\mu)$. Also, for every $i \in C$, there is some v_i such that $S_i = \mathrm{ult}(K, E_{v_i}^K)$, where $\mathrm{crit}(E_{v_i}^K) = \mathrm{crit}(E_v^K) < \mathrm{crit}(\pi)$, $v_i \geq \sup(\kappa_i^+ \cap Y)$, and $\tau(E_v^K) \leq \kappa_i$.

In order to see that $f \upharpoonright \gamma \in K$ for all $\gamma < \kappa$ it now again suffices to verify that $\dot{F}^{\mathcal{R}_i} \in K$ for all $i \in C$. Fix $i \in C$. Let

$$\sigma : S \longrightarrow_{\dot{F}^{\mathcal{R}_i}} S_i.$$

Let us also write

$$\bar{\sigma} : K \longrightarrow_{E_v^K} S,$$

and

$$\bar{\sigma}_i : K \longrightarrow_{E_{v_i}^K} S_i.$$

Standard arguments, using hull- and definability properties, show that in fact

$$\bar{\sigma}_i = \sigma \circ \bar{\sigma}.$$

Therefore,

$$\sigma(\bar{\sigma}(f)(a)) = \bar{\sigma}_i(f)(\sigma(a))$$

for the appropriate a, f. As $\kappa(S) \leq \pi(\mu) = \mathrm{crit}(\dot{F}^{\mathcal{R}_i})$, we may hence compute $\dot{F}^{\mathcal{R}_i}$ inside K.

By letting $\tilde{\mathcal{R}}$ be as in Case 1.1 we may again conclude that actually $f \in K$.

CASE 2. Clause 2 holds.

Let $i \in C$. Then $\mathcal{R}_i$ is a weasel with $\kappa(\mathcal{R}_i) = \kappa_i$ and $c(\mathcal{R}_i) = \emptyset$. This, combined with the proof of [19, Lemma 3.11], readily implies that

$$\mathcal{R}_i = \mathrm{ult}\left(K, E_{\Lambda_i}^K\right),$$

where $\tau(E_{\Lambda_i}^K) \leq \kappa_i$ and $s(E_{\Lambda_i}^K) = \emptyset$. Moreover, by the proof of [19, Lemma 3.11, Claim 2], $\mathrm{crit}(E_{\Lambda_i}^K) = \mathrm{crit}(\pi_{0\theta_i}^{\mathcal{T}}) < \mathrm{crit}(\pi)$. Let us write

$$\pi_i : K \longrightarrow_{E_{\Lambda_i}^K} \mathcal{R}_i,$$

and let us write

$$\sigma_i : \mathcal{P}_i \longrightarrow \mathcal{R}_i$$

for the canonical long ultrapower map. Notice that we must have

$$\pi_i = \sigma_i \circ \pi_{0\theta_i}^{\mathcal{T}}.$$

Furthermore, if $i \leq j \in C$, then we'll have that

$$\sigma_j \circ \pi_{0\theta_j}^T = \varphi_{ij} \circ \sigma_i \circ \pi_{0\theta_i}^T.$$

Let us define $f : \kappa \to \kappa$ as follows. We set $f(\xi) = \xi^+$ in the sense of

$$\mathrm{ult}\left(K, E_{\Lambda_i}^K \upharpoonright \xi\right),$$

where $i \in C$ is large enough so that $\xi \leq \kappa_i$. If $f(\xi)$ is independent from the choice of i then f is well-defined, $f \upharpoonright \gamma \in K$ for all $\gamma < \kappa$, and $f(\kappa_i) = \sup(Y \cap \kappa_i^+)$ for all $i \in C$.

Now let $i \leq j \in C$. We aim to verify that

$$E_{\Lambda_i}^K = E_{\Lambda_j}^K \upharpoonright \kappa_i,$$

which will prove that $f(\xi)$ is independent from the choice of i.

Well, we know that $\mathrm{crit}(E_{\Lambda_i}^K) = \mathrm{crit}(\pi_{0\theta_i}) = \mathrm{crit}(\pi_{0\theta_j}) = \mathrm{crit}(E_{\Lambda_j}^K)$; call it μ. Fix $a \in [\kappa_i]^{<\omega}$ and $X \in \mathcal{P}([\mu]^{\mathrm{Card}(a)}) \cap K$. We aim to prove that

$$X \in \left(E_{\Lambda_i}^K\right)_a \iff X \in \left(E_{\Lambda_j}^K\right)_a.$$

But we have that $X \in (E_{\Lambda_i}^K)_a$ if and only if $a \in \sigma_i \circ \pi_{0\theta_i}^T(X)$ if and only if $a \in \varphi_{ij} \circ \sigma_i \circ \pi_{0\theta_i}^T(X) = \sigma_j \circ \pi_{0\theta_j}^T(X)$ if and only if $a \in (E_{\Lambda_j}^K)_a$, as desired.

We may now finally let $\tilde{\mathcal{R}}$ be the weasel given by the direct limit of the system

$$(\mathcal{R}_i, \varphi_{ij} \mid i \leq j \in C).$$

The above arguments can then easily be adopted to show that $f \in K$.   ⊣

We have separated the arguments that $f \upharpoonright \gamma \in K$ for all $\gamma < \kappa$ from the arguments that $f \in K$, as the former ones also work for a "stable K up to κ," for which the latter ones don't make much sense.

The following is a version of Lemma 3.12 for the stable $K(X)$ up to κ.

LEMMA 3.13. *Let $\kappa > 2^{\aleph_0}$ be a limit cardinal with $\mathrm{cf}(\kappa) = \delta > \aleph_0$, and let $\vec{\kappa} = (\kappa_i \mid i < \delta)$ be a strictly increasing continuous sequence of singular cardinals below κ which is cofinal in κ with $\delta \leq \kappa_0$. Let $X \in H = \bigcup_{\theta < \kappa} H_{\theta^+}$. Suppose that for each $x \in H$ there is an X-suitable coarse premouse $\mathcal{P}$ with $x \in \mathcal{P}$. Let us further assume that $K(X)$ stabilizes in H, and let $K(X)$ denote the stable $K(X)$ up to κ. Let $\lambda = \delta \cdot \mathrm{Card}(\mathrm{TC}(X))$. Let $\mathfrak{M} = (H; \ldots)$ be a model whose type has cardinality at most λ.*

There is then a pair (Y, f) such that $(Y; \ldots) \prec \mathfrak{M}$, $\mathrm{Card}(Y) = \lambda$, $(\kappa_i \mid i < \delta) \subset Y$, $\mathrm{TC}(\{X\}) \subset Y$, $f : \kappa \to \kappa$, $f \upharpoonright \gamma \in K(X)$ for all $\gamma < \kappa$, and for all but nonstationarily many $i < \delta$, $f(\kappa_i) = \mathrm{char}_{\vec{\kappa}}^Y(\kappa_i)$.

Moreover, whenever $(Y; \ldots) \prec \mathfrak{M}$ is such that $^\omega Y \subset Y$, $\mathrm{Card}(Y) = \lambda$, $(\kappa_i \mid i < \delta) \subset Y$, and $\mathrm{TC}(\{X\}) \subset Y$, then there is an $f : \kappa \to \kappa$ such that $f \upharpoonright \gamma \in K(X)$ for all $\gamma < \kappa$, and for all but nonstationarily many $i < \delta$, $f(\kappa_i) = \mathrm{char}_{\vec{\kappa}}^Y(\kappa_i)$.

PROOF. The proof runs in much the same way as before. For each $i < \delta$, there is some $x = x_i \in H$ such that for all X-suitable coarse premice $\mathcal{P}$ with $x \in \mathcal{P}$ we have that $K(X)^{\mathcal{P}} \| \kappa_i = K(X) \| \kappa_i$. For $i < \delta$, let $\mathcal{P}_i$ be an X-suitable coarse premouse with $x_i \in \mathcal{P}$.

We may pick

$$\pi : N \cong Y \prec H,$$

where N is transitive, such that $(Y; \ldots) \prec \mathfrak{M}$, $(\kappa_i \mid i < \delta) \subset Y$, $\mathrm{TC}(\{X\}) \subset Y$, and such that simultaneously for all $i < \delta$, if one runs the proof of [18] with respect to $K(X)^{\mathcal{P}_i}$ then all the objects occuring in this proof are iterable. Let $\bar{K}(X)$ be defined over N in exactly the same way as $K(X)$ is defined over H. There is a normal iteration tree $\mathcal{T}$ on $K(X)$ such that for all $i < \delta$ there is some $\alpha_i \leq \mathrm{lh}(\mathcal{T})$ with $\bar{K}(X) \| \pi^{-1}(\kappa_i) \trianglelefteq \mathcal{M}_{\alpha_i}^{\mathcal{T}}$. (For all we know $\mathcal{T}$ might have limit length and no cofinal branch, though.)

We may then construct f in much the same way as in the proof of Lemma 3.12. If some $\gamma < \kappa$ is given with $\gamma < \kappa_i$, some $i < \delta$, then may argue inside the coarse premouse $\mathcal{P}_i$ and deduce that $f \upharpoonright \gamma \in K(X)^{\mathcal{P}} \| \kappa_i = K(X) \| \kappa_i$.

This proves the first part of Lemma 3.13. The "moreover" part of Lemma 3.13 follows from the method by which [19, Lemma 3.13] is proven.   ⊣

§4. The proofs of the main results.

PROOF OF THEOREM 1.1. Let α be as in the statement of Theorem 1.1. Set

$$\mathsf{a} = \{ \kappa \in \mathrm{Reg} \mid |\alpha|^+ \leq \kappa < \aleph_\alpha \}.$$

As $2^{|\alpha|} < \aleph_\alpha$, [2, Theorem 5.1] yields that $\max(\mathrm{pcf}(\mathsf{a})) = |\prod \mathsf{a}|$. However, $|\prod \mathsf{a}| = \aleph_\alpha^{|\alpha|}$ (cf. [14, Lemma 6.4]). Because $\aleph_\alpha^{|\alpha|} > \aleph_{|\alpha|^+}$, we therefore have that

$$\max(\mathrm{pcf}(\mathsf{a})) > \aleph_{|\alpha|^+}.$$

This in turn implies that

$$\{ \aleph_{\beta+1} \mid \alpha < \beta \leq |\alpha|^+ \} \subset \mathrm{pcf}(\mathsf{a})$$

by [2, Corollary 2.2]. Set

$$H = \bigcup_{\theta < \aleph_{|\alpha|^+}} H_{\theta^+}.$$

We aim to prove that for each $n < \omega$, H is closed under $X \mapsto M_n^\#(X)$.

To commence, let $X \in H$, and suppose that $X^\# = M_0^\#(X)$ does not exist. By [2, Theorem 6.10] there is some

$$\mathsf{d} \subset \{ \aleph_{\beta+1} \mid \alpha < \beta < |\alpha|^+ \}$$

with $\min(\mathsf{d}) > \mathrm{TC}(X)$, $|\mathsf{d}| \leq |\alpha|$, and $\aleph_{|\alpha|^++1} \in \mathrm{pcf}(\mathsf{d})$. By [3], however, we have that

$$\{f \in \textstyle\prod \mathsf{d} \mid f = \tilde{f} \restriction \mathsf{d}, \text{ some } \tilde{f} \in L[X]\}$$

is cofinal in $\prod \mathsf{d}$. As GCH holds in $L[X]$ above X, this yields $\max(\mathrm{pcf}(\mathsf{d})) \leq \sup(\mathsf{d})^+$. Contradiction!

Hence H is closed under $X \mapsto X^\# = M_0^\#(X)$.

Now let $n < \omega$ and assume inductively that H is closed under $X \mapsto M_n^\#(X)$. Fix X, a bounded subset of $\aleph_{|\alpha|^+}$. Let us assume towards a contradiction that $M_{n+1}^\#(X)$ does not exist.

Without loss of generality, $\kappa = \sup(X)$ is a cardinal of V. We may and shall assume inductively that if $\kappa \geq \aleph_2$ and if $\bar{X} \subset \kappa$ is bounded then $M_{n+1}^\#(\bar{X})$ exists.

We may use the above argument which gave that H is closed under $Y \mapsto Y^\#$ together with [24, Theorem 5.3] (rather than [3]) and deduce that for every $x \in H$ there is some (n, X)-suitable coarse premouse containing x. We claim that $K(X)$ stabilizes in H. Well, if $\kappa \leq \aleph_1$ then this follows from Theorem 3.9. On the other hand, if $\kappa \geq \aleph_2$ then this follows from Theorem 3.11 together with our inductive hypothesis according to which $M_{n+1}^\#(\bar{X})$ exists for all bounded $\bar{X} \subset \kappa$. Let $K(X)$ denote the stable $K(X)$ up to $\aleph_{|\alpha|^+}$.

Set $\lambda = |\alpha|^+ \cdot \kappa < \aleph_{|\alpha|^+}$. We aim to define a function

$$\Phi : [H]^\lambda \longrightarrow NS_{|\alpha|^+}.$$

Let us first denote by S the set of all $Y \in [H]^\lambda$ such that $Y \prec H, (\aleph_\eta \mid \eta < |\alpha|^+) \subset Y, \kappa + 1 \subset Y$, and there is a pair (C, f) such that $C \subset |\alpha|^+$ is club, $f : \aleph_{|\alpha|^+} \to \aleph_{|\alpha|^+}, f \restriction \gamma \in K(X)$ for all $\gamma < \aleph_{|\alpha|^+}$, and $f(\aleph_\eta) = \sup(Y \cap \aleph_{\eta+1})$ as well as $\aleph_{\eta+1} = (\aleph_\eta)^{+K(X)}$ for all $\eta \in C$. By Lemma 3.13, S is stationary in $[H]^\lambda$. Now if $Y \in S$ then we let (C_Y, f_Y) be some pair (C, f) witnessing $Y \in S$, and we set $\Phi(Y) = |\alpha|^+ \setminus C_Y$. On the other hand, if $Y \in [H]^\lambda \setminus S$ then we let (C_Y, f_Y) be undefined, and we set $\Phi(Y) = \emptyset$.

By Lemma 2.2, there is then some club $D \subset |\alpha|^+$ such that for all $g \in \prod_{\eta \in D} \aleph_{\eta+1}$ there is some $Y \in S$ such that $D \cap \Phi(Y) = \emptyset$ and $g(\aleph_{\eta+1}) < \sup(Y \cap \aleph_{\eta+1})$ for all $\eta \in D$. Set

$$\mathsf{d} = \{\aleph_{\eta+1} \mid \alpha \leq \eta \in D\} \subset \mathrm{pcf}(\mathsf{a}).$$

There is trivially some regular $\mu > \aleph_{|\alpha|^+}$ such that $\mu \in \mathrm{pcf}(\mathsf{d})$. (In fact, $\aleph_{|\alpha|^++1} \in \mathrm{pcf}(\mathsf{d})$.) By [2, Theorem 6.10] there is then some $\mathsf{d}' \subset \mathsf{d}$ with $|\mathsf{d}'| \leq |\alpha|$ and $\mu \in \mathrm{pcf}(\mathsf{d}')$. Set $\sigma = \sup(\mathsf{d}')$. In particular (cf. [2, Corollary 7.10]),

$$\mathrm{cf}\left(\textstyle\prod \mathsf{d}'\right) > \sigma^+.$$

However, we claim that

$$\mathcal{F} = \{f \restriction \mathsf{d}' \mid f : \sigma \to \sigma \wedge f \in K(X)\}$$

is cofinal in $\prod \mathsf{d}'$. As GCH holds in $K(X)$ above κ, $|\mathcal{F}| \leq |\sigma|^+$, which gives a contradiction!

To show that $\mathcal{F}$ is cofinal, let $g \in \prod d'$. Let $Y \in S$ be such that $D \cap \Phi(Y) = \emptyset$ and for all $\eta \in d'$, $g(\aleph_{\eta+1}) < \sup(Y \cap \aleph_{\eta+1})$. As $D \cap \Phi(Y) = \emptyset$, we have that $D \subset C_Y$. Therefore, $f_Y(\aleph_{\eta+1}) = \sup(Y \cap \aleph_{\eta+1})$ for all $\eta \in d'$, and hence $g(\aleph_{\eta+1}) < f_Y(\aleph_{\eta+1})$ for all $\eta \in d'$. Thus, if we define $f' : \sigma \to \sigma$ by $f'(\xi^{+K(X)}) = f_Y(\xi)$ for $\xi < \sigma$ then $f' \in \mathcal{F}$ and $g < f' \upharpoonright d'$. $\dashv$

PROOF OF THEOREM 1.3. Fix κ as in the statement of Theorem 1.3. Set

$$H = \bigcup_{\theta < \kappa} H_{\theta^+}.$$

We aim to prove that for each $n < \omega$, H is closed under $X \mapsto M_n^\#(X)$.

Let $C \subset \kappa$ be club. As κ is a strong limit cardinal, there is some club $\bar{C} \subset C$ such that every element of $\bar{C}$ is a strong limit cardinal. As $\{\alpha < \kappa \mid 2^\alpha = \alpha^+\}$ is co-stationary, there is some $\lambda \in \bar{C}$ with $2^\lambda \geq \lambda^{++}$. In particular, $\lambda^{\mathrm{cf}(\lambda)} > \lambda^+ \cdot 2^{\mathrm{cf}(\lambda)}$. We have shown that

$$\{\lambda < \kappa \mid \lambda^{\mathrm{cf}(\lambda)} > \lambda^+ \cdot 2^{\mathrm{cf}(\lambda)}\}$$

is stationary in κ, i.e., SCH fails stationarily often below λ.

This fact immediately implies by [3] that H is closed under $X \mapsto M_0^\#(X)$.

Now let $n < \omega$, and let us assume that H is closed under $X \mapsto M_n^\#(X)$. Let us suppose that there is some $X \in H$ such that $M_{n+1}^\#(X)$ does not exist. We are left with having to derive a contradiction.

As SCH fails stationarily often below λ, we may use [24, Theorem 5.3] and deduce that for every $x \in H$ there is some (n, X)-suitable coarse premouse containing x. By Theorem 3.9, $K(X)$ stabilizes in H. Let $K(X)$ denote the stable $K(X)$ up to κ.

Let us fix a strictly increasing and continuous sequence $\vec{\kappa} = (\kappa_i \mid i < \delta)$ of ordinals below κ which is cofinal in κ. Let us define a function

$$\Phi : [H]^{2^\delta} \longrightarrow NS_\delta.$$

Let $Y \in [H]^{2^\delta}$ be such that $Y \prec H$, $(\kappa_i \mid i < \delta) \subset Y$, $^\omega Y \subset Y$, and $TC(\{X\}) \subset Y$. By Lemma 3.13, there is a pair (C, f) such that $C \subset \delta$ is club, $f : \kappa \to \kappa$, $f \upharpoonright \gamma \in K(X)$ for all $\gamma < \kappa$, and $f(\kappa_i) = \mathrm{char}_{\vec{\kappa}}^Y(\kappa_i)$ as well as $\kappa_i^+ = \kappa_i^{+K}$ for all $i \in C$. We let (C_Y, f_Y) be some such pair (C, f), and we set $\Phi(Y) = \delta \setminus C_Y$. If Y is not as just described then we let (C_Y, f_Y) be undefined, and we set $\Phi(X) = \emptyset$.

By Lemma 2.3, there is then some club $D \subset \delta$ and some limit ordinal $i < \delta$ of D such that

$$\mathrm{cf}\left(\prod\{\kappa_j^+ \mid j \in i \cap D\}\right) > \kappa_i^+,$$

and for all $f \in \prod_{i \in D} \kappa_i^+$ there is some $Y \prec H$ such that $\mathrm{Card}(Y) = 2^\delta$, $^\omega Y \subset Y$, $D \cap \Phi(Y) = \emptyset$, and $f < \mathrm{char}_{\vec{\kappa}}^Y$. Let us write

$$\mathsf{d} = \{\kappa_j^+ \mid j \in i \cap D\}.$$

We now claim that

$$\mathcal{F} = \{f \restriction d \mid f : \kappa_i \to \kappa_i \wedge f \in K(X)\}$$

is cofinal in $\prod d$. As GCH holds in $K(X)$ above $\mathrm{TC}(X)$, $|\mathcal{F}| \leq |\kappa_i|^+$, which gives a contradiction!

To show that $\mathcal{F}$ is cofinal, let $g \in \prod d$. There is some $Y \prec H$ such that $\mathrm{Card}(Y) = 2^\delta$, ${}^\omega Y \subset Y$, $D \cap \Phi(Y) = \emptyset$, and $g(\kappa_j^+) < \sup(Y \cap \kappa_j^+)$ for all $j \in D \cap i$. As $D \cap \Phi(X) = \emptyset$, we have that $D \subset C_Y$. But now if $\kappa_j^+ \in d$ then $g(\kappa_j^+) < \sup(Y \cap \kappa_j^+) = f_Y(\kappa_j) < \kappa_j^+$. Thus, if we define $f : \kappa_i \to \kappa_i$ by $f(\xi^{+K(X)}) = f_Y(\xi)$ for $\xi < \kappa_i$ then $f \in \mathcal{F}$ and $g < f \restriction d$.   ⊣

REFERENCES

[1] U. ABRAHAM and M. MAGIDOR, *Cardinal arithmetic*, **Handbook of Set Theory** (Foreman, Kanamori, and Magidor, editors), to appear.

[2] M. BURKE and M. MAGIDOR, *Shelah's pcf theory and its applications*, **Annals of Pure and Applied Logic**, vol. 50 (1990), no. 3, pp. 207–254.

[3] K. I. DEVLIN and R. B. JENSEN, *Marginalia to a theorem of Silver*, ⊢*ISILC Logic Conference*, Lecture Notes in Mathematics, vol. 499, Springer, Berlin, 1975, pp. 115–142.

[4] W. B. EASTON, *Powers of regular cardinals*, **Annals of Pure and Applied Logic**, vol. 1 (1970), pp. 139–178.

[5] M. FOREMAN, M. MAGIDOR, and R. SCHINDLER, *The consistency strength of successive cardinals with the tree property*, **The Journal of Symbolic Logic**, vol. 66 (2001), no. 4, pp. 1837–1847.

[6] M. GITIK, *The negation of the singular cardinal hypothesis from $o(\kappa) = \kappa^{++}$*, **Annals of Pure and Applied Logic**, vol. 43 (1989), no. 3, pp. 209–234.

[7] ———, *The strength of the failure of the singular cardinal hypothesis*, **Annals of Pure and Applied Logic**, vol. 51 (1991), no. 3, pp. 215–240.

[8] ———, *Blowing up power of a singular cardinal—wider gaps*, **Annals of Pure and Applied Logic**, vol. 116 (2002), no. 1-3, pp. 1–38.

[9] ———, *Two stationary sets with different gaps of the power function*, available at http://www.math.tau.ac.il/~gitik.

[10] ———, *Introduction to prikry type forcing notions*, **Handbook of Set Theory** (Foreman, Kanamori, and Magidor, editors), to appear.

[11] M. GITIK and W. MITCHELL, *Indiscernible sequences for extenders, and the singular cardinal hypothesis*, **Annals of Pure and Applied Logic**, vol. 82 (1996), no. 3, pp. 273–316.

[12] K. HAUSER and R. SCHINDLER, *Projective uniformization revisited*, **Annals of Pure and Applied Logic**, vol. 103 (2000), no. 1-3, pp. 109–153.

[13] M. HOLZ, K. STEFFENS, and E. WEITZ, **Introduction to Cardinal Arithmetic**, Birkhäuser Verlag, Basel, 1999.

[14] T. JECH, **Set Theory**, Academic Press, New York, 1978.

[15] ———, *Singular cardinals and the PCF theory*, **The Bulletin of Symbolic Logic**, vol. 1 (1995), no. 4, pp. 408–424.

[16] M. MAGIDOR, *On the singular cardinals problem. I*, **Israel Journal of Mathematics**, vol. 28 (1977), no. 1-2, pp. 1–31.

[17] ———, *On the singular cardinals problem. II*, **Annals of Mathematics. Second Series**, vol. 106 (1977), no. 3, pp. 517–547.

[18] W. MITCHELL and E. SCHIMMERLING, *Weak covering without countable closure*, **Mathematical Research Letters**, vol. 2 (1995), no. 5, pp. 595–609.

[19] W. MITCHELL, E. SCHIMMERLING, and J. STEEL, *The covering lemma up to a Woodin cardinal*, **Annals of Pure and Applied Logic**, vol. 84 (1997), no. 2, pp. 219–255.

[20] W. MITCHELL and J. STEEL, *Fine Structure and Iteration Trees*, Lecture Notes in Logic, vol. 3, Springer-Verlag, Berlin, 1994.

[21] K. L. PRIKRY, *Changing measurable into accessible cardinals*, **Dissertationes Mathematicae (Rozprawy Matematyczne)**, vol. 68 (1970), p. 55.

[22] TH. RÄSCH and R. SCHINDLER, *A new condensation principle*, **Archive for Mathematical Logic**, vol. 44 (2005), no. 2, pp. 159–166.

[23] E. SCHIMMERLING and J. R. STEEL, *The maximality of the core model*, **Transactions of the American Mathematical Society**, vol. 351 (1999), no. 8, pp. 3119–3141.

[24] E. SCHIMMERLING and W. H. WOODIN, *The Jensen covering property*, **The Journal of Symbolic Logic**, vol. 66 (2001), no. 4, pp. 1505–1523.

[25] R. SCHINDLER, *Mutual stationarity in the core model*, **Logic Colloquium '01** (Baaz et al., editors), Lecture Notes in Logic, vol. 20, ASL, Urbana, IL, 2005, pp. 386–401.

[26] R. SCHINDLER and J. STEEL, *List of open problems in inner model theory*, available at http://wwwmath1.uni-muenster.de/logik/org/staff/rds/list.html.

[27] S. SHELAH, *Cardinal Arithmetic*, The Clarendon Press Oxford University Press, New York, 1994.

[28] ——, *Further cardinal arithmetic*, **Israel Journal of Mathematics**, vol. 95 (1996), pp. 61–114.

[29] J. SILVER, *On the singular cardinals problem*, **Proceedings of the International Congress of Mathematicians (Vancouver, B. C., 1974), Vol. 1**, Canad. Math. Congress, Montreal, Que., 1975, pp. 265–268.

[30] J STEEL, *Projectively well-ordered inner models*, **Annals of Pure and Applied Logic**, vol. 74 (1995), no. 1, pp. 77–104.

[31] J. STEEL, *The Core Model Iterability Problem*, Lecture Notes in Logic, vol. 8, Springer-Verlag, Berlin, 1996.

[32] ——, *Core models with more Woodin cardinals*, **The Journal of Symbolic Logic**, vol. 67 (2002), no. 3, pp. 1197–1226.

SCHOOL OF MATHEMATICAL SCIENCES
TEL AVIV UNIVERSITY
TEL AVIV 69978, ISRAEL
E-mail: gitik@post.tau.ac.il

INSTITUT FÜR MATHEMATISCHE LOGIK UND GRUNDLAGENFORSCHUNG
UNIVERSITÄT MÜNSTER
48149 MÜNSTER, GERMANY
E-mail: rds@math.uni-muenster.de

INSTITUTE OF MATHEMATICS
THE HEBREW UNIVERSITY OF JERUSALEM
JERUSALEM, 91904, ISRAEL
and
DEPARTMENT OF MATHEMATICS
RUTGERS UNIVERSITY
NEW BRUNSWICK, NJ 08903, USA
E-mail: shelah@rci.rutgers.edu

EMBEDDING FINITE LATTICES INTO THE COMPUTABLY ENUMERABLE DEGREES — A STATUS SURVEY

STEFFEN LEMPP, MANUEL LERMAN, AND REED SOLOMON

Abstract. We survey the current status of an old open question in classical computability theory: Which finite lattices can be embedded into the degree structure of the computably enumerable degrees? Does the collection of embeddable finite lattices even form a computable set?

Two recent papers by the second author show that for a large subclass of the finite lattices, the so-called join-semidistributive lattices (or lattices without so-called "critical triple"), the collection of embeddable lattices forms a Π_2^0-set.

This paper surveys recent joint work by the authors, concentrating on restricting the number of meets by considering "quasilattices", i.e., finite upper semilattices in which only some meets of incomparable elements are specified. In particular, we note that all finite quasilattices with one meet specified are embeddable; and that the class of embeddable finite quasilattices with two meets specified, while nontrivial, forms a computable set. On the other hand, more sophisticated techniques may be necessary for finite quasilattices with three meets specified.

§1. Introduction. One of the longstanding open questions in classical computability theory is the characterization of all finite lattices embeddable into the computably enumerable (c.e.) degrees. This problem was first raised in the late 1960's but has up to now defied many attempts at a solution. At this point, it is even unclear whether a "reasonable" (e.g., decidable, or "purely lattice-theoretic") characterization exists. Progress has been steady over the past decade but very slow. In this paper, we will try to point out an approach to solving the problem which we consider hopeful and which has led to some further, yet unpublished, partial results by the authors.

We note here that the lattice embeddings problem is currently the primary remaining obstacle toward showing the decidability of the $\forall\exists$-theory of the c.e. degrees (in the language of partial ordering) as the former obviously forms

---

2000 *Mathematics Subject Classification.* Primary: 03D25.

Key words and phrases. lattice embedding, computably enumerable degrees.

The first author's research was partially supported by NSF grant DMS-9732526 and by the Vilas Foundation of the University of Wisconsin. The second author's research was supported by the University of Connecticut Research Foundation. The third author's research was partially supported by an NSF Postdoctoral Fellowship.

Logic Colloquium '02
Edited by Z. Chatzidakis, P. Koepke, and W. Pohlers
Lecture Notes in Logic, 27

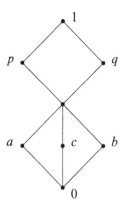

FIGURE 1. Lattice S_8.

a subproblem of the latter. If the lattice embeddings problem can be shown to have a decidable (and "reasonable") solution, then one would hope to show the remainder of the $\forall\exists$-theory also to be decidable using the techniques of Slaman and Soare [14] in their solution of the extension of embeddings problem (i.e., given two finite partial orders $\mathcal{P} \subset \mathcal{Q}$, deciding whether any embedding of $\mathcal{P}$ into the c.e. degrees can be extended to an embedding of $\mathcal{Q}$) and of Ambos-Spies, Jockusch, Shore, and Soare [1] in their work on the promptly simple degrees (i.e., those c.e. degrees not forming half of a minimal pair).

To very briefly recap the history of the lattice embeddings problem up to this point, the first lattice embeddings result is contained in the minimal pair theorem of Lachlan [6] and Yates [17], which implies that the four-element diamond lattice can be embedded into the c.e. degrees. Lerman (unpublished) and Thomason [16] extended this by showing that all finite (indeed, all countable) distributive lattices can be so embedded. Lachlan [7] found the first two examples of finite embeddable nondistributive lattices, namely the five-element lattices M_3 and N_5. Lachlan and Soare [8], following a suggestion of Lerman, exhibited the first nonembeddable finite lattice, S_8 (see Figure 1). The best possible results obtainable by the techniques of the late 1980's were presented in Ambos-Spies and Lerman [2, 3], which isolated a Nonembeddability Condition (NEC) and an Embeddability Condition (EC), respectively. The latter condition, ensuring embeddability, is very complicated and formulated in terms of trees used to carry out the construction. The former condition, however, ensuring nonembeddability, is a simple lattice-theoretic condition.

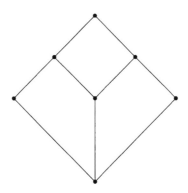

FIGURE 2. A join-semidistributive non-meet-
semidistributive lattice.

To formulate NEC, we introduce the following

DEFINITION 1. Let $\mathcal{L} = \langle L, \leq, \vee, \wedge \rangle$ be a finite lattice.

- Elements $a, b, c \in L$ form a *critical triple* if they are pairwise incomparable; $a \vee c = b \vee c$; and $a \wedge b \leq c$.
- $\mathcal{L}$ satisfies the *Nonembeddability Condition (NEC)* if there are a critical triple $a, b, c \in L$ and two additional incomparable elements $p, q \in L$ such that

(1) $$a \leq p \wedge q \leq a \vee c \leq q.$$

- $\mathcal{L}$ is *principally decomposable* if for any two elements $a > b$ in L such that a is minimal over b, the set $[0, a] - [0, b]$ has a least element (where 0 is the least element of $\mathcal{L}$).
- $\mathcal{L}$ is *join-semidistributive* if for all $a, b, c \in L$,

(2) $$a \vee c = b \vee c \quad \text{implies} \quad a \vee c = (a \wedge b) \vee c.$$

These notions are closely connected by the following easy

LEMMA 2. *A finite lattice $\mathcal{L}$ is join-semidistributive iff it is has no critical triple iff it is principally decomposable.*

From now on, we will use the term "join-semidistributive" in place of "principally decomposable" since the former is the one used by lattice theorists.

REMARK 3. (1) There is a dual, but distinct notion called "meet-semidistributive". Figure 2 shows a finite lattice which is join-semidistributive but not meet-semidistributive.

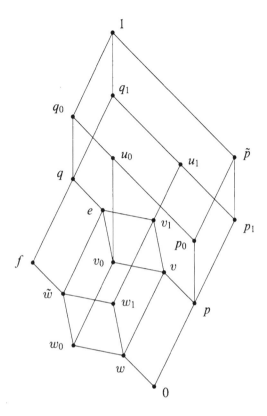

FIGURE 3. Lattice L_{20}.

(2) Join-semidistributive lattices form in some sense a class of finite lattices complementary to the modular lattices: Any finite join-semidistributive modular lattice is distributive. For more information on the lattice theory of semidistributive lattices, see Gorbunov [5].

The condition NEC thus pointed to the fact that the first step in solving the lattice embeddings problem was to consider the case of finite join-semidistributive lattices. When Downey [4] showed that there is an initial segment of c.e. degrees into which no finite non-join-semidistributive lattice can be embedded, he conjectured that on the other hand, every finite join-semidistributive lattice can be embedded into any nontrivial interval of the c.e. degrees.

This conjecture was refuted by Lempp and Lerman [9], who exhibited a finite join-semidistributive lattice, L_{20} (see Figure 3), which cannot be embedded into the c.e. degrees. Further developing the techniques used for L_{20}, Lerman [11, 12] subsequently isolated a necessary and sufficient Π_2^0-criterion for the embeddability of finite join-semidistributive lattices into the c.e. degrees. (Lerman [11] gives the partial result for finite so-called "ranked" lattices, which is then extended in Lerman [12] to the case of all finite join-semidistributive lattices.)

We defer the precise definition of Lerman's Embeddability Criterion to Section 3.5 since we need to first introduce a number of additional definitions and also provide some intuition explaining the various conditions of Lerman's criterion.

First of all, however, we would like to point out the approach we have taken over the past several years in attacking the lattice embeddings problem. As will become clearer in the intuitive discussion of lattice embedding techniques in the next section, the hardest part is to ensure that meets are preserved under the embedding. It is therefore natural to take an inductive approach towards the lattice embedding problem by restricting the number of meets to be considered, motivating our definition of quasilattices. We first remark that any finite upper semilattice carries a natural lattice structure.

REMARK 4. Any finite upper semilattice $\mathcal{L} = \langle L, \leq, \vee, 0, 1 \rangle$ (with least element 0 and greatest element 1) can be made into a lattice by defining the meet function by

$$(3) \qquad\qquad a \wedge b = \bigvee \{c \in L \mid c \leq a, b\}.$$

(The existence of 0 in L ensures that the set on the right-hand side above is always nonempty.)

We can now make the following

DEFINITION 5. A *quasilattice* $\mathcal{L} = \langle L, \leq, \vee, \wedge, 0, 1 \rangle$ (with least element 0 and greatest element 1) is an upper semilattice $\langle L, \leq, \vee \rangle$ together with a partial meet function $\wedge$ defined on some (but not necessarily all) unordered pairs of incomparable elements of L, where $a \wedge b$ (if defined) equals the meet defined by Remark 4.

Both a finite upper semilattice and a finite lattice are thus examples of quasilattices (where the meet is never or always defined, respectively). However, we will be most interested in examples where the meet is defined for a limited number of unordered pairs of incomparable elements, say, n many; we will call such a quasilattice a *quasilattice with n meets specified*.

We note that Lerman [11, 12] used closely related structures called pseudolattices. Lerman uses, in place of the partial meet functions, $(n + 1)$-ary meet relations $M_n(a_1, \ldots, a_n, b)$, denoting that $c \leq a_i$ for all i implies $c \leq b$.

The current status of the lattice embeddings problem can thus be summarized in the following

Main Statement. (1) (*Folklore*) Any finite quasilattice with no meets specified (i.e., any finite upper semilattice) is embeddable into the c.e. degrees.

(2) Any finite quasilattice with one meet specified is embeddable into the c.e. degrees.

(3) It is decidable which finite quasilattices with two meets specified are embeddable into the c.e. degrees; and not every finite quasilattice is so embeddable.

(4) It is currently unknown whether the characterization of the embeddable finite quasilattices with three meets specified is decidable. Our techniques developed for only two meets are not known to suffice to solve this problem.

The rest of this paper is devoted to explaining, at least on an intuitive level, why we believe that our approach to the lattice embeddings problem via quasilattices is the most promising one; and to discuss the various clauses of our Main Statement. Specifically, we will address clause (1) of our Main Statement in Section 2.4; clause (2) in Section 3.4; clause (3) in Section 3.7; and clause (4) in Section 3.8.

We conclude this section by remarking that a large number of variations of the lattice embeddings problem into the c.e. degrees have been studied, too numerous to cover in detail here. Firstly, one can study lattice embeddings *preserving* 0 *and* 1, i.e., mapping 0 to the degree $\mathbf{0}$ and 1 to the degree $\mathbf{0}'$, respectively. (Any currently known "plain" lattice embedding into the c.e. degrees also preserves 0.) Next, one can study lattice embeddings into initial segments or intervals of c.e. degrees. Finally, lattice embeddings into the lattice of ideals of c.e. degrees have been studied.

§2. Embedding finite upper semilattices.

We start with an intuitive description of the basic lattice embedding construction and the conflicts between the various strategies involved. Since the meet requirements are the most complicated ones, we first concentrate on the other requirements, i.e., we will present the argument for embedding finite upper semilattices (viewed as quasilattices with no meet specified). Fix a finite quasilattice $\mathcal{L}$ and its set $J_{\mathcal{L}}$ of join-irreducible elements. (Here 0 is not considered a join-irreducible element.)

Convention 6. In order to simplify notation from now on, we denote the c.e. degree which is the image of a lattice element a, say, by the corresponding bold-face letter $\mathbf{a}$, and the c.e. set representing the degree $\mathbf{a}$ by the corresponding upper-case letter A.

There are now four types of requirements to be satisfied for an embedding of $\mathcal{L}$ into the c.e. degrees:

(4) $\mathcal{C}^{a,b} : A \leq_T B$ (for $a < b$ in $\mathcal{L}$)

(5) $\mathcal{I}_{\Phi}^{a,b} : A \neq \Phi(B)$ (for $a \not\leq b$ in $\mathcal{L}$; $a \in J_{\mathcal{L}}$)

(6) $\mathcal{J}^{d,e,f} : \exists\, \Gamma^{d,e} \left(F = \Gamma^{d,e}(D \oplus E) \right)$ (for $f = d \vee e$ in $\mathcal{L}$)

              $\mathcal{M}_{\Psi}^{p,q,r} : \Psi(P) = \Psi(Q)$ is total $\Longrightarrow$

(7) $\exists\, \Delta^{p,q} \left(\Psi(P) = \Delta^{p,q}(R) \right)$ (for $r = p \wedge q$ in $\mathcal{L}$)

Here Φ and Ψ range over all possible Turing functionals. Clearly, the $\mathcal{C}$- and $\mathcal{J}$-requirements are "global", each building a single reduction (whose names we will suppress, except in the initial discussion of the $\mathcal{J}$-strategies below), whereas the $\mathcal{I}$- and $\mathcal{M}$-requirements are "local", each strategy on the tree of strategies working with a separate diagonalization witness or a separate functional $\Delta^{p,q}$, respectively.

We now gradually introduce the strategies for the four types of requirements and each time sketch the conflicts with the strategies discussed before. The discussion of the meet requirements will be deferred to the next section.

2.1. Comparability requirements $\mathcal{C}^{a,b}$. For this requirement, we simply ensure that any number x targeted for a c.e. set A, say, is first chosen at a stage $< x$, and that when (if ever) x enters A, it simultaneously enters all sets B with $b > a$. Clearly, this simple strategy ensures the comparability requirements: Given $a < b$ in $\mathcal{L}$ and a number x, we first check if x is chosen with target C (for some $c \leq a$) by stage x. If not, then $x \notin A$; otherwise, $x \in A$ iff $x \in B$.

2.2. Incomparability requirements $\mathcal{I}_{\Phi}^{a,b}$. The strategy for this requirement is simply the Friedberg-Muchnik strategy: We choose a "big" diagonalization witness x (i.e., larger than any number previously mentioned in the construction) and keep x out of A for now. We then wait for a computation $\Phi(B; x) = 0$. When (and if) such a computation appears, we enumerate x into A and preserve $\Phi(B; x)$ by restraining B up to its use.

The above two types of strategies present no serious conflict and can easily be combined to show that any finite partial order can be embedded into the c.e. degrees.

2.3. Join requirements $\mathcal{J}^{d,e,f}$. When, at a stage s, a "big" number $x > s$ is targeted to be enumerated into a set C (with $c \leq f$, and so x may also be enumerated into F), then we also define a computation $\Gamma(D \oplus E; x)$ (for the functional $\Gamma = \Gamma^{d,e}$) with "big" use $\gamma_s(x)$. We now agree that

(1) the current use $\gamma(x)$ must enter D or E by the stage at which x enters C and thus must enter F;

(2) if the current use $\gamma(x)$ enters D or E at a stage s', say, before x enters C, then, at stage s', we redefine the computation $\Gamma(D \oplus E; x)$ with new "big" use $\gamma(x)$; and

(3) the use $\gamma(x)$ of x is increased at most finitely often.

Clearly, this will ensure the join requirement since we can define $\Gamma(D \oplus E; x) = 0$ for all x which are not targeted for some set C with $c \leq f$ by stage x. In this latter case, x cannot enter F, and so $\Gamma(D \oplus E)$ is correct on those x. If x is chosen as a target by stage x, then by (3) above, $\gamma(x)$ will eventually stabilize, and $D \oplus E$ can compute the stage when this happens by (2). Now $x \in F$ iff $\gamma(x) \in D \cup E$ by (1) for the final value of the use $\gamma(x)$.

Instead of being so explicit about the join functionals, however, we will now present an alternative way to deal with the comparability and join requirements simultaneously, which will also be useful in dealing with the meet requirements later on.

2.4. Co-principal filters. We start with the following

DEFINITION 7. Let $\mathcal{L}$ be a finite quasilattice.

(1) A *filter* of $\mathcal{L}$ is any upward closed subset $F \neq L, \emptyset$ of L.
(2) The *filter generated by* a set $S \subset L$ (where $S \neq \emptyset$ and $0 \notin S$) is the upward closure of S in $\mathcal{L}$ and is denoted by (S). If $S = \{a_0, \ldots, a_n\}$, we abbreviate $(\{a_0, \ldots, a_k\})$ by $(a_0, \ldots, a_k)$ or simply by $(\overrightarrow{a})$.
(3) A *co-principal filter* of $\mathcal{L}$ is a filter F such that $L - F$ is closed under join, or equivalently, such that $L - F$ is of the form $[0, b]$ for some $b \in L - \{1\}$. For any $b \in L - \{1\}$, we denote the co-principal filter $L - [0, b]$ by $F(b)$.

Note that as long as we enumerate numbers of "roughly equal" size into all sets C (for all c in some co-principal filter F) then we can build the join functionals Γ for all possible joins as above. This observation allows one to establish clause (1) of our Main Statement, namely, to show that all finite upper semilattices can be embedded into the c.e. degrees by combining the above three types of strategies as follows: When an $\mathcal{I}^{a,b}$-strategy chooses a witness x targeted for A, it will also target the same number x for all sets C with $c \nleq b$. Now since $F(b) = L - [0, b]$ forms a co-principal filter in $\mathcal{L}$, it is easy to check that if the $\mathcal{I}^{a,b}$-strategy enumerates x into all sets C with $c \in F(b)$ when x enters A, then each $\mathcal{J}^{d,e,f}$-requirement is satisfied, since $f \in F(b)$ iff $d \in F(b)$ or $e \in F(b)$, so x is targeted, and possibly later enters, F iff x enters D or E.

2.5. Covering sequences and covering arrays. Since we will also have to consider meet requirements later on, we now restrict our attention to embedding finite join-semidistributive lattices. (This will allow us to present the Embeddability Criterion of Lerman [11, 12] in a somewhat modified and simplified form.)

First of all, when taking into consideration meet requirements, it will not always be possible to enumerate diagonalization witnesses in one step as sketched two paragraphs above; rather, we will have to "retarget" a number of times as indicated in the general description of the join functional Γ above. To simultaneously deal with all join requirements, and to simplify the description, we introduce two key notions in the following

DEFINITION 8. Let $\mathcal{L}$ be a finite lattice.

A *covering sequence* is an ordered sequence $\vec{a} = \langle a_0, \ldots, a_l \rangle$ of elements of L such that $(a_0, \ldots, a_i)$ is a co-principal filter for all $i \leq l$. (We allow $l = -1$, i.e., a covering sequence may be empty.)

Covering sequences provide an easy method to ensure the satisfaction of all comparability and join requirements intuitively as follows: When an $\mathcal{I}_\Phi^{a,b}$-strategy chooses a diagonalization witness x, then

(I) the strategy chooses a covering sequence $\langle a_0, \ldots, a_l \rangle$ with $a = a_l$ and $b \notin (a_0, \ldots, a_l)$, and chooses associated "big" numbers x_i for all $i \leq l$ where $x_l = x$;

(II) if the strategy enumerates any associated number before enumerating x into A, then, for some $l'' <$ the current l, it enumerates, for all $i \leq l''$, the currently associated numbers x_i into all sets C with $c \geq$ the current a_i, and chooses a new covering sequence $\langle a_0', \ldots, a_{l'}' \rangle$ (which contains the tail $\langle a_{l''+1}, \ldots, a_l \rangle$ of the current covering sequence as a subsequence) as well as "big" associated numbers x_i' for all "new" elements a_i' of the new covering sequence (whereas for all "old" elements a_i' of the new covering sequence, the currently associated number remains associated with a_i');

(III) if the strategy enumerates x into A then it enumerates, for all $i \leq$ the current l, the currently associated number x_i into all sets C with $c \geq$ the current a_i; and

(IV) it chooses a new covering sequence (as in (II) above) at most finitely often.

It is now not hard to see that the above (I)–(IV) will ensure the incomparability and join requirements since, by the definition of covering sequences, the associated numbers can be viewed as uses of join functionals Γ.

We make this more precise in the following

DEFINITION 9. A *covering array* consists of a sequence

(8) $$\vec{A} = \langle \vec{a_0}, \ldots, \vec{a_m} \rangle$$

of target sequences $\vec{a_j} = \langle a_{0,j}, \ldots, a_{l_j,j} \rangle$ together with a sequence of *transition maps* $\langle T_0, \ldots, T_{m-1} \rangle$ such that for each $j < m$, T_j is a map from $(l_j', l_j]$ into $[0, l_{j+1}]$ (for some $l_j' < l_j$) satisfying, for all $j < m$,

(9) $$l_j' < i < i' \leq l_j \Longrightarrow T_j(i) < T_j(i'), \text{ and}$$

(10) $$\forall i \in (l_j', l_j] \quad (a_{i,j} = a_{T_j(i),j+1}).$$

(So each T_j is an order-preserving map from a final segment of (indices of) $\overrightarrow{a_j}$ to (indices of) $\overrightarrow{a_{j+1}}$, preserving the lattice element $a_{i,j}$. Here we allow $l_j = -1$ only if $j = m$, i.e., only the last covering sequence of the covering array may be empty. We abbreviate the composition $T_{j_1-1} \cdots T_{j_0}$ (for $0 \leq j_0 \leq j_1 \leq m$) by T_{j_0,j_1}.)

We say that $\overrightarrow{A}$ is a *covering array for an* $\mathcal{I}_\Phi^{a,b}$*-requirement* if furthermore $a = a_{l_0,0}$ and $b \notin (a_{0,0}, \ldots, a_{l_0,0})$. (I.e., a is the last element of the first covering sequence in $\overrightarrow{A}$, and the set B is not initially targeted, namely, not before a computation $\Phi(B;x) = 0$ has been found. Of course, once such a computation has been found, we can target new numbers y for sets C with $c \leq b$ since such y can be chosen above the use of the computation $\Phi(B;x)$.)

The above clauses (I)–(IV) for satisfying all comparability and join requirements can now be phrased as follows: All enumerations of an $\mathcal{I}^{a,b}$-strategy, once it chooses a diagonalization witness x, correspond to a covering array $\overrightarrow{A} = \langle \overrightarrow{a_0}, \ldots, \overrightarrow{a_m} \rangle$ for this requirement in the following sense: When the $\mathcal{I}$-strategy chooses a new "big" diagonalization witness x targeted for A at a stage s_0, say, it also chooses new "big" numbers $x_{i,0}$ targeted for $A_{i,0}$. Once a computation $\Phi(B;x) = 0$ has been found, the $\mathcal{I}$-strategy enumerates into sets and chooses new numbers in m many steps: At step $j > 0$ of this process, say, at a stage s_j, any number $x_{i,j-1}$ such that $i \notin \text{dom } T_{j-1}$ is enumerated into its target $A_{i,j-1}$ (and thus into all sets C with $c \geq a_{i,j-1}$). Also, for any $i \leq l_j$, if $i \in \text{ran } T_{j-1}$ then we set $x_{i,j} = x_{T_{j-1}^{-1}(i),j-1}$; if $i \notin \text{ran } T_{j-1}$ then we choose a new "big" number $x_{i,j}$ targeted for $A_{i,j}$.

It is now easy to verify that the use of the covering array for the $\mathcal{I}_\Phi^{a,b}$-requirement will also satisfy any $\mathcal{J}^{d,e,f}$-requirement: Suppose a number $y = x_{i,j_0}$ is targeted for some set C with $c \leq f$ at some stage $s_{j_0} < y$, say. Consider the sequence

$$(11) \qquad \left\langle x_{i,j_0}, x_{T_{j_0}(i),j_0+1}, x_{T_{j_0,j_0+2}(i),j_0+2}, \ldots, x_{T_{j_0,j_1}(i),j_1} \right\rangle$$

such that $j_1 = m$ or $T_{j_0,j_1}(i) \notin \text{dom } T_{j_1}$. Since each $\overrightarrow{a_j}$ is a covering sequence, we have that for all $j \in [j_0, j_1]$, there is some $i_j < T_{j_0,j}(i)$ with $a_{i_j,j} \leq d$ or $\leq e$. So we can define $\Gamma^{d,e}(D \oplus E; x)$ with use $\gamma^{d,e}(y) = x_{i_j,j}$, and the clauses (1)–(3) of Section 2.3 will hold. (We reiterate here the remark that the above technique only works for finite join-semidistributive lattices. For finite lattices in general, without the assumption of principal decomposability, the notion of covering sequence has to be generalized, requiring that only certain, but not all, initial segments of the sequence generate co-principal filters. E.g., even in the example of arbitrary finite upper semilattices at the end of Section 2.4, it may not be possible to use covering sequences to generate co-principal filters of the form $L - [0, b]$ as outlined there, as the example of the lattice M_3, viewed as an upper semilattice, illustrates.)

§3. **Embedding finite quasilattices.** We are now ready to add the meet requirements, in the context of covering sequences and covering arrays as defined above.

3.1. Meet requirements $\mathcal{M}_{\Psi}^{p,q,r}$. The basic strategy for a meet requirement is quite simple even though its interaction with the other requirements is very complicated: As the length of agreement between $\Psi(P)$ and $\Psi(Q)$ increases, the strategy defines more and more of $\Delta(R)$ by initially setting $\Delta(R; y)$ to the common value of $\Psi(P; y)$ and $\Psi(Q; y)$ and setting the use $\delta(y) \geq$ the uses $\psi(P; y)$ and $\psi(Q; y)$. Whenever $\Delta(R; y)$ is defined for some argument y, the strategy tries to have at least one of $\Psi(P; y)$ or $\Psi(Q; y)$ defined and agreeing with $\Delta(R; y)$. If that fails, then the strategy must destroy the computation $\Delta(R; y)$ by enumerating a number $\leq \delta(y)$ into R.

The main difficulty in the above strategy is the following typical scenario: After $\Delta(R; y)$ has become defined, both computations $\Psi(P; y)$ and $\Psi(Q; y)$ may become undefined, although never at the same time. Each may now return with the same value but a larger use, so the strategy outlined in the previous paragraph sees no reason to act yet. However, this creates the so-called "dangerous interval"

(12) $$I_y = \big(\delta(y), \min\{\psi(P; y), \psi(Q; y)\}\big],$$

dangerous for $\Delta(R)$ since the enumeration of any number $z \in I_y$ into R (and thus into both P and Q by comparability requirements) will destroy both $\Psi(P; y)$ and $\Psi(Q; y)$ but not $\Delta(R; y)$. This now makes it necessary to enumerate another number $z' \leq \delta(y)$ into R to correct $\Delta(R; y)$. However, this z' may be in a dangerous interval $I_{y'}$ for some $y' < y$, possibly setting off a cascade of smaller and smaller numbers having to enter R until no dangerous interval is hit. (It is exactly this type of behavior which was at the heart of the proof of the nonembeddability of the lattice S_8 by Lachlan and Soare [8]. However, S_8 is not a join-semidistributive lattice, and a nonembeddability construction for join-semidistributive lattices, such as for the nonembeddable lattice L_{20} discovered by Lempp and Lerman [9], has to use dangerous intervals in a more subtle way.)

Since the incomparability strategies (which are, as we have now seen, the only ones *initiating* the enumeration of numbers into sets) are all finitary, we agree that we will never allow enumeration of any numbers y into dangerous intervals, but rather always directly enumerate the largest number $y' \leq y$ which does not hit a dangerous interval, thus never triggering the kind of cascade of enumerations described in the previous paragraph. This restriction will be implemented by the prohibition functions defined below in Section 3.5 where we will also take into account the interaction between dangerous intervals of different meet strategies. However, even though only the incomparability strategies will *initiate* the enumeration of numbers into sets, the meet strategies may respond to other enumerations by enumerating numbers on their own, namely so-called *correction markers* to correct their functionals Δ. We will show that this can only happen for finite join-semidistributive quasilattices with at least three meets specified. However, up to this point, we do not know if correction markers are necessary at all, or whether correction will always be automatic by numbers entering purely for coverage reasons. (In the latter case, we would have a decision procedure for a fixed number of "gates", as we will outline in Section 3.8.)

First of all, however, we will set up the machinery of pinball machine constructions and blocks which will allow us to make precise the implementation of the meet strategies.

3.2. Pinball machine constructions. Lattice embedding constructions are traditionally done using the so-called pinball machine construction introduced by Lerman [10] (see also Soare [15, Chapter VIII.5]). The rough idea is the following: Diagonalization witnesses are represented by "balls" originating from "holes" (corresponding to incomparability, i.e., $\mathcal{I}_\Phi^{a,b}$-requirements) which then have to pass by "gates" (corresponding to higher-priority meet requirements). Balls may either get permanently stuck at (i.e., be permanently restrained by) one of the gates below the hole, or they may pass by all of the finitely many gates below the hole and enter the "enumeration basket" (i.e., be enumerated into their target set). In addition to the balls corresponding to diagonalization witnesses, we need other balls (i.e., numbers) to "cover" the diagonalization witnesses (i.e., to generate a co-principal filter containing a). These other balls either originate at the same hole as the diagonalization witness (and then correspond to the elements of $\overline{a_0}$); or they originate at gates below a hole to "cover" balls currently at or above that gate now that some of the previously covering balls may have been enumerated. These latter balls correspond to "new" elements $a_{i,j}$ (for $j > 0$) of covering sequences $\overrightarrow{a_j}$ (i.e., for which $T_{j-1}^{-1}(i)$ is undefined).

Before explaining how the pinball machine construction helps satisfy the meet requirements, we first explain some of the simple mechanics of how the gates and the covering arrays interact.

3.3. Blocks. The main tool to combine the target array with the pinball machine construction is the notion of blocks. (We deviate here somewhat from the way Lerman [11, 12] defines blocks by slightly changing the definition of the functions h_k and $h_{j,k}$; however, our definition here is equivalent.)

DEFINITION 10. Fix $n > 0$. (The intuition here will be that G_0 through G_{n-1} are gates (with gate G_0 the lowest, corresponding to the highest-priority requirement) below a hole H_n. For now, however, these gates can be viewed simply as giving us indices for the block functions.)

(1) A covering sequence $\langle a_0, \ldots, a_l \rangle$, together with a function $f : [0, l] \to [0, n]$ and partial functions $h_k : [0, l] \to [0, l]$ (for $k < n$), forms a *blocked*

target sequence if for all $k < n$,

(13) $\operatorname{ran} f = \{n\}$ or $\forall i, i' \leq l \left(i < i' \Longrightarrow f(i) \leq f(i') < n \right)$;

(14) $\operatorname{dom} h_k = \{ i \leq l \mid f(i) \geq k \}$;

(15) $\forall i \in \operatorname{dom} h_k \left(i \leq h_k(i) \right)$;

(16) $\forall i, i' \in \operatorname{dom} h_k \left(i < i' \Longrightarrow h_k(i) \leq h_k(i') \right)$;

(17) $\forall i, i' \in \operatorname{dom} h_k \left(h_k(i) < h_k(i') \Longrightarrow h_k(i) < i' \right)$;

(18) $k < n - 1 \Longrightarrow \forall i, i' \in \operatorname{dom} h_k$
$\left(h_k(i) = h_k(i') \Longrightarrow h_{k+1}(i) = h_{k+1}(i') \right)$; and

(19) $\forall i, i' \in \operatorname{dom} h_k \left(h_k(i) = h_k(i') \Longrightarrow f(i) = f(i') \right)$.

(Intuitively, f indicates that the "ball" (number) x_i associated with a_i currently is at gate $G_{f(i)}$ (if $f(i) < n$), or at the hole H_n of the diagonalization requirement (if $f(i) = n$). Clause (13) now states that the balls either all reside at the hole H_n, or all reside at gates such that later balls in the sequence do not reside at lower gates. Each function h_k induces a partition of the balls at or above gate G_k into intervals called k-*blocks*, where $a_{h_k(i)}$ indicates the last element of the k-block; this is ensured by clauses (14)–(17). Clause (18) indicates that the k-blocks refine the $(k+1)$-blocks, while clause (19) states that any k-block resides at a single gate or hole. The intuition here is that each k-block consists of balls which pass gate G_k simultaneously (which explains the choice of the domain of h_k).) We denote the eth k-block of $\langle a_0, \ldots, a_l \rangle$ by B_k^e, starting with $e = 0$.

(2) A *covering array* $\overrightarrow{A} = \langle \overrightarrow{a_0}, \ldots, \overrightarrow{a_m} \rangle$ with transition functions $T_0, \ldots,$ T_{m-1} (where each of the covering sequences $\overrightarrow{a_j} = \langle a_{0,j}, \ldots, a_{l_j, j} \rangle$ is a blocked target sequence with functions f_j and $h_{j,k}$ for all $k < n$) forms a *blocked target array* if for all $k < n$,

(20) $\forall j \leq m \, \forall i \leq l_j \left(f_j(i) = n \iff j = 0 \right)$;

(21) $l_m = -1$;

(22) $\forall j < m \, \forall i \in [0, l_j] - B_{j, f_j(0)}^0$
$\left(i \in \operatorname{dom} T_j \text{ and } f_j(i) = f_{j+1}(T_j(i)) \right)$;

(23) $\forall j < m \, \forall i \in B_{j, f_j(0)}^0 \left(i \notin B_{j,0}^0 \iff i \in \operatorname{dom} T_j \right)$;

(24) $\forall j < m \, \forall i \in B_{j, f_j(0)}^0 - B_{j,0}^0$
$\left(f_{j+1}(T_j(i)) = \min\{ k - 1 \mid i \in B_{j,k}^0 \} \right)$; and

(25) $\forall j < m \, \forall i, i' \in \operatorname{dom} T_j \left(T_j(i), T_j(i') \in \operatorname{dom} h_{j+1,k} \Longrightarrow \right.$
$\left(h_{j,k}(i) = h_{j,k}(i') \iff h_{j+1,k}(T_j(i)) = h_{j+1,k}(T_j(i')) \right)$.

(Here clause (20) states that only the balls corresponding to $\overrightarrow{a_0}$ are at the hole; all balls corresponding to later $\overrightarrow{a_j}$ are at gates. Clause (21) states that only the last covering sequence $\overrightarrow{a_m}$ is empty. Clauses (22)–(24) exactly prescribe the motion of the balls in the pinball machine from step j to step $j + 1$: Set $k_j = f_j(0)$, which is the (index of the) lowest gate containing a ball at step j. Now any ball not in the first k_j-block B^0_{j,k_j} at G_{k_j} remains at the same gate at which it was by clause (22); the balls in the first 0-block $B^0_{j,0}$ at G_{k_j} are enumerated by clause (23); and the balls in $B^0_{j,k_j} - B^0_{j,0}$ move down to gate G_{k-1} iff they are in $B^0_{j,k} - B^0_{k-1}$ by clause (24). (This is more restrictive than the definition of Lerman [11, 12], but by Lerman's proof, it still gives an embeddability criterion for finite join-semidistributive lattices.) Finally, clause (25) states that k-blocks are preserved from step j to step $j + 1$ unless a ball is no longer in a k-block at step $j + 1$, i.e., is already below gate G_k.)

(3) A blocked target array is a *blocked target array for an $\mathcal{I}^{a,b}_\Phi$-requirement* if furthermore $a = a_{l_0,0}$ and $b \notin (a_{0,0}, \ldots, a_{l_0,0})$.

3.4. A single meet $p \wedge q = r$. We are now ready to consider in detail our first argument involving meet strategies. Following the philosophy of the introduction of this paper, we start by describing how to deal with a single meet in a quasilattice. Since the incomparability strategies (which are the only ones initiating the enumeration of numbers into sets) are finitary, it suffices to consider the interaction of a single $\mathcal{I}^{a,b}_\Phi$-requirement with a finite number of higher-priority $\mathcal{M}^{p,q,r}$-strategies (in the context of all comparability and join requirements). We first restrict ourselves to the case of a single gate G_0 since several gates for the same meet present no additional difficulties. We need to distinguish three cases, depending on the position of a and b relative to p, q, and r.

CASE 1. $r \nleq b$: Then there is no conflict since the enumeration of the diagonalization witness x into A can take place immediately upon finding a computation $\Phi(B; x) = 0$ and simultaneously with the enumeration of some number $x_{i,0}$ into a set $A_{i,0}$ where $a_{i,0} \leq r$ since $x_{i,0}$ can correct the meet functional $\Delta^{p,q}(R)$. The covering array can thus be chosen as $\overrightarrow{A} = \langle \overrightarrow{a_0}, \overrightarrow{a_1} \rangle$ with $\overrightarrow{a_0}$ consisting of a single 0-block and $\overrightarrow{a_1}$ empty where $r \in (a_{0,0}, \ldots, a_{l_0,0})$, i.e., all numbers $x_{i,0}$ associated with $a_{i,0}$ are enumerated immediately.

CASE 2. $p \leq b$ (or symmetrically $q \leq b$): Then again there is no conflict since the enumeration of the diagonalization witness x into A can take place immediately upon finding a computation $\Phi(B; x) = 0$ and the functional $\Psi(P)$ will not be injured. The covering array can thus be chosen as $\overrightarrow{A} = \langle \overrightarrow{a_0}, \overrightarrow{a_1} \rangle$ with $\overrightarrow{a_0}$ consisting of a single 0-block and $\overrightarrow{a_1}$ empty where $p \notin (a_{0,0}, \ldots, a_{l_0,0})$, i.e., all numbers $x_{i,0}$ associated with $a_{i,0}$ are enumerated immediately.

CASE 3. $r \leq b$ but $p, q \not\leq b$: This is the nontrivial case since the enumeration of the diagonalization witness x into A will typically require the enumeration of numbers into both P and Q while the preservation of the computation $\Phi(B; x) = 0$ does not allow the immediate enumeration into R. We resolve this problem by splitting the first covering sequence $\overrightarrow{a_0}$ with $a = a_{l_0,0}$ into 0-blocks $B_{0,0}^e$ such that for each e,

$$(26) \qquad p \notin (B_{0,0}^e) \quad \text{or} \quad q \notin (B_{0,0}^e).$$

(This can certainly be achieved by making each 0-block consist of a single element, but it probably makes more sense to use maximal 0-blocks satisfying clause (26).)

The second covering sequence $\overrightarrow{a_1}$ is now obtained from $\overrightarrow{a_0}$ by (i) deleting the first 0-block $B_{0,0}^0$ of $\overrightarrow{a_0}$; (ii) copying each subsequent 0-block $B_{0,0}^e$ of $\overrightarrow{a_0}$ (for $e > 0$) into the 0-block $B_{1,0}^{e-1}$ of $\overrightarrow{a_1}$; and (iii) adding a covering sequence generating $F(p)$ or $F(q)$ at the beginning of each 0-block $B_{1,0}^e$ depending on whether the remainder of $B_{1,0}^e$, coming from $B_{0,0}^{e+1}$, is contained in $F(p)$ or $F(q)$, respectively. The remaining covering sequences $\overrightarrow{a_j}$ (for $j > 1$) are now obtained from the previous covering sequence $\overrightarrow{a_{j-1}}$ by simply deleting the first 0-block of $\overrightarrow{a_{j-1}}$ until we end up with the empty sequence. Recall that the corresponding enumeration enumerates a number $x_{i,j}$ into $A_{i,j}$ when $T_j(i)$ is undefined, i.e., when $a_{i,j}$ is deleted from the sequence between step j and step $j + 1$.

This ensures that at each step of the enumeration, either $\Psi(P)$ or $\Psi(Q)$ will not be injured.

We illustrate the above with the example of the quasilattice in Figure 4.

EXAMPLE 11. The covering array for a hole H_1 (for an incomparability requirement $\mathcal{I}_\Phi^{a,b}$) above a gate G_0 (for a meet requirement $\mathcal{M}_\Psi^{p,q,r}$) consists of the following covering sequences

$$\langle a_0, \quad a_1, \quad a_2, \quad a \rangle$$
$$\langle p', a_1, q', a_2, p', a \rangle$$
$$\langle q', a_2, p', a \rangle$$
$$\langle p', a \rangle$$
$$\langle \, \rangle$$

where the transition maps are defined as indicated by vertical alignment. This uses that $a_1, a \in F(p) = [p', 1]$ whereas $a_0, a_2 \in F(q) = [q', 1]$.

Since the above can be generalized to n many gates G_0 through G_{n-1} for the same meet $p \wedge q = r$ by simply making all 0-blocks simultaneously also k-blocks for all $k < n$, this simple construction shows that any finite quasilattice

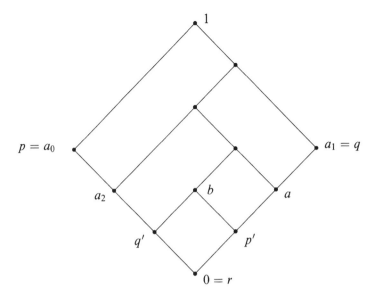

FIGURE 4. A quasilattice with one meet specified.

with only one meet specified can be embedded into the c.e. degrees. This establishes clause (2) of our Main Statement.

Before we can address the embedding of quasilattices with more than one meet specified, we need to introduce the notion of a prohibition function, which will help us identify and avoid enumeration into dangerous intervals.

3.5. Prohibition functions and Lerman's Embeddability Criterion. For each blocked target array $\overrightarrow{A} = \langle \overrightarrow{a_0}, \ldots, \overrightarrow{a_m} \rangle$, we define prohibition functions g_j (for each $j \leq m$, corresponding to each blocked target sequence $\overrightarrow{a_j}$), which associates each element $a_{i,j}$ of $\overrightarrow{a_j}$ with a subset $g_j(i)$ of $\{G_0, \ldots, G_{n-1}\}$. (Here, each g_j will depend only on $\overrightarrow{A}$, and more specifically only on $\langle \overrightarrow{a_0}, \ldots, \overrightarrow{a_j} \rangle$. Intuitively, $G_k \in g_j(i)$ will tell us that we cannot currently target new balls $\leq r_k$ at or before the k-block of $a_{i,j}$. Once the prohibition functions have been defined, we will define the notion of a *good* blocked target array, which implements this prohibition. This approach has the advantage that the various features of meet strategies in the literature ("minimal pair strategy", "Lachlan meet strategy", etc.) are unified into one single definition.)

DEFINITION 12. Fix a blocked target array $\overrightarrow{A} = \langle \overrightarrow{a_0}, \ldots, \overrightarrow{a_m} \rangle$. For each element $a_{i,j}$, we will have $g_j(i) \subseteq \{G_0, \ldots, G_{n-1}\}$, so we fix a gate G_k (corresponding to a meet $p_k \wedge q_k = r_k$, for some $k < n$) and define whether $G_k \in g_j(i)$ by induction on $j \leq m$.

Let $k_j = f_j(0)$, which is the (index of the) lowest gate containing a ball at step j. If $j > 0$ and $i \in \operatorname{dom} h_{j,k}$ then fix $i_k \le l_{j-1}$ maximal such that $T_{j-1}(i_k)$ is undefined or $\le h_{j,k}(i)$. (I.e., $a_{i_k,j-1}$ is the last element of the "preimage k-block" in $\overrightarrow{a_{j-1}}$ of the k-block of $a_{i,j}$ in $\overrightarrow{a_j}$. We leave i_k undefined if $i \notin \operatorname{dom} h_{j,k}$.)

Adding a gate: We add G_k to $g_j(i)$ iff $k < k_j$ and both $p_k, q_k \in (a_{0,j}, \ldots, a_{h_{j,k}(i),j})$ but $r_k \notin (a_{0,j}, \ldots, a_{h_{j,k}(i),j})$.

Deleting a gate: Suppose $j > 0$. Then we delete $G_k \in g_{j-1}(i_k)$ from $g_j(i)$ iff $k \le k_{j-1}$ and at least one of p_k and q_k is not in $(a_{0,j}, \ldots, a_{h_{j,k}(i),j})$.

Otherwise: If G_k is not added into, or deleted from, $g_j(i)$ by one of the above clauses, then $G_k \notin g_j(i)$ (if $j = 0$), or $G_k \in g_j(i)$ iff $G_k \in g_{j-1}(i_k)$ (if $j > 0$, respectively).

Note that whether $G_k \in g_j(i)$ depends only the k-block of $a_{i,j}$ since the definition of $g_j(i)$ only uses $h_{j,k}(i)$ but never i itself. Note also that whether $G_k \in g_j(i)$ can only change (from whether $G_k \in g_{j-1}(i_k)$) when there are no balls at G_0 through G_{k-1}, i.e., at an "expansionary stage" for G_k.

Intuitively, we add a gate G_k into $g_j(i)$ when the gates G_0 through G_k contain no balls and when the k-blocks up the k-block of $a_{i,j}$ target sets below both P_k and Q_k but not R_k, i.e., these k-blocks destroy both $\Psi_k(P_k)$ and $\Psi_k(Q_k)$ while not allowing the correction of $\Delta_k(R_k)$ at an expansionary stage for G_k when G_k should be extending the definition of $\Delta_k(R_k)$ since there are no balls at G_0 through G_k. We delete $G_k \in g_{j-1}(i_k)$ from $g_j(i)$ when the gates G_0 through G_{k-1} contain no balls and when the k-blocks up to the k-block of $a_{i,j}$ target no sets below P_k or no sets below Q_k, i.e., these k-blocks do not destroy one of $\Psi_k(P_k)$ and $\Psi_k(Q_k)$ at an expansionary stage for G_k. Otherwise, we leave $G_k \in g_j(i)$ iff $G_k \in g_{j-1}(i_k)$. This corresponds to the meet strategy outlined in Section 3.1.

DEFINITION 13. A blocked target array $\overrightarrow{A}$ is a *good blocked target array* if for any gate G_k (corresponding to a meet $p_k \wedge q_k = r_k$, for some $k < n$) with $k \le k_j$ (where $k_j = f_j(0)$ is the (index of the) lowest gate containing a ball at step j), $G_k \notin g_j(0)$.

Note that by Definition 12, if $G_k \in g_j(0)$ while $k \le k_j$ then both $p_k, q_k \in (B^0_{j,k})$, with, by Definition 12, (hereditarily) smaller numbers targeted for both P_k and Q_k than any number possibly targeted for R_k. Thus we cannot allow $G_k \in g_j(0)$ while $k \le k_j$, as stated in Definition 13.

We can now state in full detail

LERMAN'S EMBEDDABILITY CRITERION (Lerman [11, 12]). A finite join-semi-distributive lattice (or quasilattice) is embeddable into the c.e. degrees iff for any sequence of gates (corresponding to meet requirements, allowing repetition) and any hole (corresponding to an incomparability requirement), there is a good blocked target array.

Lerman's Embeddability Criterion thus provides a Π_2^0-condition for the embeddability of finite join-semidistributive lattices, with the universal quantifier ranging over sequences of gates and the existential quantifier ranging over good blocked target arrays. Bounding these two quantifiers would yield an effective condition and thus a decision procedure.

3.6. Two gates for two meets $p_0 \wedge q_0 = r_0$ **and** $p_1 \wedge q_1 = r_1$. Here we will encounter a sketch of the first nonembeddability proof, for a lattice we call L_{14}. We will also give a decidable (although rather complicated) criterion for the embeddability of finite quasilattices with two meets specified.

We first consider an incomparability strategy (hole H_2) having to deal with two higher-priority meet strategies (gates G_0 and G_1, one for each meet). We again distinguish cases, first depending on the position of a and b relative to p_1, q_1, and r_1:

CASE 1. $r_1 \not\leq b$ or $p_1 \leq b$ or $q_1 \leq b$: Then the entire initial covering sequence $\overrightarrow{a_0}$ will be a single 1-block, i.e., immediately pass by gate G_1 and go on to gate G_0, where we will proceed as in the case of one meet (i.e., as in Section 3.4).

CASE 2. $r_1 \leq b$ and $p_1, q_1 \not\leq b$: We first handle a special subcase:

CASE 2.1. $r_0 \leq b$: We begin by stating some simplifying assumptions we can make:

(1) It is useless to *add* an element $c \leq r_k$ to a block at gate G_k (for $k = 0, 1$) since this is only allowed, by Definition 13, when there is currently no element $\leq p_k$ or no element $\leq q_k$ in the blocked target sequence up to the k-block of c. Note that, by Definition 12, if such c is not added to the first k-block at G_k but to a later one, then the first k-block through the k-block of c can be combined into one k-block; but adding such c to the first k-block at G_k gives no advantage to providing coverage for the rest of the covering sequence. (However, it is possible to *add* some $c \leq r_1$ at G_0 without any restrictions since corresponding balls will be above any G_1-restraint; and to *add* some $c \leq r_0$ at G_1 as long as the goodness of the blocked target array is not violated.)

(2) Whenever there are balls at G_0, we may assume that, as in Section 3.4, they are arranged in 0-blocks each beginning with a covering sequence for all of $F(p_0)$ or $F(q_0)$. This is since, by Definition 12, this cannot violate the goodness of the covering array and cannot otherwise restrict the covering sequence at G_0 or G_1. We may also assume the 0-blocks B at G_0 to alternatingly satisfy $(B) = F(p_0)$ or $(B) = F(q_0)$.

(3) Whenever there are at least two 0-blocks, B_0 and B_1, say, at G_0 then there is no need to add new elements at G_1 since $(B_0 \cup B_1) = F(p_0) \cup F(q_0) = F(r_0)$, which will cover anything allowed at G_1.

(4) Whenever an element $c \leq r_0$ is added at G_1, then we may add all of $F(r_1)$ (which by (1) is the maximal filter we can use) together with c. This is

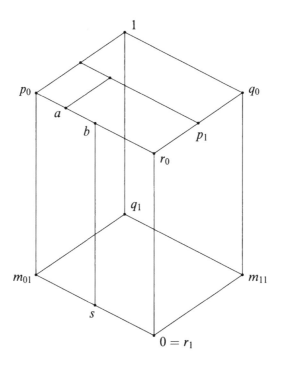

FIGURE 5. An embeddable quasilattice with two meets specified.

because once we target below r_0 at G_1 (and since we may not target below r_1 at G_1), targeting below all of $F(r_1)$ cannot violate the goodness of the blocked target array.

The above now allow us to define an effective decision procedure to decide whether, given two gates below one hole, there is a good blocked target array: There are only finitely many choices as to what to add at gate G_1 since duplicating elements at G_1 only helps if one targets below r_0; but then we can use all of $F(r_1)$ by remark (4) above. And by remark (1) above, the number of choices of what to add at gate G_0 is also effectively bounded, so there is an overall bound on the size of a potential good blocked target array, giving an effective decision procedure for its existence.

We illustrate the above with two examples, which analyze two quasilattices with two meets specified that are the same except for the position of one top p_1 of one meet. Surprisingly, this small difference makes one embeddable while making the other nonembeddable.

EXAMPLE 14. Figure 5 shows an embeddable quasilattice with the two meets $p_0 \wedge q_0 = r_0$ and $p_1 \wedge q_1 = r_1$ specified. The good blocked target array for a hole H_2 (for an incomparability requirement $\mathcal{I}_\Phi^{a,b}$) above the gates G_0 and G_1 (for the corresponding meet requirements) consists of the following blocked target sequences

$$\langle m_{11}, \quad m_{01} \qquad ; \quad p_1; \qquad a \rangle$$
$$\langle s\, m_{01}, m_{11}\, p_1; \quad p_1; \qquad a \rangle$$
$$\langle m_{11}\, p_1; \quad p_1; \qquad s\, a \rangle$$
$$\langle r_0\, p_1; m_{11}, s\, a \rangle$$
$$\langle m_{11}, s\, a \rangle$$
$$\langle s\, a \rangle$$
$$\langle\,\rangle$$

where the transition maps are defined as indicated by vertical alignment; the 0-blocks and 1-blocks are separated by commas and semicolons, respectively; and the 0-blocks $\langle s\, m_{01}, m_{11}\, p_1\rangle$, $\langle m_{11}\, p_1\rangle$, and $\langle m_{11}\rangle$ of the second, third, and fifth blocked target sequence, respectively, are at gate G_0, while all the other blocks are at gate G_1. This uses that

$$(s, m_{01}) = F(q_0),$$
$$(m_{11}, p_1) = F(p_0),$$
$$(r_0, p_1) = F(q_1),$$
$$(m_{11}, s, a) = F(p_1), \text{ and}$$
$$(s, a) = F(q_0).$$

Finally note that in the fourth blocked target sequence, we are not prohibited from inserting r_0 since $p_0 \notin (m_{11}, p_1)$.

EXAMPLE 15. Figure 6 shows the nonembeddable quasilattice L_{14} with the two meets $p_0 \wedge q_0 = r_0$ and $p_1 \wedge q_1 = r_1$ specified. (Compared to the lattice L_{20} in Figure 3 of Lempp and Lerman [9], the nonembeddability proof for L_{14} is much simpler since it requires only two meets instead of four.)

We will illustrate that there is no good blocked target array, and thus that L_{14} cannot be embedded into the c.e. degrees as follows.

The only possible blocked target sequences we can start with are the sequences $\langle m_{11}, m_{01}; m_{10}, a\rangle$ and $\langle m_{11}; m_{10}; m_{01}; a\rangle$ where the 0-blocks and 1-blocks are separated by commas and semicolons, respectively. (The first sequence could be split into more 1-blocks, at no advantage.) The proof that these two starting blocked target sequences will not yield a good blocked target array and thus will not lead to an embedding strategy are similar, so we only indicate the proof for the first sequence.

Starting from the blocked target sequence $\overrightarrow{a_0} = \langle m_{11}, m_{01}; m_{10}, a\rangle$, the second blocked target sequence $\overrightarrow{a_1}$ must contain $\langle m_{01}; m_{10}, a\rangle$ as a subsequence,

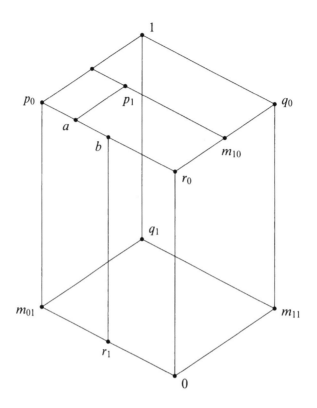

FIGURE 6. Lattice L_{14}.

with m_{01} at gate G_0 and the other elements at gate G_1. In order to have m_{01} pass by gate G_0, we must make it part of a prime filter $F \subseteq F(q_0)$, and the only choice here is $F = F(q_0)$; so the second blocked target sequence $\vec{a_1}$ must contain $\langle r_1\, m_{01}; m_{10}, a \rangle$ as a subsequence. This is still not a covering sequence since (r_1, m_{01}, m_{10}) is not a prime filter, so we need add either r_0 or m_{11} between m_{01} and m_{10}. By prohibition and since $p_0, q_0 \in (m_{01}, m_{11})$, we cannot add r_0, so the second blocked target sequence must be $\vec{a_1} = \langle r_1\, m_{01}, m_{11}; m_{10}, a \rangle$ (or some supersequence, or possibly with more 1-blocks, at no advantage) where $\langle r_1\, m_{01}, m_{11} \rangle$ is at gate G_0 and $\langle m_{10}, a \rangle$ is at G_1. Now the third blocked target sequence $\vec{a_2}$ must contain $\langle m_{11}; m_{10}, a \rangle$ as a subsequence. This is not yet a covering sequence since (m_{11}, m_{10}, a) is not a prime filter but needs r_0, r_1, or m_{01} before a. The former two are prohibited, so we need to use m_{01}, yielding $\vec{a_2} = \langle m_{11}; m_{10}; m_{01}; a \rangle$ or $\vec{a_2} = \langle m_{11}; m_{01}; m_{10}, a \rangle$ (or some supersequence thereof), i.e., we have returned to one of the two starting sequences. In this vein, one can show formally that L_{14} is not embeddable since any good covering array starting with one of the starting sequences keeps repeating a starting sequence (or some supersequence thereof) over and over.

CASE 2.2. $r_0 \not\leq b$: This case is similar to Case 2.1 except that whenever a 0-block contains an element $c \leq r_0$ from the starting sequence, that 0-block (which can extend to the end of the current 1-block) can pass by gate G_0 without problems. A simple modification of the strategy in Case 2.1 thus also shows this case to be decidable.

3.7. Two meets $p_0 \wedge q_0 = r_0$ and $p_1 \wedge q_1 = r_1$. The full argument for two meets is very similar to the case of just two gates outlined in the previous section. We sketch the argument here.

First of all, note that we can think of consecutive gates for the same meet as just one single gate for the purpose of this construction since one gate, or several gates for the same meet, will impose the same restrictions on the allowable blocked target sequences. So assume from now on that any sequence of gates alternates between the two meets specified, say, even-indexed gates work for the meet $p_0 \wedge q_0 = r_0$ and odd-indexed gates work for the meet $p_1 \wedge q_1 = r_1$.

Next, note that due to clause (1) of Case 2.1 in Section 3.6, we cannot add an element $c \leq r_l$ at a gate for the meet $p_l \wedge q_l = r_l$. Also, when we add an element $c \leq r_l$ at a gate for the meet $p_{1-l} \wedge q_{1-l} = r_{1-l}$, we can only add it where it is not prohibited.

Now observe that adding duplicate elements c neither $\leq r_0$ nor $\leq r_1$ is unnecessary since there is no additional coverage, and such elements cannot serve as correction markers. Thus we can bound the number of such elements added at any gate. Furthermore, we can argue, as in Section 3.6, that any element $c \leq r_l$ added at a gate can be replaced by the full filter $F(r_{1-l})$. This yields an effective bound on the length of the blocked target sequences, thus giving us an effective bound on the length of blocked target arrays before blocked target sequences are repeated. Therefore, the embeddings problem for finite semi-distributive quasilattices with two meets specified is decidable as claimed in clause (3) of our Main Statement.

3.8. Three meets $p_i \wedge q_i = r_i$ for $i \leq 2$. Remarks (1) and (4) in Section 3.6 and their extension to the full two-gate case tell us that for a finite quasilattice with only two meets specified, there is no need for so-called "correction markers" to correct meet functionals Δ. The reason for this is that correction markers are balls targeted for elements $c \leq r_k$ at a gate $G_{k'}$ (for some $k' > k$) specifically to correct the meet functional $\Delta_k(R_k)$ such that these elements are not needed for coverage; however, we saw that for two meets, in the only possible case, namely, targeting below r_0 at gate G_1, we will only target so when needed for coverage, and then only with the full filter $F(r_1)$.

In the three-meet case, this may no longer be so. The partial evidence we have for this is only indirect in that we cannot point to a specific finite join-semidistributive lattice which is embeddable but only using correction markers. However, we have worked out a method of "trace trees" (coding the elements needed to cover other elements), which suggests that correction markers may be necessary for some finite join-semidistributive quasilattices with three meets specified. On the other hand, if correction markers are not needed (i.e., if all numbers entering sets are for coverage and not only to correct meet functionals $\Delta^{p,q}$), then the decision procedure for two meets outlined in the previous section will bound the number of good blocked target arrays to be considered in Lerman's Embeddability Criterion. This may then lead to the decidability of the lattice embeddings problem, provided one can also bound the number of gates to be considered.

REFERENCES

[1] K. AMBOS-SPIES, C. G. JOCKUSCH, JR., RICHARD A. SHORE, and ROBERT I. SOARE, *An algebraic decomposition of the recursively enumerable degrees and the coincidence of several degree classes with the promptly simple degrees*, **Transactions of the American Mathematical Society**, vol. 281 (1984), no. 1, pp. 109–128.

[2] K. AMBOS-SPIES and MANUEL LERMAN, *Lattice embeddings into the recursively enumerable degrees*, **The Journal of Symbolic Logic**, vol. 51 (1986), no. 2, pp. 257–272.

[3] ———, *Lattice embeddings into the recursively enumerable degrees. II*, **The Journal of Symbolic Logic**, vol. 54 (1989), no. 3, pp. 735–760.

[4] RODNEY G. DOWNEY, *Lattice nonembeddings and initial segments of the recursively enumerable degrees*, **Annals of Pure and Applied Logic**, vol. 49 (1990), no. 2, pp. 97–119.

[5] VIKTOR A. GORBUNOV, *Algebraic Theory of Quasivarieties*, Siberian School of Algebra and Logic, Consultants Bureau, New York, 1998.

[6] ALISTAIR H. LACHLAN, *Lower bounds for pairs of recursively enumerable degrees*, **Proceedings of the London Mathematical Society. Third Series**, vol. 16 (1966), pp. 537–569.

[7] ———, *Embedding nondistributive lattices in the recursively enumerable degrees*, **Conference in Mathematical Logic—London '70 (Proc. Conf., Bedford Coll., London, 1970)**, Lecture Notes in Math., vol. 255, Springer, Berlin, 1972, pp. 149–177.

[8] ALISTAIR H. LACHLAN and ROBERT I. SOARE, *Not every finite lattice is embeddable in the recursively enumerable degrees*, **Advances in Mathematics**, vol. 37 (1980), no. 1, pp. 74–82.

[9] STEFFEN LEMPP and MANUEL LERMAN, *A finite lattice without critical triple that cannot be embedded into the enumerable Turing degrees*, **Annals of Pure and Applied Logic**, vol. 87 (1997), no. 2, pp. 167–185.

[10] MANUEL LERMAN, *Admissible ordinals and priority arguments*, **Cambridge Summer School in Mathematical Logic (Cambridge, 1971)**, Springer, Berlin, 1973, pp. 311–344.

[11] ———, *A necessary and sufficient condition for embedding ranked finite partial lattices into the computably enumerable degrees*, **Annals of Pure and Applied Logic**, vol. 94 (1998), no. 1-3, pp. 143–180.

[12] ———, *A necessary and sufficient condition for embedding principally decomposable finite lattices into the computably enumerable degrees*, **Annals of Pure and Applied Logic**, vol. 101 (2000), no. 2-3, pp. 275–297.

[13] ———, *Embeddings into the computably enumerable degrees*, **Computability Theory and its Applications (Boulder, CO, 1999)**, Amer. Math. Soc., Providence, RI, 2000, pp. 191–205.

[14] THEODORE A. SLAMAN and ROBERT I. SOARE, *Extension of embeddings in the computably enumerable degrees*, **Annals of Mathematics. Second Series**, vol. 154 (2001), no. 1, pp. 1–43.

[15] ROBERT I. SOARE, *Recursively Enumerable Sets and Degrees*, Perspectives in Mathematical Logic, Springer, Berlin, 1987.

[16] STEVEN K. THOMASON, *Sublattices of the recursively enumerable degrees*, *Zeitschrift für Mathematische Logik und Grundlagen der Mathematik*, vol. 17 (1971), pp. 273–280.

[17] C. E. M. YATES, *A minimal pair of recursively enumerable degrees*, *The Journal of Symbolic Logic*, vol. 31 (1966), pp. 159–168.

DEPARTMENT OF MATHEMATICS
UNIVERSITY OF WISCONSIN
MADISON, WI 53706-1388, USA
E-mail: lempp@math.wisc.edu

DEPARTMENT OF MATHEMATICS
UNIVERSITY OF CONNECTICUT
STORRS, CT 06269-3009, USA
E-mail: mlerman@math.uconn.edu
E-mail: david.solomon@uconn.edu

DIMENSION THEORY INSIDE A HOMOGENEOUS MODEL

OLIVIER LESSMANN

Abstract. Homogeneous model theory is the study of the class of (elementary) submodels of a large sufficiently homogeneous model. In this paper, we show how to develop and use dimension theory in this nonelementary context.

We focus on uncountable categoricity, $\aleph_0$-stability. superstability, and supersimplicity. We first give a Baldwin-Lachlan style theorem: If a class is categorical in some uncountable cardinal, then each uncountable model is prime and minimal over the basis of a (type)-definable pregeometry, which implies that the class is categorical in all uncountable cardinals. We then show that there is a good dependence relation in the $\aleph_0$-stable and the superstable cases; we use it to construct homogeneous models and present Main Gap theorems. Finally, we extend dimension theory to the supersimple case, without stability assumptions.

The class of free groups $F(X)$ on infinitely many generators X fits in this context but is not first order. It is categorical in all infinite cardinals, $\aleph_0$-stable (hence superstable), supersimple, and consists only of homogeneous models.

Introduction. First order model theory can be viewed as the study of the class of elementary submodels of a large sufficiently saturated model $\bar{M}$, as any (small) model of the first order theory of $\bar{M}$ embeds elementarily into $\bar{M}$. In this sense, $\bar{M}$ can be used as a universal domain. To extend the methods of first order model theory to classes of models with more complicated axiomatisations, it becomes necessary to consider weaker notions of universal domains, as saturated models will in general not be in the class.

Homogeneous model theory is concerned with the class of elementary submodels of a large (sufficiently) homogeneous model, written $\mathfrak{C}$ throughout this paper. Then, $\mathfrak{C}$ functions as a universal domain for the class of models of its first order theory, *omitting* over the empty set all the types omitted in $\mathfrak{C}$. Classes of models of a first order theory T omitting a prescribed set of types Γ are usually written $EC(T, \Gamma)$; omitting the types in Γ gives us more expressive power than first order logic. We focus here on $EC(T, \Gamma)$-classes but we could consider other infinitary classes (for example axiomatised by a sentence in $L_{\omega_1,\omega}$) without any real change in the theory.

The compactness theorem fails at this level of generality (some types are omitted). However, a weaker property follows from the homogeneity of $\mathfrak{C}$ and provides a good criterion for which complete types are realised. This is

Logic Colloquium '02
Edited by Z. Chatzidakis, P. Koepke, and W. Pohlers
Lecture Notes in Logic, 27

called *weak compactness*: A complete type p over a small subset A of $\mathfrak{C}$ is realised in $\mathfrak{C}$ if and only if $p \upharpoonright B$ is realised in $\mathfrak{C}$ for each finite subset B of A. Weak compactness is the main reason why homogeneous model theory is so well-behaved. Dimension theory is an example of this good behaviour. In this paper, we will show how to use pregeometries, bases, and dimensions, and more generally dependence relations and Morley sequences, to understand the structure of the models. The overall picture is similar to the first order case.

Homogeneous model theory provides a unifying framework for much of current model theory: First order, Robinson theories, existentially closed models, Banach space model theory, classes of models with amalgamation over sets (n-variable, infinitary). Moreover many generic constructions produce homogeneous models (predimension, beautiful pairs, generic automorphism), and furthermore, the existence of homogeneous models follows from (nonelementary) stability conditions (see Theorem 2.18 below). Finally, many mathematical structures can be dealt with in this framework; for example free groups, Hilbert spaces, and even some operator algebras like $\ell^2(X)$. Homogeneous model theory does not contain Shelah's excellent classes ([35] and [36]), but we will discuss the connections between the two contexts also.

Homogeneous model theory was initiated by Saharon Shelah and many of the results we present are due to him. This survey is also based on the works of Alexander Berenstein, Steven Buechler, Rami Grossberg, Bradd Hart, Tapani Hyttinen, H. Jerome Keisler, Boris Zilber, and myself. In the first section we describe the context of homogeneous model theory in more details, and remind the reader of some of the basic facts. In the second section, we consider the problem of categoricity in some uncountable cardinal, under the assumption that the language is countable. We present a generalisation of the Baldwin-Lachlan theorem to this context. A natural example of categoricity is the class of free groups on infinitely many generators. We also discuss $\aleph_0$-stability and the connection with excellent classes. In the third section we present the basic theory of stability and superstability. We show that in the superstable case, there is a dependence relation satisfying many (but not all) of the properties of dividing in the first order case. We give an application to the construction of homogeneous (as opposed to sufficiently homogeneous) models in all cardinals greater than or equal to the first stability cardinal. We also present Main Gap theorems for two classes of models. In the fourth section we show under which circumstances dividing has *all* the properties that it has in the first order case. We call this simplicity (in fact, we treat the supersimple case only). We give an example of Shelah showing that stability does not imply simplicity at this level of generality and also give examples of (non first order) supersimple structures. Throughout these sections, we only present some of the proofs, chosen to illustrate aspects of the general theory. Finally, in the last section we list a few open questions.

The notation is standard: sets are written with capital letters A, B, C, E (we reserve D for diagrams), finite sequences are written with small letters a,

b, c, d, finite sequences of variables x, y, and so forth. We write $a \in A$ to mean that a is a finite sequence from elements in A.

§1. Context.
In this section, we recall the main definitions and some of the basic results from [30].

A model M is *λ-homogeneous* if for any elementary map $f : M \to M$, with $|f| < \lambda$ and $a \in M$, there exists an elementary map $g : M \to M$ extending f such that $a \in \text{dom}(g)$. In this paper, we focus on elementary maps, but we could replace these elementary maps by maps preserving other types of formulas (quantifier-free, existential, n-variable, infinitary) without significant change in the theory.

A model M is *homogeneous*, if it is λ-homogeneous for $\lambda = \|M\|$. It is not difficult to see that M is homogeneous if and only if any partial elementary map $f : M \to M$, with $|f| < \|M\|$, extends to an automorphism of M. Many familiar mathematical structures are homogeneous: the countable dense linear order without endpoints, the countable random graph, any algebraically closed field, the reals (as a real-closed field), the positive integers, the ring of integers, and so on. More generally, any model of an $\aleph_0$-categorical theory is $\aleph_0$-homogeneous, and any model of an uncountably categorical first order theory is homogeneous.

Fix a model M. Let D be the *(finite) diagram* of M, namely the set of complete types over the empty set realised by finite sequences from M. We say that a set A is a *D-set* if $\text{tp}(a/\emptyset) \in D$, for each finite $a \in A$. Similarly, we say that a model is a *D-model* if its universe is a D-set.

We are interested in the class of D-models for a fixed diagram D. This is the nonelementary class of models of a first order theory omitting, over the empty set, all the complete types outside D — an $EC(T, \Gamma)$-class.

There are two extreme cases: (1) $D = D(T)$ is the set of all complete types of the first theory of M: this is the *first order* case. (2) D is the set of all complete isolated types of the first order theory of M: this is the *atomic* case. Recall that a complete type p is isolated if there is a formula $\phi \in p$ such that $\phi \vdash p$. In the atomic case, all the models are $\aleph_0$-homogeneous. The atomic case is in some way typical; for questions about the number of uncountable models of various nonelementary classes (for example axiomatised by a Scott sentence in $L_{\omega_1,\omega}$), it is enough to focus on the number of uncountable models of an atomic diagram.

DEFINITION 1.1. M is *(D, λ)-homogeneous* if M is a λ-homogeneous with diagram D.

Homogeneous model theory is the study of the class of D-models under the assumption that there exist (D, λ)-homogeneous models of size at least λ, for arbitrarily large λ. Diagrams D with this property are called *good*.

We now give some consequences of the homogeneity assumption.

Fix a diagram D. We first turn to the problem of understanding which types are realised by D-models. A basic result in first order model theory states that M is λ-saturated if and only if M is λ-homogeneous and the diagram of M consists of all complete types over the empty set. This is a general phenomenon: The identity of D completely determines which complete types can be realised by D-models.

Let M be a D-model. Let $A \subseteq M$ and p be a complete type over A. If p is realised by $c \in M$, then the set $A \cup c$ is a D-set (since it is inside a D-model). Hence, whenever p over a D-set A is realised by a D-model containing A, then $A \cup c$ is a D-set for some (all) realisations of p. The presence of (D, λ)-homogeneous models for arbitrarily large λ shows that this necessary condition is, in fact, sufficient: Let M be (D, λ)-homogeneous and let $A \subseteq M$ of size less than λ. Let p be a complete type over A and suppose that some sequence c (in some elementary extension of M which is not necessarily a D-model) realises p and is such that $A \cup c$ is a D-set. By (D, λ)-homogeneity of M, the set $A \cup c$ embeds elementarily into M. Now by λ-homogeneity of M, we may assume that this embedding is the identity on A. Hence, the image of c under this embedding is a realisation of p in M. This leads to the following definition:

DEFINITION 1.2. Let $S_D(A)$ be the collection of complete types p over A such that whenever $c \models p$ then $A \cup c$ is a D-set.

It is easy to see that if A is not a D-set, then $S_D(A)$ is empty. Also, if $A \cup c$ is a D-set for some $c \models p$, then $A \cup c'$ is a D-set for all $c' \models p$.

We have shown one direction of the following fact:

FACT 1.3. Let M be a model. The following conditions are equivalent:

(1) M is (D, λ)-homogeneous.
(2) M realises all the types in $S_D(A)$ for each $A \subseteq M$ of size less than λ.

Thus, (D, λ)-homogeneity corresponds to a notion of saturation for the appropriate set of types. *Weak compactness* is simply a restatement of this property: Let M be (D, λ)-homogeneous. Suppose p is a complete type over $A \subseteq M$ of size less than λ. Then p is realised in M if and only if $p \in S_D(A)$ if and only if

$$p \restriction B \text{ is realised in } M, \quad \text{for each finite } B \subseteq A.$$

Here are three consequences of weak compactness inside M:

FACT 1.4. Let M be (D, λ)-homogeneous. If $(p_i : i < \alpha < \lambda)$ is an increasing sequence of complete types p_i, each of which is realised in M, then the union is a complete type realised in M.

FACT 1.5. Let M be (D, λ)-homogeneous. Let $A, I \subseteq M$. If I is an infinite A-indiscernible sequence and $|I| < \lambda$ then I can be extended to an A-indiscernible sequence of size λ.

The next fact is often referred to as 'Morley's methods'. It is useful even in the first order case. The issue is that the indiscernible sequence similar to the sequence we start with exists already inside the model M.

FACT 1.6 (Morley's methods). Let M be (D, λ)-homogeneous. Let $\{a_i : i < \kappa\}$ and A be in M. If κ is suitably large compared to $|A|$, then there exists an A-indiscernible sequence $(b_n : n < \omega) \subseteq M$ such that for each $n < \omega$ there are $i_0 < \cdots < i_n < \lambda$ such that

$$\text{tp}(b_0, \ldots, b_n/A) = \text{tp}(a_{i_0}, \ldots, a_{i_n}/A).$$

Above, 'suitably large' means at least the Hanf number $\beth_{(2^{|A|+|D|})^+}$.

In order to study a good diagram D, we could simply work inside a (D, λ)-homogeneous model $\mathfrak{C}$, for a large enough λ. Then, $\mathfrak{C}$ embeds elementarily any D-set, provided it has size at most λ and the class of elementary submodels of $\mathfrak{C}$ (of size at most λ) corresponds to the class of D-models (of size at most λ). However, it is more convenient to work inside a strongly λ-homogeneous model, as the automorphism group of such models is better behaved.

DEFINITION 1.7. A model M is *strongly λ-homogeneous*, if any elementary map $f : M \to M$, with $|f| < \lambda$, extends to an automorphism of M.

Obviously, M is homogeneous if it is strongly $\|M\|$-homogeneous. Also, if M is strongly λ-homogeneous, then M is λ-homogeneous.

As in the first order case, it is easy to show the following fact:

FACT 1.8. The following conditions are equivalent.

(1) There are (D, λ)-homogeneous models of size at least λ for arbitrarily large λ.
(2) There are strongly λ-homogeneous models of size at least λ whose diagram is D, for arbitrarily large λ.

We can now make the convention that will hold throughout this paper (except for Theorem 2.18).

CONVENTION 1.9. We fix a good diagram D. We let $\mathfrak{C}$ be a strongly $\bar{\kappa}$-homogeneous model, whose diagram is D, for a suitably large $\bar{\kappa}$.

All D-sets and all D-models of size at most $\bar{\kappa}$ embed elementarily into $\mathfrak{C}$. Hence, there is no loss of generality in working inside $\mathfrak{C}$. Satisfaction is evaluated with respect to $\mathfrak{C}$ and all sets and models are inside $\mathfrak{C}$, unless specified otherwise. All sets, ordinals, and cardinals will be assumed to have size less than $\bar{\kappa}$. All types $p \in S_D(A)$ with $A \subseteq \mathfrak{C}$ of size less than $\bar{\kappa}$ are realised in $\mathfrak{C}$. Furthermore, these complete types correspond to orbits of the automorphism group of $\mathfrak{C}$.

§2. Categoricity and $\aleph_0$-stability. In this section, we assume that the language L is countable. We first consider the problem of *categoricity*: Suppose that for some cardinal λ, any two D-models of size λ are isomorphic. What can be said about the isomorphism-type of models in other cardinals? When $\lambda = \aleph_0$, nothing much can be said; the atomic case is like this and any Scott sentence in $L_{\omega_1,\omega}$ can be translated into an atomic diagram. Hence, as in the first order case, the problem is interesting mostly for $\lambda > \aleph_0$.

This problem was solved by Shelah [30] and Keisler [16] independently (in different but equivalent contexts; Keisler was considering classes axiomatisable by a sentence in $L_{\omega_1,\omega}$) in the early 1970s. We recall the definition and then state their theorem.

DEFINITION 2.1. Let λ be a cardinal. We say that D is *categorical in λ* if any two D-models of size λ are isomorphic.

THEOREM 2.2 (Keisler, Shelah). *If D is categorical in some uncountable cardinal, then D is categorical in all uncountable cardinals.*

Their proof is to show that all uncountable models are homogeneous with diagram D and therefore isomorphic if they have the same size.

In this section, we outline a Baldwin-Lachlan style proof of this result, which generalises the first order case, and exemplifies the role of dimension theory for the problem of categoricity. This theorem appears in [23]. A similar theorem with a different statement is also due to Hyttinen in [7].

THEOREM 2.3 (Lessmann). *If D is categorical in some uncountable cardinal, then each uncountable D-model is prime and minimal over the basis of a pregeometry (and its parameters), given by a quasiminimal type. It follows that D is categorical in all uncountable cardinals.*

We now introduce the relevant definitions and concepts needed and present some of the arguments. We give an outline of the full argument after Fact 2.13. We first recall Shelah's definition of λ-stability for this context [30].

DEFINITION 2.4. The diagram D is said to be *λ-stable* if $|S_D(A)| \leq \lambda$, for each D-set A of size at most λ.

We defer exploring stability to the next section. For now, we just state a fact due to Keisler [16] and Shelah [30], which is proved using Ehrenfeucht-Mostowski models, just like in the first order case.

FACT 2.5. If D is categorical in some uncountable cardinal, then D is $\aleph_0$-stable.

It follows from the previous fact that D must be countable. Note that D may be $\aleph_0$-stable, even if the first order theory is unstable:

EXAMPLE 2.6. Consider a structure M in the language $\{N, +, 0, 1\}$, where N is a predicate whose interpretation in M is the set of natural numbers $\mathbb{N}$ in the language $\{+, 0, 1\}$, and the interpretation of $\neg N$ is simply an infinite set with no structure. Let T be the first order theory of M and D its diagram. Then T is unstable, as it has the strict order property. However, D is good and categorical in all infinite cardinals: the model of size λ has a copy of $\mathbb{N}$ as the interpretation for N and the interpretation of $\neg N$ is simply a set of size λ with no structure. Hence, D is $\aleph_0$-stable.

We say that a set is *small*, *not big*, or *bounded*, if it has size less than $\bar{\kappa}$ and *not small*, *big*, or *unbounded* otherwise. We say that a formula or a type is big, if its set of realisations in $\mathfrak{C}$ is big. The previous example shows that some infinite definable sets may be small. Thus the dichotomy finite/unbounded which is at the root of strongly minimal sets cannot work. However, in the $\aleph_0$-stable case, as soon as a definable set is uncountable, it is big. To see this, we first state a useful fact concerning the existence of indiscernibles. The key ingredient of this proof is that, when D is $\aleph_0$-stable and $p \in S_D(A)$, there exists a finite subset $B \subseteq A$ such that p does not split over B. Recall that p *splits over* B, if there are $c, d \in A$ with $\text{tp}(c/B) = \text{tp}(d/B)$, and a formula $\phi(x, y)$, such that $\phi(x, c) \in p$ and $\neg\phi(x, d) \in p$. Nonsplitting is very well-behaved in the $\aleph_0$-stable case. It gives rise to a dependence relation over all the models with very good properties. We postpone this discussion to the next section though, where this will be done in more generality. The next fact is due to Shelah and proved in [30].

FACT 2.7. Let D be λ-stable. Let $I \cup A \subseteq \mathfrak{C}$. Assume that $|I| > \lambda \geq |A|$. Then there exists a subset of I of size λ^+, which is indiscernible over A.

The previous fact implies that if D is $\aleph_0$-stable and p is a type over countably many parameters which is realised uncountably many times, then p is big (it is realised by an indiscernible sequence over its parameters, and that sequence can be extended to have size $\bar{\kappa}$). We can also show:

FACT 2.8. Let D be $\aleph_0$-stable. Let p be a type over A. The following conditions are equivalent:
  (1) p is big.
  (2) For each model M containing A, there is $c \in \mathfrak{C} \setminus M$ realising p.
  (3) There is a $(D, \aleph_0)$-homogeneous model $M \prec \mathfrak{C}$ such that p is realised in $\mathfrak{C} \setminus M$.

We now look at the smallest big types. This concept can be traced back to Shelah [32] (in a different context). The name was coined by Zilber [39] in the mid-1990, in his work about complex numbers with exponentiation (see Example 2.17).

DEFINITION 2.9. A type in one-variable $q(x)$ is *quasiminimal* if it is big and there does not exist $\phi(x, b)$ such that both $q \cup \{\phi(x, b)\}$ and $q \cup \{\neg\phi(x, b)\}$ are big.

In particular, if $q(x)$ is a type over countably many parameters, q is quasi-minimal if and only if $q(\mathfrak{C})$ has size at least $\bar{\kappa}$, but for any definable set $\phi(\mathfrak{C})$, the set $q(\mathfrak{C}) \cap \phi(\mathfrak{C})$ is either countable, or has countable complement in $q(\mathfrak{C})$. Strongly minimal types are quasiminimal.

Suppose that $q(x, b)$ is quasiminimal and let $b \in A$. It follows immediately from the definition that q has a unique extension in $S_D(A)$ which is big.

The next proof that quasiminimal types always exist can be found in [24]. Initially, this was proved using the rank in [23].

PROPOSITION 2.10. *Let D be $\aleph_0$-stable. Then there exists a quasiminimal type over finitely many parameters.*

PROOF. This proof illustrates the use of weak compactness. Suppose, for a contradiction, that no quasiminimal type exists over finitely many parameters. Since $\{x = x\}$ is big, by Fact 2.7 there is a big, complete type $q_\emptyset$ over the empty set. Since $q_\emptyset$ is big but not quasiminimal, there exists a formula $\phi(x, b)$ such that both $q_\emptyset \cup \{\phi(x, b)\}$ and $q_\emptyset \cup \{\neg\phi(x, b)\}$ are big. By Fact 2.7 again, we may choose q_0 and q_1, which are big, complete over b, and extend $q_\emptyset \cup \{\phi(x, b)\}$ and $q_\emptyset \cup \{\neg\phi(x, b)\}$ respectively. We can continue in this way and construct a tree of big complete types over finitely many parameters q_η, for $\eta \in {}^{<\omega}2$, such that $q_\eta \subseteq q_\nu$ if $\eta < \nu$ and $q_{\eta^\frown 0}$ and $q_{\eta^\frown 1}$ are contradictory. By weak compactness, for each $\eta \in {}^\omega 2$, the type $q_\eta := \bigcup_{n<\omega} q_{\eta \upharpoonright n}$ is realised in $\mathfrak{C}$, by say c_η. Let $A \subseteq \mathfrak{C}$ be the countable set of parameters used in the q_ηs for $\eta \in {}^{<\omega}2$. Then $\mathrm{tp}(c_\eta/A) \in S_D(A)$, and $\mathrm{tp}(c_\eta/A) \neq \mathrm{tp}(c_\nu/A)$ for $\eta \neq \nu$. Hence, there are continuum many types in $S_D(A)$, so D is not $\aleph_0$-stable, a contradiction. ⊣

We now show that quasiminimal types induce a *pregeometry*. Recall that a *pregeometry* is a pair (W, cl), where

$$\mathrm{cl} : \mathcal{P}(W) \longrightarrow \mathcal{P}(W)$$

satisfies

(1) $x \in \mathrm{cl}(X)$, for each $x \in X$.
(2) If $x \in \mathrm{cl}(X)$, there exists a finite $Y \subseteq X$ such that $x \in \mathrm{cl}(Y)$.
(3) If $x \in \mathrm{cl}(X)$ and $X \subseteq \mathrm{cl}(Y)$, then $x \in \mathrm{cl}(Y)$.
(4) (Exchange) If $x \in \mathrm{cl}(Xy) \setminus \mathrm{cl}(X)$, then $y \in \mathrm{cl}(Xx)$.

Let (W, cl) be a pregeometry. A set $A \subseteq W$ is *closed* if $\mathrm{cl}(A) = A$. For a closed set A, we can define a *basis* for A which is the smallest $I \subseteq A$ such that $\mathrm{cl}(I) = A$, or, equivalently, the largest *independent* subset of A, i.e. the largest subset $I \subseteq A$ such that $a \notin \mathrm{cl}(I \setminus a)$, for each $a \in I$. It is easy to show, just like in linear algebra, that any two bases of the same closed set have the same size. This size is called the *dimension of A*.

What Theorem 2.3 asserts is that the dimension theory inside the pregeometry determines the isomorphism type of the D-models when D is categorical in some uncountable cardinal.

The next fact is proved in [23], and [24].

FACT 2.11. Let D be $\aleph_0$-stable. Let $q(x,b)$ be a quasiminimal type with finitely many parameters. For $a, B \subseteq q(\mathfrak{C})$, define $a \in \mathrm{cl}(B)$ if $\mathrm{tp}(a/Bb)$ extends q but is not big. Then $(q(\mathfrak{C}), \mathrm{cl})$ satisfies the axioms of a pregeometry.

The key idea is to use the property that quasiminimal types have a unique complete big extension over any set. As an illustration, we prove (2): Suppose that $a \in \mathrm{cl}(B)$. Then $\mathrm{tp}(a/Bb)$ is small, so $\mathrm{tp}(a/Bb) \neq q_B$, where q_B is the unique, big, complete type over Bb. There is $c \in B$ with $\phi(x, c, b) \in \mathrm{tp}(a/Bb)$ and $\neg\phi(x, c) \in \mathrm{tp}(a/Bb)$. By quasiminimality of q, the type $q(x, b) \cup \phi(x, b, c)$ is small, so $a \in \mathrm{cl}(c)$.

We now turn to prime models. Given a class $\mathcal{K}$ of models, we say that M is *prime over A in $\mathcal{K}$*, if $A \subseteq M \in \mathcal{K}$, and for all $N \in \mathcal{K}$, if $f : A \to N \in \mathcal{K}$ is elementary, then there is an elementary map $g : M \to N$ extending f. When $\mathcal{K}$ is the class of D-models, we simply say *prime over A*. The next fact is due to Shelah in [30]. There is a different proof using a rank in [23].

FACT 2.12. Let D be $\aleph_0$-stable. Over any D-set, there is a prime model in the class of $(D, \aleph_0)$-homogeneous models.

This is proved by showing that there are $(D, \aleph_0)$-primary models over any D-set. Recall that a model M is $(D, \aleph_0)$-*primary over A*, if $M = A \cup \{a_i : i < \lambda\}$ is a $(D, \aleph_0)$-homogeneous model and the type $\mathrm{tp}(a_i/A \cup \{a_j : j < i\})$ is implied by its restriction to some finite subset of its parameters. It is easy to see that if M is $(D, \aleph_0)$-primary over A, then M is prime over A in the class of $(D, \aleph_0)$-homogeneous models. Note that the existence of primary models (with the usual notion of isolation) is not known to follow from categoricity assumptions.

The next fact corresponds to *unidimensionality* (or *no Vaughtian pairs*). It is proved in [23] and [24].

FACT 2.13. Let D be categorical in some uncountable cardinal. Let $M \prec N$ be $(D, \aleph_0)$-homogeneous models. Let q be a quasiminimal type over finitely many parameters in M. If $q(M) = q(N)$, then $M = N$.

The previous fact is used to show that models are minimal over bases. Given a class of models $\mathcal{K}$, we say that a model $M \in \mathcal{K}$ is *minimal over A* if $A \subseteq M$ and for all $N \in \mathcal{K}$, if $A \subseteq N \prec M$, then $M = N$.

We can now finish the proof:

PROOF OF THEOREM 2.3. First, one shows that all uncountable D-models are $(D, \aleph_0)$-homogeneous (see for example [30] for a proof of this).

Let M be a D-model of size $\lambda > \aleph_0$. Then M is $(D, \aleph_0)$-homogeneous. Let $q(x, b)$ be a quasiminimal type over finitely many parameters b. By $(D, \aleph_0)$-homogeneity of M, we may assume that $b \in M$.

Notice that $q(M)$ is closed in the pregeometry $(q(\mathfrak{C}), \mathrm{cl})$: This is because if $c \in \mathrm{cl}(q(M))$, then $\mathrm{tp}(c/q(M))$ must be small, so $\mathrm{tp}(c/q(M))$ cannot be realised outside the $(D, \aleph_0)$-homogeneous model M (by Fact 2.8), so $c \in M$, and therefore $c \in q(M)$.

Since $q(M)$ is closed, choose a basis I of $q(M)$. We claim that $|I| = \lambda$ and that M is prime and minimal over I: Let M' be prime over $I \cup b$ in the class of $(D, \aleph_0)$-homogeneous models (which exists by Fact 2.12). Then $q(M') = q(M)$ (as $I \subseteq q(M') \subseteq q(M)$ and both $q(M')$ and $q(M)$ are closed). Thus $M = M'$ by Fact 2.13. This shows that M is prime over $I \cup b$ and that I has size λ. The fact that M is minimal over $I \cup b$ is shown similarly. Notice that, since I is uncountable, being prime and minimal over $I \cup b$ in the class of $(D, \aleph_0)$-homogeneous models coincides with being prime and minimal in the class of D-models.

Now, let N be another D-model of size λ. We will show that N is isomorphic to M. First, we may assume that $b \in N$. Second, N is prime and minimal over $J \cup b$, where J is a basis of $q(N)$. Since $|I| = |J| = \lambda$, there is a bijection $f : I \to J$. It is easy to see that this bijection is elementary and can be extended to the identity on b. By primeness of M over I, we can find an elementary map $g : M \to N$ extending f. But $g(M) \prec N$, both contain $J \cup b$, and N is minimal over $J \cup b$. Hence $g(M) = N$, so g is the desired isomorphism. $\dashv$

We now look at the class of infinitely generated free groups. The fact that this example fits in the context of homogeneous model theory was noticed by Keisler in [16].

EXAMPLE 2.14. Consider the class of free groups $F(X)$ on infinitely many generators X, in the language of groups with an additional predicate for the set X of generators. This class is not first order axiomatisable, but it is the class of D-models of a diagram D. We can describe both the theory and the diagram easily: The first order theory states that $F(X)$ is a group, that X is infinite, and that every element which can be written as a product of elements of X and their inverse, can be written in this way in at most one way. D is simply the set of isolated types of this theory (it is enough to omit the type of an element which cannot be written as a product of elements of X and their inverse).

Then D is categorical in all infinite cardinals and all models are homogeneous. X is the definable quasiminimal set (in fact, strongly minimal in this case), and it carries a trivial pregeometry. $F(X)$ is prime and minimal over X, since it is the definable closure of X.

The example of free groups is also instructive in that it shows the limitations of some extensions of the first order theory of stable groups. In the first order case, a theorem of Baur, Cherlin and Macintyre [1] states that any infinite group whose first order theory is categorical in all infinite cardinals is abelian by finite. Free groups have a good diagram which is categorical in all infinite cardinals and yet, all the abelian subgroups are countable. Notice that free groups do not have generics.

Another difference with the first order case is the value of the U-rank. Without going into too much details (see the last section on simplicity for more), notice that the U-rank of an element a corresponds to the size of the set $\{x_1, \ldots, x_n\}$, where $a = x_1^{\varepsilon_1} \cdots x_n^{\varepsilon_n}$, and $x_i \in X$, and $\varepsilon_i = \pm 1$. Then, all elements have finite U-rank. This is the general situation: It is possible to define a U-rank for complete types over $(D, \aleph_0)$-homogeneous models, and show that, if D is uncountably categorical, then the U-rank of any such type is finite (see [25] for details). However, the example of free groups shows that the supremum of the U-rank over all the 1-types may be ω, which is not achieved. In the first order uncountably categorical case, this does not happen: The supremum is always finite.

We now consider Keisler's question [16] concerning the existence of arbitrarily large homogeneous models. When Keisler proved his result on categoricity using homogeneous models, he asked whether this was the general case. We rephrase his question in our context. Recall that we say that D is good if there are (D, λ)-homogeneous models of size at least λ for arbitrarily large λ.

QUESTION 2.15 (Keisler). Let D be a countable diagram (which we do not assume to be good). Assume that D is categorical in some (or all) uncountable cardinal(s). Must D be good?

Shelah answered this question negatively (using an example of Marcus [27]) and proceeded to develop the theory of *excellence*. See [32], [35], and [36].
He showed:

THEOREM 2.16 (Shelah). *Let D be an atomic diagram (in a countable language).*

(1) *(GCH) If D is categorical in $\aleph_n$, for each $n < \omega$, then D is excellent.*
(2) *If D is excellent, and categorical in some uncountable cardinal, then D is categorical in all uncountable cardinals.*

Many of Zilber's examples in his work around the complex numbers with exponentiation are excellent, categorical in all infinite cardinals, even quasi-minimal, but not homogeneous. We give an example below.

Excellence is a condition on the existence of prime models over certain countable sets (which in turn, implies the existence of prime models over some uncountable sets). We can also give a Baldwin-Lachlan style proof of categoricity under the assumption of excellence (see [25]). More care has to be taken as many sets do not have prime models over them (and D is not $\aleph_0$-stable, see Theorem 2.18 below). Excellence can be developed more generally for the class of $(D, \aleph_0)$-homogeneous models of a diagram. For the proof of categoricity in this case, only the following two consequences of excellence are needed:

(1) D has the *amalgamation property*, i.e. if $M_0 \prec M_1, M_2$ are $(D, \aleph_0)$-homogeneous models, there are a $(D, \aleph_0)$-model N and elementary maps $f_\ell : M_\ell \to N$, such that $f_1 \upharpoonright M_0 = f_2 \upharpoonright M_0$.

(2) If M is a $(D, \aleph_0)$-homogeneous model and $c \models p \in S_D(M)$, then there is a $(D, \aleph_0)$-primary model over $M \cup c$.

A consequence of (2) is that any type $p \in S_D(M)$ is realised in a D-model extending M (hence, the natural notion of semantic types given by the amalgamation property and the syntactic notion given by $S_D(M)$ coincide). Together with (1), we can form a universal domain $\bar{M}$, which realizes all types in $S_D(M)$ for $M \prec \bar{M}$ of smaller size (such models are called *full*, they are not homogeneous in general).

We now give one of Zilber's pseudo-analytic, excellent, quasi-minimal structures.

EXAMPLE 2.17. This is a summary of [40] written with $EC(T, \Gamma)$-classes, rather than $L_{\omega_1, \omega}$. Consider

$$0 \longrightarrow \mathbb{Z} \longrightarrow^i H \longrightarrow^{ex} F^* \longrightarrow 1,$$

where F^* is the multiplicative group of an algebraically closed field F of characteristic 0, H is a torsion-free, divisible, abelian group, and the sequence is exact.

The canonical example, the simplest among Zilber's pseudo-analytic structures [39], is when $F = \mathbb{C}$, $H = \mathbb{C}^*$, and *ex* is the exponentiation map

$$\exp : \mathbb{C}_+ \longrightarrow \mathbb{C}^*.$$

Zilber represents this as a one-sorted structure H, whose universe is the universe of the torsion-free divisible abelian group H, in the language of abelian groups $+$. He adds two basic relations: a basic equivalence relation E, whose interpretation is

$$E(h_1, h_2) \quad \text{if and only if} \quad ex(h_1) = ex(h_2),$$

and a ternary relation S with interpretation

$$S(h_1, h_2, h_3) \quad \text{if and only if} \quad ex(h_1) + ex(h_2) = ex(h_3).$$

So the nonzero elements of the field F are the equivalence classes h/E, for $h \in H$. Multiplication in F can then be defined via E and $+$ and addition in F is defined via S. The kernel of ex, which is simply the class corresponding to the unit 1 of the field, is then definable.

This class can be axiomatised as follows:

- H is a torsion free, divisible, abelian group,
- $H/E \cup \{0\}$ is an algebraically closed field of characteristic 0.

We also omit a type to express:

- For all x_1 and x_2 in the kernel of ex, there are $z_1, z_2 \in \mathbb{Z} \setminus \{0\}$ such that $z_1 x_1 + z_2 x_2 = 0$.

This class is categorical in all infinite cardinals, i.e. the fields are isomorphic, the abelian groups are isomorphic, and these isomorphisms commute with ex. It follows from categoricity that all the models have the same complete first order theory in this language. Zilber further proves that every uncountable model is $(D, \aleph_0)$-homogeneous, where D is the set of complete types over the empty set realised by the models in the class and that the class is excellent.

In this example, the universe H is quasiminimal and $a \in \operatorname{cl}(B)$ if $ex(a) \in acl(ex(B))$, where acl is the algebraic closure operator in the sense of the field structure (on $ex(H) \cup \{0\}$).

This example does not belong to homogeneous model theory. A variation, due to Zilber, where $\mathbb{Z}$ is replaced by its completion in the profinite topology, does belong to homogeneous model theory.

In the remainder of this section, we discuss the difference between homogeneity and excellence.

We have seen that if D is categorical in some uncountable cardinal, then if D is good, then D is $\aleph_0$-stable and all uncountable models are $(D, \aleph_0)$-homogeneous.

We can, in fact, show the converse. In the next theorem, we do *not* assume that D is good, as it is the conclusion. Also, the language may be uncountable. The next result is proved in [25]. It generalises results of Shelah in [30] and Kudaibergenov in [21].

THEOREM 2.18 (Lessmann). *Assume that D is λ-stable and has a (D, λ)-homogeneous model of size greater than λ. Then D has (D, μ)-homogeneous models of size at least μ, for arbitrarily large μ.*

Let us look at the case when $\lambda = \aleph_0$: If D is $\aleph_0$-stable and has an uncountable $(D, \aleph_0)$-homogeneous model, then D is good. (The requirement of uncountability of the $(D, \aleph_0)$-homogeneous model cannot be weakened, as is shown by the example of the natural numbers.) It follows from this, that if D is a countable diagram which is uncountably categorical, then D is good if and only if there is a $(D, \aleph_0)$-primary model over any countable D-set.

This shows precisely the difference between homogeneity and excellence in the uncountably categorical case: it lies in the strength of the $\aleph_0$-stability assumption, or, equivalently, in the existence of prime models. In excellent, uncountably categorical classes, there may be countable D-sets A over which $S_D(A)$ is uncountable, and these are precisely the sets over which there are no prime models!

Another particular case of Theorem 2.18 is that if M is an $\aleph_0$-homogeneous, uncountable model of an $\aleph_0$-stable first order *theory* T, then there are arbitrarily large homogeneous models with the same diagram as M (this particular case was proved earlier by Kudaibergenov in [21]; it can be seen as an omitting type theorem). This fact cannot be weakened to superstable T: Example 2.17 has superstable first order theory. This is therefore the simplest, nonhomogeneous possibility for excellence.

§3. Stability and superstability. In this section, the language L may be uncountable. We extend $\aleph_0$-stability to the more general context of superstability and stability.

We first state several results due to Shelah in the stable context: The stability spectrum [30], the equivalence between instability and the existence of unbounded orders [31], and the homogeneity spectrum [33]. An exposition of these results with different proofs can be found in [5].

We then focus on superstability and state the central dimension-theoretic fact that, in the superstable case, there exists a dependence relation on the subsets of the models which satisfies many of the properties of dependence inside a pregeometry. As an application, we use this theorem to prove a particular case of the homogeneity spectrum: When D is superstable, there exists a (D, λ)-homogeneous model of size λ, whenever λ is at least the first stability cardinal.

We also discuss the Main Gap for two classes of models, which is another dramatic illustration of how dimension theory allows us to understand the isomorphism type of models.

We already defined λ-stability. Shelah defines *stability* in this context in [30]:

DEFINITION 3.1. D is *stable* if D is λ-stable for some cardinal λ.

In the previous section, we saw that uncountable categoricity of a countable D implied $\aleph_0$-stability. We could show in the same way that uncountable categoricity implies that D is λ-stable, for all infinite λ. In fact, the $\aleph_0$-stability of D already implies the λ-stability of D in all the infinite cardinals. This transfer phenomenon, by which stability in one cardinal implies stability in other cardinals, holds in much more generality. In addition to [30] and [5], another proof of Shelah's stability spectrum theorem is due to Hyttinen and Shelah in [13].

THEOREM 3.2 (Stability spectrum theorem). *If D is stable, there are cardinals $\kappa(D) \leq \lambda(D) < \beth_{(2^{|D|})^+}$, such that D is μ-stable if and only if $\mu \geq \lambda(D)$ and $\mu^{<\kappa(D)} = \mu$.*

The cardinal $\beth_{(2^{|D|})^+}$ follows from the use of Morley's methods. The cardinal $\lambda(D)$ is the first cardinal λ such that D is stable in λ. The cardinal $\kappa(D)$ is the smallest cardinal κ such that for $p \in S_D(A)$ there exists $B \subseteq A$ of size less than $\kappa(D)$ such that p *does not split strongly over* B, i.e. whenever $(a_i : i < \omega)$ is a B-indiscernible sequence with $a_0, a_1 \in A$, then for any formula $\phi(x, y)$, if $\phi(x, a_0) \in p$, then $\phi(x, a_1) \in p$. Strong splitting is better behaved than splitting in general, and has good properties when the diagram D is stable (see the paragraph following Theorem 3.12). It is a coarser version of dividing (see the section on simplicity)

DEFINITION 3.3. We have three contexts:

- D is $\aleph_0$-*stable* if $\lambda(D) = \kappa(D) = \aleph_0$.
- D is *superstable* if $\lambda(D) < \infty$ and $\kappa(D) = \aleph_0$.
- D is *stable* if $\lambda(D) < \infty$.

When the diagram D is uncountable and categorical in some cardinal greater than $|D|$, then D is superstable (but not $\aleph_0$-stable in general).

We now turn to the equivalence between order and instability. We saw in Example 2.6 a stable diagram D with an infinite order. However, the order there was bounded. We do have the dichotomy between order and stability for unbounded orders:

THEOREM 3.4 (Shelah). *The following conditions are equivalent:*

(1) D *is not stable.*

(2) *There is a formula* $\phi(x, y)$ *and an indiscernible sequence* $(a_i : i < \omega)$ *such that*

$$\mathfrak{C} \models \phi(a_i, a_j) \quad \textit{if and only if} \quad i < j < \omega.$$

(3) *There exists a formula* $\phi(x, y)$ *such that for arbitrarily large* λ, *there is* $\{a_i : i < \lambda\}$ *such that*

$$\mathfrak{C} \models \phi(a_i, a_j) \quad \textit{if and only if} \quad i < j < \lambda.$$

The equivalence of (2) and (3) is a simple application of Morley's methods and the fact that infinite indiscernible sequences can be extended. The crucial point is the equivalence of (1) and (2): (2) implies (1) uses homogeneity in the fact that indiscernible sequences can be extended. (1) implies (2) is purely combinatorial and uses no homogeneity.

This shows, for example, that we can use the natural numbers, or the integers, or the rationals, or the reals in the axiomatisations of our structures without necessarily destroying stability, as all these mathematical structures are bounded when axiomatised with a finite diagram.

We now focus on *dependence relations*. To motivate the properties that we will require of a dependence relation, suppose that D is $\aleph_0$-stable and let $q(x)$ be a quasiminimal type. Inside the pregeometry $q(\mathfrak{C})$, we can define what we mean by A *is free from* C *over* B, which we will write $A \underset{B}{\downarrow} C$, by

$$\dim(A'/B) = \dim(A'/B \cup C), \quad \text{for any finite } A' \subseteq A.$$

It is not difficult to see that $A \not\perp_B C$, if and only if there exists $a_0, \ldots, a_n \in A$ such that

$$a_0 \in \text{cl}(B \cup C \cup a_1 \cdots a_n) \setminus \text{cl}(Ba_1 \cdots a_n).$$

Intuitively, $A \not\perp_B C$ if C has information over A that B does not have.

We can easily establish the following eight properties, using the definition of the pregeometry, and the fact that quasiminimal sets have a unique big extension over any set. We attach to these properties their name, as we will use them later:

FACT 3.5. Let D be $\aleph_0$-stable. Let q be quasiminimal. Let $A \perp_B C$ be defined as above inside $q(\mathfrak{C})$, i.e. all sets and elements in the list below are in $q(\mathfrak{C})$. Then, the following properties hold:

(1) (Local Character) For any a and C, there exists a finite $B \subseteq C$ such that $a \perp_B C$.

(2) (Extension) For any a, B and C containing B, with $\dim(C) < \dim(q(\mathfrak{C}))$, there exists $a' \models \text{tp}(a/B)$ such that $a' \perp_B C$.

(3) (Symmetry) $A \perp_B C$ if and only if $C \perp_B A$.

(4) (Transitivity) If $B \subseteq C \subseteq E$, then $A \perp_B C$ and $A \perp_C E$ imply $A \perp_B E$.

(5) (Monotonicity) If $A' \subseteq A$, $B \subseteq B_1 \subseteq B \cup C$ and $A \perp_B C$, then $A' \perp_{B_1} C$.

(6) (Finite Character) $A \perp_B C$ if and only if $A' \perp_B C'$ for all finite $A' \subseteq A$ and finite $C' \subseteq C$.

(7) (Invariance) $A \perp_B C$ if and only if $f(A) \perp_{f(B)} f(C)$ for any automorphism f of $\mathfrak{C}$.

(8) (Stationarity) If $\text{tp}(a_1/B) = \text{tp}(a_2/B)$ and $a_\ell \perp_B C$ for $\ell = 1, 2$, then

$$\text{tp}(a_1/C) = \text{tp}(a_2/C).$$

In the first order case, any superstable theory has a dependence relation defined everywhere satisfying (1)–(8), except that B in (8) is a model in general. In the context of homogeneous model theory, even in the $\aleph_0$-stable case, B is a $(D, \aleph_0)$-homogeneous model not only in (8), but in (2) and (3), as well (see [24]). There are ways of going around using a model for (3), but (1) has to be changed (see [13] or [10] for details).

We are going to state what is known in the superstable case. For this, we need to introduce *Lascar strong types*. Lascar strong types in the context of stable diagrams were introduced by Hyttinen and Shelah [13] in the context of stability. We will use them also in the next section, where stability is not assumed. The proof of Facts 3.7, 3.8, and 3.10, concerning Lascar strong types without assuming that D is stable can be found in [3].

Recall that an equivalence relation $E(x, y)$ on $\mathfrak{C}$ is said to be *A-invariant* if $E(a, b)$ implies $E(f(a), f(b))$ for any $f \in \text{Aut}(\mathfrak{C}/A)$. As usual, we say that E has a *bounded* number of equivalence classes if it has fewer than $\bar{\kappa}$ equivalence classes.

DEFINITION 3.6. Two sequences $b, c \in \mathfrak{C}$ of the same length have *the same Lascar strong type over A*, written $\text{Lstp}(b/A) = \text{Lstp}(c/A)$ if $E(b, c)$ holds for any *A*-invariant equivalence relation E with a bounded number of equivalence classes.

It is easy to see that $\text{Lstp}(x/A) = \text{Lstp}(y/A)$ is itself an *A*-invariant equivalence relation with a bounded number of equivalence classes, so it is the finest such equivalence relation.

We gather a few facts about the behaviour of Lascar strong types.

FACT 3.7. If $\text{tp}(b/A)$ is not big and $\text{Lstp}(c/A) = \text{Lstp}(b/A)$, then $b = c$.

Suppose that b, c belong to an infinite *A*-indiscernible sequence. Then, since infinite indiscernible sequences can be extended, we must have $\text{Lstp}(b/A) = \text{Lstp}(c/A)$. The following fact clarifies this connection with indiscernibles.

FACT 3.8. Suppose that $\text{tp}(b/A)$ is big. Then, the following conditions are equivalent:

(1) $\text{Lstp}(b/A) = \text{Lstp}(c/A)$.
(2) There exist $n < \omega$ and $a_0, \ldots, a_n$ such that $a_0 = b$, $a_n = c$, and for each $\ell < n$, the elements $a_\ell, a_{\ell+1}$ belong to an infinite *A*-indiscernible sequence.

We can introduce *strong automorphisms*.

DEFINITION 3.9. An automorphism $f \in \text{Aut}(\mathfrak{C}/A)$ is *strong* if

$$\text{Lstp}(a/A) = \text{Lstp}(f(a)/A),$$

for any $a \in \mathfrak{C}$.

We denote the set of strong automorphisms over A by $\text{Aut}_f(\mathfrak{C}/A)$. Then, $\text{Aut}_f(\mathfrak{C}/A)$ is easily seen to be a normal subgroup of $\text{Aut}(\mathfrak{C}/A)$. The next fact shows that, just like complete types can be construed as orbits of the automorphism group, so too can Lascar strong types be construed as orbits of the group of strong automorphisms:

FACT 3.10. The following conditions are equivalent:

(1) $\text{Lstp}(b/A) = \text{Lstp}(c/A)$.
(2) There exists $f \in \text{Aut}_f(\mathfrak{C}/A)$ such that $f(b) = c$.

The definition of a-saturated model is due to Hyttinen and Shelah and appears in [13].

DEFINITION 3.11. Let D be stable. We say that M is *a-saturated* if M realises every Lascar strong type $\mathrm{Lstp}(a/A)$, for each subset $A \subseteq M$ of size less than $\kappa(D)$.

When D is superstable, M is a-saturated if M realises every Lascar strong types over finite subsets of M.

In the superstable case, Hyttinen and myself [10] introduced a rank R to measure the complexity of type-definable sets. Notice that the rank must measure the complexity of the type-definable sets as they appear in $\mathfrak{C}$, and not in a saturated extension of $\mathfrak{C}$.

If we set $A \underset{B}{\downarrow} C$, by

$$R(\mathrm{tp}(a/B \cup C)) = R(\mathrm{tp}(a/B)),$$

for any finite $a \in A$, then we can prove:

THEOREM 3.12 (Hyttinen-Lessmann). *Let D be superstable. Then the relation $A \underset{B}{\downarrow} C$ satisfies* (1)–(8) *in the above properties, except that B needs to be replaced by an a-saturated model in Extension, Symmetry, and Stationarity.*

I had previously shown in [23] that Extension, Symmetry, and Stationarity hold over any $(D, \aleph_0)$-homogeneous model, when D is $\aleph_0$-stable, using a rank which is bounded in this case. The above theorem is a generalisation, as $(D, \aleph_0)$-homogeneous models are a-saturated in the $\aleph_0$-stable case. Hyttinen and Shelah have proved a version of this theorem under stability [13] assumptions for strong splitting: The main difference is in the statement of Transitivity, which requires a-saturated models as well.

An example due to Shelah which we give in the next section (Example 4.10) shows that we cannot expect better properties (like those in the first order case) in general.

We are now going to give an application of Theorem 3.12. For this, we first need to define *averages*. Averages were introduced by Shelah in [33].

The next fact follows immediately from the definition of $\kappa(D)$ in terms of strong splitting.

FACT 3.13. Let D be stable. Let I be an indiscernible sequence of size at least $\kappa(D)$. Let $b \in \mathfrak{C}$. Then all but fewer than $\kappa(D)$ elements of I realise the same type over b.

Given an indiscernible sequence I of size at least $\kappa(D)$ and a set A, we can define *the average of I over A*:

$$Av(I/A) = \{\phi(x, a) : \text{At least } \kappa(D) \text{ elements of } I \text{ realise } \phi(x, a)\}.$$

Fact 3.13 shows that it is well-defined when D is stable. Furthermore $Av(I/A)$ $\in S_D(A)$: To see this, extend I to an indiscernible $J \subseteq \mathfrak{C}$ with $|J| \geq |A|^+ + \kappa(D)$. Then $Av(I/A) = Av(J/A)$, and by using the previous fact, we have that all but fewer than $|A|^+ + \kappa(D)$ elements of J realise $Av(J/A)$. By choice of J, some elements of J will realise $Av(I/A)$, so $Av(I/A) \in S_D(A)$. Averages are very useful tools; together with the dependence relation, they allow us to construct homogeneous models. Shelah proved:

THEOREM 3.14 (Homogeneity Spectrum). *There exists a* (D, λ)-*homogeneous model of size* λ *if and only if* $\lambda \geq |D|$, *and either* D *is stable in* λ *or* $\lambda^{<\lambda} = \lambda$.

There are two issues: (1) To construct homogeneous models in those cardinals when D is stable. (2) To show that homogeneous models do not exist in the other cases. Not surprisingly, the inexistence results are much more combinatorial in flavour. We prove the positive part in the superstable case only, but we follow the ideas of the general proof:

PROPOSITION 3.15. *If* D *is superstable then there is a* (D, λ)-*homogeneous model of size* λ, *for each* $\lambda \geq \lambda(D)$.

PROOF. Let $\lambda \geq \lambda(D)$. By superstability, D is λ-stable. Construct an increasing and continuous sequence of D-models of size λ

$$(M_i : i < \lambda)$$

such that M_{i+1} realises all the Lascar strong types over M_i.

This is possible: Since D is λ-stable, there are at most λ distinct Lascar strong types over each set of size λ (if $I = \{a_i : i < \lambda^+\}$ realises distinct strong types over a set A of size λ, then by Fact 2.7, there is $J \subseteq I$ of size λ^+ which is indiscernible over A, and hence, all the $a_i \in J$ realise the same Lascar strong type over A, a contradiction).

Now let $M = \bigcup_{i<\lambda} M_i$. Then, M has size λ and we claim that M is (D, λ)-homogeneous.

For $\lambda = \aleph_0$ this is obvious, so suppose that λ is uncountable. Let $p \in S_D(A)$, for $A \subseteq M$ of size less than λ. Let $c \in \mathfrak{C}$ realise p and consider $q = \mathrm{tp}(c/M)$. If $c \in M$, we are done, so we may assume that $c \notin M$. By the Local Character of superstability, we have that

$$c \underset{B}{\downarrow} M, \quad \text{for some finite set } B \subseteq M.$$

Without loss of generality, we may assume that $B \subseteq M_0$. We construct a sequence

$$I = (a_i : i < \lambda, i \text{ a limit ordinal})$$

such that $a_i \in M_{i+1}$ realises $q \upharpoonright M_i$. Then $|I| = \lambda$, since $c \notin M$. Notice that when i is a limit, M_i must be a-saturated. Hence, by Stationarity over a-saturated models, it is not difficult to see that I is an indiscernible sequence. It then follows from the definition that $Av(I/A) = p$. Furthermore, all but fewer than $|A|^+ + \kappa(D)$ elements of I realise $Av(I/A)$ by Fact 3.13. But, $\lambda(D) \geq |A|^+ + \kappa(D)$ so there is $c' \in I$ realising p. $\dashv$

In the rest of this section, we will consider Main Gaps. For this, we first need to consider pregeometries generalising quasiminimal types:

DEFINITION 3.16. A type $p \in S_D(B)$ is *regular* if the relation $a \in \mathrm{cl}(C)$ given by $a \not\downarrow_B C$ induces a pregeometry on the set of realisations of p.

The next result is a generalisation of a theorem in [23], using the methods of [10] or [14].

THEOREM 3.17. *Let D be superstable. Let $M \subseteq N$ be a-saturated models, and $M \neq N$. Then there exists $a \in N \setminus M$ such that $\mathrm{tp}(a/M)$ is a regular type.*

With these ingredients, it is possible to develop *orthogonality calculus*, which essentially studies the relations between different regular types from the point of view of independence. We first consider the $\aleph_0$-stable case.

DEFINITION 3.18. Assume D is $\aleph_0$-stable. D has *NDOP* if whenever M_ℓ are $(D, \aleph_0)$-homogeneous models for $\ell = 0, 1, 2$ satisfying

$$M_1 \underset{M_0}{\downarrow} M_2,$$

then there exists a $(D, \aleph_0)$-primary model M_3 over $M_1 \cup M_2$, which is minimal in the class of $(D, \aleph_0)$-homogeneous models. We say that D has DOP if D fails to have NDOP.

In the first order case, the Main Gap for the class of $\aleph_\varepsilon$-saturated models (the first order counterpart of a-saturated models here) was proved by Shelah (see also [37] for the proof of the Main Gap for the class of all models of a countable first order theory). The first Main Gap for a nonelementary class of models (atomic and excellent) was due to Grossberg and Hart and appears in [4] using NDOP. Here also NDOP is the main dividing line.

THEOREM 3.19 (Grossberg-Lessmann). *Let D be $\aleph_0$-stable.*

- *If D has NDOP, then each $(D, \aleph_0)$-homogeneous model is prime and minimal (in the class of $(D, \aleph_0)$-homogeneous models) over a tree of height at most ω consisting of independent countable models.*
- *If D has DOP, then there are 2^λ nonisomorphic $(D, \aleph_0)$-homogeneous models of size λ, for cofinally many cardinals λ.*
- *The Main Gap holds for the class of $(D, \aleph_0)$-homogeneous models: Either there are $2^{\aleph_\alpha}$ nonisomorphic $(D, \aleph_0)$-homogeneous models of size $\aleph_\alpha$ for cofinally many cardinals $\aleph_\alpha$ or there are fewer than $\beth_{\omega_1}(|\alpha + \aleph_0|)$ nonisomorphic $(D, \aleph_0)$-homogeneous models of size $\aleph_\alpha$, for each cardinal $\aleph_\alpha$.*

The ingredients of the proof are: (1) the existence of $(D, \aleph_0)$-prime models, which we have already established; (2) the good dependence relation over the subsets of the models given by Theorem 3.12; (3) the existence of the more general pregeometries given by regular types in Theorem 3.17; and finally (4) a good understanding of the relationships between the regular types, the so-called orthogonality calculus that we mentioned earlier. Having established this, the proof is similar to [4] and [37], and the details are in [22]:

Let M be a $(D, \aleph_0)$-homogeneous model. The models M_s, for s in the tree of the decomposition of M under NDOP are thus: If $s = \emptyset$, then $M_s \prec M$ is the $(D, \aleph_0)$-primary model over the empty set. If s is a successor of t, then $M_s \prec M$ is a $(D, \aleph_0)$-primary model over $M_t \cup a_s$, where $p = \text{tp}(a_s/M_t) \in S_D(M_t)$ is regular and a_s belongs to a basis of $p(M)$. Furthermore, the regular types involved over a given model M_t are chosen so that if $\text{tp}(a_{s_1}/M_t) \neq \text{tp}(a_{s_2}/M_t)$, where t is the predecessor of s_1 and s_2, then

$$a_{s_1} \underset{M_t}{\downarrow} a_{s_2}.$$

Clearly, if $a_{s_1} \neq a_{s_2}$ belong to a basis of the same regular type, then the same relation holds. It follows that

$$M_{s_1} \underset{M_t}{\downarrow} M_{s_2}, \quad \text{if } t \text{ is a predecessor of } s_1, s_2 \text{ and } s_1 \neq s_2.$$

Using the properties of independence, the entire tree is independent. NDOP guarantees that the maximal such tree that can be formed inside M exhausts all the elements of M.

We have encountered this idea in a simpler form before. In the uncountably categorical case, we have a tree of height one for each uncountable model M: Having constructed M_0, there is essentially only one regular $p \in S_D(M_0)$ realised in M (any other regular types are nonorthogonal to p). Let I be a basis for $p(M)$ and let $M_a \prec M$ be $(D, \aleph_0)$-primary over $M_0 \cup a$, for $a \in I$. Then M is prime and minimal over the independent tree $(M_0, M_a : a \in I)$.

The weakness of Theorem 3.19 is that $\aleph_0$-stability is not a dividing line, i.e. the failure of $\aleph_0$-stability does not imply the existence of many models. On the other hand, superstability is a dividing line. This was proved by Hyttinen [8, 6] and also by Hyttinen and Shelah [12] for various classes of models. One should therefore establish Main Gap theorems under the assumption of superstability, rather than $\aleph_0$-stability. The natural generalisation is to consider the class of a-saturated models. Unfortunately, the conclusion is not known in this case: the obstacle is that we do not know whether there exist prime models in the class of a-saturated models. However, Hyttinen and Shelah were able to prove the Main Gap for the class of *locally saturated models* [14], which is a subclass of the class of a-saturated models.

DEFINITION 3.20. A D-model M is *locally λ-saturated* if for each finite $A \subseteq M$ there exists a (D, λ)-homogeneous model N such that $A \subseteq N \prec M$. When D is superstable, M is called *locally saturated* if M is locally $\lambda(D)$-saturated.

Hyttinen and Shelah [14] proved that when D is not superstable, for each cardinal μ, there are cofinally many cardinals λ, with 2^λ nonisomorphic locally

μ-saturated models of size λ. Furthermore, under the appropriate generalisation of NDOP for the class of locally saturated models when D is superstable, they proved:

THEOREM 3.21 (Hyttinen-Shelah). *Let D be superstable.*

- *If D has NDOP, then each locally saturated model is prime and minimal (in the class of locally saturated models) over a tree of height at most ω consisting of independent models of size at most $\lambda(D)$.*
- *If D has DOP, then there are 2^{λ} nonisomorphic locally saturated models of size λ, for cofinally many cardinals λ.*
- *The Main Gap holds for the class of locally saturated models: Either there are $2^{\aleph_\alpha}$ nonisomorphic locally saturated models of size $\aleph_\alpha$ for cofinally many cardinals $\aleph_\alpha$ or there are fewer than $\beth_{(|D|^{\aleph_0})^+}(|\alpha + \lambda(D)|)$ nonisomorphic locally saturated models of size $\aleph_\alpha$ for each cardinal $\aleph_\alpha$.*

§4. Simplicity and supersimplicity. In this section, L may be uncountable. We study dependence relations without assuming any stability. We want to have a theory of dependence on the subsets of a model, which shares all the properties of dividing in the first order case. The main motivation is that several interesting mathematical examples exhibit such a good behaviour (see Example 4.12 and Example 4.13).

We focus on *dividing*, which is a dependence relation defined syntactically. The chief reason is that one can show that if a dependence relation has all the required properties, it has to be dividing (see for example [10] or [20], and [18] in the first order case). The idea of *simplicity* is to see if we can derive all the properties of dividing (among the list of (1)–(8)), from the knowledge that some of them hold. In this section, we focus on supersimplicity (for simplicity, one needs to assume Finite Character and relax Local Character). All the results in this section are due to Buechler and myself in [3]. They are generalisations of the first order results due to Shelah [34], Kim [17] and Kim-Pillay [18]. See also their exposition in [19].

We first define dividing. The definition we give is equivalent to Shelah's original definition when $D = D(T)$, that is, in the first order case.

DEFINITION 4.1. Let $p(x, c) \in S_D(Ac)$. We say that $p(x, c)$ *divides over B* if there exists an infinite B-indiscernible sequence I containing c such that

$$\bigcup_{d \in I} p(x, d)$$

is not realized in $\mathfrak{C}$.

Although we have used the forking symbol before, all the definitions coincide in the context where they have been defined, so this should not be a problem. Hence, we will write

$$A \underset{B}{\downarrow} C$$

if $\mathrm{tp}(a/Bc)$ does not divide over B, for every finite $a \in A$ and $c \in C$.
We now define supersimplicity. See [3] for simplicity.

DEFINITION 4.2. We say that D is *supersimple* if

(1) (Local Character) For all a and C we have

$$a \underset{B}{\downarrow} C, \quad \text{for some finite } B \subseteq C.$$

(2) (Extension) Let $B \subseteq C$ and a be given. Then

$$a' \underset{B}{\downarrow} C, \quad \text{for some } a' \in \mathfrak{C} \text{ realising } \mathrm{tp}(a/B).$$

The requirement for Extension involves realising partial types in $\mathfrak{C}$. This is a problematic question inside a homogeneous model in general and may fail (see Example 4.10).

We have the following *algebraic* properties of dividing:

THEOREM 4.3 (Buechler-Lessmann). *Let D be supersimple.*

(1) (*Symmetry*)

$$A \underset{B}{\downarrow} C \quad \text{if and only if} \quad C \underset{B}{\downarrow} A.$$

(2) (*Transitivity*) Let $B \subseteq C \subseteq E$.

$$A \underset{B}{\downarrow} C \text{ and } A \underset{C}{\downarrow} E \quad \text{implies} \quad A \underset{B}{\downarrow} E.$$

It is easy to see that Monotonicity, Finite Character and Invariance already follow.

The key property is the use of *Morley sequences*. A Morley sequence is an infinite indiscernible independent sequence:

DEFINITION 4.4. Let D be supersimple. Let $p \in S_D(A)$ and let $B \subseteq A$. Let X be an infinite linear order. The sequence $(a_i : i \in X)$ is a *Morley sequence for p over B* if

(1) Each a_i realises p.
(2) $(a_i : i \in X)$ is indiscernible over A.
(3) For each $i_1 < \cdots < i_m < j_1 < \cdots < j_n \in X$, we have

$$a_{j_1} \cdots a_{j_n} \underset{B}{\downarrow} a_{i_1} \cdots a_{i_m}.$$

When $B = A$, we simply say Morley sequence for p.

We have used Morley sequences twice already: In the proof of Theorem 2.3: The basis I for $q(M)$ is a Morley sequence for the quasiminimal type q. In the proof of Proposition 3.15: The sequence $I = (a_i : i < \lambda, i \text{ a limit})$ is a Morley sequence for $q \upharpoonright M_0$.

We will outline the proof of Theorem 4.3 to illustrate the role that Morley sequences play in the proof.

FACT 4.5. Let $p(x, b) = \text{tp}(a/Cb) \in S_D(Cb)$. The following conditions are equivalent:

(1) $p(x, b)$ does not divide over C.
(2) For each infinite C-indiscernible sequence $I \ni b$, there is an automorphism f of $\mathfrak{C}$ fixing Cb pointwise such that $f(I)$ is indiscernible over Ca.

From this, it follows quickly that:

FACT 4.6. If $\text{tp}(a/Cb)$ does not divide over C and $\text{tp}(c/Cba)$ does not divide over C, then $\text{tp}(ac/Cb)$ does not divide over C.

We now prove that Morley sequences exist.

PROPOSITION 4.7. *Let D be supersimple. Suppose $p \in S_D(C)$ is big and let $B \subseteq C$. Let X be an infinite linear order. There is a Morley sequence $(a_i : i \in X)$ for p over B.*

PROOF. By Local Character of supersimplicity, there exists $B' \subseteq C$ finite such that p does not divide over B'. Thus, by making B bigger if necessary, we may assume that p does not divide over B.

Let λ be suitably large compared to B in the sense of Fact 1.6. By Extension, we can define $(b_i : i < \lambda)$ such that $b_i \models \text{tp}(a/C)$ and

$$b_{i+1} \underset{B}{\bigcup} \{b_j : j < i\}, \quad \text{for } i < \lambda.$$

Using Morley's methods (Fact 1.6 and a property similar to Fact 1.5), we can find $(a_i : i \in X)$ indiscernible over A such that for $i_0 < \cdots < i_n \in X$, there are $j_1 < \cdots < j_n < \lambda$ with $\text{tp}(a_{i_0}, \ldots, a_{i_n}/A) = \text{tp}(b_{i_0}, \ldots, b_{i_n}/A)$, and it follows that

$$a_{i_n} \underset{B}{\bigcup} a_{i_1} \cdots a_{i_{n-1}}, \quad \text{for each } n < \omega.$$

Using Fact 4.6 inductively, we can show that $(a_i : i \in X)$ is a Morley sequence (i.e. satisfies (3) in the previous definition).   ⊣

The next proposition was proved by Kim [17] in the first order case. It holds also at this level of generality and we give the proof, as it is different from Kim's and very short.

PROPOSITION 4.8 (Buechler-Lessmann). *Assume that D is supersimple. Let $p(x, b) \in S_D(Cb)$. The following conditions are equivalent:*

(1) $p(x, b)$ *divides over* C.

(2) *There is a Morley sequence* $(b_i : i \in \mathbb{Z})$ *for* $\mathrm{tp}(b/C)$ *such that* $\bigcup_{i \in \mathbb{Z}} p(x, b_i)$ *is not realized in* $\mathfrak{C}$.

(3) *For each Morley sequence* $(b_i : i \in \mathbb{Z})$ *for* $\mathrm{tp}(b/C)$, *the set of formulas* $\bigcup_{i \in \mathbb{Z}} p(x, b_i)$ *is not realized in* $\mathfrak{C}$.

PROOF. (2) implies (1) is by definition. Also, since D is supersimple, Morley sequences exists and (3) implies (2). Hence, it is enough to show that (1) implies (3): Let $(b_i : i \in \mathbb{Z})$ be a Morley sequence for $\mathrm{tp}(b/C)$ over C. Suppose that $p(x, b)$ divides over C. We must show that $\bigcup_{i \in \mathbb{Z}} p(x, b_i)$ is not realized in $\mathfrak{C}$.

We first claim and if $i_0 < i_1 < \cdots < i_n \in \mathbb{Z}$, then $p(x, b_{i_0})$ divides over $Cb_{i_1} \ldots b_{i_n}$.

By (1) $p(x, b)$ divides over C, so certainly $p(x, b_{i_0})$ divides over C (by using an automorphism of $\mathfrak{C}$ fixing C pointwise and sending b to b_{i_0}). Let J be a C-indiscernible sequence containing b_{i_0} witnessing the fact that $p(x, b_{i_0})$ divides over C. Since

$$b_{i_1} \cdots b_{i_n} \underset{C}{\downarrow} b_{i_0},$$

by definition of a Morley sequence, we have $\mathrm{tp}(b_{i_1}, \ldots, b_{i_n}/Cb_{i_0})$ does not divide over C. Since $b_{i_0} \in J$, by Fact 4.5, there is an automorphism f of $\mathfrak{C}$ fixing Cb_{i_0} such that $f(J)$ is indiscernible over $Cb_{i_1} \ldots b_{i_n}$. Then $f(J)$ demonstrates that $p(x, b_{i_0})$ divides over $Cb_{i_1} \ldots b_{i_n}$.

To continue the proof, suppose that $\bigcup_{i \in \mathbb{Z}} p(x, b_i)$ was realized by some element $a \in \mathfrak{C}$. Consider the type $\mathrm{tp}(a/C \cup \{b_i : i \in \mathbb{Z}\})$ extending $\bigcup_{i \in \mathbb{Z}} p(x, b_i)$. By Local Character of supersimplicity, there are $i_1 < \cdots < i_n$ such that

$$a \underset{Cb_{i_1} \ldots b_{i_n}}{\downarrow} C \cup \{b_i : i \in \mathbb{Z}\}.$$

Choose $i_0 \in \mathbb{Z}$ such that $i_0 < i_1$. Then $p(x, b_{i_0})$ is realized by a by definition and

$$a \underset{Cb_{i_1} \ldots b_{i_n}}{\downarrow} b_{i_0},$$

by Monotonicity. But then $p(x, b_{i_0})$ does not divide over $Cb_{i_1} \ldots b_{i_n}$, which contradicts our claim.   ⊣

We can now give the proof of Symmetry:

PROOF OF THEOREM 4.3(1). Suppose that $p(x, b) = \mathrm{tp}(a/Cb)$. Suppose that $\mathrm{tp}(a/Cb)$ does not divide over C. If $\mathrm{tp}(a/C)$ is small, then $\mathrm{tp}(b/Ca)$ does not divide over C. If $\mathrm{tp}(a/C)$ is big, then also $\mathrm{tp}(a/Cb)$ is big; let $(a_i : i \in \mathbb{Z})$ be a Morley sequence for $\mathrm{tp}(a/Cb)$ over C, which exists by Proposition 4.7. In particular, $(a_i : i \in \mathbb{Z})$ is Cb-indiscernible and is a Morley sequence for $\mathrm{tp}(a/C)$ over C. So a_i realizes $p(x, b)$ for each $i \in \mathbb{Z}$ by indiscernibility over C. So $b \in \mathfrak{C}$ realizes $\bigcup_{i \in \mathbb{Z}} p(a_i, y)$. Since $(a_i : i \in \mathbb{Z})$ is a Morley sequence, we have $\mathrm{tp}(b/Ca)$ does not divide over C by Proposition 4.8.   ⊣

This finishes the proof for the algebraic properties. We now focus on the *multiplicity properties* of dividing, which are extensions of Stationarity. Stationarity is a strong requirement about the compatibility of free extensions, which can only hold in the stable case (see below). The Independence Theorem is a weaker compatibility requirement. In the first order case, it is due to Kim and Pillay. The proof used in [3] is closer to Pillay's proof in the existentially closed case [28], and involves ideas from Shami [29].

THEOREM 4.9 (Buechler-Lessmann). (*The Independence Theorem*) *Let D be supersimple. Let a_1, a_2, b_1, b_2 and A be such that*

(1) $b_1 \underset{A}{\downarrow} b_2$;
(2) $\mathrm{Lstp}(a_1/A) = \mathrm{Lstp}(a_2/A)$;
(3) $a_\ell \underset{A}{\downarrow} b_\ell$, *for $\ell = 1, 2$.*

Then there is $a \in \mathfrak{C}$ realising $\mathrm{Lstp}(a_\ell/Ab_\ell)$, for $\ell = 1, 2$, such that $a \underset{A}{\downarrow} b_1 b_2$.

We now describe an example of Shelah which can be found in [10]. It shows that, at this level of generality, stability does not imply simplicity, and that Local Character does not imply Extension (both of these statements are consequences of compactness in the first order case).

EXAMPLE 4.10. The language contains a binary relation symbol $E_n(x, y)$, for each $n < \omega$. The first order theory asserts that each $E_n(x, y)$ is an equivalence relation, with an infinite number of equivalence classes, all of which are infinite. It also asserts that E_n partitions each E_{n+1}-equivalence class into infinitely many E_n-classes.

This theory is complete, has quantifier elimination, and is $\aleph_0$-stable. Let $\bar{M}$ be the saturated monster model for this first order theory. Pick an element $a \in \bar{M}$, and consider $\mathfrak{C} = \bigcup_{n < \omega} a/E_n$. Then $\mathfrak{C}$ is homogeneous, not saturated (the type $\{\neg E_n(x, y) : n < \omega\}$ is omitted). Let D be the diagram of $\mathfrak{C}$. Since T is $\aleph_0$-stable, then D is $\aleph_0$-stable.

It is not difficult to see that Local Character holds by using the supersimplicity of the first order theory. However, if $p(x) \in S_D(A)$ and $A \neq \emptyset$, then p divides over $\emptyset$. Thus Extension fails.

We now focus on the connections with stability. The stability of D implies Local Character, but the previous example shows that Extension may fail. This is why the simplicity assumption has to be added in the next theorem.

THEOREM 4.11. *Let D be supersimple. The following conditions are equivalent:*

(1) *D is stable;*
(2) *D is superstable;*

(3) (*Lascar strong types are stationary*) If $\text{Lstp}(b_1/A) = \text{Lstp}(b_2/A)$ *and*
$b_\ell \underset{A}{\bigcup} C$ *for* $\ell = 1, 2$ *then*

$$\text{tp}(b_1/C) = \text{tp}(b_2/C).$$

We finish this section with two examples of supersimplicity.

EXAMPLE 4.12. Consider again the class of free groups $F(X)$, on infinitely many generators X. Let $A \subseteq F(X)$. Let $A' \subseteq X$ be the set of elements needed to generate all the elements of A. It is not difficult to see that A' is uniquely determined by A and that any automorphism f of $F(X)$ commutes with $'$. The reader can check easily that

$$A \underset{B}{\bigcup} C \quad \text{if and only if} \quad A' \underset{B'}{\bigcup} C',$$

where the primed relation coincides with independence in the sense of the trivial pregeometry in X. It follows that the class of free groups is supersimple.

We can go back to the U-rank: Inside any supersimple structure, the partial ordering $<$ defined on complete types $p \in S_D(B)$ and $q \in S_D(A)$ by

$$p < q \quad \text{if} \quad p \subseteq q \text{ and there is } a \models q \text{ with } a \underset{B}{\not\bigcup} A$$

is well-founded. The foundational rank for this ordering is called the SU-rank. We showed in the section of categoricity that the SU-rank of the complete type of any element of the free group over the empty set is an integer, and the supremum of the SU-rank over such 1-types is ω.

The model theory of Hilbert spaces (and more generally Banach space structures) has been extensively studied by Henson and Iovino in a framework that preserves some of the compactness (see [15] for an exposition). The study of Hilbert spaces in the context of homogeneous model theory was started in [3], and in parallel by Berenstein [2] and Berenstein and Buechler. Hilbert spaces are $\aleph_1$-stable and simple [3]. Berenstein observed that in this context nondividing corresponds to orthogonality, and that Hilbert spaces (considered as a group) do not have generics. We reproduce below some examples of operator algebras which can be found in [2].

EXAMPLE 4.13. Consider first the operator algebra $(\ell^2(X), (,), +, \times)$ of the square integrable functions from X into the reals. We consider this as a two-sorted structure with the reals as one sort (which we do not write explicitly) and $\ell^2(X)$ as the other sort; $(,)$ is a function symbol for the scalar product and $+$ and $\times$ for addition and product of functions. Berenstein shows that this structure is strongly $|X|$-homogeneous, $\aleph_1$-stable and simple (this means here that each type does not divide over a countable subset).

The structure $(\ell^2_{fin}(X), (,), +, \times)$, which consists of those square integrable functions generated by functions with finite support is stable and supersimple when $|X| > 2^{\aleph_0}$.

There are more examples in [2]: Hilbert spaces with generic subsets, generic automorphisms, self-adjoint operators, etc.

§5. **Some open questions.** In this section, we list some related basic questions.

In the first order uncountably categorical case, the Baldwin-Lachlan theorem extends to describe the structure of the countable models; there is either only one, or ω-many ordered in an elementary increasing chain $(M_i : i \leq \omega)$, where M_i has dimension i in some strongly minimal set defined over M_0, the prime model over the empty set.

QUESTION 5.1. What is the structure of the countable D-models, for an uncountably categorical, good, countable diagram D?

This question is linked to the existence of prime models over countable sets, as well as the definability of the property '$p(x)$ is big'.

Another question is on the bound for $\kappa(D)$: In the first order case, one can show that $\kappa(T) \leq |T|^+$. Here, it is only known to be less than $\beth_{(2^{|D|})^+}$.

QUESTION 5.2. Let D be stable. Is there a better bound on $\kappa(D)$ than $\lambda(D)$?

QUESTION 5.3. Can one prove the Main Gap for the class of a-saturated models of a good, superstable diagram D?

This was answered positively by Hyttinen and myself (in an unpublished result), under the additional assumption that D is (super)simple. As we pointed out in the body of the paper, this question is related to the next one, which was asked by Hyttinen and Shelah in [13].

QUESTION 5.4. Let D be superstable. Are there prime models in the class of a-saturated models? If so, over which sets A?

This, in turn, is related to the behaviour of Lascar strong types: For example, under which conditions are increasing unions of Lascar strong types realised? When is equality between Lascar strong types type-definable?

Also, it is not difficult to see that the Lascar group $\Gamma(D)$, which is obtained by taking the quotient of the automorphism group of $\mathfrak{C}$ with the normal subgroup of strong automorphisms, is an invariant of D. In the first order case, the Lascar group can be made into a topological group. This is not known at this level of generality:

QUESTION 5.5. When can $\Gamma(D)$ be made into a topological group?

QUESTION 5.6. Is it possible to develop a good substitute for $\mathfrak{C}^{eq}$, i.e. work in a natural expansion of $\mathfrak{C}$ in which types have canonical bases and the stability/simplicity properties of $\mathfrak{C}$ are preserved?

A major open question is also the participation of groups in the theory of good diagrams. We have the following broad questions.

QUESTION 5.7. When do groups arise? What is the structure of these groups? How are they definable (first order, infinitary, invariant)? How can they be used?

Several partial answers exist. There is a group configuration theorem for the quasiminimal case in [26], there are group configurations and binding group results in the simple stable case in [2], and groups arising from nonorthogonality in the superstable or excellent case in [11]. Some generic group actions are studied in [9], and a description of which groups appear under the assumption that the bounded closure coincides with the algebraic closure can be found in [38]. There is a good understanding of groups when they have generics [2], but not in the general case. Here is a very modest questions without an answer:

QUESTION 5.8. Assume that G is an infinite group, whose diagram is good and categorical in all infinite cardinals. Does G have an infinite abelian subgroup?

A larger abelian subgroup cannot exist in general, as free groups demonstrate.

Finally, a question on the existence of a Hanf number for homogeneity. The answer is positive in the stable case (it follows from the stability spectrum and Theorem 2.18).

QUESTION 5.9. Let D be given. Is there a cardinal λ, such that if D has a (D, λ)-homogeneous model of size at least λ, then D is good?

REFERENCES

[1] WALTER BAUR, GREGORY CHERLIN, and ANGUS MACINTYRE, *Totally categorical groups and rings*, **Journal of Algebra**, vol. 57 (1979), no. 2, pp. 407–440.

[2] ALEXANDER BERENSTEIN, **Dependence Relations on Homogeneous Groups and Homogeneous Expansions of Hilbert Spaces**, Ph.D. thesis, Notre Dame, 2002.

[3] STEVE BUECHLER and OLIVIER LESSMANN, *Simple homogeneous models*, **Journal of the American Mathematical Society**, vol. 16 (2003), pp. 92–121.

[4] RAMI GROSSBERG and BRADD HART, *The classification theory of excellent classes*, **The Journal of Symbolic Logic**, vol. 54 (1989), pp. 1359–1381.

[5] RAMI GROSSBERG and OLIVIER LESSMANN, *Shelah's stability and homogeneity spectrum in finite diagrams*, **Archive for Mathematical Logic**, vol. 41 (2002), no. 1, pp. 1–31.

[6] TAPANI HYTTINEN, *On nonstructure of elementary submodels of an unsuperstable homogeneous structure*, **Mathematical Logic Quarterly**, vol. 43 (1997), pp. 134–142.

[7] ——, *Generalizing Morley's theorem*, **Mathematical Logic Quarterly**, vol. 44 (1998), pp. 176–184.

[8] ——, *On non-structure of elementary submodels of a stable homogeneous structure*, **Fundamenta Mathematicae**, vol. 156 (1998), pp. 167–182.

[9] ——, *Generic group actions*, preprint.

[10] TAPANI HYTTINEN and OLIVIER LESSMANN, *A rank for the class of elementary submodels of a superstable homogeneous structure*, **The Journal of Symbolic Logic**, vol. 67 (2002), no. 4, pp. 1469–1482.

[11] TAPANI HYTTINEN, OLIVIER LESSMANN, and SAHARON SHELAH, *Interpreting groups and fields in some nonelementary classes*, **Journal of Mathematical Logic**, vol. 4 (2005), pp. 1–47.

[12] TAPANI HYTTINEN and SAHARON SHELAH, *On the number of elementary submodels of an unsuperstable homogeneous structure*, **Mathematical Logic Quaterly**, vol. 44 (1998), no. 3, pp. 354–358.

[13] ———, *Strong splitting in stable homogeneous models*, **Annals of Pure and Applied Logic**, vol. 103 (2000), pp. 201–228.

[14] ———, *Main gap for locally saturated elementary submodels of a homogeneous structure*, **The Journal of Symbolic Logic**, vol. 66 (2001), no. 3, pp. 1286–1302.

[15] JOSÉ IOVINO, *A Quick Introduction to Banach Space Model Theory*, Lectures notes. Carnegie Mellon University.

[16] H. JEROME KEISLER, **Model Theory for Infinitary Logic**, North-Holland, Amsterdam, 1971.

[17] BYUNGHAN KIM, *Forking in simple unstable theories*, **Journal of the London Mathematical Society**, vol. 57 (1998), no. 2, pp. 257–267.

[18] BYUNGHAN KIM and ANAND PILLAY, *Simple theories*, **Annals of Pure and Applied Logic**, vol. 88 (1997), pp. 149–164.

[19] ———, *From stability to simplicity*, **The Bulletin of Symbolic Logic**, vol. 4 (1998), no. 1, pp. 17–36.

[20] ALEXEI KOLESNIKOV, *Dependence relations in nonelementary classes*, preprint.

[21] KANAT KUDAIBERGENOV, *Homogeneous models of stable theories*, **Siberian Advances in Mathematics**, vol. 3 (1993), no. 3, pp. 1–33.

[22] OLIVIER LESSMANN, **Dependence Relations in Some Nonelementary Classes**, Ph.D. thesis, Carnegie Mellon University, 1998.

[23] ———, *Ranks and pregeometries in finite diagrams*, **Annals of Pure and Applied Logic**, vol. 106 (2000), no. 1, pp. 49–83.

[24] ———, *Homogeneous model theory; existence and categoricity*, **Logic and Algebra** (Yi Zhang, editor), Contemporary Mathematics, vol. 302, AMS, 2002, pp. 149–164.

[25] ———, *Categoricity and U-rank in excellent classes*, **The Journal of Symbolic Logic**, vol. 68 (2003), no. 4, pp. 1317–1336.

[26] ———, *Abstract group configuration*, preprint.

[27] LEO MARCUS, *A prime minimal model with an infinite set of indiscernibles*, **Israel Journal of Mathematics**, vol. 11 (1972), pp. 180–183.

[28] ANAND PILLAY, *Forking in existentially closed structures*, **Connections between Model Theory and Algebraic and Analytic Geometry** (Angus Macintyre, editor), quaderni di matematica, University of Naples, 2001.

[29] ZIV SHAMI, *A natural finite equivalence relation definable in low theories*, **The Journal of Symbolic Logic**, vol. 65 (2000), no. 4, pp. 1481–1490.

[30] SAHARON SHELAH, *Finite diagrams stable in power*, **Annals of Pure and Applied Logic**, vol. 2 (1970), pp. 69–118.

[31] ———, *A combinatorial problem, stability and order in infinitary languages*, **Pacific Journal of Mathematics**, vol. 41 (1972), pp. 246–261.

[32] ———, *Categoricity in $\aleph_1$ of sentences in $L_{\omega_1\omega}(Q)$*, **Israel Journal of Mathematics**, vol. 20 (1975), pp. 127–148.

[33] ———, *The lazy model theorist's guide to stability*, **Proceedings of a Symposium in Louvain** (P. Henrand, editor), Logique et Analyse, no. 71, March 1975, pp. 241–308.

[34] ———, *Simple unstable theories*, **Annals of Mathematical Logic**, vol. 19 (1980), pp. 177–203.

[35] ———, *Classification theory for nonelementary classes. I. the number of uncountable models of $\psi \in L_{\omega_1,\omega}$. part A*, **Israel Journal of Mathematics**, vol. 46 (1983), pp. 212–240.

[36] ——, *Classification theory for nonelementary classes. I. the number of uncountable models of $\psi \in L_{\omega_1,\omega}$. part B*, **Israel Journal of Mathematics**, vol. 46 (1983), pp. 241–273.

[37] ——, **Classification Theory**, revised ed., North-Holland, Amsterdam, 1990.

[38] Boris Zilber, *Hereditarily transitive groups and quasi-urbanik structures*, **American Mathematical Society Translations**, vol. 195 (1999), no. 2, pp. 165–186.

[39] ——, *Analytic and pseudo-analytic structures*, **Logic Colloquium 2000** (R. Cori et al., editors), Lecture Notes in Logic, vol. 19, ASL and AK Peters, 2005, pp. 392–408.

[40] ——, *Covers of the multiplicative group of an algebraically closed field of characteristic 0*, preprint.

MATHEMATICAL INSTITUTE
OXFORD UNIVERSITY
OXFORD, OX1 3LB, UK
E-mail: lessmann@maths.ox.ac.uk

REALS WHICH COMPUTE LITTLE

ANDRÉ NIES

Abstract. We investigate combinatorial lowness properties of sets of natural numbers (reals). The real A is super-low if $A' \leq_{tt} \emptyset'$, and A is jump-traceable if the values of $\{e\}^A(e)$ can be effectively approximated in a sense to be specified. We investigate those properties, in particular showing that super-lowness and jump-traceability coincide within the r.e. sets but none of the properties implies the other within the ω-r.e. sets. Finally we prove that, for any low r.e. set B, there is a K-trivial set $A \not\leq_T B$.

§1. Introduction. In computability theory, one measures and compares the computational complexity of sets of natural numbers (also called *reals*). The first question one is interested in is whether the real is computable. Reals which come close to being computable are therefore of particular interest. A *lowness property* of a real A says that, in some sense, A has low computational power when used as an oracle (and therefore A is close to being computable). To qualify as a lowness property, we require that the property be downward closed under Turing reducibility $\leq_T$, and that each real A with that property is generalized low, namely $A' \leq_T A \oplus \emptyset'$. In this paper we study and compare two lowness properties, being super-low and being jump-traceable.

Superlow reals. Recall that a real A is *low* if its jump A' is Turing-below the halting problem $\emptyset'$, or, equivalently, $A'(e) = \lim_s g(e, s)$ for a computable $0, 1$-valued g. The following concept is more restrictive.

DEFINITION 1.1. The real A is super-low if $A' \leq_{tt} \emptyset'$. Equivalently, $A'(e) = \lim_s g(e, s)$ for a computable $0, 1$-valued g such that $g(e, s)$ changes at most $b(e)$ times, for a computable function b.

This notion goes back to work of Mohrherr [8], and an unpublished manuscript of Bickford and Mills [1] (where only super-low r.e. sets are studied, called "abject" there). The canonical construction of a low simple set (see [10, Theorem VII.1.1]) produces in fact a super-low set: one satisfies lowness requirements

$$L_e : \exists^\infty s \ \{e\}^A(e) \downarrow [s\text{-}1] \implies \{e\}^A(e) \downarrow.$$

Key words and phrases. Lowness, r.e. degrees, reals, traceability, K-triviality.

Logic Colloquium '02
Edited by Z. Chatzidakis, P. Koepke, and W. Pohlers
Lecture Notes in Logic, 27

L_e is injured at most e times by requirements

$$P_i : |W_i| = \infty \Longrightarrow W_i \cap A \neq \emptyset,$$

$i < e$, which enumerate a number $x \geq 2i$ such that $x \in W_{i,s}$ into A at a stage s if $W_{i,s} \cap A_{s-1} = \emptyset$. Then $\{e\}^A(e)$ can become undefined at most e times. Thus, if we let $g(e, s) = 1$ when $\{e\}^A(e)$ converges at stage s and $g(e, s) = 0$ otherwise, then g is an approximation as in Definition 1.1, where $b(e) = e$.

The low basis theorem of Jockusch and Soare [6] can also be strengthened to "super-low": each non-empty Π_1^0 class has a super-low member (Proposition 3.1 below).

Jump-traceable reals. We write $J^A(e)$ for $\{e\}^A(e)$, the jump at argument e. While lowness and super-lowness restrict the *domain* $A' = \{e : J^A(e) \downarrow\}$ of J^A, jump traceability expresses that $J^A(e)$ has few possible *values*. Given $T \subseteq \mathbb{N}$, let $T^{[x]} = \{y : \langle y, x \rangle \in T\}$.

DEFINITION 1.2.   (i) An r.e. set $T \subseteq \mathbb{N}$ is a TRACE if for some computable h, $\forall n |T^{[n]}| \leq h(n)$. We say that h is a BOUND for T.
   (ii) The real A is JUMP-TRACEABLE if there is a trace T such that

$$\forall e \; J^A(e) \downarrow \Longrightarrow J^A(e) \in T^{[e]}.$$

This modifies the property of being recursively traceable, used in [11] to give a characterization of the reals that are low for Schnorr tests. We will see below that, because of the universality of the jump, jump traceability of A actually restricts the possible values of any partial A-recursive function via a trace.

Both super-lowness and jump traceability are closed downward under $\leq_T$ and imply GL_1. Thus they satisfy our criteria for being lowness properties. Super-low reals A are ω-r.e., that is, $A \leq_{tt} \emptyset'$. On the other hand, we will see that there is a perfect Π_1^0-class of jump-traceable reals. Among our main results are:

- *super-lowness and jump-traceability coincide within the r.e. sets*
- *none of the properties implies the other within the ω-r.e. sets.*

We also prove that jump traceability is Σ_3^0 on the ω-r.e. sets, namely, if $(\Theta_e)_{e \in \mathbb{N}}$ is an effective listing of all tt-reduction procedures defined on an initial segment of $\mathbb{N}$, then $\{e : \Theta_e(\emptyset') \text{ jump-traceable}\}$ is Σ_3^0. The same result follows for the r.e. sets. Recall that $\{e : W_e \text{ low}\}$ is Σ_4^0-complete [10, Corollary XII 4.7]. Since our two properties coincide for r.e. sets, super-lowness is strictly stronger than lowness even for the r.e. sets.

Our "combinatorial" lowness properties can be used to study very interesting lowness properties related to randomness and prefix Kolmogorov complexity. We first recall some definitions. For each real A, we want to define $K^A(y)$, the length of a shortest prefix description of y using oracle A. An *oracle machine* is a partial recursive functional $M : 2^\omega \times 2^{<\omega} \mapsto 2^{<\omega}$. We write $M^A(x)$ for $M(A, x)$. M is an *oracle prefix machine* if the domain of M^A is an antichain under inclusion of strings, for each A. Let $(M_d)_{d \in \mathbb{N}^+}$ be an effective listing of all oracle prefix machines. The universal oracle prefix machine U is given by

$$U^A(0^d 1\sigma) = M_d^A(\sigma).$$

Let $K^A(y) = \min\{|\sigma| : U^A(\sigma) = y\}$. If $A = \emptyset$, we simply write $U(\sigma)$ and $K(y)$. $U_s(\sigma) = y$ indicates that $U(\sigma) = y$ and the computation takes at most s steps.

The real A is K-*trivial* if the K-complexity of its initial segments is as low as possible, up to a constant c, namely $\forall n \ K(X \restriction n) \leq K(n) + c$. Let $\mathcal{K}$ denote this class of reals. $\mathcal{K}$ contains nonrecursive r.e. sets and is closed under $\oplus$ (see [3] for proofs and more references). Here we show that,

- *for each low r.e. B, there is an r.e. $A \in \mathcal{K}$ such that $A \not\leq_T B$.*

In [9] I prove that $\mathcal{K}$ is closed downward under Turing reducibility, and each $A \in \mathcal{K}$ is truth table-below some r.e. $D \in \mathcal{K}$. Thus $\mathcal{K}$ is an example of a lowness property which is an ideal in the ω-r.e. reals, generated by its r.e. members. In contrast, the super-low r.e. sets do not form an ideal, since there are super-low r.e. sets A_0, A_1 such that $A_0 \oplus A_1$ is Turing complete (see [1] or Theorem 3.3 below).

We also show in Nies [9] that each $A \in \mathcal{K}$ is superlow. Since the construction in [3] produces a noncappable $A \in \mathcal{K}$, the class of super-low r.e. degrees is downward dense in the nonrecursive r.e. degrees.

The notion of jump traceability can be used to characterize reals which are computationally weak in the following sense.

DEFINITION 1.3. Let $p : \mathbb{N} \mapsto \mathbb{N}$ be a non-decreasing computable function such that $\lim_n p(n) - n = \infty$. A real A is p-*low* if $\forall y \ K(y) \leq p(K^A(y) + c_0) + c_1$ for some constants $c_0, c_1 \in \mathbb{N}$.

Thus, for such A, $K^A(y)$ is not much smaller than $K(y)$. Let $\mathcal{M}[p]$ denote this class of reals. In the last section of [4] we show that

$$A \text{ jump traceable} \Longleftrightarrow \exists \ p \text{ computable} \quad A \in \mathcal{M}[p].$$

PRELIMINARIES. If $f : \mathbb{N} \to \mathbb{N}$, then we say f is ω-r.e. if $f \leq_{wtt} \emptyset'$, that is f can be computed from $\emptyset'$ with recursively bounded use. This is easily seen to be equivalent to $f(e) = \lim_s g(e, s)$, where g is a computable function such that $g(e, s)$ changes at most $b(e)$ times, for a computable function b. For instance, if T is a trace in the sense of Definition 1.2, then $f(n) = \max T^{[n]}$ is ω-r.e., via $g(n, s) = \max T_s^{[n]}$ and $b(n) = h(n)$.

The following notation is also useful. A Δ_2^0-*approximation* $(A_r)_{r \in \mathbb{N}}$ of a real A is an effective sequence of finite sets such that $A(x) = \lim_r A_r(x)$. Then $A \leq_{tt} \emptyset'$ iff $A \leq_{wtt} \emptyset'$ iff the number of changes in such an approximation is recursively bounded.

Recall that we write $J^A(e)$ for $\{e\}^A(e)$. If A is given by a Δ_2^0-approximation, we write $J^A(e)[s]$ for $\{e\}_s^{A_s}(e)$. The use of the computation $J^A(e)$ is denoted $j(A, e)$, and the use of $J^A(e)[s]$ is denoted $j(A, e)[s]$.

Recall that a *partial recursive functional* is an r.e. set Ψ of "axioms" $\langle \sigma, e, v \rangle$, $\sigma \in 2^{<\omega}$ such that if $\langle \sigma, e, v \rangle, \langle \sigma', e, v' \rangle \in \Psi$ and σ, σ' are compatible, then $v = v'$. Given Δ_2^0-approximation (A_s), to *define $\Psi^A(e) = v$ with use u at stage s* means to put the axiom $\langle A_s \restriction u, e, v \rangle$ into Ψ.

While the proof of the following fact is not hard, it depends on the particular implementation of the universal machine.

FACT 1.4. From a partial recursive functional Ψ, one can effectively obtain a primitive recursive function α, called a reduction function for Ψ, such that

$$\forall X \; \forall e \; \Psi^X(e) \simeq J^X(\alpha(e)).$$

§2. **Jump-traceability.** In this section we collect some basic facts on jump-traceability, prove existence of a perfect class of jump-traceable reals, and place this notion in context.

Jump traceable reals at large.

FACT 2.1. *If A is jump-traceable via T, then there is a trace S such that, for each partial recursive functional Ψ,*

$$\text{a.e. } m \big[\Psi^A(m) \downarrow \Rightarrow \Psi^A(m) \in S^{[m]} \big].$$

For, define S by $S^{[m]} = \bigcup_{i \leq m} T^{[\alpha_i(m)]}$, where (α_i) is a listing of all primitive recursive unary functions. Then S is a trace which is as required by Fact 1.4.

PROPOSITION 2.2. *Let A be any jump-traceable real. Then A is generalized low$_1$, namely $A' \leq_T A \oplus \emptyset'$. The reduction procedure can be obtained effectively from an r.e. index for the trace T. In this reduction, the use on $\emptyset'$ is recursively bounded, and the use on A is ω-r.e.*

PROOF. Consider the partial A-recursive functional

$$\Psi^A(e) = \mu s \big[J^A(e) \downarrow \text{ in } s \text{ steps} \big].$$

Choose a reduction function α by Fact 1.4. To see if $e \in A'$, first compute $t = \max T^{[\alpha(e)]}$, using $\emptyset'$ as an oracle. Then, using $A \restriction t$, check whether $J^A(e) \downarrow$ in $\leq t$ steps. If so, answer YES, otherwise No.

Since T is a trace, the use on $\emptyset'$ to compute t is recursively bounded.   ⊣

Recall that a Π_1^0-class P is a subset of 2^ω given as the set of paths $[B]$ through a recursive subtree of B of $2^{<\omega}$.

THEOREM 2.3. *There is a perfect Π_1^0-class of reals which are jump-traceable via a fixed trace T, whose bound is $h(e) = 2 \cdot 4^e$.*

PROOF. We will define an effective sequence of 1-1-maps (F_s) such that $F_s : 2^{<\omega} \mapsto 2^{<\omega}$ preserving the ordering and compatibility relations, and for each $\alpha \in 2^{<\omega}$, $F_s(\alpha) \subseteq F_{s+1}(\alpha)$ and $\lim_s F_s(\alpha)$ exists. If we let $B = \{\rho : \forall s \ \exists \alpha[|\alpha| \leq |\rho| \ \& \ \rho \subseteq F_s(\alpha)]\}$, then $Q = [B]$ is a perfect Π_1^0-class. We define $F_s(\alpha)$, $|\alpha| = e$ in a way to minimize the number of values $J^{F(\alpha)}(e)$. Any such value we see at some stage needs to be enumerated into $T^{[e]}$.

CONSTRUCTION. Let $F_0(\alpha) = \alpha$ for each α. At stage $s + 1$ look for the length-lexicographically least α such that, where $|\alpha| = e$, there is $\beta \succeq \alpha$ such that $y = J^{F(\beta)}(e) \downarrow$ and $y \notin T^{[e]}$. If there is such an α, enumerate y into $T^{[e]}$. Define $F_{s+1}(\alpha) = F_s(\beta)$. Moreover, for all $\rho \neq \lambda$ define $F_{s+1}(\alpha\rho) = F_s(\beta\rho)$.

This ends the construction. A value $F_s(\alpha)$, $|\alpha| = e$, changes at most $2^{e+1} - 1$ times, and causes the enumeration of at most that many elements into $T^{[e]}$. Thus $|T^{[e]}| \leq 2 \cdot 4^e$. $\dashv$

This construction could be massaged a bit to obtain a bound close to 2^e. However, it is unknown if there is a perfect class for much smaller bounds. Below we show that there is a fixed C such that, if A is low for K via b (i.e., $\forall y \ K^A(y) \geq K(y) - b$), then A is jump traceable via the bound $C2^b i \log i$ (see Proposition 5.9 below). However, each low for K set is Δ_2^0. In [4] we construct a non-computable r.e. A which is *strongly jump traceable*, namely, jump traceable via each unbounded montonic computable h.

Jump traceable ω-r.e. sets. Next we determine the index set complexity of jump traceability on the ω-r.e. sets. Recall that $(\Theta_e)_{e \in \mathbb{N}}$ is an effective listing of all (possibly partial) tt-reduction procedures defined on an initial segment of $\mathbb{N}$. Thus $\Theta_e(x)$ can be viewed as a truth table. Then we obtain an effective listing $(V_e)_{e \in \mathbb{N}}$ of the ω-r.e. sets by letting $V_{e,s}(x) = \Theta_e(\emptyset'; x)[s]$, which is interpreted as 0 if $\Theta_e(x)[s]$ is undefined. Now let $V_e(x) = \lim_s V_{e,s}(x)$.

PROPOSITION 2.4. $\{e : V_e \text{ jump-traceable}\}$ is Σ_3^0-complete. Similarly, $\{e : W_e \text{ jump-traceable}\}$ is Σ_3^0-complete.

PROOF. V_e is jump-traceable iff $\exists T \subseteq \mathbb{N} \ \exists h$ total

$$\left(\forall n |T^{[n]}| \leq h(n) \ \& \ \forall x \forall s \exists t \geq s \left[J^{V_e}(x)[t] \uparrow \lor J^{V_e}(x)[t] \downarrow \in T_t^{[x]}\right]\right).$$

The direction "$\Rightarrow$" is clear. For the other direction, note that if $J^{V_e}(x) \downarrow$ then the condition implies $J^{V_e}(x) \in T^{[x]}$.

For each $e \in \mathbb{N}$ we can effectively obtain $\widehat{e}$ such that $W_e = V_{\widehat{e}}$. This proves the second index set is Σ_3^0. For Σ_3^0-hardness it suffices to consider the r.e. case. But it is easy to show that *any* nontrivial Σ_3^0-class of r.e. sets which is closed under finite differences and contains the computable sets has a Σ_3^0-complete index set. $\dashv$

Digression: r.e. traceable reals. A real A is *r.e. traceable* if there is a trace S such that $\forall \Gamma(\Gamma^A \text{ total} \Rightarrow \text{a.e. } m \ \Gamma^A(m) \in S^{[m]})$. This was studied in (Ishmukhametov [5]), who used the term *weakly recursive*. By Fact 2.1 each jump traceable real is r.e. traceable. Since the function $g(x) = \max T^{[x]}$ is ω-r.e., a weakly recursive real is array recursive [2], which means that there is an ω-r.e. function which eventually dominates any A-computable function. For r.e. sets A, the converse implication holds by [5], a fact which can be proved in the same style as the proof of Theorem 4.1 below.

The r.e. traceable Δ_2^0 reals have an interesting uniformity property. Recall that a real A is low$_2$ if Tot$^A = \{e : \{e\}^A \text{ total}\} \in \Sigma_3^0$.

PROPOSITION 2.5. *The r.e. traceable Δ_2^0-reals (and hence the jump traceable Δ_2^0-reals) are uniformly low$_2$. Thus, from a Δ_2^0 approximation (A_s) to A one can effectively obtain a Σ_3^0-index for Tot^A.*

The point is that the Δ_2^0 approximation alone suffices, in case it actually approximates a r.e. traceable real, to obtain the Σ_3^0-index.

PROOF. Using the same argument as in [11, Fact 1], if a real A is weakly recursive, then it is irrelevant what the actual bound h for the trace is, as long as $\lim_n h(n) = \infty$. Thus there is a trace S such that $|S^{[m]}| \leq m$. Let S_i be a u.r.e. list of all traces with bound $h(m) = m$, and let $V^{[e]} = \bigcup_{i \leq e} S_i^{[e]}$, so that V is a trace which works for all r.e. traceable reals. Let $g(m, s) = \max V_s^{[m]}$. Then $\{e\}^A$ is total iff

$$\exists x \exists s \, \forall t \geq s \left(\forall z < x \, e^A(z) \downarrow [t] \, \& \, \forall z \geq x \exists v \geq t \, u(A; e, z)[v] \leq g(z, v) \right).$$

The direction from left to right holds since $u(A; e, z)$, the use of $\{e\}^A(z)$ is an A-recursive function. The converse direction holds because, for each z, there are only finitely many possibilities for $\{e\}^A(z)[v]$.

The right hand side gives a Σ_3^0 index for Tot^A, which was obtained uniformly in the Δ_2^0-approximation to A.   ⊣

§3. Super-low reals.

Jockusch and Soare [6] proved that each non-empty Π_1^0 class has a low member. An analisis of their proof yields

PROPOSITION 3.1. *Each non-empty Π_1^0 class has a super-low member.*

PROOF. Suppose $P = [B]$, where B is an infinite recursive subtree of $2^{<\omega}$. For each finite set F, let $B_F = \{\sigma \in B : \forall e \in F \, J^\sigma(e) \uparrow\}$. Since being finite is a Σ_1^0-property of recursive trees, there is a computable g defined on (strong indices for) finite subsets of $\mathbb{N}$ such that

$$B_F \text{ finite} \iff g(F) \in \emptyset'.$$

As in [6], let $B = B_{F_0} \supseteq B_{F_1} \supseteq \cdots$ be a sequence of recursive trees defined as follows: let $F_0 = \emptyset$, and $F_{i+1} = F_i$ if $B_{F_i \cup \{i\}}$ is finite, and $F_{i+1} = F_i \cup \{i\}$ else. Then one can compute F_i from $\emptyset'$, where the use is bounded by the computable function $\max\{g(F) + 1 : F \subseteq \{0, \ldots, i\}\}$.

By compactness, there is a (unique) path $A \in \bigcap_i [B_{F_i}]$. This path satisfies $J^A(e) \uparrow \Leftrightarrow e \in F_{e-1}$. Thus $A' \leq_{wtt} \emptyset'$ and hence $A' \leq_{tt} \emptyset'$.   ⊣

COROLLARY 3.2. *There is a Martin-Löf random super-low real.*

PROOF. This follows since the set of random reals forms a union of Π_1^0-classes (given by a universal Martin-Löf test). $\dashv$

In contrast, a Martin-Löf random real R is not of n-r.e. degree unless $R \equiv_T \emptyset'$. This is because there is a fixed point free $f \leq_T R$ (i.e., $\forall e \ W_e \neq W_{f(e)}$), and the Arslanov completeness criterion applies to n-r.e. sets (see [10, p. 277]).

Recall that, by the Sacks Splitting Theorem, there are low r.e. sets A_0, A_1 such that $\emptyset' \leq_T A_0 \oplus A_1$. Again, we strengthen this to super-low. This was first proved in [1].

THEOREM 3.3. [1] *There are super-low r.e. sets A_0, A_1 such that $\emptyset' \leq_T A_0 \oplus A_1$.*

PROOF. We enumerate A_0, A_1 and also build a Turing functional Γ such that $\emptyset' = \Gamma(A_0 \oplus A_1)$. The use of $\Gamma(A_0 \oplus A_1; p)$ is denoted $\gamma(A_0 \oplus A_1; p)$ (and pictured as a movable marker). For the duration of this proof, k, l denote numbers in $\{0, 1\}$, p, q denote numbers in $\mathbb{N}$ and $[p, k]$ stands for $2p + k$.

To avoid that $J^{A_k}(p)$ change too often, we ensure that at each stage s, for each p, k such that $[p, k] \leq s$,

$$(1) \qquad J^{A_k}(p)[s] \downarrow \Longrightarrow \gamma(A_0 \oplus A_1)([p, k]) > j(A_k, p)[s]$$

CONSTRUCTION. At stage s, define $\Gamma(A_0 \oplus A_1; s)$ with large use, and do the following.

a) If there is $[p, k]$ such that $J^{A_k}(p)[s - 1] \uparrow$ and $J^{A_k}(p) \downarrow$ at the beginning of stage s, choose $[p, k]$ minimal such. Put $\gamma(A_0 \oplus A_1; [p, k])$ into A_{1-k} and redefine $\Gamma(A_0 \oplus A_1; q)$, $s \geq q \geq [p, k]$, with the correct value and large use.

b) If $n \in \emptyset'_s - \emptyset'_{s-1}$, then put $\gamma(A_0 \oplus A_1; n)$ into A_{1-k}.

A typical set-up looks like this:

$$j(A_0, p) \qquad \gamma(2p) \qquad\qquad j(A_1, p) \qquad \gamma(2p + 1)$$

(Here, $J^{A_1}(p)$ converged after $J^{A_0}(p)$.)

VERIFICATION. We first check (1) by induction on s. The condition holds for $s = 0$. If $s > 0$, we may suppose there is $[p, k]$ minimal such that $J^{A_k}([p, k])[s - 1] \uparrow$ and $J^{A_k}(p)[s] \downarrow$ (else there is nothing to prove), in which case we put $v = \gamma(A_0 \oplus A_1; [p, k])$ into A_{1-k}.

• If $[q, l] < [p, k]$, then $v > \gamma(A_0 \oplus A_1; [q, l]) \geq j(A_l, q)[s]$ by inductive hypothesis, so that (1) remains true for $J^{A_l}(q)[s]$.
• Since we enumerate v into the "other" side, $J^{A_k}(p)[s]$ remains convergent, so we ensure $\gamma(A_0 \oplus A_1; [p, k]) > j(A_k, p)[s]$.
• For $[q, l] > [p, k]$, computations $J^{A_l}(q)[s]$ have their use below the new value of $\gamma([q, l])$.

Next we show that both A_0 and A_1 are super-low. As in the standard construction of a (super-)low simple set, let $g_k(p,s) = 1$ if $J^{A_k}(p)[s] \downarrow$ and let $g_k(e,s) = 0$ otherwise. We define a computable function c such that $c([p,k])$ is a bound on the number of times $J^{A_k}(p)$ can become defined. Then $b_k(p) = 2c([p,k]) + 1$ bounds how often $g_k(p,s)$ changes.

By (1), $J^{A_0}(p)$ becomes undefined at most $2p$ times due to change of $\emptyset' \restriction 2p$. Otherwise, $J^{A_k}(p)$ becomes undefined only when some computation $J^{A_{1-k}}([q, 1-k])$ becomes defined, where $[q, 1-k] < [p,k]$. Thus the recursive function c given by $c(0) = 1$, $c([p,k]) = 2p + \sum\{c([q, 1-k]) : [q, 1-k] < [p,k]\}$ is as desired.

As a consequence, each marker $\gamma(A_0 \oplus A_1; m)$ reaches a limit. Thus $\emptyset' = \Gamma(A_0 \oplus A_1)$.   ⊣

In contrast to the case of the Sacks Splitting theorem, we cannot achieve that Γ above is a *wtt*-reduction. Bickford and Mills [1, Theorem 4.1] show that, in fact, no super-low r.e. set is cuppable in the r.e. *wtt*-degrees.

§4. Traceability versus super lowness.

THEOREM 4.1. *Let A be r.e. Then the following are equivalent.*

(i) *A is jump traceable*
(ii) *A is super-low.*

Both directions are uniform.

PROOF. (i) $\Rightarrow$ (ii). Suppose A is jump traceable via a trace T with bound h. By convention, for each s, $T_s \subseteq [0, s)$. Consider the following partial A-recursive function:

$$q(e) = \mu s(J^A(e)[s] \downarrow \ \& \ A_s \restriction j(A_s, e, s) = A \restriction j(A_s, e, s)).$$

By Fact 1.4 there is a total computable α such that, for all e, $q(e) \simeq J^A(\alpha(e))$. Then, for each s,

$$(2) \qquad \left(J^A(e)[s] \uparrow \ \& \ J^A(e) \downarrow\right) \Longrightarrow J^A(\alpha(e)) \geq s,$$

since $J^A(\alpha(e)) < s$ implies that $J^A(e)$ has reached a final value by stage s.

We define computable functions $g(e,s), b(e)$ as in Definition 1.1 witnessing that A is super-low. Let $g(e, 0) = 0$. For $t > 0$, if $J^A(e)[t] \uparrow$ then let $g(e,t) = 0$. Now suppose $J^A(e)[t] \downarrow$. If $g(e, t-1) = 1$ then let $g(e, t) = 1$. If $g(e, t-1) = 0$, then we first test the stability of the computation $J^A(e)[t]$ before allowing $g(e,t) = 1$: let $s < t$ be the greatest stage such that $J^A(e)[s] \uparrow$. If $v = J^A(\alpha(e))[t] \downarrow$, $s \leq v$ and $v \in T_t^{[e]}$ then let $g(e,t) = 1$, otherwise $g(e,t) = 0$.

We claim that $g(e,t)$ changes at most $2h(\alpha(e)) + 2$ times. It suffices to show that $g(e,t)$ changes from 1 to 0 and back to 1 at most $h(\alpha(e))$ times. Thus, suppose $s > 0$, $g(e, s-1) = 1$, $g(e,s) = 0$ (so that $J^A(e)[s] \uparrow$) and

$t > s$ is least such that $g(e, t) = 1$. Then $v = J^A(\alpha(e))[t] \downarrow$ and $s \leq v$. Since $T_s \subseteq [0, s)$ and $v \in T_t^{[e]}$, $T_t^{[e]} - T_s^{[e]} \neq \emptyset$. This can happen at most $h(\alpha(e))$ times.

$$J^A(e) \uparrow \qquad\qquad\qquad\qquad\qquad\qquad v \in T_t^{[e]}$$

$$s \qquad\qquad\qquad v = J^A(\alpha(e)) \qquad\qquad t$$

$$g(e, s - 1) = 1, \, g(e, s) = 0 \qquad\qquad\qquad\qquad g(e, t) = 1$$

It remains to be shown that $A'(e) = \lim_s g(e, s)$. If $J^A(e) \uparrow$, then $g(e, s) = 0$ for infinitely many s, so $\lim_s g(e, s) = 0$. Now suppose $J^A(e) \downarrow$. Let s be greatest such that $J^A(e)[s] \uparrow$. Since $J^A(\alpha(e)) \downarrow$, there is a $t \geq s$ such that the computation $v = J^A(\alpha(e))$ is stable and $v \in T_t^{[e]}$. Then $s \leq v$ by (2). So we define $g(e, t') = 1$ for each $t' \geq t$.

Note that we have obtained g and b effectively in the trace T and its bound.

(ii) $\Rightarrow$ (i). Suppose A is super-low. Thus A' is ω-r.e. via some functions g, b. We enumerate a trace T to show A is jump traceable, and also define an auxiliary partial recursive functional Ψ, which copies computation of the jump J with some delay. We assume a partial recursive functional $\widetilde{\Psi}$ is given, and let α be the reduction function for $\widetilde{\Psi}$ according to Fact 1.4. Since we produce Ψ effectively from α, by the Recursion Theorem we can assume that $\widetilde{\Psi} = \Psi$, so that α is also a reduction function for Ψ.

Given e, let $\widehat{e} = \alpha(e)$. At stage $s = 0$, Ψ is totally undefined. For $s > 0$, we distinguish two cases.

a) $g(\widehat{e}, s) = 0$. If $\Psi^A(e)[s - 1] \uparrow$ and $J^A(e)[s] \downarrow = v$, define $\Psi^A(e)[s] = v$ with use $j(A_s, e, s)$.

b) $g(\widehat{e}, s) = 1$. If $\Psi^A(e)[s] \downarrow$ then enumerate $y = J^A(e)[s]$ into $T^{[e]}$.

Note that, since Ψ just copies computations of J at a later stage, when a new computation $J^A(e)[s]$ appears, then no computation $\Psi^A(e)[t]$ which was defined at $t < s$ still applies at stage s.

Suppose $J^A(e) = z$, and let s be the least stage where this (final) computation appears. We show $z \in T^{[e]}$. At a stage $t \geq s$, we may only define a new computation $\Psi^A(e)[t]$ in case $g(\widehat{e}, t) = 0$. Since $\Psi^A(e)[t]$ remains undefined till this happens, by the definition of α, in fact there must be such a stage $t \geq s$. Since the use for $\Psi^A(e)[t]$ is $j(A_s, e, s)$ and $A_s \upharpoonright j(A_s, e, s)$ is stable, $\Psi^A(e) \downarrow$. Hence $g(\widehat{e}, r) = 1$ for some $r > t$, at which point we enumerate z into $T^{[e]}$.

Next we show T is a trace with bound $h(e) = \lfloor \frac{1}{2} b(\alpha(e)) \rfloor$. Suppose $q < r$ are stages where distinct elements y, z are enumerated into $T^{[e]}$. Then $y = J^A(e)[q]$, $z = J^A(e)[r]$, and $g(\widehat{e}, q) = g(\widehat{e}, r) = 1$. Since $A_q \upharpoonright j(A_q, e, q) \neq A_r \upharpoonright j(A_q, e, q)$, no definition $\Psi^A(e)[q']$ issued at a stage $q' \leq q$ is valid at stage r. (Here is where we need that A is r.e.) So we must have made a new definition $\Psi^A(e)[t]$ at a stage t, $q < t < r$, whence $g(\widehat{e}, t) = 0$. Since $g(\widehat{e}, s)$ can change from 1 to 0 and back at most $h(e)$ times, this proves that $|T^{[e]}| \leq h(e)$.

Using the Recursion Theorem with indices for g and b as parameters, we obtain T and h effectively in those indices.   ⊣

We obtain an interesting consequence which is not obvious from the definition.

COROLLARY 4.2. $\{e : W_e \text{ super-low}\}$ is Σ_3^0-complete.

PROOF. This follows from the corresponding fact for jump-traceability, Proposition 2.4.   ⊣

THEOREM 4.3. There is a super-low real A which is not jump-traceable.

PROOF. In [7] we show that no r.e. traceable set is diagonally non-computable. Since a ML-random set is diagonally non-computable, the Martin-Löf random real obtained in Corollary 3.2 is not jump-traceable.   ⊣

THEOREM 4.4. There is an ω-r.e. jump-traceable real A which is not super-low.

Notice however that A is low by Proposition 2.2.

PROOF. Fix an effective listing $(g_e, b_e)_{e \in \mathbb{N}}$ of all pairs consisting of a binary and a unary partial recursive function, such that for all w, $\{q : g_e(w, q) \downarrow\}$ is an initial segment of $\mathbb{N}$. Then we can assume the same property for the approximation at a stage s, $g_e(w, q)[s]$.

To ensure A is not super-low, we meet the requirements

$$P_e : g_e, b_e \text{ total } \& \forall x \ g_e(x, q) \text{ changes at most } b_e(x) \text{ times} \implies$$
$$\exists y \ \neg A'(y) = \lim_q g_e(y, q).$$

We define an auxiliary binary p.r. functional Ψ. As usual, by the Recursion Theorem, we are given a reduction function α such that $\Psi^X(e, y) = J^X(\alpha(\langle e, y \rangle))$. The strategy for P_e is as follows.

1. Pick a fresh candidate y at stage t. Let $\widetilde{y} = \alpha(e, y)$. Wait till $b_e(\widetilde{y}) \downarrow$ at a stage t.
2. Pick a fresh number z (thus, $z \geq b_e(\widetilde{y})$), called the *parameter* of P_e. From now on, ensure that

$$\Psi^A(e, y) \downarrow \Longleftrightarrow z \in A.$$

To do so, for all strings σ of length z, define $\Psi^{\sigma 1}(e, y) = 1$. This is allowed, since there have been no definitions with arguments e, y so far.

Do the following at most $b_e(\widetilde{y})$ times at stages $s \geq t$: Whenever $g_e(\widetilde{y}, q)[s-1] \uparrow$, $g_e(\widetilde{y}, q)[s] \downarrow$, and $g_e(\widetilde{y}, q-1) \neq g_e(\widetilde{y}, q)$, then declare $A_s(z) = 1 - g_e(\widetilde{y}, q)$. Otherwise $A_s(z) = A_{s-1}(z)$.

Then, if the hypothesis of P_e is satisfied,

$$\widetilde{y} \in A' \Longleftrightarrow \Psi^A(e, y) \downarrow \Longleftrightarrow \lim_s g_e(\widetilde{y}, s) = 0.$$

Moreover, $A_s(z)$ changes at most z times (since $z \geq b_e(\widetilde{y})$), so that A is ω-r.e. To ensure A is jump traceable, we enumerate a trace T. We meet the requirements

$$Q_e : |T^{[e]}| \leq h(e) \ \& \ \left(v = J^A(e) \downarrow \Longrightarrow v \in T^{[e]}\right),$$

where $h(e)$ is a recursive bound to be determined below.

The priority ordering of requirements is $Q_0 < P_0 < Q_1 < \cdots$. The strategy for Q_e is simple: whenever a computation $J^A(e) = v$ appears at stage s which has not been seen before, then

1. put v into $T^{[e]}$,
2. initialize the requirements P_i, $i \geq e$.

We say that Q_e *acts*. In that case, $A(z)$ retains its value, for any parameter z of a lower priority requirement P_j. Therefore, unless also a higher priority P_i is initialized, for $t \geq s$, $A_t \restriction j(A_t, e, t)$ only depends on the values $A(z)$, where z is the parameter of a higher priority P_i, which gives at most 2^e possibilities for $A_t \restriction j(A_t, e, t)$ (here we need that P_i only needs to change $A(z)$ for a single number z, which would fail if we had to make A r.e.).

Construction. Let $A_0 = \emptyset$. At stage $s > 0$, go through the requirements $Q_0, P_0, \ldots, Q_s, P_s$ and let them carry out one step of their strategies. At the end, if $y \leq s$ and no value has been assigned yet to $A_s(y)$, retain the value at stage $s - 1$.

Verification. Let $h(0) = 1$ and, for $e > 0$, let $h(e) = h(e - 1)(2^e + 1)$.

LEMMA 4.5. *Let* $e \geq 0$. *Then*

(i) Q_e *is met,*

(ii) P_e *is initialized at most* $h(e)$ *times and met.*

PROOF. For $e = 0$, (i) and (ii) hold, since P_0 is initialized at most once, when $J^A(0)$ converges for the first time. Assume $e > 0$.

(i) While P_{e-1} is not initialized, the requirements P_i, $i < e$ pick at most one number z. If F_s is the set of such numbers at a stage s, then there are at most 2^e possibilities for $A_s \cap F_s$. Hence Q_e enumerates at most 2^e numbers into $T^{[e]}$ before P_{e-1} is initialized another time. Hence, by inductive hypothesis (ii) for $e - 1$, $|T^{[e]}| \leq 2^e h(e - 1) \leq h(e)$.

(ii) If P_e is initialized, then either P_{e-1} is also initialized, or Q_e acts. So P_e is initialized at most $h(e)$ times. Once it is no more initialized, P_e diagonalizes successfully.   ⊣

§5. A construction of a K-trivial r.e. real.
The following theorem was considered in discussions with Downey and Hirschfeldt.

THEOREM 5.1. *For each low r.e. set B there is an r.e. K-trivial set A such that $A \not\leq_T B$.*

PROOF. Let $\mathbb{N}^{\langle e \rangle}$ denote the set of numbers of the form $\langle y, e \rangle$. We meet the requirements

$$P_e : A \neq \{e\}^B,$$

by enumerating numbers $x \in \mathbb{N}^{\langle e \rangle}$ into A. To ensure A is K-trivial, we apply the criterion implicit in [3, Theorem 3.1] in the form presented in [9, Proposition 4.1]. This is actually a characterization of $\mathcal{K}$, as proved in [9, Theorem 7.1]. We refer to those papers for motivation, and to [9] for a proof.

Note that $K(y) = \lim_s K_s(y)$, where $K_s(y) = \min\{|\sigma| : U_s(\sigma) = y\}$. One uses the "cost function"

$$c(x, s) = 1/2 \sum_{x < y \leq s} 2^{-K_s(y)},$$

which bounds the cost of changing $A(x)$ at stage s. Note that $c(x, s)$ is nondecreasing in s, $\lim_s c(x, s) \leq 1/2$ for each x, and $\lim_x \lim_s c(x, s) = 0$ by the definition of prefix Kolmogorov complexity.

FACT 5.2. [9] Suppose that $A(x) = \lim_s A_s(x)$ for a Δ_2^0-approximation (A_s) such that

(3) $\quad S = \sum\{c(x, s) : s > 0 \ \& \ x \text{ is minimal s.t. } A_{s-1}(x) \neq A_s(x)\} \leq 1/2.$

Then A is K-trivial.

To meet the requirements P_e, we use a Robinson type procedure, using the lowness of B to "certify" computations $\{e\}^B(x)[s] = 0$. We may ask a $\Sigma_1^0(B)$-question about the enumeration of A, and we have a Δ_2^0-approximation to the answer. But which enumeration? We may assume that it is given, by the recursion theorem. Formally, an *enumeration* is an index for a partial recursive function $\mathcal{A}$ defined on an initial segment of $\mathbb{N}$ such that, where $\mathcal{A}(t)$, is interpreted as a strong index for the part of A enumerated by stage t, $\mathcal{A}(s) \subseteq \mathcal{A}(s+1)$ for each s. We write A_t for $\mathcal{A}(t)$. Given any (possibly partial) enumeration $\widetilde{\mathcal{A}}$, we effectively produce an enumeration $\mathcal{A}$, asking $\Sigma_1^0(B)$-questions about the given enumeration $\widetilde{\mathcal{A}}$. We must show that $\mathcal{A}$ is total in the interesting case that $\mathcal{A} = \widetilde{\mathcal{A}}$ (by the recursion theorem), where these questions are actually about $\mathcal{A}$.

Here is the $\Sigma_1^0(B)$-question for requirement P_e:

Is there a stage s and an $x \in \mathbb{N}^{\langle e \rangle}$ such that $\widetilde{\mathcal{A}}$ is defined up to $s - 1$, and

- $\{e\}^B(x) = 0[s]$, where $B_s \upharpoonright u(B_s, e, x, s) = B \upharpoonright u(B_s, e, x, s)$ (B is correct on the use of the computation), and
- $c(x, s) \leq 2^{-(e+n+3)}$,

where $n = |\mathbb{N}^{\langle e \rangle} \cap \tilde{A}(s-1)|$ is the number of enumerations for the sake of P_e prior to s.

Since B is low, there is a total computable function $g(e, s)$ such that $\lim_s g(e, s) = 1$ if the answers is YES, and $\lim_s g(e, s) = 0$ otherwise. (The function $g(e, s)$ actually depends on a further argument which we supress, an index for $\tilde{A}$.)

CONSTRUCTION. We define $\mathcal{A}_s$, assuming $\mathcal{A}_{s-1}$ has been defined or $s = 0$. For each $e < s$, if there is an $x < s$, $x \in \mathbb{N}^{\langle e \rangle}$ satisfying

$$\{e\}^B(x) = 0[s] \ \& \ c(x, s) \le 2^{-(e+n+3)},$$

where $n = |\mathbb{N}^{\langle e \rangle} \cap \mathcal{A}_{s-1}|$, then choose x least and search for the least $t \ge s$ such that $g(e, t) = 1$, or $B_t \upharpoonright u \ne B_s \upharpoonright u$, where $u = u(B_s, e, x, s)$ is the use at s. In the first case, enumerate x into A (at the current stage s). If the search does not end for some $e < s$, then we leave $\mathcal{A}_s$ undefined.

VERIFICATION. We may assume $A = \tilde{A}$ by the recursion theorem.

LEMMA 5.3. A *is total.*

PROOF. Assume that $\mathcal{A}_{s-1}$ is defined or $s = 0$. Since $A = \tilde{A}$ and by the correctness of $\lim_t g(e, t)$, the search at stage s ends for each e. So we define $\mathcal{A}_s$.   ⊣

LEMMA 5.4. A *is K-trivial.*

PROOF. We apply the Fact 5.2. At stage s, suppose x is minimal s.t. $\mathcal{A}_{s-1}(x) \ne \mathcal{A}_s(x)$. We enumerate x for the sake of some requirement P_e, which so far has enumerated n numbers. Then $c(x, s) \le 2^{-(e+n+3)}$, hence $S \le \sum_{0 \le e, n} 2^{-(e+n+3)} = 1/2$.   ⊣

LEMMA 5.5. *Each requirement P_e is met.*

PROOF. Suppose for a contradiction that $A = \{e\}^B$. First assume $\lim_s g(e, s) = 1$. Choose witnesses x, s for the affirmative answer to the $\Sigma_1^0(B)$ question for P_e. Since $B \upharpoonright u$ does not change after s where $u = u(B_s, e, x, s)$, we search for t till we see $g(e, t) = 1$. Then P_e enumerates x at stage s.

Now consider the case $g(e, s) = 0$ for all $s \ge s_0$. Then we do not enumerate numbers for the sake of P_e after stage s_0. Then there is n such that P_e puts just n numbers into A. Since $A = \{e\}^B$, there is $x \in \mathbb{N}^{\langle e \rangle}$ and $s \ge s_0$ such that $\{e\}^B(x) = 0[s]$ and $c(x, s) \le 2^{-(e+n+3)}$, where $n = |\mathbb{N}^{\langle e \rangle} \cap A|$. So the answer to the $\Sigma_1^0(B)$ question for P_e is YES, contradiction.   ⊣
                                            ⊣

Note that the action of P_e may be infinitary, which is harmless here, but could be avoided by refining the $\Sigma_1^0(B)$ question.

Also note that the argument in the proof of Lemma 5.5, in the case $\lim_s g(e, s) = 1$ breaks down if B is merely Δ_2^0. The opponent can now present the correct computation $\{e\}^B(x) = 0$ at a stage s where the limit

$\lim_s g(e, s)$ has not yet been reached. Then he temporarily changes B below the use at stage $t > s$ while keeping $g(e, t) = 0$, and we do not put x into A at s. At a later stage where the old computation $\{e\}^B(x) = 0$ comes back, he has increased the cost function above $2^{-(e+n+3)}$. Thus the following question remains:

QUESTION 5.6. Does Theorem 5.1 hold for Δ_2^0 low sets B?

We may replace B by a u.r.e. sequence of uniformly low sets B_i and obtain a stronger result, which is proved by making the appropriate notational changes in the proof of Theorem 5.1.

COROLLARY 5.7. *For each u.r.e. sequence of uniformly low sets B_i, there is an r.e. K-trivial set A such that $A \not\leq_T B_i$ for each i.*

We apply this to a class first studied by Andrei Muchnik (1998).

DEFINITION 5.8. [9] A is low for K via a constant b if

$$\forall y \ K(y) \leq K^A(y) + b.$$

Let $\mathcal{M}$ denote this class of reals.

In [9] it is proved as a main result that $\mathcal{K} = \mathcal{M}$.

Note that $\mathcal{M} \subseteq \mathcal{M}[p]$ for reach p as in Definition 1.3. Thus each $A \in \mathcal{M}$ is jump traceable. Here is a uniform version of this.

PROPOSITION 5.9. *There is a fixed C such that, if A is low for K via b, then A is jump-traceable, where the trace T_b is obtained effectively in b and has bound $C 2^b i \log i$.*

PROOF. Up to constants, for each i such that $J^A(i)$ is defined,

$$K(i) \geq K^A(i) \geq K^A(J^A(i)) \geq K(J^A(i)) - b,$$

and hence $J^A(i) \in \{y : K(y) \leq K(i) + b + d\}$, for some fixed d. Since $K(i) \leq \log i + 2\log\log i + d'$ for some fixed constant d', the trace T_b given by $T_b^{[i]} = \{y : K(y) \leq \log i + 2\log\log i + b + d + d'\}$ is as required.   ⊣

The class $\mathcal{M}$ is Σ_3^0 on both the ω-r.e. and the r.e. sets. Since it includes all finite sets, there is a u.r.e. listing of the r.e. sets in $\mathcal{M}$. However, there is no way to determine a constant for being low for K:

THEOREM 5.10. *There is no effective sequence (B_i, b_i) of pairs of an r.e. set and a constant such that each B_i is low for K via b_i, and for each r.e. set A,*

$$A \in \mathcal{M} \implies \exists i \ A \leq_T B_i.$$

In particular, there is no such sequence listing all the r.e. sets in $\mathcal{M}$.

PROOF. By Proposition 5.9, each B_i is jump-traceable, where the trace and its bound are obtained effectively in b_i. By Theorem 4.1, we obtain a witness for the (super)-lowness of B_i, effectively in i. The result follows by Corollary 5.7.   ⊣

We close with a further question.

QUESTION 5.11. Find an elementary property which distinguishes the classes of low and super-low r.e. degrees. For instance, is there a noncappable (hence, low cuppable) degree which does not cup with a super-low r.e. degree to $0'$? Is there a degree which is not the supremum of two super-low degrees?

It would also be interesting to see to what extent Theorem 4.1 holds for d.r.e. sets.

Acknowledgment. The author thanks Frank Stephan and Sebastiaan Terwijn for helpful comments.

REFERENCES

[1] M. BICKFORD and F. MILLS, *Lowness properties of r.e. sets*, Manuscript, UW Madison, 1982.

[2] R. G. DOWNEY, CARL G. JOCKUSCH, JR., and M. STOB, *Array nonrecursive sets and multiple permitting arguments*, **Recursion Theory Week, Oberwolfach 1989** (K. Ambos-Spies, G. H. Müller, and Gerald E. Sacks, editors), Lecture Notes in Mathematics, vol. 1432, Springer–Verlag, Heidelberg, 1990, pp. 141–174.

[3] ROD G. DOWNEY, DENIS R. HIRSCHFELDT, ANDRÉ NIES, and FRANK STEPHAN, *Trivial reals*, **Proceedings of the 7th and 8th Asian Logic Conferences**, Singapore Univ. Press, Singapore, 2003, pp. 103–131.

[4] S. FIGUEIRA, A. NIES, and F. STEPHAN, *Combinatorial lowness properties*, **Annals of Pure and Applied Logic**, to appear.

[5] SHAMIL ISHMUKHAMETOV, *Weak recursive degrees and a problem of Spector*, **Recursion Theory and Complexity (Kazan, 1997)**, de Gruyter Series in Logic and its Applications, vol. 2, de Gruyter, Berlin, 1999, pp. 81–87.

[6] CARL G. JOCKUSCH, JR. and ROBERT I. SOARE, Π_1^0 *classes and degrees of theories*, **Transactions of the American Mathematical Society**, vol. 173 (1972), pp. 33–56.

[7] B. KJOS-HANSSEN, A. NIES, and F. STEPHAN, *Lowness for the class of Schnorr random sets*, **SIAM Journal on Computing**, to appear.

[8] JEANLEAH MOHRHERR, *A refinement of low n and high n for the r.e. degrees*, **Zeitschrift für Mathematische Logik und Grundlagen der Mathematik**, vol. 32 (1986), no. 1, pp. 5–12.

[9] A. NIES, *Lowness properties and randomness*, **Advances in Mathematics**, vol. 197 (2005), pp. 274–305.

[10] R. SOARE, **Recursively Enumerable Sets and Degrees**, Perspectives in Mathematical Logic, Omega Series, Springer–Verlag, Heidelberg, 1987.

[11] S. TERWIJN and D. ZAMBELLA, *Algorithmic randomness and lowness*, **The Journal of Symbolic Logic**, vol. 66 (2001), pp. 1199–1205.

DEPARTMENT OF COMPUTER SCIENCE
UNIVERSITY OF AUCKLAND
PRIVATE BA5 92019, AUCKLAND, NEW ZEALAND
URL: www.cs.auckland.ac.nz/~nies
E-mail: andre@cs.auckland.ac.nz

BISIMULATION INVARIANCE AND FINITE MODELS

MARTIN OTTO

Abstract. We study bisimulation invariance over finite structures. This investigation leads to a new, quite elementary proof of the van Benthem-Rosen characterisation of basic modal logic as the bisimulation invariant fragment of first-order logic. The ramification of this characterisation for the finer notion of global two-way bisimulation equivalence is based on bisimulation respecting constructions of models that recover in finite models some of the desirable properties of the usually infinite bisimilar unravellings.

§1. Introduction and preliminaries. Model theory—understood in the broad sense of the study of the expressive power of logical languages and definability of structural properties, or the interplay between syntax (*language*) and semantics (*structural properties*)—has applications in many areas. Many of these fall outside the realm of 'classical' logic and model theory, because any reasonably mature application area comes with its own pre-defined notion of what are its *relevant structures* and which are the *relevant structural properties* for its purposes. Decisions about syntax and semantics should be subordinate to the mathematical modelling that captures the essence of the application by delineating the class of structural properties that matter.

First and foremost these decisions concern the class of structures to be considered (certain classes of algebras or relational structures, possibly just finite structures of a certain kind, etc.); and a notion of equivalence between structures (isomorphism, elementary equivalence, bisimulation equivalence, etc.). While the class of structures depends on *what* is to be modelled, the accompanying notion of equivalence depends on the *coarseness* of this modelling, or its level of abstraction. Classical algebra, for instance, would typically consider the class of all algebras of a certain type, and analyse them up to isomorphism. Classical model theory is concerned with classes of all structures of a certain type, considered up to elementary equivalence. In fact, classical model

---

Key words and phrases. Finite model theory, modal logic, guarded fragment, bisimulation, preservation and characterisation theorems.

Research carried out at the Computer Science Department, University of Wales Swansea, United Kingdom; partially supported by EPSRC grant GR/R11896/01.

Logic Colloquium '02
Edited by Z. Chatzidakis, P. Koepke, and W. Pohlers
Lecture Notes in Logic, 27

theory to a large extent can be understood as an exploration of elementary equivalence (first-order logic) and its relationship to isomorphism (algebra). Finite model theory on the other hand restricts its domain to just the finite structures of certain type, under a variety of equivalences that are coarser than elementary equivalence. Elementary equivalence is of no particular interest since over finite structures it coincides with isomorphism. Apart from equivalences like bounded-variable elementary equivalence, which are partly motivated as appropriate technical substitutes for elementary equivalence in this setting, other notions of equivalence are motivated by specific requirements from application domains in the above sense. One such domain is the theoretical study of the behaviour of computational processes, which has long been a fruitful domain of interaction between logic and computer science.

Transition systems are edge and vertex coloured directed graphs, i.e. relational structures $\mathfrak{A} = (A, E^{\mathfrak{A}}, \ldots, P^{\mathfrak{A}}, \ldots)$ in a vocabulary consisting of unary predicates P and binary relations E. Intuitively the elements of the universe A of $\mathfrak{A}$ model the states, the unary P describe atomic properties of states, and the binary edge relations E formalise directed transitions from state to state.[1] In as far as possible *behaviours* are concerned, one is interested in transition systems not up to isomorphism but up to bisimulation equivalence. In fact bisimulation equivalence was, apparently independently, isolated as an adequate notion of equivalence both in the context of process analysis [11, 20] (where the term bisimulation was coined), and in the Ehrenfeucht-Fraïssé analysis of Kripke structures for propositional modal logics [23, 24] (under the name of zig-zag equivalence). The common domain of *modal model theory* is that of transition systems (or Kripke structures) up to bisimulation equivalence, or the *model theory of bisimulation invariance*. Of course this can be further ramified, according to whether, for instance, we want to consider the domain of all transition systems, or restrict to the subclass of finite transition systems (process behaviours in finite state systems), or other subclasses (e.g., connected, finite and connected, etc). Generally one can expect any such domain to have its own characteristic flavour of model theory, equally dependent on the class of structures as on the notion of equivalence. The modal domain and its ramifications provide a particularly successful example of specific model theoretic techniques and results, both classically (all structures) and in the sense of finite model theory (just finite structures). An aspect particularly relevant to computer science applications is that within the modal domain there are logics whose expressive power goes far beyond that of first-order logic while still being algorithmically very tractable. Tractability here mainly concerns decidability of the satisfiability problem and feasible model

---

[1] Formally there is no distinction between this setting and that of relational formalisations of Kripke structures for propositional modal logic with atomic propositions P and accessibility relations E between possible worlds $a \in A$.

checking complexities. Indeed, bisimulation invariance itself and some of its immediate model theoretic consequences, like the tree model property, can account for these good algorithmic features.

A fundamental issue in every ramification of model theory for a particular choice of structures and equivalence is the quest for *expressive completeness*. A logic is expressively complete for a class of structural properties if it precisely expresses just these structural properties. Basic modal logic is expressively complete for the class of bisimulation invariant first-order properties over the class of all transition systems (van Benthem's theorem, Theorem 2.1 below). In other words, basic modal logic is the 'first-order logic for the modal domain'. One of the most important process logics, the modal μ-calculus which extends basic modal logic by means of least and greatest fixed points is similarly known to be expressively complete for precisely the bisimulation invariant monadic second-order properties, by a well known theorem of Janin and Walukiewicz [14].

On the technical side, a model theory for a specific domain will always have its characteristic techniques for constructing and transforming models, in accordance with the underlying equivalence. Bisimulation preserving constructions in the domain of all, finite and infinite, transition systems mostly center on tree models that are naturally obtained as bisimilar unravellings of arbitrary transitions systems. Most of our technical contributions here revolve around bisimulation preserving model constructions in finite models, where unravellings no longer work. It is a characteristic feature of finite model theory in general that model constructions and transformations tend to be combinatorially more involved simply because the result is required to remain finite. This is also true of modal finite model theory.

In the rest of this section, we provide some of the basic definitions and fix some notation. Much of it can safely be skipped by readers familiar with the fundamentals of Ehrenfeucht-Fraïssé games and equivalences, bisimulation games and equivalences, and with basic modal logic.

We typically write $\mathfrak{A} = (A, E^{\mathfrak{A}}, \ldots, P^{\mathfrak{A}}, \ldots)$ for a structure with universe A, binary predicates $E^{\mathfrak{A}}, \ldots$, and unary predicates $P^{\mathfrak{A}}, \ldots$; superscripts $^{\mathfrak{A}}$ are dropped where inessential. Often a distinguished node or parameter $a \in A$ is indicated as in $\mathfrak{A}, a$.

We want to see bisimulation equivalence both as a notion of behavioural equivalence between transition systems or—more classically—as the equivalence induced by the infinite Ehrenfeucht-Fraïssé game whose moves capture the relativised pattern of modal quantification.

1.1. Ehrenfeucht-Fraïssé games. Recall the classical Ehrenfeucht-Fraïssé characterisation of elementary equivalence (see [7, 6, 21] for textbook treatments). The game is played by two players over the two structures, $\mathfrak{A}$ and $\mathfrak{B}$, that are to be compared. In each round of the game the first player marks an element of one of the structures (his choice) and the second player responds by selecting an element in the opposite structure. After k rounds, k elements will have been individually marked in each structure: we denote the resulting configuration as $\mathfrak{A}; a_1, \ldots, a_k$ versus $\mathfrak{B}; b_1, \ldots, b_k$, where (a_i, b_i) is the pair of elements marked in round i. It is the second player's task to maintain similarity between the two sides: she loses as soon as the correspondence $\{(a_1, b_1), \ldots, (a_k, b_k)\}$ stops being a local isomorphism.[2]

The well known Ehrenfeucht-Fraïssé theorem says that the second player has a winning strategy for q rounds in this game iff $\mathfrak{A}$ and $\mathfrak{B}$ cannot be distinguished by FO-sentences of quantifier rank q, $\mathfrak{A} \equiv_q \mathfrak{B}$. Algebraically or combinatorially the existence of a winning strategy for the second player in the q round game, $\mathfrak{A} \simeq_q \mathfrak{B}$, is captured in a *back-and-forth system* of depth q of local isomorphisms.

A winning strategy in the unbounded game with an infinite succession of rounds in which the second player needs to maintain a partial isomorphism characterises equivalence in the infinitary logic $L_{\infty\omega}$, $\mathfrak{A} \equiv_{\infty\omega} \mathfrak{B}$ (Karp's theorem). The corresponding combinatorial notion of partial isomorphy between structures, $\mathfrak{A} \simeq_{\text{part}} \mathfrak{B}$, precisely captures the existence of such a winning strategy as a single back-and-forth system of local isomorphisms. All these notions naturally extend to structures with parameters, which may be thought of as elements marked before the start of the game.

It is natural to introduce bisimulation equivalence in this game context.

1.2. Bisimulation games. The *bisimulation game* on transition systems $\mathfrak{A}$, a and $\mathfrak{B}$, b is played by two players as above, but now only one marked element in each structure is kept. Initially the marked elements are the distinguished nodes a and b. In each round the first player selects one of the structures and one of the transitions available from the marked element in that structure and moves the marker along this transition to its target element; the second player has to respond by moving the marker in the opposite structure along a corresponding transition, and she loses if she cannot move (for lack of an appropriate transition) or if the marked elements do not satisfy exactly the same unary predicates. Clearly the existence of a winning strategy for the second player can be captured by a back-and-forth system of pairs in $A \times B$. More precisely, for the unbounded game the existence of a strategy corresponds to one system $Z \subseteq A \times B$ with $(a, b) \in Z$ (for initialisation in the distinguished elements), with $a' \in P^{\mathfrak{A}} \Leftrightarrow b' \in P^{\mathfrak{B}}$ for all $(a', b') \in Z$ and all unary predicates P, and satisfying the usual back-and-forth conditions:

forth: for all binary E and for all $(a', b') \in Z$ and all a'' such that $(a', a'') \in E^{\mathfrak{A}}$ there is some b'' such that $(b', b'') \in E^{\mathfrak{B}}$ and $(a'', b'') \in Z$.

---

[2]We follow [21] in using the more natural name of *local isomorphisms* for what the older literature calls *partial isomorphisms*: partial maps that are isomorphisms between the substructures induced on their domain and range.

back: for all binary E and for all $(a', b') \in Z$ and all b'' such that $(b', b'') \in E^{\mathfrak{B}}$ there is some a'' such that $(a', a'') \in E^{\mathfrak{A}}$ and $(a'', b'') \in Z$.

DEFINITION 1.1. Two structures are bisimilar, $\mathfrak{A}, a \sim \mathfrak{B}, b$, if and only if the second player has a winning strategy in the unbounded game starting from $\mathfrak{A}$, a and $\mathfrak{B}, b$; equivalently, if there is a back-and-forth system Z with $(a, b) \in Z$.

Similarly we may consider an ℓ-round bisimulation game. The formalisation of a winning strategy takes the form of a depth ℓ stratified back-and-forth system, $(Z_i)_{0 \leqslant i \leqslant \ell}$, with $(a, b) \in Z_\ell$ and back-and-forth conditions that correspondingly guarantee the existence of $(a'', b'') \in Z_{i-1}$ for $(a', b') \in Z_i$. We write $\mathfrak{A}, a \sim^\ell \mathfrak{B}, b$ if there is such, i.e., if the second player has a strategy to survive for ℓ rounds in the game on $\mathfrak{A}, a$ and $\mathfrak{B}, b$.

DEFINITION 1.2. We write $\mathfrak{A}, a \sim^\ell \mathfrak{B}, b$ and speak of ℓ-*bisimulation equivalence* if the second player has a winning strategy in the ℓ-round bisimulation game from $\mathfrak{A}, a$ and $\mathfrak{B}, b$; or, equivalently if there is a stratified back-and-forth systems of depth ℓ, with $(a, b) \in Z_\ell$.

Some variations of basic bisimulation equivalence will be considered in Section 4.

1.3. Basic modal logic. When comparing the bisimulation game with the usual FO Ehrenfeucht-Fraïssé game, we see that the severe restriction of moves in the bisimulation game corresponds to restricted access to elements through quantification. Any move is bound to the current position via a transition along some specific binary edge predicate E, in precisely the way that is captured by the modal operators $\langle E \rangle$ and their duals $[E]$. These may be transcribed into relativised first-order quantifiers according to

(1)
$$(\langle E \rangle \varphi)(x) \equiv \exists y \big(Exy \wedge \varphi(y) \big),$$
$$([E] \varphi)(x) \equiv \forall y \big(Exy \to \varphi(y) \big).$$

For us, basic modal logic ML is exactly the fragment of FO consisting of all FO-formulae in a single free variable in which all quantifications take the form of $\langle E \rangle$ and $[E]$ for binary predicates E.

DEFINITION 1.3. We denote basic modal logic as ML, and regard it as a fragment of first-order logic, syntactically generated from atomic formulae Px, where P is unary, as the closure under Boolean connectives and modal quantification according to (1) for all binary predicates E.

1.4. Unravellings. By the *modal domain* we mean the world of transition systems considered up to bisimulation equivalence. One may pass from any given transition system to one of its bisimilar companions that is most convenient for the task at hand. For a wide variety of such tasks *tree models*—transition systems based on trees rather than arbitrary graphs—are very suitable and it is not hard to obtain bisimilar companion structures that are trees for any given transition system, by a simple process of unravelling.

The unravelling of $\mathfrak{A}, a$ has as its universe all directed finite paths emanating from a in $\mathfrak{A}$; unary predicates are interpreted so as to put a path $a_0, a_1, \ldots, a_\ell$, where $a_0 = a$, into P iff $a_\ell \in P^{\mathfrak{A}}$; and there is an E-edge from $a_0, a_1, \ldots, a_\ell$ precisely to all $a_0, a_1, \ldots, a_\ell, a_{\ell+1}$ for which $(a_\ell, a_{\ell+1}) \in E^{\mathfrak{A}}$. The natural projection that associates $a_0, a_1, \ldots, a_\ell \mapsto a_\ell$ induces a bisimulation with $\mathfrak{A}, a$.

Thanks to this universal availability of bisimilar companions that are trees, any bisimulation invariant model theoretic issue about arbitrary transition systems is immediately reduced to the model theory of trees. One can hardly overestimate the scope and success of this approach. To mention but the most obvious application, we may consider satisfiability issues for logics that are preserved under bisimulation (e.g., basic modal logic ML, but also computation tree logic or the modal μ-calculus—the latter two fragments of monadic second-order logic that reach outside FO). By bisimulation invariance a formula is satisfiable if and only if it is satisfiable in a tree model: any bisimulation invariant logic enjoys the *tree model property*. Satisfiability in a tree model can be formalised as a monadic second-order fact about trees, for each of the logics mentioned. Hence, Rabin's theorem on the decidability of the monadic second-order theory of trees immediately yields decidability results. Quite often the complexity of the satisfiability problem for bisimulation invariant logics can moreover be pinpointed with translations into emptiness problems for suitable tree-automata. As Vardi argues in [25] the tree model property (rather than the finite model property) accounts for the robust decidability of logics for the modal domain (also compare [9]).

Unravellings, however, rarely work for the purpose of finite model theory, simply because any unravelling of a transition system with directed cycles will be infinite; indeed, acyclicity and finiteness are mutually exclusive in bisimilar companions of any system with directed cycles. In Section 4 we shall specifically deal with certain qualified analogues of unravellings that work in finite model theory.

1.5. Partial unravellings. One rather simplistic local analogue that can also be useful is provided by partial unravellings that unravel the given structure to a certain depth from its distinguished node, merged into isomorphic copies of the given structure. The *depth ℓ unravelling* of $\mathfrak{A}, a$ provides a structure bisimilar with $\mathfrak{A}, a$ which is tree-like inside a radius ℓ from the distinguished node. We restrict the full unravelling of $\mathfrak{A}, a$ to all nodes whose distance is at most ℓ from the root, and identify any node of the form $a = a_0, a_1, \ldots, a_\ell = b$ with the node corresponding to b in a new isomorphic copy of $\mathfrak{A}, b$. Note that the resulting bisimilar companion structure is finite for finite $\mathfrak{A}$, and acyclic at least up to depth ℓ from a.

§2. The van Benthem-Rosen characterisation. The classical version of the characterisation theorem that links ML with bisimulation invariance is the following, due to van Benthem [23, 24]. We choose a formulation that highlights the harder direction of the equivalence, namely the converse of the semantic preservation property.

THEOREM 2.1 (van Benthem). FO/$\sim$ $\equiv$ ML: *any first-order formula $\varphi(x)$ that is invariant under bisimulation is equivalent to a formula of basic modal logic, and vice versa.*

Clearly this theorem can be read in two different ways. We may take it as a semantic characterisation of the modal fragment inside FO, in the spirit of the classical model theoretic preservation theorems (noting that as with those, the preservation statement is not the interesting direction of the stated equivalence). Alternatively, we may see it as a provision of effective syntax for the bisimulation invariant FO-properties. The full set of FO-formulae that are bisimulation invariant cannot be the answer, since bisimulation invariance of FO formulae is not decidable. The essence of the characterisation theorem, therefore, is that ML is a logic that is *expressively complete* for bisimulation invariant FO.

The finite model theory version of this characterisation result (cf. Theorem 2.4 below) is not an immediate consequence of the classical version, since there are first-order formulae that are bisimulation invariant over finite structures without being bisimulation invariant over all structures. Trivial examples can be generated with the use of some infinity axiom. Let for instance ψ be the first-order sentence that asserts that the binary relation R is a linear ordering without maximal element. Then any formula of the form $\psi \wedge \varphi(x)$ is trivially bisimulation invariant over finite structures, but not bisimulation invariant over infinite structures unless $\varphi(x)$ is unsatisfiable in any model of ψ.

Interestingly, many classical characterisation theorems fail in the sense of finite model theory. For instance, the classical characterisation of FO^2, seemingly a close relative to modal logics [10], as the 2-pebble game invariant fragment of first-order logic, fails in finite model theory. Indeed, the first-order sentence (in three variables) that says of a binary relation R that it is a linear order of the universe, is invariant under 2-pebble game equivalence in restriction to finite structures (but not in general)—and it is easy to see that no first-order sentence with just two variables is equivalent to it over all finite structures. Compare also [3], and, e.g., Example 1.12 in [16]. Failures like this are only to be expected of course, since the typical tools of classical model theory and most notably the compactness theorem for FO fail in restriction to finite structures [6].

2.1. Generic classical proof. The classical proof of van Benthem's theorem makes use of compactness and saturation techniques that essentially involve infinite models. Suppose more generally that we are dealing with any equivalence relation $\leftrightharpoons$ between structures (like some unbounded Ehrenfeucht-Fraïssé game equivalence) with finitary approximations $\leftrightharpoons^\ell$ (the corresponding ℓ-round approximants). Let $\mathcal{L} \subseteq$ FO be a syntactic fragment of first-order logic stratified according to $\mathcal{L} = \bigcup_\ell \mathcal{L}_\ell$, with each $\mathcal{L}_\ell$ closed under disjunction, and related to the $\leftrightharpoons^\ell$ and $\leftrightharpoons$ such that

  (i) the $\leftrightharpoons^\ell$ form a chain of successive refinements $\leftrightharpoons^0 \supseteq \leftrightharpoons^1 \supseteq \cdots \supseteq \leftrightharpoons$,

  (ii) each $\mathcal{L}_\ell$ is preserved under $\leftrightharpoons^\ell$,

  (iii) $\leftrightharpoons^\ell$ has finite index and each $\leftrightharpoons^\ell$ class is definable by a formula of $\mathcal{L}_\ell$.

Typically, the $\mathcal{L}_\ell$ would correspond to levels of $\mathcal{L}$ defined in terms of some suitable notion of quantifier rank, and the formulae for (iii) are the Hintikka formulae associated with the Ehrenfeucht-Fraïssé analysis. In fact (ii) and (iii) imply that any $\mathcal{L}$-fomula is equivalent to a finite disjunction over such Hintikka formulae, thus providing a syntactic normal form.

For $\varphi \in$ FO consider the following two statements:

(2) $\varphi \leftrightharpoons$ invariant $\Longleftrightarrow \varphi$ expressible in $\mathcal{L}$,

(3) $\varphi \leftrightharpoons$ invariant $\Longrightarrow \varphi \leftrightharpoons^\ell$ invariant for some ℓ.

FO$/\leftrightharpoons \equiv \mathcal{L}$ precisely means that (2) holds for all φ.

PROPOSITION 2.2. *Given* $\leftrightharpoons$, $\leftrightharpoons^\ell$, $\mathcal{L}$ *and* $\mathcal{L}_\ell$ *with* (i)–(iii), (2) *and* (3) *are equivalent for every* $\varphi \in$ FO.

PROOF. (2) $\Rightarrow$ (3) follows from (ii). For the converse implication we only need (3) to establish the crucial direction in (2), that $\leftrightharpoons$ invariance implies $\mathcal{L}$-definability. But if, by (3), the given φ is in fact $\leftrightharpoons^\ell$ invariant for some ℓ, then we may use closure under disjunction and (iii) to equivalently express φ as a disjunction over the defining $\mathcal{L}_\ell$ formulae of all those $\leftrightharpoons^\ell$-classes that satisfy φ.   ⊣

Indeed, the characterisation (2) follows outright if $\leftrightharpoons$ and its approximants $\leftrightharpoons^\ell$ further satisfy the following condition (iv), which is best seen as a weak *convergence* of the sequence of the $\leftrightharpoons^\ell$ to $\leftrightharpoons$. Let $\leftrightharpoons^\omega$ stand for the common refinement $\bigcap_\ell \leftrightharpoons^\ell$ of all the finite levels $\leftrightharpoons^\ell$.

  (iv) $\leftrightharpoons^\omega$ coincides with $\leftrightharpoons$ in restriction to ω-saturated models.

This condition is naturally satisfied for equivalence relations derived from Ehrenfeucht-Fraïssé games in the indicated manner.

PROPOSITION 2.3. *Let* $\leftrightharpoons$, $\leftrightharpoons^\ell$, *and* $\mathcal{L} = \bigcup_\ell \mathcal{L}_\ell$ *be as above with* (i)–(iv). *Then* FO$/\leftrightharpoons \equiv \mathcal{L}$.

PROOF. We show (3). Assuming, for the sake of contradiction, that φ were not $\leftrightharpoons^\ell$ invariant for any ℓ, we obtain a sequence of models $\mathfrak{A}_\ell \models \varphi$ and $\mathfrak{B}_\ell \models \neg\varphi$ with $\mathfrak{A}_\ell \leftrightharpoons^\ell \mathfrak{B}_\ell$. A simple compactness argument (or a straightforward ultraproduct construction) allows us to construct a pair of models $\mathfrak{A} \models \varphi$

and $\mathfrak{B} \models \neg\varphi$ where $\mathfrak{A} \leftrightharpoons^\ell \mathfrak{B}$ for all finite ℓ. For instance one can argue for finite satisfiability of the FO-theory $\{\chi^A \leftrightarrow \chi^B : \chi \in \mathcal{L}\} \cup \{\varphi^A, \neg\varphi^B\}$, where superscripts A and B refer to relativisation to new unary predicates A and B that delineate sub-universes for the $\mathfrak{A}$- and $\mathfrak{B}$-parts, respectively. The substructures induced on the A- and B-parts of any model of this theory will provide $\mathfrak{A}$ and $\mathfrak{B}$ as desired.

Passing to ω-saturated elementary extensions $\mathfrak{A}_\infty \succcurlyeq \mathfrak{A}$ and $\mathfrak{B}_\infty \succcurlyeq \mathfrak{B}$, respectively, we have $\mathfrak{A}_\infty \models \varphi$, $\mathfrak{B}_\infty \models \neg\varphi$ even though $\mathfrak{A}_\infty \leftrightharpoons \mathfrak{B}_\infty$ by (iv), contradicting $\leftrightharpoons$ invariance of φ. ⊣

The equivalence of (2) and (3) remains valid when interpreted in the sense of finite model theory. Moreover, (iv) holds in particular of finite structures (which all are ω-saturated): $\leftrightharpoons^\omega$ coincides with $\leftrightharpoons$ in finite model theory.

Incidentally, this allows us to reformulate the finite model theory version of the characterisation $\mathcal{L} \equiv \mathrm{FO}/\leftrightharpoons$ as an instance of compactness in finite model theory, as follows. In finite model theory, and by the above, $\leftrightharpoons$ invariance of φ is captured by

$$\{\chi^A \leftrightarrow \chi^B : \chi \in \mathcal{L}\} \models \varphi^A \longleftrightarrow \varphi^B.$$

Here we could moreover restrict the formulae χ to the Hintikka formulae characterising the $\leftrightharpoons^\ell$ classes (see condition (iii) above), for all finite ℓ. Similarly, $\leftrightharpoons^\ell$ invariance of φ, for fixed level ℓ, is captured (both classically and in finite model theory) by

$$\{\chi^A \leftrightarrow \chi^B : \chi \in \mathcal{L}_\ell\} \models \varphi^A \longleftrightarrow \varphi^B.$$

Therefore, the crucial implication (2) $\Rightarrow$ (3) in finite model theory becomes

(4)
$$\{\chi^A \leftrightarrow \chi^B : \chi \in \mathcal{L}\} \models \varphi^A \longleftrightarrow \varphi^B$$
$$\implies \text{ for some finite } \ell: \{\chi^A \leftrightarrow \chi^B : \chi \in \mathcal{L}_\ell\} \models \varphi^A \longleftrightarrow \varphi^B.$$

Compactness, of course, fails as a general principle in finite model theory. If we look at the (indirect) proof of Proposition 2.3, the structures $\mathfrak{A}$ and $\mathfrak{B}$ obtained there would typically have to be infinite, even if based on sequences of finite $\mathfrak{A}_\ell$ and $\mathfrak{B}_\ell$. And since $\leftrightharpoons$ invariance can now only be assumed in restriction to finite models, it cannot be brought to bear.

The modal characterisation theorem itself, however, does go through in finite model theory, as shown by Rosen [22]. In other words, with basic modal logic of nesting depth ℓ for $\mathcal{L}_\ell$ and $\mathcal{L} = \mathrm{ML}$, (4) is a valid instance of compactness in finite model theory.

THEOREM 2.4 (Rosen). *Any first-order formula $\varphi(x)$ that is invariant under bisimulation over finite structures is equivalent over finite structures to a formula of basic modal logic, and vice versa.*

While the generic classical proof is rather smooth, it really tells us nothing about the finite model theory of the matter. The rather more constructive argument given by Rosen, however, does equally apply to the classical version, thus providing a new proof there as well. The same is true of our new proof of this theorem and of those ramifications of the van Benthem-Rosen characterisation that will concern us in Section 4. Those ramifications in particular will give an exemplary insight into ways in which finite model theory techniques can provide surprising alternatives to classical arguments.

§3. The van Benthem-Rosen theorem reproved. This section provides a self-contained exposition of an elementary proof of the van Benthem-Rosen characterisation, based solely on playing Ehrenfeucht-Fraïssé games. An extended version of the argument can be found in [18].

DEFINITION 3.1. Let $\mathfrak{A}$ be a relational structure. The *Gaifman graph* $G(\mathfrak{A})$ associated with $\mathfrak{A}$ has universe A and edges (a, a') for any pair of distinct elements that coexist in some relational ground atom of $\mathfrak{A}$. *Gaifman distance* d on $\mathfrak{A}$ is the metric induced by ordinary graph distance in $G(\mathfrak{A})$: $d(a, a') = \ell$ if ℓ is the minimal length of a path between a and a' in $G(\mathfrak{A})$.

DEFINITION 3.2. Let $\mathfrak{A}$ be a relational structure. The ℓ-*neighbourhood* of a in $\mathfrak{A}$ is the subset $U^\ell(a) = \{a' \in A : d(a, a') < \ell\}$.

DEFINITION 3.3. A formula $\psi(x)$ is ℓ-*local* if it is logically equivalent to its relativisation to $U^\ell(x)$.

Gaifman distance d is first-order definable, in the sense that for every ℓ there is a first-order formula saying that $d(x, y) \leqslant \ell$. For a vocabulary consisting of just unary and binary predicates in particular, as is the case for transition systems, $d(x, y) \leqslant 1$ is just a disjunction over $x = y$ and all $Exy \vee Eyx$ for binary E; inductively, $d(x, y) \leqslant \ell$ is expressible as $\exists z (d(x, z) \leqslant \ell_1 \wedge d(z, y) \leqslant \ell_2)$ for any $\ell = \ell_1 + \ell_2$, $\ell_i \geqslant 1$. Generally, Gaifman neighbourhoods are FO-definable, and ℓ-locality has an obvious syntactic counterpart in FO. For a fuller account of decompositions of FO formulae into local formulae see Section 5.1, where Gaifman's theorem (Theorem 5.3) will feature prominently.

DEFINITION 3.4. For two structures $\mathfrak{A}$ and $\mathfrak{B}$ we say that a in $\mathfrak{A}$ and b in $\mathfrak{B}$ are ℓ-*locally* q-*equivalent* if $\mathfrak{A} \restriction U^\ell(a)$, a and $\mathfrak{B} \restriction U^\ell(b)$, b are q-elementarily equivalent, i.e., cannot be distinguished by first-order formulae of quantifier rank q. Symbolically we write

$$\mathfrak{A}, a \equiv_q^{(\ell)} \mathfrak{B}, b.$$

Of course $\mathfrak{A}, a \equiv_q^{(\ell)} \mathfrak{B}, b$ may also be characterised by saying that $\mathfrak{A}, a$ and $\mathfrak{B}, b$ agree on all ℓ-local FO-formulae of quantifier rank up to q.

LEMMA 3.5. *Let* $\varphi(x) \in$ FO *be invariant under disjoint sums. Then* φ *is* ℓ-*local for* $\ell = 2^{qr(\varphi)}$.

PROOF. Let $q := \mathrm{qr}(\varphi)$. It suffices to show that for $\ell := 2^q$

$$\mathfrak{A}, a \equiv_q^{(\ell)} \mathfrak{B}, b \quad \Longrightarrow \quad \left(\mathfrak{A} \models \varphi[a] \iff \mathfrak{B} \models \varphi[b]\right).$$

Let $\mathfrak{A}, a \equiv_q^{(\ell)} \mathfrak{B}, b$. As φ is invariant under disjoint sums the desired preservation of φ follows from the following:

$$(5) \quad q \cdot \mathfrak{A} + q \cdot \mathfrak{B} + \mathfrak{A} \models \varphi[a] \quad \Longleftrightarrow \quad q \cdot \mathfrak{A} + q \cdot \mathfrak{B} + \mathfrak{B} \models \varphi[b].$$

Here $q \cdot \mathfrak{A} + q \cdot \mathfrak{B}$ stands for the disjoint sum of q copies of $\mathfrak{A}$ and $\mathfrak{B}$ each, also disjoint form the original copies of $\mathfrak{A}$ and $\mathfrak{B}$ in which we take a, respectively b, to live. Let $\mathfrak{C} := q \cdot \mathfrak{A} + q \cdot \mathfrak{B}$ denote the common part of the two structures.

We establish (5) by exhibiting a strategy for the second player in the q-round Ehrenfeucht-Fraïssé game on $\mathfrak{C} + \mathfrak{A}, a$ versus $\mathfrak{C} + \mathfrak{B}, b$.

After k moves in the game, positions are of the form $\mathfrak{C} + \mathfrak{A}, a, a_1, \ldots, a_k$ versus $\mathfrak{C} + \mathfrak{B}, b, b_1, \ldots, b_k$. Some of the moves will have been played according to a strategy in the game on $\mathfrak{A} \upharpoonright U^\ell(a), a$ versus $\mathfrak{B} \upharpoonright U^\ell(b), b$, which is available as $\mathfrak{A}, a \equiv_q^\ell \mathfrak{B}, b$. Some other moves will have been made according to an isomorphic copying strategy into appropriate isomorphic partner structures in $\mathfrak{C}$. Let us call moves of the first kind (played according to $\mathfrak{A}, a \upharpoonright U^\ell(a)$ versus $\mathfrak{B}, b \upharpoonright U^\ell(b)$ in the original copies of $\mathfrak{A}$ and $\mathfrak{B}$) close, and also refer to the elements a_i and b_i marked in those rounds as close. Note that in this terminology a_i is close iff b_i is. We let $(a_i)_{a_i \text{ close}}$ and $(b_i)_{b_i \text{ close}}$, respectively, stand for the subtuples consisting of just those elements that were played 'close' in this sense, in either structure. A critical distance related to 'closeness' in round k of the game will be $\ell_k := 2^{q-k}$; note that $\ell_0 = \ell$. Our strategy will be such that the following conditions are maintained. After the k-th round:

(i) all close elements have been played within distance $\ell - \ell_k$ of a or b, respectively,

(ii) $\mathfrak{A} \upharpoonright U^\ell(a), a, (a_i)_{a_i \text{ close}} \equiv_{q-k} \mathfrak{B} \upharpoonright U^\ell(b), b, (b_i)_{b_i \text{ close}}$,

(iii) if a_i and b_i are not close, then their distance from a (respectively b) or any other close element is greater than ℓ_k,

(iv) if a_i and b_i are not close then there is an isomorphism ρ_i between the copy of $\mathfrak{A}$ or $\mathfrak{B}$ in which a_i lives and the copy in which b_i lives, such that $\rho_i(a_i) = b_i$,

(v) if a_i and a_j (b_i and b_j) are both not close, and $d(a_i, a_j) \leq \ell_k$ (or $d(b_i, b_j) \leq \ell_k$) then $\rho_i = \rho_j$.

It is clear that (i)–(v) are satisfied at the start of the game, for $k = 0$. We show how the second player can maintain these conditions in response to any next move by the first player. If these conditions are maintained through all q rounds, then the second player wins: note in particular that the non-close elements are still at a distance greater than $\ell_q = 1$ from the close ones, and hence have no relational edges in common with those, while amongst the close or non-close elements the validity of the strategy inside $\mathfrak{A}, a \upharpoonright U^\ell(a)$

and $\mathfrak{B}, b \upharpoonright U^\ell(b)$, or the existence of isomorphisms according to (iv) and (v) guarantee that $\{(a, b), (a_1, b_1), \ldots, (a_q, b_q)\}$ is a local isomorphism.

So, how can (i)–(v) be maintained? W.l.o.g. assume that the first player makes his $(k + 1)$st move into the a-structure, choosing a_{k+1}. In order to advise the second player on her response we distinguish several cases.[3]

Case (A): a_{k+1} is at distance greater than ℓ_{k+1} from a and all previously played elements; a_{k+1} and b_{k+1} are not going to be close. Consider the copy of $\mathfrak{A}$ or $\mathfrak{B}$ in which a_{k+1} lives in the a-structure. As there are q copies of $\mathfrak{A}$ and $\mathfrak{B}$ in $\mathfrak{C}$, at least one hitherto unused copy of the respective kind is still available in the b-structure. Pick an isomorphism onto such a fresh copy and pick b_{k+1} to be the image of a_{k+1} under this isomorphism. (i)–(v) are clearly maintained.

Case (B): a_{k+1} is within distance ℓ_{k+1} of some previously played element that is not close. Again, a_{k+1} and b_{k+1} are not going to be close, but b_{k+1} needs to be chosen according to requirement (v). Let b_{k+1} be the image of a_{k+1} under the isomorphism that works for all the (non-close) previously played elements within distance ℓ_{k+1} of a_{k+1}. (i)–(v) are clearly maintained.

Case (C): a_{k+1} is within distance ℓ_{k+1} of some previously played close element; a_{k+1} and b_{k+1} are going to be close. Note that condition (i) will be satisfied for a_{k+1}: $d(a, a_{k+1}) \leqslant \ell - \ell_k + \ell_{k+1} = \ell - \ell_{k+1}$. b_{k+1} can be selected according to the strategy in $\mathfrak{A} \upharpoonright U^\ell(a), a; (a_i)_{a_i \text{ close}} \equiv_{q-k}^{(\ell)} \mathfrak{B} \upharpoonright U^\ell(b), b; (b_i)_{b_i \text{ close}}$. It remains to argue that condition (i) also applies to this b_{k+1}. But the game for $\mathfrak{A} \upharpoonright U^\ell(a), a; (a_i)_{a_i \text{ close}} \equiv_{q-k}^{(\ell)} \mathfrak{B} \upharpoonright U^\ell(b), b; (b_i)_{b_i \text{ close}}$ automatically preserves distances up to ℓ_{k+1} in its first round, as $d(x, y) \leqslant 2^m$ is expressible in quantifier rank m. Again, (i)–(v) are seen to be maintained.

$\dashv$

COROLLARY 3.6. *Let $\varphi(x) \in$ FO be invariant under bisimulation (over finite structures). Then φ is equivalent (over finite structures) to an ML-formula of nesting depth less than $2^{\operatorname{qr}(\varphi)}$. This bound is tight: FO can be exponentially more succinct than ML in expressing bisimulation invariant properties.*

PROOF. Note that bisimulation invariance in particular implies invariance under disjoint sums. So $\varphi(x)$ is ℓ-local for $\ell = 2^{\operatorname{qr}(\varphi)}$, by the previous lemma. We now want to show that

$$\mathfrak{A}, a \sim^{\ell-1} \mathfrak{B}, b \quad \Longrightarrow \quad (\mathfrak{A} \models \varphi[a] \Longleftrightarrow \mathfrak{B} \models \varphi[b]).$$

By bisimulation invariance (in finite structures) we may w.l.o.g. assume that $\mathfrak{A} \upharpoonright U^\ell(a)$ and $\mathfrak{B} \upharpoonright U^\ell(b)$ are trees: if they are not we pass to depth ℓ unravellings [in the classical case we may of course fully unravel even if the resulting

---

[3]The three cases are mutually exclusive; for (B) and (C) this comes from (iii).

structures are infinite]. With this proviso observe that then

$$
\begin{array}{rcccl}
 & \mathfrak{A}, a & \sim^{\ell-1} & \mathfrak{B}, b \\
\Longleftrightarrow & \mathfrak{A}{\upharpoonright}U^\ell(a), a & \sim^{\ell-1} & \mathfrak{B}{\upharpoonright}U^\ell(b), b \\
\Longrightarrow & \mathfrak{A}{\upharpoonright}U^\ell(a), a & \sim & \mathfrak{B}{\upharpoonright}U^\ell(b), b \\
\Longrightarrow & \mathfrak{A}{\upharpoonright}U^\ell(a) \models \varphi[a] & \Longleftrightarrow & \mathfrak{B}{\upharpoonright}U^\ell(b) \models \varphi[b] \\
\Longrightarrow & \mathfrak{A} \models \varphi[a] & \Longleftrightarrow & \mathfrak{B} \models \varphi[b].
\end{array}
$$

The equivalence between the first two lines, as well the implication to the third are due to the fact that the ℓ-neighbourhoods encompass all elements that are reachable from the source node in $\ell - 1$ rounds of the bisimulation game. The next two implications use first bisimulation invariance, then ℓ-locality of φ.

For the tightness claim it suffices to consider the property

$$
\mathrm{Red}_q(x) : \begin{cases} \text{a red element is reachable from } x \\ \text{on an } E\text{-path of length less than } 2^q \end{cases}
$$

in structures with binary E and unary R for *red*. Observe that $\mathrm{Red}_q(x)$ is expressible by an FO-formula $\varphi_q(x)$ of quantifier rank q. As auxiliary formulae we use variants of the distance formulae considered in connection with Definition 3.1 above, but now for distances along (directed) E-paths: let $\delta(x, y) \leqslant 2^q$ be shorthand for the quantifier rank q formulae generated from $\delta(x, y) \leqslant 2^0 := x = y \vee Exy$ according to $\delta(x, y) \leqslant 2^{q+1} := \exists z (\delta(x, z) \leqslant 2^q \wedge \delta(z, y) \leqslant 2^q)$. Now let $\varphi_0(x) = \mathrm{Red}(x)$ and inductively $\varphi_{q+1}(x) := \exists y (d(x, y) \leqslant 2^q \wedge \varphi_q(y))$. Clearly $\varphi_q(x)$ is invariant under bisimulation (in fact under $\sim^\ell$ for $\ell = 2^q - 1$), but not invariant under $\sim^\ell$ for any $\ell < 2^q - 1$, hence not expressible in ML at nesting depth less than $2^q - 1$.   ⊣

§4. Locally acyclic covers.

We look at stronger variants of bisimulation equivalence and in particular at global two-way bisimulations. A bisimulation between $\mathfrak{A}$, a and $\mathfrak{B}$, b is called a *two-way bisimulation* if the back-and-forth conditions are also satisfied w.r.t. to the inverse edge relations (backward transitions) $E^{-1} = \{(y, x) : (x, y) \in E\}$. A bisimulation is *global* if it pairs every node in $\mathfrak{A}$ with some node in $\mathfrak{B}$, and vice versa.

DEFINITION 4.1. Two-way global bisimulation equivalence, $\mathfrak{A}, a \approx \mathfrak{B}, b$, is defined like bisimulation equivalence, but on the basis of back-and-forth conditions w.r.t. both forward and backward moves along E-edges, and with the additional requirement that for every a' in $\mathfrak{A}$ there is some b' from $\mathfrak{B}$ such that the second player has a winning strategy in the game from $\mathfrak{A}, a'$ and $\mathfrak{B}, b'$, and vice versa.

Two-way global ℓ-bisimulation, $\approx^\ell$, is the corresponding approximation, defined in terms of a winning strategy in the ℓ-round game, or in terms of a stratified back-and-forth system of depth ℓ.

Note that $\approx$ and $\approx^\ell$ are meaningful also without distinguished nodes, as in $\mathfrak{A} \approx \mathfrak{B}$, due to the global nature of these bisimulations.

The passage from ordinary bisimulation to two-way and/or global bisimulation corresponds to the introduction of moves in which markers are moved backwards along edges, and moves in which markers can be freely relocated to any element of their structure. In Ehrenfeucht-Fraïssé terms these extra moves capture the power of inverse and universal modalities, respectively. In first-order terms, unconstrained moves correspond to universal and existential quantification, $\forall x \varphi(x)$ and $\exists x \varphi(x)$, which in the classical modal context may be associated with modal operators $[U]$ and $\langle U \rangle$ where $U = A \times A$ is the full binary relation on $\mathfrak{A}$. Inverse (backward) modalities take the form of

$$\left(\langle E^{-1} \rangle \varphi \right)(x) \equiv \exists y \left(Eyx \wedge \varphi(y) \right),$$
$$\left([E^{-1}] \varphi \right)(x) \equiv \forall y \left(Eyx \rightarrow \varphi(y) \right).$$

DEFINITION 4.2. We denote as $\mathrm{ML}^{-\vee}$ the extension of basic modal logic by global and inverse modalities.

A characterisation theorem analogous to van Benthem's obtains. Indeed, we shall see that this also can be proved in ways that show the same characterisation to be valid in finite model theory: both classically and in finite model theory, the $\approx$ invariant fragment of first-order logic is precisely captured by $\mathrm{ML}^{-\vee}$.

As global quantification clearly allows us to jump anywhere inside the given structure, the tight localisation to a neighbourhood of the distinguished parameter (or free variable) is lost. For model constructions this means that a more global approach, which treats all nodes and their local neighbourhoods on the same footing, becomes necessary. This is where approximate, at least locally good, finite approximations to unravellings come into play. As these considerations are of some technical interest in their own right, we firstly isolate these model construction issues and then return to the proof of the above-mentioned characterisation result as an application.

DEFINITION 4.3. For an onto mapping $\pi\colon \hat{\mathfrak{A}} \to \mathfrak{A}$ between transition systems of the same type:

π is a *cover* of $\mathfrak{A}$ by $\hat{\mathfrak{A}}$, if
– for every unary predicate P: $\hat{a} \in P^{\hat{\mathfrak{A}}} \Leftrightarrow \pi(\hat{a}) \in P^{\mathfrak{A}}$,
– for every binary relation E, π provides *lifts*: if $(a, a') \in E^{\mathfrak{A}}$ then there are E-edges in $\hat{\mathfrak{A}}$ out of every node $\hat{a} \in \pi^{-1}(a)$ and into every node $\hat{a}' \in \pi^{-1}(a')$.
π is a *faithful cover* if all these lifts are unique.
π is a *bisimilar cover* if its graph $\{(\hat{a}, a)\colon \pi(\hat{a}) = a\}$ is a global two-way bisimulation between $\hat{\mathfrak{A}}$ and $\mathfrak{A}$.

Note that covers need not be homomorphisms. Unlike bisimilar companions, covers merely provide a kind of one-sided simulation, satisfying

the back-requirements, but not necessarily the forth-requirements of bisimulations. Bisimilar covers, however, are homomorphisms by virtue of the
forth-requirements. Even bisimilar covers need not be faithful. Faithful covers always allow unique lifts of (undirected) paths to any node in the fibre
above any given node in the path.

Recall the bisimilar unravellings discussed in Section 1.4. Consider the
unravelling of $\mathfrak{A}$, a, based on the set of all paths from a in $\mathfrak{A}$; let π be the natural
projection from this unravelling to $\mathfrak{A}$, which maps a path to the last vertex on
that path. This projection induces a bisimulation, though typically not a two-
way bisimulation, nor a global one unless every node of $\mathfrak{A}$ is reachable from a.
Instead of the ordinary (directed) unravelling, however, we may consider a
two-way unravelling based on all *undirected* paths, in which edges may be
traversed backwards as well as forwards. The corresponding projection π
induces a two-way bisimulation. Taking the disjoint union of such two-way
unravellings from at least one element from each connected component of
$\mathfrak{A}$, π becomes a bisimilar cover; and even a faithful bisimilar cover if we
only admit paths that do not traverse the same edge in opposite directions in
consecutive steps. In this sense, two-way unravellings generally can provide
faithful bisimilar covers, albeit typically infinite ones.

A fuller justification for the following proviso is given, with explicit en-
coding prescriptions, in [17, 19]. The essential steps of the constructions
discussed below remain the same but become somewhat more transparent in
the restricted setting. The assumption laid out in the proviso can essentially
be made without loss of generality, as the construction of bisimilar covers is
compatible with suitable encodings of arbitray edge relations in structures that
do conform to the proviso.

PROVISO 4.4. *For the remainder of this section we assume that there is just
one single, strictly asymmetric edge relation.*

We are interested in bisimilar covers of finite transition systems by finite
transition systems that are at least locally acyclic, i.e., do not have short
(undirected) cycles. An *undirected cycle* of length $k \geq 3$ in $\mathfrak{A}$ is a sequence of
nodes $a_0, a_1, \ldots, a_{k-1}$ (cyclically indexed by $\mathbb{Z}_k$) such that for every i either
(a_i, a_{i+1}) or (a_{i+1}, a_i) is an edge and $a_i \neq a_{i+2}$.

DEFINITION 4.5.   (i) The *girth* of a vertex a in a transition system $\mathfrak{A}$,
girth$(a, \mathfrak{A})$, is the minimal length of an undirected cycle through a.
(ii) The *girth* of a transition system $\mathfrak{A}$, girth$(\mathfrak{A})$, is the minimal length of any
cycle in $\mathfrak{A}$.
(iii) $\mathfrak{A}$ is *k-acyclic* if its girth is at least k.

Note that k-acyclicity is related to *local acyclicity* in the sense that the restrictions of the Gaifman graph of $\mathfrak{A}$ to ℓ-neighbourhoods of any $a \in \mathfrak{A}$ will
be acyclic if $\mathfrak{A}$ is k-acyclic for some $k \geq 2\ell$: ℓ-locally one will not see any

(undirected) cycles. It is necessary for our purposes to consider undirected cycles, precisely because we shall need local acyclicity in the Gaifman graph, which corresponds to the symmetrised version of the transition relations. It may be worth pointing out that short directed cycles are much more easily avoided. In fact one would merely have to use a truncated partial unravelling to sufficient depth and introduce new edges from the frontier of the unravelled part to suitable nodes closer to the root. The point of the rather more elaborate proofs required for the following theorem is that we want to avoid even short undirected cycles, or short cycles in the Gaifman graph of the desired companion structure. Both of the constructions given below supersede the construction from [17], which is hyper-exponential in nature.

THEOREM 4.6. *For any fixed $k \geqslant 3$: Any finite transition system $\mathfrak{A}$ possesses a faithful bisimilar cover $\pi \colon \hat{\mathfrak{A}} \to \mathfrak{A}$ by a finite k-acyclic transition system $\hat{\mathfrak{A}}$. For fixed k, the size of $\hat{\mathfrak{A}}$ can be polynomially bounded in terms of the size of $\mathfrak{A}$.*

It is clear that the size of any cover as stated in the theorem has to be exponential in k: if $\ell \leqslant k/2$ then the ℓ-neighbourhood of any $\hat{a}$ in $\hat{\mathfrak{A}}$ is isomorphic with the ℓ-neighbourhood of $\pi(a)$ in a two-way unravelling of $\mathfrak{A}$ from $\pi(a)$ (since both neighbourhoods are acyclic and faithful bisimilar covers of each other); the size of the latter is in general exponential in ℓ.

4.1. A game oriented construction. We reduce the task of finding a bisimilar cover $\hat{\mathfrak{A}}$ to that of finding faithful covers $\mathfrak{A}^*$ that are suitable locally in the vicinity of individual nodes. This construction is inspired by the powerset construction in the determinisation of non-deterministic automata. In the associated bisimilar cover $\hat{\mathfrak{A}}$ we keep track of all possible lifts of paths to the faithful cover $\mathfrak{A}^*$ to make sure that a path in $\hat{\mathfrak{A}}$ cannot cycle back as long as at least one of the lifts into $\mathfrak{A}^*$ does not. In other words we need only make sure that at least one 'leaf' of the faithful cover $\mathfrak{A}^*$ above every node is locally acyclic. This construction is exponential and does not yield a polynomial size cover; the alternative group theoretic construction indicated below, however, will yield a polynomial size cover.

Let $\rho \colon \mathfrak{A}^* \to \mathfrak{A}$ be a faithful cover. It is easy to see that the cardinality of the fibres $\rho^{-1}(a)$ must be constant within each connected component of $\mathfrak{A}$. Indeed, the set of all lifts to $\mathfrak{A}^*$ of a single edge (a, a') of $\mathfrak{A}$ induces a bijection between $\rho^{-1}(a)$ and $\rho^{-1}(a')$. For the rest of the construction we assume w.l.o.g. that $\mathfrak{A}$ is weakly connected. Let $\rho \colon \mathfrak{A}^* \to \mathfrak{A}$ be a faithful cover of finite multiplicity N. We obtain a bisimilar cover for $\mathfrak{A}$ as follows.

$$\hat{A} := \big\{ (s \colon N \to A^*) \colon s \text{ one-to-one, } \rho \circ s \text{ constant} \big\}.$$

The conditions on s say that it is a bijection between N and some fibre $\rho^{-1}(a)$; intuitively each s can be used as an instantaneous description in tracking all *possible* positions in a one-sided simulation of $\mathfrak{A}$-behaviours in $\mathfrak{A}^*$. We let $\pi \colon \hat{A} \to A$ be the natural projection induced by ρ.

In $\hat{\mathfrak{A}}$ we now put an edge from s to s' iff there is an edge from $a := \pi(s)$ to $a' := \pi(s')$ and if

$$s' \circ s^{-1} \colon \rho^{-1}(a) \longrightarrow \rho^{-1}(a')$$

is the bijection induced by the lifts of (a, a') to $\mathfrak{A}^*$. It is now easy to check that $\pi \colon \hat{\mathfrak{A}} \to \mathfrak{A}$ is a faithful bisimilar cover.

Moreover, the girth of an element s of $\hat{\mathfrak{A}}$ above $a = \pi(s)$ can be bounded as follows. If $s_0, s_1, \ldots, s_{k-1}$ forms an undirected cycle of length k at $s_0 = s$ in $\hat{\mathfrak{A}}$, then $s_0(m), s_1(m), \ldots, s_{k-1}(m)$ is an undirected cycle in $\mathfrak{A}^*$, for every $m \in N$. This is because each edge $(s_i(m), s_{i+1}(m))$ (or its inverse) is a lift of the edge $(\pi(s_i), \pi(s_{i+1}))$ (or its inverse) to $\mathfrak{A}^*$. As the $s_0(m)$ list all the elements in $\pi^{-1}(a)$, the girth of s in $\hat{\mathfrak{A}}$ is at lesat as large as the girth of any of the elements of $\pi^{-1}(s)$ in $\mathfrak{A}^*$:

$$\text{girth}(s, \hat{\mathfrak{A}}) \geqslant \max \left\{ \text{girth}(a^*, \mathfrak{A}^*) \colon \rho(a^*) = \pi(s) \right\}.$$

It remains to obtain faithful covers $\mathfrak{A}^*$ that possess for each $a \in \mathfrak{A}$ at least one covering element a^* above a whose ℓ-neighbourhood is acyclic (where $\ell = \lceil k/2 \rceil$ to make $\hat{\mathfrak{A}}$ k-acyclic). Since we are only after covers, not bisimilar covers, it suffices to cover any complete tournament of the same size as $\mathfrak{A}$ in the required manner—this cover will automatically also be good for $\mathfrak{A}$, up to re-direction of edges where required. We therefore fix $\mathfrak{A}_n$ to be a complete tournament on n vertices, $\mathfrak{A}_n = (n = \{0, \ldots, n-1\}, E = <)$.

LEMMA 4.7. *For every ℓ, $\mathfrak{A}_n$ admits a finite faithful cover in which every vertex is covered by at least one vertex whose ℓ-neighbourhood is acyclic.*

PROOF. Let $\mathfrak{A} := \mathfrak{A}_n$ and fix ℓ. For every $i \in n$ let $\mathfrak{A}^{(i)}$ be the restriction of the two-way unravelling of $\mathfrak{A}$ from i to the $(\ell + 1)$-neighbourhood of the root i. We let $\rho \colon \mathfrak{A}^{(i)} \to \mathfrak{A}$ be the natural projection. By the *rim* of $\mathfrak{A}^{(i)}$ we mean the set of vertices at distance ℓ from the root vertex i. The desired $\mathfrak{A}^*$ will simply be the disjoint union of the $\mathfrak{A}^{(i)}$ for $i \in n$ with additional edges joining rim vertices. This immediately guarantees that the ℓ-neighbourhood of the root i in $\mathfrak{A}^{(i)}$ is isomorphic to the ℓ-neighbourhood of i in $\mathfrak{A}^*$, and therefore acyclic. We also denote by ρ the union of the individual ρ, defined on the union of the $\mathfrak{A}^{(i)}$.

The missing edges on the rim can be filled in arbitrarily in such a way that for every $i < j$: every rim element in $\rho^{-1}(i)$ which is not already linked (inside its own $\mathfrak{A}^{(i)}$ that is) to an element in $\rho^{-1}(j)$, is joined to precisely one other rim element in $\rho^{-1}(j)$ which is not already linked to an element in $\rho^{-1}(i)$. This is possible as there are exactly the same number of rim elements of each kind across all the $\mathfrak{A}^{(i)}$, by symmetry. This turns $\rho \colon \mathfrak{A}^* \to \mathfrak{A}$ into a faithful cover.   ⊣

4.2. A simple group theoretic construction. Instead of adherence to a (bi-)simulation game intuition we may directly construct a bisimilar cover as a bundle over $\mathfrak{A}$ for whose fibres we use an arbitrary abstract group. Using Cayley groups of large girth will produce covers of large girth. This construction has the benefit of simplicity and, using groups constructed by Margulis and Imrich, can also be kept polynomial for each fixed value of k.

Let G be any group and suppose the edge set E of $\mathfrak{A}$ is injectively embedded into a subset of G not containing both g and g^{-1} for any $g \in G$. For simplicity we identify E with a corresponding subset $E \subseteq G$. Let

$$\hat{A} := A \times G, \quad \pi \colon \hat{A} \longrightarrow A \text{ the natural projection.}$$

In $\hat{\mathfrak{A}}$ we now put an edge from (a, g) to (a', g') iff $(a, a') = e \in E$ and $g' = g \cdot e$. It is obvious that $\pi \colon \hat{\mathfrak{A}} \to \mathfrak{A}$ is a faithful bisimilar cover. The girth of $\hat{\mathfrak{A}}$ cannot be less than the girth of the Cayley graph associated with $E \subseteq G$. This is the graph whose vertices are elements of the subgroup $H \subseteq G$ generated by E in G, with an edge between g and g' if $g' = g \cdot h$ for some $h \in E \cup E^{-1}$. Again, this relationship between the girths stems from the fact that any (undirected) cycle in $\hat{\mathfrak{A}}$ induces a cycle in the Cayley graph.

So in this case it remains to provide Cayley graphs of large girth, an issue which has independently been studied in the context of algebraic methods in combinatorial graph theory [1]. Alon [1] attributes to Biggs [4] the following simple construction of n-regular Cayley graphs of exponential size for fixed girth. For our purposes, n is the number of generators and corresponds to $|E|$. Let T_n^ℓ be the complete symmetric tree graph of degree n and depth ℓ, with a node 0 at the root. Let the edges of T_n^ℓ be coloured $1, \ldots, n$ such that every internal node is incident with exactly one edge of each colour. With colour i associate a permutation $i \colon T_n^\ell \to T_n^\ell$, which swaps all vertex pairs (t, t') that form an edge of colour i. Let G be the permutation group generated by these n generators. Any non degenerate composition of ℓ generators takes the root to a leaf, whence any non-degenerate generator sequence that fixes the root must have length greater than 2ℓ. In other words, the girth of the Cayley graph of G with respect to generators $1, \ldots, n$ is greater than 2ℓ.

Slightly more involved explicit group theoretic constructions, due to Imrich [13] and Margulis [15], yield smaller Cayley graphs of given degree and girth. In fact, Imrich's construction realises an asymptotically near optimal dependence of the girth on degree and size. In our context, these Cayley graphs give rise to bisimilar covers whose size is polynomial in the size of the given structure, for any fixed acyclicity k.

THEOREM 4.8 (Margulis, Imrich). *For every n and ℓ there are n-regular Cayley graphs of girth greater than 2ℓ in size $\mathcal{O}(n^{c\ell})$, c a fixed constant.*

§5. Characterising $\approx$ invariant first-order logic.

THEOREM 5.1. *Both in the sense of classical and of finite model theory:* $\mathrm{FO}/{\approx} \equiv \mathrm{ML}^{-\forall}$. *I.e., for any $\varphi(x) \in \mathrm{FO}$: $\varphi(x)$ is invariant under global* *two-way bisimulation* (*over finite transition systems*) *if and only it is equivalently expressible in* $ML^{-\forall}$.

While this characterisation result is exactly as (classically) expected, its proof illustrates several interesting features of the approach inspired by finite model theory: the more combinatorial and also rather more constructive nature of the argument; and a strategy that is exactly orthogonal to the one in the generic classical proof (as outlined in Section 2.1 above; that proof applies verbatim to the classical case of this theorem).

5.1. Gaifman locality. Recall the definition of Gaifman distance, Gaifman neighbourhoods and ℓ-local formulae from Definitions 3.1, 3.2, and 3.3.

DEFINITION 5.2. A subset of $\mathfrak{A}$ is *ℓ-scattered* if the mutual distance between any two distinct members of the set is at least 2ℓ.

A *basic local sentence*—for some locality radius ℓ, an ℓ-local formula $\psi(x)$, and some $n \geqslant 1$—is a sentence that asserts the existence of an ℓ-scattered set of size n whose elements all satisfy ψ.

The following classical result of Gaifman [8] says that every first-order formula is essentially local. See [6] for a textbook treatment.

THEOREM 5.3 (Gaifman). *Any first-order formula $\varphi(x)$ is equivalent to one that is a Boolean combination of local formulae and basic local sentences.*

Let us say that a first-order formula $\varphi(x)$ has *locality rank* ℓ, *local quantifier rank* q, and *scattering rank* n, if it has a representation according to Gaifman's theorem in which all constituent local formulae are ℓ-local, of quantifier rank q, and all its basic local sentences speak about k-scattered sets of sizes m, where $m \leqslant n$ and $k \leqslant \ell$.

Note that Lemma 3.5, in this terminology, says that any $\varphi(x)$ that is invariant under disjoint sums has locality rank 2^q, where $q = \mathrm{qr}(\varphi)$ ($=$ local quantifier rank), and scattering rank 0 (i.e., no basic local sentences are required).

Global (two-way) bisimulation, however, does not imply invariance under disjoint sums. But it still implies invariance under disjoint copies, i.e., $q \cdot \mathfrak{A}, a \approx \mathfrak{A}, a$, where as above $q \cdot \mathfrak{A}$ stands for the q-fold sum of (isomorphic copies of) $\mathfrak{A}$. The following is the natural ramification of Gaifman's theorem for this setting. A detailed proof can be found in [17, 19]; it applies Gaifman's theorem and shows how to eliminate constituents that make assertions about non-trivial scattered sets.

PROPOSITION 5.4. *Both in the sense of classical and finite model theory: if $\varphi(x) \in$ FO is invariant under disjoint copies, then φ has scattering rank 1, i.e., the only basic local sentences needed are plain existentially quantified local formulae.*

In the following paragraph we use Gaifman locality of a given $\approx$ invariant formula $\varphi(x)$ to capture an approximate, local level of elementary equivalence strong enough to preserve φ.

5.2. Upgrading ℓ-bisimulation towards elementary equivalence. Let $\doteq$ be some equivalence relation between structures, coarser than elementary equivalence $\equiv$ and serving as an approximation to $\equiv$. We say that $\leftrightarrows^\ell$ can be *upgraded* to $\doteq$ modulo $\leftrightarrows$, if for any (finite) $\mathfrak{A}, a \leftrightarrows^\ell \mathfrak{B}, b$ there are (finite) $\tilde{\mathfrak{A}}, a \leftrightarrows \mathfrak{A}, a$ and $\tilde{\mathfrak{B}}, b \leftrightarrows \mathfrak{B}, b$ such that $\tilde{\mathfrak{A}}, a \doteq \tilde{\mathfrak{B}}, b$:

$$
\begin{array}{ccc}
\mathfrak{A}, a & \xrightarrow{\;\;\leftrightarrows^\ell\;\;} & \mathfrak{B}, b \\[2pt]
\Big\updownarrow{\scriptstyle\leftrightarrows} & & \Big\updownarrow{\scriptstyle\leftrightarrows} \\[2pt]
\tilde{\mathfrak{A}}, a & \xrightarrow{\;\;\doteq\;\;} & \tilde{\mathfrak{B}}, b
\end{array}
$$

If $\doteq$ is sufficiently strong to preserve $\varphi(x)$, then the diagram shows that $\leftrightarrows$ invariance of φ implies $\leftrightarrows^\ell$ invariance of φ. This is exactly as required for Theorem 5.1 according to the discussion in Section 1; see in particular the crucial 'compactness' properties (3) and (4) there.

The approximation levels of $\equiv$ needed in our applications are the following. The parameters ℓ, q, and n in $\equiv^{(\ell)}_{q,n}$ are modelled after the levels of locality rank, local quantifier rank, and scattering rank of formulae in Gaifman form.

DEFINITION 5.5. $\mathfrak{A}, a \equiv^{(\ell)}_{q,n} \mathfrak{B}, b$ if for every $k \leqslant \ell$, for every k-local formula $\psi(x)$ of quantifier rank q, and for every $m \leqslant n$:
(i) $\mathfrak{A} \models \psi[a] \Leftrightarrow \mathfrak{B} \models \psi[b]$.
(ii) $\mathfrak{A}$ has a k-scattered subset of size m for ψ iff $\mathfrak{B}$ has.

Note that $\equiv^{(\ell)}_q$ of Definition 3.4 is recovered in the special case of $\equiv^{(\ell)}_{q,0}$. Since our $\varphi(x)$ is $\approx$ invariant, and hence invariant under disjoint copies, we may use Proposition 5.4 to work with scattering rank 1, and in fact with $\equiv^{(\ell)}_{q,1}$ where ℓ is the locality rank, q the local quantifier rank of φ.

LEMMA 5.6. *Both classically and in finite models, $\approx^\ell$ can be upgraded modulo $\approx$ to $\equiv^{(\ell)}_{q,1}$, for any q.*

PROOF. The main step in upgrading is achieved by using ℓ-locally acyclic covers, which guarantee acyclicity of the relevant ℓ-neighbourhoods in both covers. There remains an issue of multiplicities, however: a first-order formula of quantifier rank q can distinguish multiplicities less than q. We therefore precede the acyclic cover construction by a process that boosts all multiplicities to at least q.

Let $\mathfrak{A}, a \approx^\ell \mathfrak{B}, b$ be given. Let $q \otimes \mathfrak{A}$ be the transition system over universe $q \times A$ with $((i, a'), (j, a'')) \in E^{q \otimes \mathfrak{A}}$ iff $(a', a'') \in E^{\mathfrak{A}}$. Clearly $q \otimes \mathfrak{A} \approx \mathfrak{A}$. Moreover, all in-degrees and out-degrees in $q \otimes \mathfrak{A}$ are multiples of q, and this carries over to any faithful ℓ-acyclic bisimilar cover $\tilde{\mathfrak{A}}$ of $q \otimes \mathfrak{A}$, $\tilde{\pi} \colon \tilde{\mathfrak{A}} \to q \otimes \mathfrak{A}$. Let $\pi \colon \tilde{\mathfrak{A}} \to \mathfrak{A}$ be the composition of $\tilde{\pi}$ with the natural projection from $q \otimes \mathfrak{A}$ to $\mathfrak{A}$; we identify a with some element from $\pi^{-1}(a)$ and clearly have $\mathfrak{A}, a \approx \tilde{\mathfrak{A}}, a$. If $\tilde{\mathfrak{B}}$ is similarly obtained from $\mathfrak{B}$, we claim that $\tilde{\mathfrak{A}}, a \equiv_{q,1}^{(\ell)} \tilde{\mathfrak{B}}, b$. Indeed, it suffices to show that $\tilde{\mathfrak{A}}, a' \approx^{\ell} \tilde{\mathfrak{B}}, b'$ implies that $\tilde{\mathfrak{A}} {\restriction} U^{\ell}(a'), a' \equiv_q \tilde{\mathfrak{B}} {\restriction} U^{\ell}(b'), b'$. But this follows with a simple Ehrenfeucht-Fraïssé argument. $\tilde{\mathfrak{A}} \restriction U^{\ell}(a')$ and $\tilde{\mathfrak{B}} {\restriction} U^{\ell}(b')$ are acyclic with multiplicities greater than q, and the second player will always find suitable branches (new where needed) to respond for q rounds. $\dashv$

We remark that explicit appeal to Proposition 5.4 can be avoided in the proof of Theorem 5.1, in favour of a stronger upgrading as follows.

LEMMA 5.7. *For any n, $\approx^{\ell}$ can be upgraded to $\equiv_{q,n}^{(\ell)}$ modulo $\approx$, classically and in finite models.*

PROOF. Let $\mathfrak{A}, a \approx^{\ell} \mathfrak{B}, b$. With Lemma 5.6 we find (finite) $\hat{\mathfrak{A}} \approx \mathfrak{A}$ and $\hat{\mathfrak{B}} \approx \mathfrak{B}$ such that $\hat{\mathfrak{A}}, a \equiv_{q,1}^{(\ell)} \hat{\mathfrak{B}}, b$. It follows that $\tilde{\mathfrak{A}}, a \equiv_{q,n}^{(\ell)} \tilde{\mathfrak{B}}, b$, if we let $\tilde{\mathfrak{A}} := n \cdot \hat{\mathfrak{A}}$ and $\tilde{\mathfrak{B}} := n \cdot \hat{\mathfrak{B}}$. Clearly still $\tilde{\mathfrak{A}}, a \approx \mathfrak{A}, a$ and $\tilde{\mathfrak{B}}, b \approx \mathfrak{B}, b$. $\dashv$

It is interesting to observe how the constructive proof of our characterisation theorem lined out above is truly orthogonal to the generic classical proof as sketched in Section 2.1. Both proofs establish that for any first-order $\varphi(x)$:

$$\varphi \approx \text{invariant} \implies \varphi \approx^{\ell} \text{invariant for some } \ell,$$

cf. (3) in Section 2. The classical proof would achieve this (indirectly) through a compactness argument that upgrades the sequence of the $\approx^{\ell}$, first to its limit $\approx^{\omega}$ and then to $\approx$, and preserves φ in the process on the basis of its first-order nature. The alternative proof which also works in finite model theory, on the other hand, upgrades $\approx^{\ell}$ (where ℓ is the locality rank of φ) to an approximation of elementary equivalence that is sufficient to preserve φ, and preserves φ in the process because the entire construction fully respects $\approx$.

§6. Outlook and further ramifications.

The proof technique illustrated above for the characterisation of the first-order logic of properties invariant under two-way global bisimulation invariance, can be adapted to work for other related characterisation results.

Almost as it stands, it can be used to infer a similar characterisation of properties of (finite) transition systems invariant under *guarded bisimulation*. As shown in [17], the classical characterisation theorem for the *guarded fragment*, due to Andréka, van Benthem and Németi [2], is also valid over finite transition systems. It remains open, however, whether a similar finite model theory characterisation obtains also in the setting of general relational structures with predicates of arities greater than 2.

In the modal setting, some modifications are necessary in order to obtain a similar characterisation for the interesting class of properties that are invariant under global (but forward) bisimulation. As one might expect, these are captured precisely by the extension of basic modal logic by a universal modality. The proof however has to take into account the apparent mismatch between the inherently symmetric notion of locality in Gaifman's theorem and the forward direction of the notion of bisimulation. This extension is presented in [19].

Another natural class of transition systems is that of (possibly just finite) connected structures $\mathfrak{A}$, a, i.e., those in which every node is reachable from the source node a. In restriction to such structures, the distinction between global and ordinary bisimulation equivalence is lost. Correspondingly, even the extension of ML by global quantification is preserved under ordinary bisimulation over connected structures. A number of related characterisation theorems—also w.r.t. other natural frame conditions, and in particular transitivity—are explored in [5].

The construction of certain kinds of finite bisimilar covers has also been successfully carried out in the general guarded scenario, where at least one aspect of the usually infinite guarded unravellings of relational structured is recovered in a finite model construction. It remains open how much acyclicity (in the hypergraph theoretical rather than the graph theoretical sense) can be achieved in finite guarded covers. The construction given in [12] breaks up cliques in the Gaifman graph exactly as an unravelling would. See [12] also for interesting applications of that construction not just to the finite model theory of guarded logics, but also to issues of classical model theory to do with extension properties for partial isomorphisms in finite structures.

REFERENCES

[1] N. ALON, *Tools from higher algebra*, **Handbook of Combinatorics** (R. Graham et al., editors), vol. II, North-Holland, 1995, pp. 1749–1783.

[2] H. ANDRÉKA, J. VAN BENTHEM, and I. NÉMETI, *Modal languages and bounded fragments of predicate logic*, **Journal of Philosophical Logic**, vol. 27 (1998), pp. 217–274.

[3] J. BARWISE and J. VAN BENTHEM, *Interpolation, preservation, and pebble games*, **The Journal of Symbolic Logic**, vol. 64 (1999), pp. 881–903.

[4] N. BIGGS, *Cubic graphs with large girth*, **Annals of the New York Academy of Sciences** (G. Blum et al., editors), vol. 555, New York Acad. Sci., New York, 1989, pp. 56–62.

[5] A. DAWAR and M. OTTO, *Modal characterisation theorems over special classes of frames*, **Proceedings of 20th Annual IEEE Symposium on Logic in Computer Science LICS'05**, 2005, pp. 21–30.

[6] H.-D. EBBINGHAUS and J. FLUM, **Finite Model Theory**, 2nd ed., Springer, Berlin, 1999.

[7] H.-D. EBBINGHAUS, J. FLUM, and W. THOMAS, **Mathematical Logic**, Springer, New York, 1994.

[8] H. GAIFMAN, *On local and nonlocal properties*, **Logic Colloquium '81** (J. Stern, editor), North Holland, Amsterdam, 1982, pp. 105–135.

[9] E. GRÄDEL, *Why are modal logics so robustly decidable?*, **Bulletin of the European Association for Theoretical Computer Science**, vol. 68 (1999), pp. 90–103.

[10] E. GRÄDEL and M. OTTO, *On logics with two variables*, **Theoretical Computer Science**, vol. 224 (1999), pp. 73–113.

[11] M. HENNESSY and R. MILNER, *Algebraic laws of indeterminism and concurrency*, **Journal of the ACM**, vol. 32 (1985), pp. 137–162.

[12] I. HODKINSON and M. OTTO, *Finite conformal hypergraph covers and Gaifman cliques in finite structures*, **The Bulletin of Symbolic Logic**, vol. 9 (2003), pp. 387–405.

[13] W. IMRICH, *Explicit construction of regular graphs without small cycles*, **Combinatorica**, vol. 4 (1984), pp. 53–59.

[14] D. JANIN and I. WALUKIEWICZ, *On the expressive completeness of the propositional mu-calculus with respect to monadic second order logic*, **Proceedings of 7th International Conference on Concurrency Theory CONCUR '96**, Lecture Notes in Computer Science, vol. 1119, Springer-Verlag, Berlin, 1996, pp. 263–277.

[15] G. MARGULIS, *Graphs without short cycles*, **Combinatorica**, vol. 2 (1982), pp. 71–78.

[16] M. OTTO, **Bounded Variable Logics and Counting**, Lecture Notes in Logic, vol. 9, Springer, Berlin, 1997.

[17] ———, *Modal and guarded characterisation theorems over finite transition systems*, **Proceedings of 17th Annual IEEE Symposium on Logic in Computer Science LICS '02**, 2002, pp. 371–380.

[18] ———, *An elementary proof of the van Benthem–Rosen characterisation theorem*, TUD online preprint no. 2342, 2004.

[19] ———, *Modal and guarded characterisation theorems over finite transition systems*, **Annals of Pure and Applied Logic**, vol. 130 (2004), pp. 173–205. extended journal version of [17].

[20] D. PARK, *Concurrency and automata on infinte sequences*, **Proceedings of 5th GI Conference**, Springer-Verlag, 1981, pp. 176–183.

[21] B. POIZAT, **A Course in Model Theory**, Springer-Verlag, New York, 2000.

[22] E. ROSEN, *Modal logic over finite structures*, **Journal of Logic, Language and Information**, vol. 6 (1997), pp. 427–439.

[23] J. VAN BENTHEM, **Modal Correspondence Theory**, Ph.D. thesis, University of Amsterdam, 1976.

[24] ———, **Modal Logic and Classical Logic**, Bibliopolis, Napoli, 1983.

[25] M. VARDI, *Why is modal logic so robustly decidable?*, **Descriptive Complexity and Finite Models** (N. Immerman, P. Kolaitis. and P. Deussen, editors), DIMACS Series in Discrete Mathematics and Theoretical Computer Science, vol. 31. AMS, Providence, 1997, pp. 149–184.

FACHBEREICH MATHEMATIK
TECHNISCHE UNIVERSITÄT DARMSTADT
64289 DARMSTADT, GERMANY
E-mail: otto@mathematik.tu-darmstadt.de
URL: www.mathematik.tu-darmstadt.de/~otto

CHOICE PRINCIPLES IN CONSTRUCTIVE AND CLASSICAL SET THEORIES

MICHAEL RATHJEN

Abstract. The objective of this paper is to assay several forms of the axiom of choice that have been deemed constructive. In addition to their deductive relationships, the paper will be concerned with metamathematical properties effected by these choice principles and also with some of their classical models.

§1. Introduction. Among the axioms of set theory, the axiom of choice is distinguished by the fact that it is the only one that one finds ever mentioned in workaday mathematics. In the mathematical world of the beginning of the 20th century, discussions about the status of the axiom of choice were important. In 1904 Zermelo proved that every set can be well-ordered by employing the axiom of choice. While Zermelo argued that it was self-evident, it was also criticized as an excessively non-constructive principle by some of the most distinguished analysts of the day. At the end of a note sent to the *Mathematische Annalen* in December 1905, Borel writes about the axiom of choice:

> *It seems to me that the objection against it is also valid for every reasoning where one assumes an arbitrary choice made an uncountable number of times, for such reasoning does not belong in mathematics.*
> ([9, pp. 1251–1252]; translation by H. Jervell, cf. [20, p. 96].)

Borel canvassed opinions of the most prominent French mathematicians of his generation — Hadamard, Baire, and Lebesgue — with the upshot that Hadamard sided with Zermelo whereas Baire and Lebesgue seconded Borel. At first blush Borel's strident reaction against the axiom of choice utilized in Cantor's new theory of sets is surprising as the French analysts had used and continued to use choice principles routinely in their work. However, in the context of 19th century classical analysis only the Axiom of Dependent

The research reported in this paper was supported by United Kingdom Engineering and Physical Sciences Research Council Grant GR/R 15856/01.

Logic Colloquium '02
Edited by Z. Chatzidakis, P. Koepke, and W. Pohlers
Lecture Notes in Logic, 27

Choices, **DC**, is invoked and considered to be natural, while the full axiom of choice is unnecessary and even has some counterintuitive consequences.

Unsurprisingly, the axiom of choice does not have a unambiguous status in constructive mathematics either. On the one hand it is said to be an immediate consequence of the constructive interpretation of the quantifiers. Any proof of $\forall x \in A \, \exists y \in B \, \phi(x, y)$ must yield a function $f : A \to B$ such that $\forall x \in A \, \phi(x, f(x))$. This is certainly the case in Martin-Löf's intuitionistic theory of types. On the other hand, it has been observed that the full axiom of choice cannot be added to systems of extensional constructive set theory without yielding constructively unacceptable cases of excluded middle (see [10] and Proposition 3.2). In extensional intuitionistic set theories, a proof of a statement $\forall x \in A \, \exists y \in B \, \phi(x, y)$, in general, provides only a function F, which when fed a proof p witnessing $x \in A$, yields $F(p) \in B$ and $\phi(x, F(p))$. Therefore, in the main, such an F cannot be rendered a function of x alone. Choice will then hold over sets which have a canonical proof function, where a constructive function h is a canonical proof function for A if for each $x \in A$, $h(x)$ is a constructive proof that $x \in A$. Such sets having natural canonical proof functions "built-in" have been called *bases* (cf. [36, p. 841]).

The objective of this paper is to assay several forms of the axiom of choice that have been deemed constructive. In addition to their deductive relationships, the paper will be concerned with metamathematical properties effected by these choice principles and also with some of their classical models. The particular form of constructivism adhered to in this paper is Martin-Löf's intuitionistic type theory (cf. [21, 22]). Set-theoretic choice principles will be considered as constructively justified if they can be shown to hold in the interpretation in type theory. Moreover, looking at set theory from a type-theoretic point of view has turned out to be valuable heuristic tool for finding new constructive choice principles.

The plan for the paper is as follows: After a brief review of the axioms of constructive Zermelo-Fraenkel set theory, **CZF**, in the second section, the third section studies the implications of full **AC** on the basis of **CZF**. A brief Section 4 addresses two choice principles which have always featured prominently in constructive accounts of mathematics, namely the axioms of countable choice and dependent choices. A stronger form of choice is the *presentation axiom*, **PAx** (also known as the *existence of enough projective sets*). **PAx** is the topic of Section 5. It asserts that every set is the surjective image of a set over which the axiom of choice holds. It implies countable choice as well as dependent choices. **PAx** is validated by various realizability interpretations and also by the interpretation of **CZF** in Martin-Löf type theory. On the other hand, **PAx** is usually not preserved under sheaf constructions. Moerdijk and Palmgren in their endeavour to find a categorical counterpart for constructive type theory, formulated a categorical form of an axiom of choice which they christened the *axiom of multiple choice*. The pivotal properties of this axiom are that it is preserved under the construction of sheaves and that it encapsulates "enough choice" to allow for the construction of categorical models of *CZF* plus the regular extension axiom, *REA*. Section 6 explores a purely set-theoretic version of the axiom of multiple choice, notated *AMC*, due to Peter Aczel and Alex Simpson. The main purpose of this section is to show that "almost all" models of *ZF* satisfy *AMC*. Furthermore, it is shown that in *ZF*, *AMC* implies the existence of arbitrarily large regular cardinals.

In the main, the corroboration for the constructivenes of *CZF* is owed to its interpretation in Martin-Löf type theory given by Aczel (cf. [1, 2, 3]). This interpretation is in many ways canonical and can be seen as providing *CZF* with a standard model in type theory. It will be recalled in Section 7. Except for the general axiom of choice, all the foregoing choice principles are validated in this model and don't add any proof-theoretic strength. In Section 8 it will be shown that an axiom of subcountability, which says that every set is the surjective image of a subset of ω, is also validated by this type of interpretation.

It is a natural desire to explore whether still stronger version of choice can be validated through this interpretation. Aczel has discerned several new principles in this way, among them are the $\Pi\Sigma - AC$ and $\Pi\Sigma W - AC$. In joint work with S. Tupailo we have shown that these are the strongest choice principles validated in type theory in the sense that they imply all the "mathematical" statements that are validated in type theory. Roughly speaking, the "mathematical statements" encompass all statements of workaday mathematics, or more formally, they are statements wherein the quantifiers are bounded by sets occurring in the cumulative hierarchy at levels $< \omega + \omega$. Section 9 reports on these findings. They also have metamathematical implications for the theories $CZF + \Pi\Sigma - AC$ and $CZF + REA + \Pi\Sigma W - AC$ such as the disjunction property and the existence property for mathematical statements, as will be shown in Section 10.

§2. Constructive Zermelo-Fraenkel set theory.

Constructive set theory grew out of Myhill's endeavours (cf. [26]) to discover a simple formalism that relates to Bishop's constructive mathematics as *ZFC* relates to classical Cantorian mathematics. Later on Aczel modified Myhill's set theory to a system which he called *Constructive Zermelo-Fraenkel Set Theory*, *CZF*.

DEFINITION 2.1 (Axioms of *CZF*). The language of *CZF* is the first order language of Zermelo-Fraenkel set theory, *LST*, with the non logical primitive symbol $\in$. *CZF* is based on intuitionistic predicate logic with equality. The set theoretic axioms of axioms of *CZF* are the following:

1. *Extensionality* $\forall a \, \forall b \, (\forall y \, (y \in a \leftrightarrow y \in b) \to a = b)$.
2. *Pair* $\forall a \, \forall b \, \exists x \, \forall y \, (y \in x \leftrightarrow y = a \lor y = b)$.

3. *Union* $\forall a \, \exists x \, \forall y \, (y \in x \leftrightarrow \exists z \in a \, y \in z)$.
4. *Restricted Separation scheme* $\forall a \, \exists x \, \forall y \, (y \in x \leftrightarrow y \in a \wedge \varphi(y))$,
   for every *restricted* formula $\varphi(y)$, where a formula $\varphi(x)$ is restricted, or
   Δ_0, if all the quantifiers occurring in it are restricted, i.e., of the form
   $\forall x \in b$ or $\exists x \in b$.
5. *Subset Collection scheme*

$$\forall a \, \forall b \, \exists c \, \forall u \, \big(\forall x \in a \, \exists y \in b \, \varphi(x, y, u) \rightarrow$$
$$\exists d \in c \, (\forall x \in a \, \exists y \in d \, \varphi(x, y, u) \wedge \forall y \in d \, \exists x \in a \, \varphi(x, y, u))\big)$$

   for every formula $\varphi(x, y, u)$.
6. *Strong Collection scheme*

$$\forall a \, \big(\forall x \in a \, \exists y \, \varphi(x, y) \rightarrow$$
$$\exists b \, (\forall x \in a \, \exists y \in b \, \varphi(x, y) \wedge \forall y \in b \, \exists x \in a \, \varphi(x, y))\big)$$

   for every formula $\varphi(x, y)$.
7. *Infinity*

$$\exists x \forall u \big[u \in x \leftrightarrow \big(0 = u \vee \exists v \in x (u = v \cup \{v\})\big)\big]$$

   where $y + 1$ is $y \cup \{y\}$, and 0 is the empty set, defined in the obvious
   way.
8. *Set Induction scheme*

$$(IND_\in) \qquad \forall a \, (\forall x \in a \, \varphi(x) \rightarrow \varphi(a)) \longrightarrow \forall a \, \varphi(a),$$

   for every formula $\varphi(a)$.

From Infinity, Set Induction, and Extensionality one can deduce that there
exists exactly one set x such that $\forall u[u \in x \leftrightarrow (0 = u \vee \exists v \in x(u = v \cup \{v\}))]$;
this set will be denoted by ω.

The Subset Collection scheme easily qualifies for the most intricate axiom
of **CZF**. It can be replaced by a single axiom in the presence of Strong
Collection.

DEFINITION 2.2. [1] For sets A, B let $^A B$ be the class of all functions with
domain A and with range contained in B. Let $mv(^A B)$ be the class of all sets
$R \subseteq A \times B$ satisfying $\forall u \in A \, \exists v \in B \, \langle u, v \rangle \in R$. A set C is said to be *full in*
$mv(^A B)$ if $C \subseteq mv(^A B)$ and

$$\forall R \in mv(^A B) \, \exists S \in C \, S \subseteq R.$$

The expression $mv(^A B)$ should be read as the class of *multi-valued functions*
(or *multi functions*) from the set A to the set B.
   Additional axioms we consider are:
Fullness: For all sets A, B there exists a set C such that C is full in $mv(^A B)$.

Exponentiation Axiom: This axiom (abbreviated **Exp**) postulates that for sets A, B the class of all functions from A to B forms a set.

$$\forall a \forall b \exists c \, \forall f \, \left[f \in c \leftrightarrow (f : a \rightarrow b) \right].$$

THEOREM 2.3. *Let* **CZF⁻** *be* **CZF** *without Subset Collection.*

(i) (**CZF⁻**) *Subset Collection and Fullness are equivalent.*

(ii) (**CZF⁻**) *Fullness implies Exponentiation.*

(iii) (**CZF⁻**) *The Power Set axiom implies Subset Collection.*

PROOF. (ii) is obvious. (iii) is obvious in view of (i). For (i) see [1] or [4, Theorem 3.12].   ⊣

On the basis of classical logic and basic set-theoretic axioms, **Exp** implies the Power Set Axiom. However, the situation is radically different when the underlying logic is intuitionistic logic.

In what follows we shall use the notions of proof-theoretic equivalence of theories and proof-theoretic strength of a theory whose precise definitions can be found in [29].

THEOREM 2.4. *Let* **KP** *be Kripke-Platek Set Theory* (*with the Infinity Axiom*) (*see* [5]).

(i) **CZF** *and* **CZF⁻** *are of the same proof-theoretic strength as* **KP** *and the classical theory* **ID**$_1$ *of non-iterated positive arithmetical inductive definitions. These systems prove the same* Π_2^0 *statements of arithmetic.*

(ii) *The system* **CZF** *augmented by the Power Set axiom is proof-theoretically stronger than classical Zermelo Set theory,* **Z** (*in that it proves the consistency of* **Z**).

(iii) **CZF** *does not prove the Power Set axiom.*

PROOF. Let **Pow** denote the Power Set axiom. (i) follows from [28, Theorem 4.14]. Also (iii) follows from [28] Theorem 4.14 as one easily sees that 2-order Heyting arithmetic has a model in **CZF** + **Pow**. Since second-order Heyting arithmetic is of the same strength as classical second-order arithmetic it follows that **CZF** + **Pow** is stronger than classical second-order arithmetic (which is much stronger than **KP**). But more than that can be shown. Working in **CZF** + **Pow** one can iterate the power set operation $\omega + \omega$ times to arrive at the set $V_{\omega+\omega}$ which is readily seen to be a model of intuitionistic Zermelo Set Theory, **Z**^{i}. As **Z** can be interpreted in **Z**^{i} by means of a double negation translation as was shown in [14, Theorem 2.3.2], we obtain (ii).   ⊣

In view of the above, a natural question to ask is whether **CZF** proves that $mv(^A B)$ is a set for all sets A and B. This can be answered in the negative as the following the result shows.

PROPOSITION 2.5. *Let* $\mathcal{P}(x) := \{u : u \subseteq x\}$, *and* **Pow** *be the Power Set axiom, i.e.,* $\forall x \exists y \; y = \mathcal{P}(x)$.

(i) (**CZF⁻**) $\forall A \forall B \, (mv(^A B) \text{ is a set}) \leftrightarrow$ **Pow**.

(ii) **CZF** *does not prove that* $mv(^A B)$ *is set for all sets* A *and* B.

PROOF. (i): We argue in CZF^-. It is obvious that Power Set implies that $mv(^A B)$ is a set for all sets A, B. Henceforth assume the latter. Let C be an arbitrary set and $D = mv(^C\{0, 1\})$, where $0 := \emptyset$ and $1 := \{0\}$. By our assumption, D is a set. To every subset X of C we assign the set $X^* := \{\langle u, 0 \rangle \mid u \in X\} \cup \{\langle z, 1 \rangle \mid z \in C\}$. As a result, $X^* \in D$. For every $S \in D$ let $pr(S)$ be the set $\{u \in C \mid \langle u, 0 \rangle \in S\}$. We then have $X = pr(X^*)$ for every set $X \subseteq C$, and thus

$$\mathcal{P}(C) = \{pr(S) \mid S \in D\}.$$

Since $\{pr(S) \mid S \in D\}$ is a set by Strong Collection, $\mathcal{P}(S)$ is a set as well.

(ii) follows from (i) and Theorem 2.4(iii).   ⊣

The first large set axiom proposed in the context of constructive set theory was the *Regular Extension Axiom*, *REA*, which was introduced to accommodate inductive definitions in *CZF* (cf. [1], [3]).

DEFINITION 2.6. A set c is said to be *regular* if it is transitive, inhabited (i.e., $\exists u\, u \in c$) and for any $u \in c$ and set $R \subseteq u \times c$ if $\forall x \in u\, \exists y\, \langle x, y \rangle \in R$ then there is a set $v \in c$ such that

$$\forall x \in u\, \exists y \in v\, \langle x, y \rangle \in R \wedge \forall y \in v\, \exists x \in u\, \langle x, y \rangle \in R.$$

We write $Reg(a)$ for 'a is regular'.

REA is the principle

$$\forall x\, \exists y\, (x \in y \wedge Reg(y)).$$

THEOREM 2.7. *Let KPi be Kripke-Platek Set Theory plus an axiom asserting that every set is contained in an admissible set (see [5]).*

(i) *$CZF + REA$ is of the same proof-theoretic strength as KPi and the subsystem of second-order arithmetic with Δ_2^1-comprehension and Bar Induction.*

(ii) *$CZF + REA$ does not prove the Power Set axiom.*

PROOF. (i) follows from [28, Theorem 5.13]. (ii) is a consequence of (i) and Theorem 2.4.   ⊣

DEFINITION 2.8. Another familiar intuitionistic set theory is *Intuitionistic Zermelo-Fraenkel Set Theory*, *IZF*, which is obtained from *CZF* by adding the Power Set Axiom and the scheme of Separation for all formulas.

What is the constructive notion of set that constructive set theory claims to be about? An answer to this question has been provided by Peter Aczel in a series of three papers on the type-theoretic interpretation of *CZF* (cf. [1, 2, 3]). These papers are based on taking Martin-Löf's predicative type theory as the most acceptable foundational framework of ideas to make precise the constructive approach to mathematics. The interpretation shows how the elements of a particular type V of the type theory can be employed to interpret the sets of set theory so that by using the Curry-Howard 'formulae as types' paradigm the theorems of constructive set theory get interpreted as provable propositions. This interpretation will be recalled in Section 6.

§3. *CZF* **plus general choice.** The *Axiom of Choice*, *AC*, asserts that for all sets I, whenever $(A_i)_{i \in I}$ is family of inhabited sets (i.e., $\forall i \in I \, \exists y \in A_i$), then there exists a function f with domain I such that $\forall i \in I \, f(i) \in A_i$.

A set I is said to be a *base* if the axioms of choice holds over I, i.e., whenever $(A_i)_{i \in I}$ is family of inhabited sets (indexed) over I, then there exists a function f with domain I such that $\forall i \in I \, f(i) \in A_i$.

Diaconescu [10] showed that the full Axiom of Choice implies certain forms of excluded middle. On the basis of *IZF*, *AC* implies excluded middle for all formulas, and hence *IZF* + *AC* = *ZFC*. As will be shown shortly that, on the basis of *CZF*, *AC* implies the restricted principle of excluded middle, *REM*, that is the scheme $\theta \vee \neg\theta$ for all restricted formulas. Note also that, in the presence of Restricted Separation, *REM* is equivalent to the decidability of $\in$, i.e., the axiom $\forall x \forall y \, (x \in y \vee x \notin y)$. To see that the latter implies *REM* let $a = \{x \in 1 : \phi\}$, where ϕ is Δ_0. Then $0 \in a \vee 0 \notin a$ by decidability of $\in$. $0 \in a$ yields ϕ while $0 \notin a$ entails $\neg\phi$.

LEMMA 3.1. *CZF* + *REM* ⊢ *Pow*.

PROOF. Let $0 := \emptyset$, $1 := \{0\}$, and $2 := \{0, 1\}$. Let B be a set. In the presence of *REM*, the usual proof that there is a one-to-one correspondence between subsets of B and the functions from B to 2 works. Thus, utilizing Exponentiation and Strong Collection, $\mathcal{P}(B)$ is a set.   ⊣

PROPOSITION 3.2.   (i) *CZF* + *full Separation* + *AC* = *ZFC*.
(ii) *CZF* + *AC* ⊢ *REM*.
(iii) *CZF* + *AC* ⊢ *Pow*.

PROOF. (i) and (ii) follow at once from Diaconescu [10] but for the readers convenience proofs will be given below. The proofs also slightly differ from [10] in that they are phrased in terms of equivalence classes and quotients.

(i): Let ϕ be an arbitrary formula. Define an equivalence relation $\sim_\phi$ on 2 by

$$a \sim_\phi b :\Longleftrightarrow a = b \vee \phi,$$
$$[a]_{\sim_\phi} := \{b \in 2 : a \sim_\phi b\},$$
$$2/_{\sim_\phi} := \{[0]_{\sim_\phi}, [1]_{\sim_\phi}\}.$$

Note that $[0]_{\sim_\phi}$ and $[1]_{\sim_\phi}$ are sets by full Separation and thus $2/_{\sim_\phi}$ is a set, too. One easily verifies that $\sim_\phi$ is an equivalence relation.

We have

$$\forall z \in 2/_{\sim_\phi} \, \exists k \in 2 \, (k \in z).$$

Using AC, there is a choice function f defined on $2/\sim_\phi$ such that

$$\forall z \in 2/\sim_\phi \left[f(z) \in 2 \wedge f(z) \in z\right],$$

in particular, $f([0]_{\sim_\phi}) \in [0]_{\sim_\phi}$ and $f([1]_{\sim_\phi}) \in [1]_{\sim_\phi}$. Next, we are going to exploit the important fact

(1) $\forall n, m \in 2 \, (n = m \vee n \neq m).$

As $\forall z \in 2/\sim_\phi [f(z) \in 2]$, we obtain

$$f([0]_{\sim_\phi}) = f([1]_{\sim_\phi}) \vee f([0]_{\sim_\phi}) \neq f([1]_{\sim_\phi})$$

by (1). If $f([0]_{\sim_\phi}) = f([1]_{\sim_\phi})$, then $0 \sim_\phi 1$ and hence ϕ holds. So assume $f([0]_{\sim_\phi}) \neq f([1]_{\sim_\phi})$. As ϕ would imply $[0]_{\sim_\phi} = [1]_{\sim_\phi}$ (this requires Extensionality) and thus $f([0]_{\sim_\phi}) = f([1]_{\sim_\phi})$, we must have $\neg\phi$. Consequently, $\phi \vee \neg\phi$.

(i) follows from the fact that **CZF** plus the schema of excluded middle for all formulas has the same provable formulas as **ZF**.

(ii): If ϕ is restricted, then $[0]_{\sim_\phi}$ and $[1]_{\sim_\phi}$ are sets by Restricted Separation. The rest of the proof of (i) then goes through unchanged.

(iii) follows from (ii) and Lemma 3.1,(i).   ⊣

REMARK 3.3. It is interesting to note that the form of AC responsible for **EM** is reminiscent of that used by Zermelo in his well-ordering proof of $\mathbb{R}$. AC enables one to pick a representative from each equivalence class in $2/\sim_\phi$. Being finitely enumerable and consisting of subsets of $\{0, 1\}$, $2/\sim_\phi$ is a rather small set, though. Adopting a pragmatic constructive stance on AC, one might say that choice principles are benign as long as they don't imply the decidability of $\in$ and don't destroy computational information. From this point of view, Borel's objection against Zermelo's usage of AC based on the size of the index set of the family is a non sequitur. As we shall see later, it makes constructive sense to assume that ${}^\omega\omega$ is a base, i.e., that inhabited families of sets indexed over the set of all functions from ω to ω possess a choice function. Indeed, as will be detailed in Section 9, this applies to any index set generated by the set formation rules of Martin-Löf type theory. The axiom of choice is (trivially) provable in Martin-Löf type theory on account of the propositions-as-types interpretation. (Allowing for quotient types, though, would destroy this feature.) The interpretation of set theory in Martin-Löf type theory provides an illuminating criterion for singling out the sets for which the axiom of choice is validated. Those are exactly the sets which have an *injective presentation* (see Definition 7.4 and Section 9) over a type. The canonical and, in general, non-injective presentation of $2/\sim_\phi$ is the function $\wp$ with domain $\{0, 1\}$, $\wp(0) = [0]_{\sim_\phi}$, and $\wp(1) = [1]_{\sim_\phi}$.

What is the strength of $CZF + AC$? From Theorem 2.4 and Proposition 3.2 it follows that $CZF + AC$ and $CZF + REM$ are hugely more powerful than CZF. In $CZF + AC$ one can show the existence of a model of $Z + AC$. Subset Collection is crucial here because $CZF^- + AC$ is not stronger than CZF^-. To characterize the strength of $CZF + AC$ we introduce an extension of Kripke-Platek set theory.

DEFINITION 3.4. Let $KP(\mathcal{P})$ be Kripke-Platek Set Theory (with Infinity Axiom) formulated in a language with a primitive function symbol $\mathcal{P}$ for the power set operation. The notion of Δ_0 formula of $KP(\mathcal{P})$ is such that they may contain the symbol $\mathcal{P}$. In addition to the Δ_0-Separation and Δ_0-Collection schemes for this expanded language, $KP(\mathcal{P})$ includes the defining axiom

$$\forall x \, \forall y \, \left[y \subseteq x \leftrightarrow y \in \mathcal{P}(x) \right]$$

for $\mathcal{P}$.

THEOREM 3.5.   1. CZF, CZF^-, $CZF^- + REM$, $CZF^- + AC$, and KP are of the same proof-theoretic strength. These systems prove the same Π_2^0 statements of arithmetic.

2. $CZF + REM$ and $KP(\mathcal{P})$ are of the same proof-theoretic strength, while $CZF + AC$ is proof-theoretically reducible to $KP(\mathcal{P}) + V = L$.

3. The strength of $CZF + AC$ and $CZF + REM$ resides strictly between Zermelo Set Theory and ZFC.

PROOF. (1): In view of Theorem 2.4 it suffices to show that $CZF^- + AC$ can be reduced to KP. This can be achieved by making slight changes to the formulae-as-classes interpretation of CZF in KP as presented in [28, Theorem 4.11]. The latter is actually a realizability interpretation, where the underlying computational structure (aka partial combinatory algebra or applicative structure) is the familiar Kleene structure, where application is defined in terms of indices of partial recursive functions, i.e., $App_{Kl}(e, n, m) := \{e\}(n) \simeq m$. Instead of the Kleene structure, one can use the applicative structure of Σ_1 partial (class) functions. By a Σ_1 partial function we mean an operation (not necessarily everywhere defined) given by relations of the form $\exists z \, \phi(e, x, y, z)$ where e is a set parameter and ϕ is a bounded formula (of set theory) not involving other free variables. It is convenient to argue on the basis of $KP + V = L$ which is a theory that is Σ_1-conservative over KP. Then there is a universal Σ_1 relation that parametrizes all Σ_1 relations. We the help of the universal Σ_1 relation and the Δ_1 wellordering $<_L$ of the constructible universe one defines the Σ_1 partial recursive set function with index e (see [5]). The formulae-as-classes interpretation of $CZF^- + AC$ in $KP + V = L$ is obtained by using indices of Σ_1 partial recursive set functions rather than partial recursive functions on the integers. Since these indices form a proper class it is no longer possible to validate Subset Collection. All the axioms of CZF^- can still be validated, and in addition, AC is validated because if an object with a certain property exists, the hypothesis $V = L$ ensures that a search along the ordering $<_L$ finds the $<_L$-least such. Space limitations (and perhaps laziness) prevent us from giving all the details.

(2): Again, the complete proof is too long to be included in this paper. However, a sketch may be sufficient. Let us first address the reduction of **CZF + REM** to **KP**($\mathcal{P}$). In **KP**($\mathcal{P}$) one can develop the theory of power E-recursive functions, which are defined by the same schemata as the E-recursive functions except that the function $\mathcal{P}(x) := \{u : u \subseteq x\}$ is thrown in as an initial function (cf. [25]). The next step is to mimic the recursive realizability interpretation of **CZF** in **KP** as given in [28]. In that interpretation, formulas of **CZF** were interpreted as types (mainly) consisting of indices of partial recursive functions and realizers were indices of partial recursive functions, too. Crucially, for the realizability interpretation at hand one has to use classes of indices of power E-recursive functions for the modelling of types and indices of power E-recursive functions as realizers. The details of the interpretation can be carried out in **KP**($\mathcal{P}$), and are very similar to the E-realizability techniques employed in [32]. Thereby it is important that the types associated to restricted set-theoretic formulas are interpreted as sets.

The interpretation of **CZF + AC** in **KP**($\mathcal{P}$) $+ V = L$ proceeds similarly to the foregoing, however, here we use indices of partial functions Σ_1-definable in the power set function as realizers. To realize **AC** we need this collection of functions to be closed under a search operator. This is were the hypothesis $V = L$ is needed.

For the reduction of **CZF + REM** to **KP**($\mathcal{P}$) one can use techniques from the ordinal analysis of **KP**. First note that the ordinals are linearly ordered in **CZF + REM**. The ordinal analysis of **KP** requires an ordinal representation system in which one can express the Bachmann-Howard ordinal. For reference purposes let this be the representation system $T(\Omega)$ of [35]. The first step is to develop a class size analogue of this representation system, notated OR, where the role of Ω is being played by the class of ordinals. Such a class size ordinal representation system has been developed in [27], Section 4. The next step consists in finding an analogue $RS(OR)$ of the infinitary proof system $RS(\Omega)$ of [27]. Let's denote the element of OR which has all ordinals as predecessors by $\hat{\Omega}$. Similarly as the case of $RS(\Omega)$ one uses the ordinals (i.e., the elements of OR preceding $\hat{\Omega}$) to build a hierarchy of set terms. The main difference here is that rather than modelling this hierarchy on the constructible hierarchy, one uses the cumulative hierarchy V_α to accommodate the power set function $\mathcal{P}$. The embedding of **KP**($\mathcal{P}$) $+$ **REM** into $RS(OR)$, the cut elimination theorems and the collapsing theorem for $RS(OR)$ are proved in much the same way as for $RS(\Omega)$ in [27]. Finally one has to code infinitary $RS(OR)$ derivations in **CZF + REM** and prove a soundness theorem similar to [27, Theorem 3.5] with L_α being replaced by V_α.

(2): We have already indicated that $V_{\omega+\omega}$ provides a set model for Z in $CZF + REM$. Using the reflection theorem of ZF one can show in ZF that there exists a cardinal κ such that $L_\kappa \models KP(\mathcal{P}) + V = L$. $\dashv$

REMARK 3.6. Employing Heyting-valued semantics, N. Gambino also showed (cf. [15, Theorem 5.1.4]) that $CZF^- + REM$ is of the same strength as CZF^-.

The previous results show that the combination of CZF and the general axiom of choice has no constructive justification in Martin-Löf type theory.

§4. **Old acquaintances.** In many a text on constructive mathematics, axioms of countable choice and dependent choices are accepted as constructive principles. This is, for instance, the case in Bishop's constructive mathematics (cf. [7] as well as Brouwer's intuitionistic analysis (cf. [36, Chapter 4, Section 2]). Myhill also incorporated these axioms in his constructive set theory [26].

The weakest constructive choice principle we shall consider is the *Axiom of Countable Choice*, AC_ω, i.e., whenever F is a function with domain ω such that $\forall i \in \omega\ \exists y \in F(i)$, then there exists a function f with domain ω such that $\forall i \in \omega\ f(i) \in F(i)$.

Let xRy stand for $\langle x, y \rangle \in R$. A mathematically very useful axiom to have in set theory is the *Dependent Choices Axiom*, DC, i.e., for all sets a and (set) relations $R \subseteq a \times a$, whenever

$$(\forall x \in a)\,(\exists y \in a)\,xRy$$

and $b_0 \in a$, then there exists a function $f : \omega \to a$ such that $f(0) = b_0$ and

$$(\forall n \in \omega)\ f(n)Rf(n+1).$$

Even more useful in constructive set theory is the *Relativized Dependent Choices Axiom*, RDC.[1] It asserts that for arbitrary formulae ϕ and ψ, whenever

$$\forall x \big[\phi(x) \to \exists y(\phi(y) \wedge \psi(x, y))\big]$$

and $\phi(b_0)$, then there exists a function f with domain ω such that $f(0) = b_0$ and

$$(\forall n \in \omega)\big[\phi(f(n)) \wedge \psi(f(n), f(n+1))\big].$$

A restricted form of RDC where ϕ and ψ are required to be Δ_0 will be called $\Delta_0\text{-}RDC$.

---

[1] In [2], RDC is called the dependent choices axiom and DC is dubbed the axiom of limited dependent choices. We deviate from the notation in [2] as it deviates from the usage in classical set theory texts.

The *Bounded Relativized Dependent Choices Axiom*, **bRDC**, is the following schema: For all Δ_0-formulae θ and ψ, whenever

$$(\forall x \in a)[\theta(x) \to (\exists y \in a)(\theta(y) \wedge \psi(x, y))]$$

and $b_0 \in a \wedge \phi(b_0)$, then there exists a function $f : \omega \to a$ such that $f(0) = b_0$ and

$$(\forall n \in \omega)[\theta(f(n)) \wedge \psi(f(n), f(n+1))].$$

Letting $\phi(x)$ stand for $x \in a \wedge \theta(x)$, one sees that **bRDC** is a consequence of Δ_0-**RDC**.

Here are some other well known relationships.

PROPOSITION 4.1 (**CZF$^-$**).

(i) **DC** *implies* **AC**$_\omega$.
(ii) **bRDC** *and* **DC** *are equivalent.*
(iii) **RDC** *implies* **DC**.

PROOF. (i): If z is an ordered pair $\langle x, y \rangle$ let $1^{st}(z)$ denote x and $2^{nd}(z)$ denote y.

Suppose F is a function with domain ω such that $\forall i \in \omega \, \exists x \in F(i)$. Let $A = \{\langle i, u \rangle \mid i \in \omega \wedge u \in F(i)\}$. A is a set by Union, Cartesian Product and restricted Separation. We then have

$$\forall x \in A \, \exists y \in A \, xRy,$$

where $R = \{\langle x, y \rangle \in A \times A \mid 1^{st}(y) = 1^{st}(x) + 1\}$. Pick $x_0 \in F(0)$ and let $a_0 = \langle 0, x_0 \rangle$. Using **DC** there exists a function $g : \omega \to A$ satisfying $g(0) = a_0$ and

$$\forall i \in \omega \, [g(i) \in A \wedge 1^{st}(g(i+1)) = 1^{st}(g(i)) + 1].$$

Letting f be defined on ω by $f(i) = 2^{nd}(g(i))$ one gets $\forall i \in \omega \, f(i) \in F(i)$.

(ii) We argue in **CZF$^-$** + **DC** to show **bRDC**. Assume

$$\forall x \in a \, [\phi(x) \to \exists y \in a(\phi(y) \wedge \psi(x, y))]$$

and $\phi(b_0)$, where ϕ and ψ are Δ_0. Let $\theta(x, y)$ be the formula $\phi(x) \wedge \phi(y) \wedge \psi(x, y)$ and $A = \{x \in a \mid \phi(x)\}$. Then θ is Δ_0 and A is a set by Δ_0 Separation. From the assumptions we get $\forall x \in A \, \exists y \in A \, \theta(x, y)$ and $b_0 \in A$. Thus, using **DC**, there is a function f with domain ω such that $f(0) = b_0$ and $\forall n \in \omega \, \theta(f(n), f(n+1))$. Hence we get $\forall n \in \omega \, [\phi(n) \wedge \psi(f(n), f(n+1))]$. The other direction is obvious.

(iii) is obvious.   ⊣

AC$_\omega$ does not imply **DC**, not even on the basis of **ZF**.

PROPOSITION 4.2. **ZF** + **AC**$_\omega$ *does not prove* **DC**.

PROOF. This was shown by Jensen [19].   ⊣

An interesting consequence of **RDC** which is not implied by **DC** is the following:

PROPOSITION 4.3 (**CZF$^-$ + RDC**). *Suppose* $\forall x \exists y \phi(x, y)$. *Then for every set d there exists a transitive set A such that $d \in A$ and*

$$\forall x \in A \, \exists y \in A \, \phi(x, y).$$

Moreover, for every set d there exists a transitive set A and a function $f : \omega \to A$ such that $f(0) = d$ and $\forall n \in \omega \, \phi(f(n), f(n+1))$.

PROOF. The assumption yields that $\forall x \in b \, \exists y \phi(x, y)$ holds for every set b. Thus, by Collection and the existence of the transitive closure of a set, we get

$$\forall b \, \exists c \, [\theta(b, c) \wedge \mathrm{Tran}(c)],$$

where $Tran(c)$ means that c is transitive and $\theta(b, c)$ is the formula $\forall x \in b \, \exists y \in c \, \phi(x, y)$. Let B be a transitive set containing d. Employing **RDC** there exists a function g with domain ω such that $g(0) = B$ and $\forall n \in \omega \, \theta(g(n), g(n+1))$. Obviously $A = \bigcup_{n \in \omega} g(n)$ satisfies our requirements.

The existence of the function f follows from the latter since **RDC** entails **DC**.   ⊣

§5. The Presentation Axiom.

The *Presentation Axiom*, *PAx*, is an example of a choice principle which is validated upon interpretation in type theory. In category theory it is also known as the *existence of enough projective sets*, *EPsets* (cf. [8]). In a category $\mathbb{C}$, an object P in $\mathbb{C}$ is *projective* (in $\mathbb{C}$) if for all objects A, B in $\mathbb{C}$, and morphisms $A \xrightarrow{f} B$, $P \xrightarrow{g} B$ with f an epimorphism, there exists a morphism $P \xrightarrow{h} A$ such that the following diagram commutes

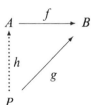

It easily follows that in the category of sets, a set P is projective if for any P-indexed family $(X_a)_{a \in P}$ of inhabited sets X_a, there exists a function f with domain P such that, for all $a \in P$, $f(a) \in X_a$.

PAx (or *EPsets*), is the statement that every set is the surjective image of a projective set.

Alternatively, projective sets have also been called *bases*, and we shall follow that usage henceforth. In this terminology, AC_ω expresses that ω is a base whereas *AC* amounts to saying that every set is a base.

PROPOSITION 5.1 (**CZF$^-$**). *PAx implies DC*.

PROOF. See [1] or [8, Theorem 6.2].   ⊣

The preceding implications cannot be reversed, not even on the basis of ZF.

PROPOSITION 5.2. $ZF + DC$ does not prove PAx.

PROOF. To see this, note that there are symmetric models $\mathcal{M}$ of $ZF + DC + \neg AC$, as for instance the one used in the proof of [18, Theorem 8.3]. A *symmetric model* (cf. [18, Chapter 5]) of ZF is specified by giving a ground model M of ZFC, a complete Boolean algebra B in M, an M-generic filter G in B, a group $\mathcal{G}$ of automorphisms of B, and a normal filter of subgroups of $\mathcal{G}$. The symmetric model consists of the elements of $M[G]$ that hereditarily have symmetric names. By [8, Theorem 4.3 and Theorem 6.1], $\mathcal{M}$ is not a model of PAx.   ⊣

REMARK 5.3. Very little is known about the classical strength of PAx. It is an open problem whether ZF proves that PAx implies AC (cf. [8, p. 50]). In P. Howard's and J. E. Rubin's book on consequences of the axiom of choice [17], this problem appears on page 322. Intuitionistically, however, PAx is much weaker than AC. In CZF, AC implies restricted excluded middle and thus the power set axiom. Moreover, in IZF, AC implies full excluded middle, whereas PAx does not yield any such forms of excluded middle.

PROPOSITION 5.4. $IZF + PAx$ does not prove AC. More specifically, $IZF + PAx$ does not prove that $^{\omega}\omega$ is a base.

PROOF. The realizability model $V(Kl)$ of [23] validates PAx [23, Theorem 7.6] as well as Church's thesis [23, Theorem 3.1]. It is well known that $^{\omega}\omega$ being a basis is incompatible with Church's thesis.   ⊣

§6. The axiom of multiple choice. A special form of choice grew out of Moerdijk's and Palmgren's endeavours to find a categorical counterpart for constructive type theory. In [24] they introduced a candidate for a notion of "predicative topos", dubbed *stratified pseudotopos*. The main results of [24] are that any stratified pseudotopos provides a model for CZF and that the sheaves on an internal site in a stratified pseudotopos again form a stratified pseudotopos. Their method of obtaining CZF models from stratified pseudo-toposes builds on Aczel's original interpretation of CZF in type theory. They encountered, however, a hindrance which stems from the fact that Aczel's interpretation heavily utilizes that Martin-Löf type theory satisfies the axiom of choice for types. In categorical terms, the latter amounts to the principle of the existence of enough projectives. As this principle is usually not preserved under sheaf constructions, Moerdijk and Palmgren altered Aczel's construction and thereby employed a category-theoretic choice principle dubbed the *axiom of multiple choice*, AMC. Rather than presenting the categorical axiom of multiple choice, the following will be concerned with a purely set-theoretic version of it which was formulated by Peter Aczel and Alex Simpson (cf. [4]).

DEFINITION 6.1. If X is a set let $MV(X)$ be the class of all multi-valued functions R with domain X, i.e., the class of all sets R of ordered pairs such that $X = \{x : \exists y \langle x, y \rangle \in R\}$. A set C *covers* $R \in MV(X)$ if

$$\forall x \in X \exists y \in C \left[(x, y) \in R\right] \quad \wedge \quad \forall y \in C \exists x \in X \left[(x, y) \in R\right].$$

A class $\mathcal{Y}$ is a *cover base* for a set X if every $R \in MV(X)$ is covered by an image of a set in $\mathcal{Y}$. If $\mathcal{Y}$ is a set then it is a *small cover base* for X.

We use the arrow $\twoheadrightarrow$ in $g : A \twoheadrightarrow B$ to convey that g is a surjective function from A to B. If $g : A \twoheadrightarrow B$ we also say that g is an *epi*.

PROPOSITION 6.2. $\mathcal{Y}$ *is a cover base for X iff for every epi $f : Z \twoheadrightarrow X$ there is an epi $g : Y \twoheadrightarrow X$, with $Y \in \mathcal{Y}$, that factors through $f : Z \to X$.*

PROOF. This result is due to Aczel and Simpson; see [4].   ⊣

DEFINITION 6.3. $\mathcal{Y}$ is a (*small*) *collection family* if it is a (small) cover base for each of its elements.

DEFINITION 6.4.

Axiom of Multiple Choice (*AMC*): Every set is in some small collection family.

H-axiom: For every set A there is a smallest set $H(A)$ such that if $a \in A$ and $f : a \to H(A)$ then $\mathbf{ran}(f) \in H(A)$.

THEOREM 6.5 (*CZF*).

1. *PAx implies AMC.*
2. *AMC plus H-axiom implies REA.*

PROOF. These results are due to Aczel and Simpson; see [4].   ⊣

PROPOSITION 6.6. *ZF does not prove that AMC implies PAx.*

PROOF. This will follow from Corollary 6.11.   ⊣

ZF models of AMC.

DEFINITION 6.7. There is a weak form of the axiom of choice, which holds in a plethora of *ZF* universes. The *axiom of small violations of choice*, *SVC*, has been studied by A. Blass [8]. It says in some sense, that all failure of choice occurs within a single set. *SVC* is the assertion that there is a set S such that, for every set a, there exists an ordinal α and a function from $S \times \alpha$ onto a.

LEMMA 6.8. (i) *If X is transitive and $X \subseteq B$, then $X \subseteq H(B)$.*
(ii) *If $2 \in B$ and $x, y \in H(B)$, then $\langle x, y \rangle \in H(B)$.*

PROOF. (i): By Set Induction on a one easily proves that $a \in X$ implies $a \in H(B)$.
(ii): Suppose $2 \in B$ and $x, y \in H(B)$. Let f be the function $f : 2 \to H(B)$ with $f(0) = x$ and $f(1) = y$. Then $\mathbf{ran}(f) = \{x, y\} \in H(B)$. By repeating the previous procedure with $\{x\}$ and $\{x, y\}$ one gets $\langle x, y \rangle \in H(B)$.   ⊣

THEOREM 6.9 (*ZF*). *SVC implies AMC and REA*.

PROOF. Let M be a ground model that satisfies *ZF* + *SVC*. Arguing in M let S be a set such that, for every set a, there exists an ordinal α and a function from $S \times \alpha$ onto a.

Let $\mathbb{P}$ be the set of finite partial functions from ω to S, and, stepping outside of M, let G be an M-generic filter in $\mathbb{P}$. By the proof of [8, Theorem 4.6], $M[G]$ is a model of *ZFC*.

Let A be an arbitrary set in M. Let $B = \bigcup_{n \in \omega} F(n)$, where

$$F(0) = TC(A \cup \mathbb{P}) \cup \omega \cup \{A, \mathbb{P}\},$$

$$F(n+1) = \left\{ b \times \mathbb{P} : b \in \bigcup_{k \leq n} F(k) \right\}.$$

Then $B \in M$. Let $Z = (H(B))^M$. Then $A \in Z$. First, we show by induction on n that $F(n) \subseteq Z$. As $F(0)$ is transitive, $F(0) \subseteq Z$ follows from Lemma 6.8, (i). Now suppose $\bigcup_{k \leq n} F(k) \subseteq Z$. An element of $F(n+1)$ is of the form $b \times \mathbb{P}$ with $b \in \bigcup_{k \leq n} F(k)$. If $x \in b$ and $p \in \mathbb{P}$ then $x, p \in Z$, and thus $\langle x, p \rangle \in Z$ by Lemma 6.8 since $2 \in B$. So, letting *id* be the identity function on $b \times \mathbb{P}$, we get $id : b \times \mathbb{P} \to Z$, and hence $ran(id) = b \times \mathbb{P} \in Z$. Consequently we have $F(n+1) \subseteq Z$. It follows that $B \subseteq Z$.

We claim that

(2) $M \models Z$ *is a small collection family*.

To verify this, suppose that $x \in Z$ and $R \in M$ is a multi-valued function on x. x being an element of $(H(B))^M$, there exists a function $f \in M$ and $a \in B$ such that $f : a \to x$ and $ran(f) = x$. As $M[G]$ is a model of *AC*, we may pick a function $\ell \in M[G]$ such that $dom(\ell) = x$ and $\forall v \in x \, uR\ell(v)$. We may assume $x \neq \emptyset$. So let $v_0 \in x$ and pick d_0 such that $v_0 R d_0$. Let $\dot{\ell}$ be a name for ℓ in the forcing language. For any $z \in M$ let $\check{z}$ be the canonical name for z in the forcing language. Define $\chi : a \times \mathbb{P} \to M$ by

$$\chi(u, p) := \begin{cases} w & \text{iff } f(u)Rw \text{ and } p \Vdash [\dot{\ell} \text{ is a function } \wedge \dot{\ell}(\check{f}(\check{u})) = \check{w}], \\ d_0 & \text{otherwise.} \end{cases}$$

For each $u \in a$, there is a $w \in Z$ such that $f(u)Rw$ and $\ell(f(u)) = w$, and then there is a $p \in G$ that forces that $\dot{\ell}$ is a function and $\dot{\ell}(\check{f}(\check{u})) = \check{w}$, so w is in the range of χ. χ is a function with domain $a \times \mathbb{P}$, $\chi \in M$, and $ran(\chi) \subseteq ran(R)$. Note that $a \times \mathbb{P} \in B$, and thus we have $a \times \mathbb{P} \in Z$. As a result, with $C = ran(\chi)$ we have $\forall v \in x \, \exists y \in C \, vRy \, \wedge \, \forall y \in C \, \exists v \in x \, vRy$, confirming the claim.   ⊣

From the previous theorem and results in [8] it follows that *AMC* and *REA* are satisfied in all permutation models and symmetric models. A *permutation model* (cf. [18, Chapter 4]) is specified by giving a model V of *ZFC* with atoms in which the atoms form a set A, a group $\mathcal{G}$ of permutations of A, and a normal filter $\mathcal{F}$ of subgroups of $\mathcal{G}$. The permutation model then consists of the hereditarily symmetric elements of V.

A *symmetric model* (cf. [18, Chapter 5]), is specified by giving a ground model M of **ZFC**, a complete Boolean algebra B in M, an M-generic filter G in B, a group $\mathcal{G}$ of automorphisms of B, and a normal filter of subgroups of $\mathcal{G}$. The symmetric model consists of the elements of $M[G]$ that hereditarily have symmetric names.

If B is a set then $\mathrm{HOD}(B)$ denotes the class of sets hereditarily ordinal definable over B.

COROLLARY 6.10. *The usual models of classical set theory without choice satisfy **AMC** and **REA**. More precisely, every permutation model and symmetric model satisfies **AMC** and **REA**. Furthermore, if V is a universe that satisfies **ZF**, then for every transitive set $A \in V$ and any set $B \in V$ the submodels $L(A)$ and $\mathrm{HOD}(B)$ satisfy **AMC** and **REA**.*

PROOF. This follows from Theorem 6.9 in conjunction with [8, Theorems 4.2, 4.3, 4.4, 4.5]. ⊣

COROLLARY 6.11 (**ZF**). ***AMC** and **REA** do not imply the countable axiom of choice, AC_ω. Moreover, **AMC** and **REA** do not imply any of the mathematical consequences of **AC** of [18, Chapter 10]. Among those consequences are the existence of a basis for any vector space and the existence of the algebraic closure of any field.*

PROOF. This follows from Corollary 6.9 and [18, Chapter 10]. ⊣

There is, however, one result known to follow from **AMC** that is usually considered a consequence of the axiom of choice.

PROPOSITION 6.12 (**ZF**). ***AMC** implies that, for any set X, there is a cardinal κ such that X cannot be mapped onto a cofinal subset of κ.*

PROOF. From [33, Proposition 4.1], it follows that $ZF \vdash H$-*axiom*, and thus, by Theorem 6.5, we get $ZF + AMC \vdash REA$. The assertion then follows from [33, Corollary 5.2]. ⊣

COROLLARY 6.13 (**ZF**). ***AMC** implies that there are class many regular cardinals.*

PROOF. If α is an ordinal then by the previous result there exists a cardinal κ such that α cannot be mapped onto a cofinal subset of κ. Let π be the cofinality of κ. Then π is a regular cardinal $> \alpha$. ⊣

The only models of $ZF + \neg AMC$ known to the author are the models of

$$ZF + \text{All uncountable cardinals are singular}$$

given by Gitik [16] who showed the consistency of the latter theory from the assumption that

$$ZFC + \forall \alpha \, \exists \kappa > \alpha \, (\kappa \text{ is a strongly compact cardinal})$$

is consistent. This large cardinal assumption might seem exaggerated, but it is known that the consistency of all uncountable cardinals being singular cannot be proved without assuming the consistency of the existence of some large cardinals. For instance, it was shown in [11] that if $\aleph_1$ and $\aleph_2$ are both singular one can obtain an inner model with a measurable cardinal.

It would be very interesting to construct models of $ZF + \neg AMC$ that do not hinge on large cardinal assumptions.

§7. Interpreting set theory in type theory. The basic idea of the interpretation CZF in Martin-Löf type theory is easily explained. The type V that is to be the universe of sets in type theory consists of elements of the form $\sup(A, f)$, where $A : U$ and $f : A \to U$. $\sup(A, f)$ may be more suggestively written as $\{f(x) : x \in A\}$. The elements of V are constructed inductively as families of sets indexed by the elements of a small type.

ML_1V is the extension of Martin-Löf type theory with one universe, ML_1, by Aczel's set of iterative sets V (cf. [1]). To be more precise, ML_1 is a type theory which has one universe U and all the type constructors of [22] except for the W-type. Indicating discharged assumptions by putting brackets around them, the natural deduction rules pertaining to V are as follows:[2]

(V-formation) V set

(V-introduction) $$\frac{A : U \quad f : A \to V}{\sup(A, f) : V}$$

(V-elimination)
$$\frac{c : V \quad \begin{array}{c} [A : U, f : A \to V] \\ {}[z : (\Pi v : A)C(f(v))] \\ d(A, f, z) : C(\sup(A, f)) \end{array}}{T_V(c, (A, f, z)d) : C(c)}$$

$$\frac{B : U \quad g : B \to V \quad \begin{array}{c} [A : U, f : A \to V] \\ {}[z : (\Pi v : A)C(f(v))] \\ d(A, f, z) : C(\sup(A, f)) \end{array}}{T_V(\sup(B, g), (A, f, z)d) = d(B, g, (\lambda v)T_V(g(v), (A, f, z)d)) : C(\sup(B, g))},$$

where the last rule is called (V-equality).

In ML_1V there are one-place functions assigning $\bar{\alpha} : U$ and $\tilde{\alpha} : \bar{\alpha} \to V$ to $\alpha : V$ such that if $\alpha = \sup(A, f) : V$, where $A : U$ and $f : A \to V$, then $\bar{\alpha} = A : U$ and $\tilde{\alpha} = f : A \to V$. Moreover, if $\alpha : V$ then

$$\sup(\bar{\alpha}, \tilde{\alpha}) = \alpha : V.$$

---

[2]To increase readability and intelligibility, the formulation à la Russell is used here for type theories with universes (cf. [22]). Rather than '$a \in A$', we shall write '$a : A$' for the type-theoretic elementhood relation to distinguish it from the set-theoretic one.

In the formulae-as-types interpretation of CZF in ML_1V we shall assume that CZF is formalized with primitive bounded quantifiers $(\forall x \in y)\phi(x)$ and $(\exists x \in y)\phi(x)$. Each formula $\psi(x_1, \ldots, x_n)$ of CZF (whose free variables are among $x_1, \ldots, x_n$) will be interpreted as a dependent type $\|\psi(a_1, \ldots, a_n)\|$ for $a_1 : V, \ldots, a_n : V$, which is also small, i.e., in U, if ψ contains only bounded quantifiers. The definition of $\|\psi(\alpha_1, \ldots, \alpha_n)\|$ for a non-atomic formula ψ proceeds by recursion on the build-up of ψ and is as follows:[3]

$$\|\varphi(\vec{\alpha}) \wedge \vartheta(\vec{\alpha})\| \text{ is } \|\varphi(\vec{\alpha})\| \times \|\vartheta(\vec{\alpha})\|,$$
$$\|\varphi(\vec{\alpha}) \vee \vartheta(\vec{\alpha})\| \text{ is } \|\varphi(\vec{\alpha})\| + \|\vartheta(\vec{\alpha})\|,$$
$$\|\varphi(\vec{\alpha}) \supset \vartheta(\vec{\alpha})\| \text{ is } \|\varphi(\vec{\alpha})\| \to \|\vartheta(\vec{\alpha})\|,$$
$$\|\neg\varphi(\vec{\alpha})\| \text{ is } \|\varphi(\vec{\alpha})\| \to N_0,$$
$$\|(\forall x \in \alpha_k)\vartheta(\vec{\alpha}, x)\| \text{ is } (\Pi i : \overline{\alpha_k})\|\vartheta(\vec{\alpha}, \widetilde{\alpha_k}(i))\|,$$
$$\|(\exists x \in \alpha_k)\vartheta(\vec{\alpha}, x)\| \text{ is } (\Sigma i : \overline{\alpha_k})\|\vartheta(\vec{\alpha}, \widetilde{\alpha_k}(i))\|,$$
$$\|\forall x \vartheta(\vec{\alpha}, x)\| \text{ is } (\Pi \beta : V)\|\vartheta(\vec{\alpha}, \beta)\|,$$
$$\|\exists x \vartheta(\vec{\alpha}, x)\| \text{ is } (\Sigma \beta : V)\|\vartheta(\vec{\alpha}, \beta)\|.$$

To complete the above interpretation it remains to provide the types $\|\alpha = \beta\|$ and $\|\alpha \in \beta\|$ for $\alpha, \beta : V$. To this end one defines by recursion on V a function $\doteq : V \times V \to U$ such that

$$\alpha \doteq \beta \text{ is } \left[(\Pi i : \bar{\alpha})(\Sigma j : \bar{\beta}) \, \tilde{\alpha}(i) \doteq \tilde{\beta}(j)\right] \times \left[(\Pi j : \bar{\beta})(\Sigma i : \bar{\alpha}) \, \tilde{\alpha}(i) \doteq \tilde{\beta}(j)\right].$$

Finally let $\|\alpha = \beta\|$ be $\alpha \doteq \beta$ and let $\|\alpha \in \beta\|$ be $(\Sigma j : \bar{\beta}) \, \alpha \doteq \tilde{\beta}(j)$.

The interpretation theorem can now be stated in a concise form.

THEOREM 7.1. *Let θ be a sentence of set theory. If $CZF \vdash \theta$ then there exists a term t_θ such that $ML_1V \vdash t_\theta : \|\theta\|$.*

PROOF. [1, Theorem 4.2]. ⊣

A strengthening of ML_1V is the type theory $ML_{1W}V$ in which the universe U is also closed under the W-type but the W-type constructor cannot be applied to families of types outside of U.

THEOREM 7.2. *Let θ be a sentence of set theory. If $CZF + REA \vdash \theta$ then there exists a term t_θ such that $ML_{1W}V \vdash t_\theta : \|\theta\|$.*

PROOF. [3]. ⊣

A further strengthening of the properties of U considered by Aczel is U-induction which asserts that U is inductively defined by its closure properties (cf. [2, 1.10]).

THEOREM 7.3. (i) *CZF, ML_1V, and $ML_1V + U$-induction are of the same strength.*

(ii) *$CZF + REA$, $ML_{1W}V$, and $ML_{1W}V + U$-induction are of the same strength.*

---

[3]Below $\supset$ denotes the conditional and N_0 denotes the empty type.

PROOF. This follows from [28, Theorem 4.14 and Theorem 5.13].   ⊣

Several choice principles are also validated by the interpretations of Theorems 7.1 and 7.2. The sets that are bases in the interpretation can be characterized by the following notion.

DEFINITION 7.4. $\alpha : V$ is *injectively presented* if for all $i, j : \bar{\alpha}$, whenever $\|\bar{\alpha}(i) = \bar{\alpha}(j)\|$ is inhabited, then $i = j : \bar{\alpha}$.

Section 9 will describe large collections of sets that have injective presentations.

§8. **Subcountability.** Certain classical set theories such as Kripke-Platek set theory possess models wherein all sets are (internally) countable, and thus a particular strong form of the axiom of choice obtains. Although **CZF** has the same proof-theoretic strength as **KP**, **CZF** refutes the statement that every set is countable. However, a weaker form of countability, dubbed *subcountability*, is not only compatible with **CZF** and **CZF + REA** but doesn't increase the proof-theoretic strength of these theories.

DEFINITION 8.1. We use the arrow $\twoheadrightarrow$ in $g : A \twoheadrightarrow B$ to convey that g is a surjective function from A to B.

$$X \text{ is } countable \quad \text{iff} \quad \exists u\, u \in X \longrightarrow \exists f\, (f : \omega \twoheadrightarrow X),$$
$$X \text{ is } subcountable \quad \text{iff} \quad \exists A \subseteq \omega\, \exists f\, (f : A \twoheadrightarrow X).$$

Let **EC** be the statement that every set is countable and let **ESC** be the statement that every set is subcountable. **ESC** has also been called the *Axiom of Enumerability* by Myhill (cf. [26]).

Obviously, in (rather weak) classical set theories countability and subcountability amount to the same. This is, however, far from being provable in intuitionistic set theories. Letting $^{\omega}\omega$ be the set of functions from ω to ω, it is, for instance, known that the theories **IZF** and **IZF** + $^{\omega}\omega$ *is subcountable* are equiconsistent, while **CZF** refutes the countability of $^{\omega}\omega$. The former fact is an immediate consequence of the equiconsistency of **IZF** and **IZF** augmented by Church's Thesis (cf. [23, Theorem 3.1]).

PROPOSITION 8.2. $\mathbf{CZF} \vdash \neg\mathbf{EC}$ *and* $\mathbf{IZF} \vdash \neg\mathbf{ESC}$.

PROOF. These facts are well-known, but it won't do much harm to repeat them here. To refute **EC** in **CZF** suppose $g : \omega \twoheadrightarrow {}^{\omega}\omega$. Define $h : \omega \to \omega$ by $h(n) = (g(n)(n)) + 1$. Then $h = g(m)$ for some $m \in \omega$. As a result, $g(m)(m) = h(m) = (g(m)(m)) + 1$, which is impossible.

To refute **ESC** in **IZF** suppose that $A \subseteq \omega$ and $f : A \twoheadrightarrow \mathcal{P}(\omega)$. Then let $D = \{n \in A : n \notin f(n)\}$. As $D \subseteq \omega$ there exists $k \in A$ such that $D = f(k)$. But this is absurd since $k \in D$ iff $k \notin D$.   ⊣

THEOREM 8.3. *ESC may be consistently added to **CZF** and **CZF** + **REA**. More precisely, **CZF** and **CZF** + **ESC** are of the same proof-theoretic strength. The same holds for **CZF** + **REA** and **CZF** + **REA** + **ESC**. Furthermore, the previous assertions allow for improvements in that one can strengthen **ESC** to the following statement*:

(3)   Every set is the surjective image of a base which is also a subset of ω.

PROOF. This follows from close inspection of the proofs of [28, Theorem 4.14 and 5.13]. The essence of those proofs is that the systems ML_IV and $ML_{IW}V$ of Martin-Löf type theory (of the foregoing section) with one universe U and the type V can be interpreted in KP and KPi, respectively. Recall that KPi denotes Kripke-Platek set theory plus an axiom saying that every set is contained in an admissible set. The theories CZF and $CZF + REA$ can in turn be interpreted in ML_IV and $ML_{IW}V$, respectively, conceiving of V as a universe of sets and using the formulae-as-types interpretation. U is modelled in KP via an inductively defined subclass of ω whose elements code subsets of ω. Each element α of V is a pair of natural numbers n and e, denoted $\sup(n, e)$, with $n \in U$ and e being the index of a partial recursive function $\tilde{\alpha}$ total on $\bar{\alpha}$ such that $\tilde{\alpha}(u) \in V$ for all $u \in \bar{\alpha}$, where $\bar{\alpha}$ denotes the subset of ω coded by n. Let $\omega_v \in V$ be the internalization of ω in V. Then $\overline{\omega_v} = \omega$. Also, for $\alpha, \beta \in V$ let $\langle \alpha, \beta \rangle_v \in V$ denote the internal ordered pair. The assignment $\alpha, \beta \mapsto \langle \alpha, \beta \rangle_v$ is a partial recursive function on $V \times V$.

Now, given any $\alpha \in V$ with $\alpha = \sup(n, e)$, let $\beta := \sup(n, e^*)$, where e^* is an index of the partial recursive function $\tilde{\beta}$ with $\tilde{\beta}(u) := \widetilde{\omega_v}(u)$ for $u \in \bar{\alpha}$. Also, define $\gamma := \sup(n, e^{\#})$, where $e^{\#}$ is an index of the partial recursive function $\tilde{\gamma}$ with

$$\tilde{\gamma}(u) := \langle \widetilde{\omega_v}(u), \tilde{\alpha}(u) \rangle_v$$

for $u \in \bar{\alpha}$. e^* and $e^{\#}$ are both effectively computable from α and so are β and γ. One then has to verify that, internally in V, β is a subset of ω_v and γ is a function that maps β onto α (in the sense of V). Here one utilizes that $\omega_v \in V$ is injectively represented, that is, whenever $\widetilde{\omega_v}(k) \dot{=} \widetilde{\omega_v}(k')$ (where $\dot{=}$ stands for the bi-simulation relation on V which interprets set-theoretic equality defined in the previous section) then $k = k'$.

To show the stronger statement (3), note first that the interpretations of [28, Theorem 4.14 and 5.13] also validate the presentation axiom PAx. Further note that every set that is in one-to-one correspondence with a base is a base, too. Thus, arguing in $CZF + PAx + ESC$, it suffices to show that every base is in one-to-one correspondence with a subset of ω. Let B be a base. Then there exist $A \subseteq \omega$ and $f : A \twoheadrightarrow B$. But B being a base, f can be inverted, that is, there exists $g : B \to A$ such that $f(g(u)) = u$ for all $u \in B$. g is an injective function and thus B is in one-to-one correspondence with $\{g(u) : u \in B\} \subseteq \omega$.   ⊣

§9. "Maximal" choice principles. The interpretation of constructive set theory in type theory not only validates all the theorems of *CZF* (resp. *CZF + REA*) but many other interesting set-theoretic statements. Ideally, one would like to have a characterization of these statements and determine an extension *CZF** of *CZF* (resp. *CZF + REA*) which deduces exactly the set-theoretic statements validated in *ML₁V* (resp. *ML₁wV*). It will turn out that the search for *CZF** amounts to finding the "strongest" version of the axiom of choice that is validated in *ML₁V*.

The interpretation of set theory in type theory gave rise to a plethora of new choice principles which will be described next.

DEFINITION 9.1 (*CZF*). If A is a set and B_x are classes for all $x \in A$, we define a class $\prod_{x \in A} B_x$ by:

$$(4) \qquad \prod_{x \in A} B_x := \left\{ f : A \to \bigcup_{x \in A} B_x \mid \forall x \in A(f(x) \in B_x) \right\}.$$

If A is a class and B_x are classes for all $x \in A$, we define a class $\sum_{x \in A} B_x$ by:

$$(5) \qquad \sum_{x \in A} B_x := \{ \langle x, y \rangle \mid x \in A \wedge y \in B_x \}.$$

If A and B are classes, we define a class $I(A, B)$ by:

$$(6) \qquad I(A, B) := \{ z \in 1 \mid A = B \}.$$

If A is a class and for each $a \in A$, B_a is a set, then

$$W_{a \in A} B_a$$

is the smallest class Y such that whenever $a \in A$ and $f : B_a \to Y$, then $\langle a, f \rangle \in Y$.

LEMMA 9.2 (*CZF*). *If A and B are sets and B_x is a set for all $x \in A$, then $\prod_{x \in A} B_x, \sum_{x \in A} B_x$ and $I(A, B)$ are sets.*

PROOF.   ⊣

LEMMA 9.3 (*CZF + REA*). *If A is a set and B_x is a set for all $x \in A$, then $W_{a \in A} B_a$ is a set.*

PROOF. This follows from [3, Corollary 5.3].   ⊣

LEMMA 9.4 (*CZF*). *There exists a smallest $\Pi\Sigma$-closed class ($\Pi\Sigma I$-closed class), i.e., a smallest class Y such that the following holds:*

(i) $n \in Y$ for all $n \in \omega$;
(ii) $\omega \in Y$;
(iii) $\prod_{x \in A} B_x \in Y$ and $\sum_{x \in A} B_x \in Y$ whenever $A \in Y$ and $B_x \in Y$ for all $x \in A$.
(iv) *If $A \in Y$ and $a, b \in A$, then $I(a, b) \in Y$.*

PROOF. See [34, Lemma 1.2].   ⊣

LEMMA 9.5 (*CZF* + *REA*). *There exists a smallest* $\Pi\Sigma W$-*closed class* ($\Pi\Sigma WI$-*closed class*), *i.e., a smallest class* **Y** *such that the following holds:*

(i) $n \in Y$ *for all* $n \in \omega$;

(ii) $\omega \in Y$;

(iii) $\prod_{x\in A} B_x \in Y$ *and* $\sum_{x\in A} B_x \in Y$ *whenever* $A \in Y$ *and* $B_x \in Y$ *for all* $x \in A$.

(iv) $W_{a\in A} B_a \in Y$ *whenever* $A \in Y$ *and* $B_x \in Y$ *for all* $x \in A$.

(v) *If* $A \in Y$ *and* $a, b \in A$, *then* $I(a,b) \in Y$.

PROOF. See [3].   ⊣

DEFINITION 9.6. The $\Pi\Sigma$-generated sets are the sets in the smallest $\Pi\Sigma$-closed class. Similarly one defines the $\Pi\Sigma I$, $\Pi\Sigma W$ and $\Pi\Sigma WI$-generated sets.

$\Pi\Sigma - AC$ is the statement that every $\Pi\Sigma$-generated set is a base. Similarly one defines the axioms $\Pi\Sigma I - AC$, $\Pi\Sigma WI - AC$, and $\Pi\Sigma W - AC$.

COROLLARY 9.7.   (i) (*CZF*) $\Pi\Sigma - AC$ *and* $\Pi\Sigma I - AC$ *are equivalent.*

(ii) (*CZF* + *REA*) $\Pi\Sigma W - AC$ *and* $\Pi\Sigma WI - AC$ *are equivalent.*

PROOF. [3, Theorem 3.7 and Theorem 5.9].   ⊣

The axioms $\Pi\Sigma - AC$ and $\Pi\Sigma W - AC$ may be added to the theories on the left hand side in Theorems 7.1 and 7.2, respectively. Below we shall show that these are in some sense the strongest axioms of choice that may be added.

DEFINITION 9.8. The *mathematical set terms* are a collection of class terms inductively defined by the following clauses:

1. ω is a mathematical set term.

2. If S and T are mathematical set terms then so are

$$\bigcup S := \{u : \exists x \in S\ u \in x\},$$
$$\{S, T\} := \{u : u = S \lor u = T\}$$

3. If S and T are mathematical set terms then so are

$$S + T := \{\langle 0, x\rangle : x \in S\} \cup \{\langle 1, x\rangle : x \in T\},$$
$$S \times T := \{\langle x, y\rangle : x \in S \land y \in T\},$$
$$S \to T := \{f : f : S \to T\}.$$

4. If $S, T_1, \ldots, T_n$ are mathematical set terms and $\psi(x, y_1, \ldots, y_n)$ is a restricted formula (of set theory) then

$$\{x \in S : \psi(x, T_1, \ldots, T_n)\}$$

is a mathematical set term.

5. If $S, T_1, \ldots, T_n, P_1, \ldots, P_k$ are mathematical set terms and

$$\psi(x, y_1, \ldots, y_n, z_1, \ldots, z_k)$$

is a bounded formula (of set theory) then

$$\{u : u = \{x \in S : \psi(x, y_1, \ldots, y_n, \vec{P})\} \wedge y_1 \in T_1 \wedge \cdots \wedge y_n \in T_n\}$$

is a mathematical set term, where $\vec{P} = P_1, \ldots, P_k$.

The *generalized mathematical set terms* are defined by the clauses for mathematical set terms plus the following clause:

6. If T is a generalized mathematical set term then so is $H(A)$, where $H(A)$ denotes the smallest class Y such that $ran(f) \in Y$ whenever $a \in A$ and $f : a \to Y$.

A *mathematical formula* (*generalized mathematical formula*) is a formula of the form $\psi(T_1, \ldots, T_n)$, where $\psi(x_1, \ldots, x_n)$ is bounded and $T_1, \ldots, T_n$ are mathematical set terms (generalized mathematical set terms). A *mathematical sentence* (*generalized mathematical sentence*) is a mathematical formula (generalized mathematical formula) without free variables.

REMARK 9.9.   1. From the point of view of **ZFC**, the mathematical set terms denote sets of rank $< \omega + \omega$ in the cumulative hierarchy while the generalized mathematical set terms denote sets of rank $< \aleph_\omega$.

2. The idea behind mathematical set terms is that they comprise all sets that one is interested in ordinary mathematics. E.g., with the help of Definition 9.8, clauses (1) and (3) one constructs the set of natural numbers, integers, rationals, and arbitrary function space as, e.g, $\mathbb{N} \to \mathbb{Q}$.

The main applications of clause (5) are made in constructing quotients: if $R \subseteq S \times S$ are set terms and R is an equivalence relation on S, then (5) permits to form the set term

$$S/R = \{[a]_R : a \in S\},$$

where $[a]_R = \{x \in S : aRx\}$.

Using clause (4) one obtains the set of Cauchy sequences of rationals from $\mathbb{N} \to \mathbb{Q}$, and finally by employing clause (5) one can define the set of equivalence classes of such Cauchy sequences, i.e., the set of reals.

3. Definition 9.8 clause (5) is related to the abstraction axiom of Friedman's system **B** in [12].

LEMMA 9.10.   1. (**CZF**) *Every mathematical set term is a set.*
2. (**CZF + REA**) *Every generalized mathematical set term is a set.*

PROOF. [34].   ⊣

THEOREM 9.11. *Let ψ be a mathematical sentence and let θ be a generalized mathematical sentence. Then there are closed terms t_ψ and t_θ of ML_1V and $ML_{1w}V$, respectively such that*

(i)   **CZF** $+ \Pi\Sigma - AC \vdash \psi$ *if and only if* $ML_1V \vdash t_\psi : \|\psi\|$.
(ii)   **CZF** $+$ **REA** $+ \Pi\Sigma W - AC \vdash \theta$ *if and only if* $ML_{1w}V \vdash t_\theta : \|\theta\|$.
(iii) *The foregoing results also hold if one adds U-induction to the type theories.*

PROOF. The "only if" parts are due to Rathjen and Tupailo [34, Theorem 7.6]. The "if" parts are due to [2, 3] and are proved by showing that the $\Pi\Sigma$- and $\Pi\Sigma W$-generated sets are injectively presentable (see Definition 7.4).   ⊣

[2, 3] feature several more choice principles. The main reason for their omission is that these axioms have no impact on the preceding result. This will be made precise below.

DEFINITION 9.12. Let $\Pi\Sigma - PAx$ be the assertion that every $\Pi\Sigma$-generated set is a base and every set is an image of a $\Pi\Sigma$-generated set. Similarly, one defines $\Pi\Sigma W - PAx$.

Let BCA_Π be the statement that whenever A is a base and B_a is a base for each $a \in A$, then $\prod_{x \in A} B_x$ is a base.

Let BCA_I be the statement that whenever A is a base then $I(b, c)$ is a base for all $b, c \in A$.

THEOREM 9.13. *Let ψ be a mathematical sentence and let θ be a generalized mathematical sentence. Then the following obtain:*

(i) $CZF + \Pi\Sigma - AC \vdash \psi$ *if and only if*

$$CZF + \Pi\Sigma - AC + \Pi\Sigma - PAx + BCA_\Pi + BCA_I + DC \vdash \psi.$$

(ii) $CZF + REA + \Pi\Sigma W - AC \vdash \theta$ *if and only if*

$$CZF + REA + \Pi\Sigma W - PAx + BCA_\Pi + BCA_I + DC \vdash \theta.$$

PROOF. [34].   ⊣

§10. The existence property. It is often considered a hallmark of intuitionistic systems that they possess the disjunction and existential definability properties.

DEFINITION 10.1. Let T be a theory whose language, $L(T)$, encompasses the language of set theory. Moreover, for simplicity, we shall assume that $L(T)$ has a constant ω denoting the set of von Neumann natural numbers and for each n a constant $\bar{n}$ denoting the n-th natural number.

1. T has the *disjunction property*, **DP**, if whenever $T \vdash \psi \vee \theta$ then $T \vdash \psi$ or $T \vdash \theta$.
2. T has the *numerical existence property*, **NEP**, if whenever
$$T \vdash (\exists x \in \omega)\phi(x)$$
then $T \vdash \phi(\bar{n})$ for some n.
3. T has the *existence property*, **EP**, if whenever $T \vdash \exists x \phi(x)$ then $T \vdash \exists! x [\vartheta(x) \wedge \phi(x)]$ for some formula ϑ.

Of course, above we assume that the formulas have no other free variables than those exhibited.

Slightly abusing terminology, we shall also say that T enjoys any of these properties if this holds only for a definitional extension of T rather than T.

ZF and **ZFC** are known not to have the existence property. But even classical set theories can have the **EP**. Kunen observed that an extension of **ZF** has the **EP** if and only if it proves that all sets are ordinal definable, i.e., $V = OD$. Going back to intuitionistic set theories, let IZF_R result from **IZF** by replacing Collection with Replacement, and let **CST** be Myhill's constructive set theory of [26]. CST^- denotes Myhill's **CST** without the axioms of countable and dependent choice.

THEOREM 10.2 (Myhill). *IZF_R and CST^- have the* **DP**, **NEP**, *and the* **EP**. *CST has the* **DP** *and the* **NEP**.

PROOF. [26].   ⊣

THEOREM 10.3 (Beeson). *IZF has the* **DP** *and the* **NEP**.

PROOF. [6].   ⊣

THEOREM 10.4. *CZF and CZF + REA have the* **DP** *and the* **NEP**. *This also holds when one adds any combination of the choice principles AC_ω, DC, RDC, or PAx to these theories.*

PROOF. [30] and [31].   ⊣

THEOREM 10.5 (Friedman, Ščedrov). *IZF does not have the* **EP**.

PROOF. [13].   ⊣

The question of whether **CZF** enjoys the **EP** is currently unsolved. The proof of the failure of **EP** for **IZF** given in [13] seems to single out Collection as the culprit. It appears unlikely that proof can be adapted to **CZF** because the refutation utilizes existential statements of the form

$$\exists b \left[\forall u \in a \, \exists y \, \varphi(u, y) \;\rightarrow\; \forall u \in a \, \exists y \in b \, \varphi(u, y)\right],$$

that are always deducible in **IZF** by employing first full Separation and then Collection, but, in general, are not deducible in **CZF**. We conjecture that the **EP** fails for **CZF** on account of Subset Collection (and maybe Collection). There are, however, positive answers available for $CZF + \Pi\Sigma - AC$ and $CZF + REA + \Pi\Sigma W - AC$. These theories have the pertaining properties for mathematical and generalized mathematical statements, respectively.

THEOREM 10.6. *Let θ_1, θ_2 be mathematical sentences and let $\psi(x)$ be a mathematical formula with at most x free. Then we have the following*:

(i) *If $CZF + \Pi\Sigma - AC \vdash \theta_1 \vee \theta_2$ then $CZF + \Pi\Sigma - AC \vdash \theta_1$ or $CZF + \Pi\Sigma - AC \vdash \theta_2$.*

(ii) *If $CZF + \Pi\Sigma - AC \vdash \exists x \psi(x)$ then there is a formula $\vartheta(x)$ (with at most x free) such that $CZF + \Pi\Sigma - AC \vdash \exists! x[\vartheta(x) \wedge \psi(x)]$.*

PROOF. (i) and (ii) are stated in [34] as Theorem 8.2. (i) and (ii) follow from results in [32] and [30].   ⊣

THEOREM 10.7. *Let θ_1, θ_2 be generalized mathematical sentences and let $\psi(x)$ be a generalized mathematical formula with at most x free. Then we have the following*:

(i) *If* $CZF + REA + \Pi\Sigma W - AC \vdash \theta_1 \vee \theta_2$ *then* $CZF + REA + \Pi\Sigma W - AC \vdash \theta_1$ *or* $CZF + REA + \Pi\Sigma W - AC \vdash \theta_2$.

(ii) *If* $CZF + REA + \Pi\Sigma W - AC \vdash \exists x \psi(x)$ *then there exists a formula* $\vartheta(x)$ *(with at most x free) such that* $CZF + REA + \Pi\Sigma W - AC \vdash \exists! x [\vartheta(x) \wedge \psi(x)]$.

PROOF. (i) and (ii) follow from results in [32] and [30]. They are stated in [34, Theorem 8.4]. ⊣

REFERENCES

[1] P. ACZEL, *The type theoretic interpretation of constructive set theory*, **Logic Colloquium '77** (A. MacIntyre, L. Pacholski, and J Paris, editors), North-Holland, Amsterdam, 1978, pp. 55–66.

[2] ———, *The type theoretic interpretation of constructive set theory: choice principles*, **The L. E. J. Brouwer Centenary Symposium** (A. S. Troelstra and D. van Dalen, editors), North-Holland, Amsterdam, 1982, pp. 1–40.

[3] ———, *The type theoretic interpretation of constructive set theory: inductive definitions*, **Logic, Methodology and Philosophy of Science, VII** (R. B. Marcus et al., editors), North-Holland, Amsterdam, 1986, pp. 17–49.

[4] P. ACZEL and M. RATHJEN, *Notes on constructive set theory*, Technical Report 40, Institut Mittag-Leffler, The Royal Swedish Academy of Sciences, 2001, http://www.ml.kva.se/preprints/archive2000-2001.php.

[5] J. BARWISE, **Admissible Sets and Structures**, Springer-Verlag, Berlin, 1975.

[6] M. BEESON, *Continuity in intuitionistic set theories*, **Logic Colloquium '78** (M Boffa, D. van Dalen, and K. McAloon, editors), North-Holland, Amsterdam, 1979, pp. 1–52.

[7] E. BISHOP and D. BRIDGES, **Constructive Analysis**, Springer-Verlag, Berlin, 1985.

[8] A. BLASS, *Injectivity, projectivity, and the axiom of choice*, **Transactions of the American Mathematical Society**, vol. 255 (1979), pp. 31–59.

[9] É BOREL, *Œuvres de Émil Borel*, Centre National de la recherche Scientifique, Paris, 1972.

[10] R. DIACONESCU, *Axiom of choice and complementation*, **Proceedings of the American Mathematical Society**, vol. 51 (1975), pp. 176–178.

[11] A. DODD and R. JENSEN, *The core model*, **Annals of Mathematical Logic**, vol. 20 (1981), no. 1, pp. 43–75.

[12] H. FRIEDMAN, *Set theoretic foundations for constructive analysis*, **Annals of Mathematics. Second Series**, vol. 105 (1977), no. 1, pp. 1–28.

[13] H. FRIEDMAN and S. ŠČEDROV, *The lack of definable witnesses and provably recursive functions in intuitionistic set theories*, **Advances in Mathematics**, vol. 57 (1985), no. 1, pp. 1–13.

[14] N. GAMBINO, **Types and Sets: A Study on the Jump to Full Impredicativity**, Laurea Dissertation, Department of Pure and Applied Mathematics, University of Padua, 1999.

[15] ———, **Heyting-Valued Interpretations for Constructive Set Theory**, Department of Computer Science, Manchester University, 2002.

[16] M. GITIK, *All uncountable cardinals can be singular*, **Israel Journal of Mathematics**, vol. 35 (1980), no. 1-2, pp. 61–88.

[17] P. HOWARD and J. E. RUBIN, **Consequences of the Axiom of Choice**, Mathematical Surveys and Monographs, vol. 59, American Mathematical Society, Providence, 1998.

[18] T. JECH, **The Axiom of Choice**, North-Holland, Amsterdam, 1973.

[19] R. B. Jensen, *Independence of the axiom of dependent choices from the countable axiom of choice (abstract)*, **The Journal of Symbolic Logic**, vol. 31 (1966), p. 294.

[20] H. Jervell, *From the axiom of choice to choice sequences*, **Nordic Journal of Philosophical Logic**, vol. 1 (1996), no. 1, pp. 95–98.

[21] P. Martin-Löf, *An intuitionistic theory of types: predicative part*, **Logic Colloquium '73** (H. E. Rose and J. Sheperdson, editors), North-Holland, Amsterdam, 1975, pp. 73–118.

[22] ———, *Intuitionistic Type Theory*, Bibliopolis, Naples, 1984.

[23] D. C. McCarty, *Realizability and Recursive Mathematics*, Ph.D. thesis, Oxford University, 1984.

[24] I. Moerdijk and E. Palmgren, *Type theories, toposes and constructive set theory: predicative aspects of AST*, **Annals of Pure and Applied Logic**, vol. 114 (2002), no. 1-3, pp. 155–201.

[25] L. Moss, *Power set recursion*, **Annals of Pure and Applied Logic**, vol. 71 (1995), no. 3, pp. 247–306.

[26] J. Myhill, *Constructive set theory*, **The Journal of Symbolic Logic**, vol. 40 (1975), no. 3, pp. 347–382.

[27] M. Rathjen, *Fragments of Kripke-Platek set theory with infinity*, **Proof Theory** (P. Aczel, H. Simmons, and S. Wainer, editors), Cambridge University Press, Cambridge, 1992, pp. 251–273.

[28] ———, *The Strength of Some Martin-Löf Type Theories*, Department of Mathematics, preprint, Ohio State University, 1993.

[29] ———, *The realm of ordinal analysis*, **Sets and Proofs** (S. B. Cooper and J. K. Truss, editors), Cambridge University Press, Cambridge, 1999, pp. 219–279.

[30] ———, *The disjunction and other properties for constructive Zermelo-Fraenkel set theory*, **The Journal of Symbolic Logic**, vol. 70 (December 2005), no. 4, pp. 1233–1254.

[31] ———, *The disjunction and other properties for constructive Zermelo-Fraenkel set theory with choice principles*, in preparation.

[32] ———, *The formulae-as-classes interpretation of constructive set theory*, **Proof Technology and Computation**, IOS Press, Amsterdam. Proceedings of the International Summer School Marktoberdorf 2003, to appear.

[33] M. Rathjen and R. Lubarsky, *On the regular extension axiom and its variants*, **Mathematical Logic Quarterly**, vol. 49 (2003), no. 5, pp. 511–518.

[34] M. Rathjen and S. Tupailo, *Characterizing the interpretation of set theory in Martin-Löf type theory*, to appear in Annals of Pure and Applied Logic.

[35] M. Rathjen and A. Weiermann, *Proof-theoretic investigations on Kruskal's theorem*, **Annals of Pure and Applied Logic**, vol. 60 (1993), no. 1, pp. 49–88.

[36] A. S. Troelstra and D. van Dalen, *Constructivism in Mathematics. Vol. I, II*, North-Holland, Amsterdam, 1988.

DEPARTMENT OF PURE MATHEMATICS
UNIVERSITY OF LEEDS
LEEDS LS2 9JT, UNITED KINGDOM
E-mail: rathjen@amsta.leeds.ac.uk

ASH'S THEOREM FOR ABSTRACT STRUCTURES

IVAN N. SOSKOV AND VESSELA BALEVA

Abstract. We introduce and study the class of relatively α-intrinsic sets on abstract structures. The main results are the Abstract Inversion Theorem and the Normal Form Theorem for the relatively α-intrinsic sets.

§1. Introduction. In this paper we are going to prove an analog of Ash's Theorem [1] for abstract structures. We shall consider structures

$$\mathfrak{A} = (\mathbb{N}; R_1, \ldots, R_k),$$

where $\mathbb{N}$ is the set of all natural numbers, each R_i is a subset of $\mathbb{N}^{r_i}$ and the equality "$=$" and the inequality "$\neq$" are among the predicates $R_1, \ldots, R_k$.

A total mapping from $\mathbb{N}$ onto $\mathbb{N}$ is called *enumeration of* $\mathfrak{A}$.

Let $\langle ., . \rangle$ be a recursive coding of the ordered pairs of natural numbers.

Given an enumeration f of $\mathfrak{A}$ and a subset of A of $\mathbb{N}^a$, let

$$f^{-1}(A) = \{\langle x_1, \ldots, x_a \rangle : (f(x_1), \ldots, f(x_a)) \in A\}.$$

We denote by $f^{-1}(\mathfrak{A})$ the set $f^{-1}(R_1) \oplus \cdots \oplus f^{-1}(R_k)$. In particular, if $f = \lambda x \cdot x$, then $f^{-1}(\mathfrak{A})$ will be denoted by $D(\mathfrak{A})$.

Next we define for every recursive ordinal α the relatively α-intrinsic sets. Given a set D of natural numbers and a recursive ordinal α, by $D_e^{(\alpha)}$ we shall denote the α-th enumeration jump of D. The exact definition will be given in the next section.

DEFINITION 1.1. Let α be a recursive ordinal and let $A \subseteq \mathbb{N}^a$. The set A is relatively α-intrinsic on the structure $\mathfrak{A}$ if for every enumeration f of $\mathfrak{A}$ the set $f^{-1}(A)$ is enumeration reducible to $(f^{-1}(\mathfrak{A}))_e^{(\alpha)}$.

Our definition is similar to the definition of the relatively intrinsically Σ_α^0 sets given in [2] and [3] but not the same. Actually the relatively intrinsically $\Sigma_{\alpha+1}^0$ sets on $\mathfrak{A}$ coincide with the relatively α-intrinsic sets on the structure $\mathfrak{A}^+ = (\mathbb{N}; R_1, \ldots, R_k, R_1^-, \ldots, R_k^-)$, where $R_i^- = \{\overline{x} : \overline{x} \notin R_i\}$.

2000 *Mathematics Subject Classification.* 03D35, 03D30.

Key words and phrases. abstract computability, enumeration reducibility, enumeration jump.

Logic Colloquium '02
Edited by Z. Chatzidakis, P. Koepke, and W. Pohlers
Lecture Notes in Logic, 27

Roughly speaking we could say that both notions define a kind of reducibility of sets to structures and while the definition of [2] and [3] corresponds to "r.e. in", our definition corresponds to the enumeration reducibility.

Next step in generalizing the enumeration reducibility is in the spirit of ASH [1]. Consider two subsets A and B of $\mathbb{N}$ and fix two recursive ordinals α and β. Say that A is relatively α, β-intrinsic on $\mathfrak{A}$ with respect to B if for all enumerations f such that $f^{-1}(B)$ is enumeration reducible to $f^{-1}(\mathfrak{A})^{(\beta)}$, $f^{-1}(A)$ is enumeration reducible to $f^{-1}(\mathfrak{A})^{(\alpha)}$.

In other words, in the definition above we take into account not all enumerations of $\mathfrak{A}$ but only those enumerations which "assume" that B is relatively β-intrinsic.

More generally, consider a sequence $\{B_\gamma\}_{\gamma \leq \zeta}$, where each B_γ is a subset of $\mathbb{N}^{a_\gamma}$, ζ is a recursive ordinal and there exists a recursive function ρ such that $\rho(\gamma) = a_\gamma$ for all $\gamma \leq \zeta$.

DEFINITION 1.2. Let $\alpha < \omega_1^{CK}$. A subset A of $\mathbb{N}^a$ is *relatively α-intrinsic on $\mathfrak{A}$ with respect to the sequence* $\{B_\gamma\}_{\gamma \leq \zeta}$ if for every enumeration f of $\mathfrak{A}$ such that $(\forall \gamma \leq \zeta)(f^{-1}(B_\gamma) \leq_e (f^{-1}(\mathfrak{A}))^{(\gamma)})$ uniformly in γ, the set $f^{-1}(A)$ is enumeration reducible to $(f^{-1}(\mathfrak{A}))^{(\alpha)}$.

In what follows we are going to present an explicit internal characterization of the relatively α-intrinsic with respect to the sequence $\{B_\gamma\}$ sets.

The proofs are based on a forcing construction and make use of our Jump Inversion Theorem [9].

For the sake of simplicity we shall assume that all sets B_γ are subsets of $\mathbb{N}$, i.e. $(\forall \gamma \leq \zeta)(a_\gamma = 1)$. The proofs in the general case are similar.

§2. Preliminaries.

2.1. Ordinal notations. We shall consider only recursive ordinals α which are below a fixed recursive ordinal η. We shall suppose that a notation $e \in \mathcal{O}$ for η is fixed and the notations for the ordinals $\alpha < \eta$ are the elements a of $\mathcal{O}$ such that $a <_o e$. For the definitions of the set $\mathcal{O}$ and the relation "$<_o$" the reader may consult [7] or [8].

We shall identify every ordinal with its notation and denote the ordinals by the letters α, β and γ. In particular we shall write $\alpha < \beta$ instead of $\alpha <_o \beta$. If α is a limit ordinal then by $\{\alpha(p)\}_{p \in \mathbb{N}}$ we shall denote the unique strongly increasing sequence of ordinals with limit α, determined by the notation of α, and write $\alpha = \lim \alpha(p)$.

2.2. The enumeration jump. Given two sets of natural numbers A and B, we say that A is enumeration reducible to B ($A \leq_e B$) if $A = \Gamma(B)$ for some enumeration operator Γ. In other words, using the notation D_v for the finite set having canonical code v and $W_0, \ldots, W_z, \ldots$ for the Gödel enumeration of the r.e. sets, we have

$$A \leq_e B \iff \exists z \forall x (x \in A \iff \exists v (\langle v, x \rangle \in W_z \ \& \ D_v \subseteq B)).$$

The relation $\leq_e$ is reflexive and transitive and induces an equivalence relation $\equiv_e$ on all subsets of $\mathbb{N}$. The respective equivalence classes are called enumeration degrees. For an introduction to the enumeration degrees the reader might consult COOPER [5].

Given a set A denote by A^+ the set $A \oplus (\mathbb{N} \setminus A)$. The set A is called *total* iff $A \equiv_e A^+$. Clearly A is recursively enumerable in B iff $A \leq_e B^+$ and A is recursive in B iff $A^+ \leq_e B^+$. Notice that the graph of every total function is a total set.

The enumeration jump operator is defined in COOPER [4] and further studied by McEVOY [6]. Here we shall use the following definition of the e-jump which is m-equivalent to the original one, see [6]:

DEFINITION 2.1. Given a set A, let $K_A^0 = \{\langle x, z \rangle : x \in \Gamma_z(A)\}$. Define the *e-jump* A_e' of A to be the set $(K_A^0)^+$.

The following properties of the enumeration jump are proved in [6]:
Let A and B be sets of natural numbers. Set $B_e^{(0)} = B$ and $B_e^{(n+1)} = (B_e^{(n)})_e'$.

(J1) If $A \leq_e B$, then $A_e' \leq_e B_e'$.

(J2) A is Σ_{n+1}^0 relatively to B iff $A \leq_e (B^+)_e^{(n)}$.

Let α be a recursive ordinal. To define the α-th enumeration jump of a set A we are going to use a construction very similar to that used in the definition of the α-th Turing jump. For every recursive ordinal α we define the set E_α^A by means of transfinite recursion on α:

DEFINITION 2.2.

(i) $E_0^A = A$.

(ii) $E_{\beta+1}^A = (E_\beta^A)_e'$.

(iii) If $\alpha = \lim \alpha(p)$, then $E_\alpha^A = \{\langle p, x \rangle : x \in E_{\alpha(p)}^A\}$.

From now on $A_e^{(\alpha)}$ will stand for E_α^A. Of course the definition of the set $A_e^{(\alpha)}$ depends on the fixed notation of the ordinal α. On the other hand, it is easy to see by a minor modification of the proof of the respective theorem of Spector for the H_α^A sets, see [7] or [8], that if α_1 and α_2 are two notations of the same recursive ordinal, then $A_e^{(\alpha_1)} \equiv_e A_e^{(\alpha_2)}$.

The following properties of the transfinite iterations of the enumeration jump follow easily from the definition:

(E1) If $\beta \leq \alpha$ are recursive ordinals, then $A_e^{(\beta)} \leq_e A_e^{(\alpha)}$ uniformly in β and α.

(E2) If $A \leq_e B$, then for every recursive ordinal α, $A_e^{(\alpha)} \leq_e B_e^{(\alpha)}$.

(E3) If $\alpha > 0$, then $A_e^{(\alpha)}$ is a total set.

Finally, we have that for total sets the α-th enumeration jump and the α-th Turing jump are equivalent. Namely the following is true:

PROPOSITION 2.3. *Let A be a total set of natural numbers. Then for every recursive ordinal α, $E_\alpha^A \equiv_e (H_\alpha^A)^+$ uniformly in α.*

Since we are going to consider only e-jumps here, from now on we shall omit the subscript e in the notation of the enumeration jump. So for every recursive ordinal α by $A^{(\alpha)}$ we shall denote the α-th enumeration jump of A.

2.3. The jump set of a sequence of sets. Let ζ be a recursive ordinal and let $\{B_\gamma\}_{\gamma \leq \zeta}$ be a sequence of sets of natural numbers. For every recursive ordinal α we define the *jump set $\mathcal{P}_\alpha$ of the sequence $\{B_\gamma\}$* by means of transfinite recursion on α:

DEFINITION 2.4.

(i) $\mathcal{P}_0 = B_0$.
(ii) Let $\alpha = \beta + 1$. Then let

$$\mathcal{P}_\alpha = \begin{cases} \mathcal{P}_\beta' \oplus B_\alpha & \text{if } \alpha \leq \zeta, \\ \mathcal{P}_\beta' & \text{otherwise.} \end{cases}$$

(iii) Let $\alpha = \lim \alpha(p)$. Then set $\mathcal{P}_{<\alpha} = \{\langle p, x \rangle : x \in \mathcal{P}_{\alpha(p)}\}$ and let

$$\mathcal{P}_\alpha = \begin{cases} \mathcal{P}_{<\alpha} \oplus B_\alpha & \text{if } \alpha \leq \zeta, \\ \mathcal{P}_{<\alpha} & \text{otherwise.} \end{cases}$$

Notice that if the sequence $\{B_\gamma\}$ contains only one member, i.e. $\zeta = 0$, then for every recursive α, $\mathcal{P}_\alpha = B_0^{(\alpha)}$.

The properties of the jump sets $\mathcal{P}_\alpha$ are similar to the properties of the enumeration jumps. Again we have that if α_1 and α_2 are two notations of the same recursive ordinal, then $\mathcal{P}_{\alpha_1} \equiv_e \mathcal{P}_{\alpha_2}$.

We shall use the following properties of the jump sets which follow easily from the definition:

($\mathcal{P}$1) If $\beta \leq \alpha$, then $\mathcal{P}_\beta \leq_e \mathcal{P}_\alpha$ uniformly in β and α.
($\mathcal{P}$2) If $\gamma \leq \min(\alpha, \zeta)$, then $B_\gamma \leq_e \mathcal{P}_\alpha$ uniformly in γ and α.
($\mathcal{P}$3) Let $(\forall \gamma \leq \min(\alpha, \zeta))(B_\gamma \leq_e A^{(\gamma)}$ uniformly in γ). Then $\mathcal{P}_\alpha \leq_e A^{(\alpha)}$.
($\mathcal{P}$4) If α is a limit ordinal, then the set $\mathcal{P}_{<\alpha}$ is total.
($\mathcal{P}$5) If $\zeta < \alpha$, then the set $\mathcal{P}_\alpha$ is total.

We conclude the preliminaries by a jump inversion theorem proved in [9]:

THEOREM 2.5. *Let $A \subseteq \mathbb{N}$ and $\{B_\gamma\}_{\gamma \leq \zeta}$ be a sequence of sets of natural numbers. Suppose that $\alpha < \zeta$ is a recursive ordinal such $A \not\leq_e \mathcal{P}_\alpha$. Let Q be a total subset of $\mathbb{N}$ such that $\mathcal{P}_\zeta \leq_e Q$ and $A^+ \leq_e Q$. Then there exists a total set F having the following properties:*

1. *For all $\gamma \leq \zeta$, $B_\gamma \leq_e F^{(\gamma)}$ uniformly in γ.*

2. *For all* $\gamma \leq \zeta$ *if* $\gamma = \beta + 1$, *then* $F^{(\gamma)} \equiv_e F \oplus \mathcal{P}'_\beta$ *uniformly in* γ.

3. *For all limit* $\gamma \leq \zeta$, $F^{(\gamma)} \equiv_e F \oplus \mathcal{P}_{<\gamma}$ *uniformly in* γ.

4. $F^{(\zeta)} \equiv_e Q$.

5. $A \not\leq_e F^{(\alpha)}$.

Here we shall use the following obvious corollary of the above theorem.

THEOREM 2.6. *Let* $\{B_\gamma\}_{\gamma \leq \zeta}$ *be a sequence of sets of natural numbers. Let* α *be a recursive ordinal and* $\xi = \max(\alpha + 1, \zeta)$. *Suppose that* $A \subseteq \mathbb{N}$ *and* $A \not\leq_e \mathcal{P}_\alpha$, Q *is a total set and* $\mathcal{P}_\xi \oplus A^+ \leq_e Q$. *Then there exists a total set* F *with the following properties*:

1. *For all* $\gamma \leq \zeta$, $B_\gamma \leq_e F^{(\gamma)}$ *uniformly in* γ.

2. $F^{(\xi)} \equiv_e Q$.

3. $A \not\leq_e F^{(\alpha)}$.

PROOF. For every γ such that $\zeta < \gamma \leq \xi$ set $B_\gamma = \emptyset$. Apply the previous theorem for the sequence $\{B_\gamma\}_{\gamma \leq \xi}$. $\dashv$

Throughout the rest of the paper we shall suppose fixed a partial structure $\mathfrak{A} = (\mathbb{N}; R_1, \ldots, R_k)$ and a sequence $\{B_\gamma\}_{\gamma \leq \zeta}$ of sets of natural numbers.

§3. **Forcing fundamentals.** Let f be an enumeration of $\mathfrak{A}$. For every recursive ordinal α by $\mathcal{P}^f_\alpha$ we shall denote the α-th jump set of the sequence $f^{-1}(\mathfrak{A}) \oplus f^{-1}(B_0), f^{-1}(B_1), \ldots, f^{-1}(B_\zeta)$.

For every α, e and x in $\mathbb{N}$ we define the relations $f \models_\alpha F_e(x)$ and $f \models_\alpha \neg F_e(x)$ as follows:

(i) $f \models_0 F_e(x)$ iff there exists a v such that $\langle v, x \rangle \in W_e$ and for all $u \in D_v$ either

    a) $u = \langle 0, \langle i, x_1^u, \ldots, x_{r_i}^u \rangle \rangle$ & $(f(x_1^u), \ldots, f(x_{r_i}^u)) \in R_i$ or

    b) $u = \langle 2, x_u \rangle$ & $f(x_u) \in B_0$.

(ii) Let $\alpha = \beta + 1$. Then

    a) if $\alpha \leq \zeta$, then

$$f \models_\alpha F_e(x) \iff (\exists v)(\langle v, x \rangle \in W_e \ \& \ (\forall u \in D_v)$$
$$((u = \langle 0, e_u, x_u \rangle \ \& \ f \models_\beta F_{e_u}(x_u)) \vee$$
$$(u = \langle 1, e_u, x_u \rangle \ \& \ f \models_\beta \neg F_{e_u}(x_u)) \vee$$
$$(u = \langle 2, x_u \rangle \ \& \ f(x_u) \in B_\alpha)));$$

    b) if $\zeta < \alpha$, then

$$f \models_\alpha F_e(x) \iff (\exists v)(\langle v, x \rangle \in W_e \ \& \ (\forall u \in D_v)$$
$$((u = \langle 0, e_u, x_u \rangle \ \& \ f \models_\beta F_{e_u}(x_u)) \vee$$
$$(u = \langle 1, e_u, x_u \rangle \ \& \ f \models_\beta \neg F_{e_u}(x_u)))).$$

(iii) Let $\alpha = \lim \alpha(p)$. Then
    a) if $\alpha \leq \zeta$, then

$$f \models_\alpha F_e(x) \iff (\exists v)(\langle v, x \rangle \in W_e \ \& \ (\forall u \in D_v)$$
$$((u = \langle 0, p_u, e_u, x_u \rangle \ \& \ f \models_{\alpha(p_u)} F_{e_u}(x_u)) \vee$$
$$(u = \langle 2, x_u \rangle \ \& \ f(x_u) \in B_\alpha)));$$

    b) if $\zeta < \alpha$, then

$$f \models_\alpha F_e(x) \iff (\exists v)(\langle v, x \rangle \in W_e \ \& \ (\forall u \in D_v)$$
$$((u = \langle 0, p_u, e_u, x_u \rangle \ \& \ f \models_{\alpha(p_u)} F_{e_u}(x_u)))).$$

(iv) $f \models_\alpha \neg F_e(x) \iff f \not\models_\alpha F_e(x)$.

An immediate corollary of the definitions above is the following:

LEMMA 3.1. *Let $A \subseteq \mathbb{N}$ and let $\alpha \leq \zeta$. Then $A \leq_e \mathcal{P}_\alpha^f$ iff there exists an e such that $A = \{x : f \models_\alpha F_e(x)\}$.*

The forcing conditions, which we shall call *finite parts*, are arbitrary finite mappings of $\mathbb{N}$ into $\mathbb{N}$. We shall denote the finite parts by the Greek letters τ, ρ and δ.

For every $\alpha \leq \zeta$, e and x in $\mathbb{N}$ and every finite part τ we define the forcing relations $\tau \Vdash_\alpha F_e(x)$ and $\tau \Vdash_\alpha \neg F_e(x)$ following the definition of "$\models$":

(i) $\tau \Vdash_0 F_e(x)$ iff there exists a v such that $\langle v, x \rangle \in W_e$ and for all $u \in D_v$ either
    a) $u = \langle 0, \langle i, x_1^u, \ldots, x_{r_i}^u \rangle \rangle$, $x_1^u, \ldots, x_{r_i}^u \in \mathrm{dom}(\tau)$ and $(\tau(x_1^u), \ldots, \tau(x_{r_i}^u)) \in R_i$ or
    b) $u = \langle 2, x_u \rangle$, $x_u \in \mathrm{dom}(\tau)$ and $\tau(x_u) \in B_0$.

(ii) Let $\alpha = \beta + 1$. Then
    a) if $\alpha \leq \zeta$, then

$$\tau \Vdash_\alpha F_e(x) \iff (\exists v)(\langle v, x \rangle \in W_e \ \& \ (\forall u \in D_v)$$
$$((u = \langle 0, e_u, x_u \rangle \ \& \ \tau \Vdash_\beta F_{e_u}(x_u)) \vee$$
$$(u = \langle 1, e_u, x_u \rangle \ \& \ \tau \Vdash_\beta \neg F_{e_u}(x_u)) \vee$$
$$(u = \langle 2, x_u \rangle \ \& \ \tau(x_u) \in B_\alpha)));$$

    b) if $\zeta < \alpha$, then

$$\tau \Vdash_\alpha F_e(x) \iff (\exists v)(\langle v, x \rangle \in W_e \ \& \ (\forall u \in D_v)$$
$$((u = \langle 0, e_u, x_u \rangle \ \& \ \tau \Vdash_\beta F_{e_u}(x_u)) \vee$$
$$(u = \langle 1, e_u, x_u \rangle \ \& \ \tau \Vdash_\beta \neg F_{e_u}(x_u)))).$$

(iii) Let $\alpha = \lim \alpha(p)$. Then
    a) if $\alpha \leq \zeta$, then

$$\tau \Vdash_\alpha F_e(x) \iff (\exists v)(\langle v, x \rangle \in W_e \ \& \ (\forall u \in D_v)$$
$$((u = \langle 0, p_u, e_u, x_u \rangle \ \& \ \tau \Vdash_{\alpha(p_u)} F_{e_u}(x_u)) \vee$$
$$(u = \langle 2, x_u \rangle \ \& \ \tau(x_u) \in B_\alpha)));$$

b) if $\zeta < \alpha$, then

$$\tau \Vdash_\alpha F_e(x) \iff (\exists v)(\langle v, x \rangle \in W_e \,\&\, (\forall u \in D_v)$$
$$((u = \langle 0, p_u, e_u, x_u \rangle \,\&\, \tau \Vdash_{\alpha(p_u)} F_{e_u}(x_u)))).$$

(iv) $\tau \Vdash_\alpha \neg F_e(x) \iff (\forall \rho \supseteq \tau)(\rho \nVdash_\alpha F_e(x)).$

For every recursive ordinal α, $e, x \in \mathbb{N}$ set $X^\alpha_{\langle e,x \rangle} = \{\rho : \rho \Vdash_\alpha F_e(x)\}.$

DEFINITION 3.2. An enumeration f of $\mathfrak{A}$ is α-generic if for every $\beta < \alpha$, $e, x \in \mathbb{N}$ the following condition holds:

(1) $(\forall \tau \subseteq f)(\exists \rho \in X^\beta_{\langle e,x \rangle})(\tau \subseteq \rho) \implies (\exists \tau \subseteq f)(\tau \in X^\beta_{\langle e,x \rangle}).$

The following standard properties of the forcing relation follow immediately from the definitions:

LEMMA 3.3.   1. *Let α be a recursive ordinal, $e, x \in \mathbb{N}$ and let $\tau \subseteq \rho$ be finite parts. Then*

$$\tau \Vdash_\alpha (\neg) F_e(x) \implies \rho \Vdash_\alpha (\neg) F_e(x).$$

2. *Let f be an α-generic enumeration. Then*

$$f \models_\alpha F_e(x) \iff (\exists \tau \subseteq f)(\tau \Vdash_\alpha F_e(x)).$$

3. *Let f be an $\alpha + 1$-generic enumeration. Then*

$$f \models_\alpha \neg F_e(x) \iff (\exists \tau \subseteq f)(\tau \Vdash_\alpha \neg F_e(x)).$$

Finally we would like to estimate an upper bound of the complexity of the forcing relation.

Given a sequence $\{X_n\}$ of sets of natural numbers, say that $\{X_n\}$ is e-reducible to the set P if there exists a recursive function g such that for all n we have that $X_n = \Gamma_{g(n)}(P)$. The sequence $\{X_n\}$ is T-reducible to P, if the function $\lambda n, x \cdot \chi_{X_n}(x)$ is recursive in P.

From the definition of the enumeration jump it follows immediately that if $\{X_n\}$ is e-reducible to P, then $\{X_n\}$ is T-reducible to P'.

For every recursive ordinal α let $\mathcal{P}_\alpha$ be the α-th jump set of the sequence $B_0 \oplus D(\mathfrak{A}), B_1, \ldots, B_\zeta$.

LEMMA 3.4. *For every α the sequence $\{X^\alpha_n\}$ is uniformly in α e-reducible to $\mathcal{P}_\alpha$ and hence it is uniformly in α T-reducible to $\mathcal{P}'_\alpha$.*

PROOF. Using effective transfinite recursion and following the definition of the forcing, one can define a recursive function $g(\alpha, n)$ such that for every α, $X^\alpha_n = \Gamma_{g(\alpha,n)}(\mathcal{P}_\alpha).$ ⊣

§4. An abstract jump inversion theorem.

DEFINITION 4.1. Let $A \subseteq \mathbb{N}$ and let α be a recursive ordinal. The set A is *forcing α-definable* on $\mathfrak{A}$ if there exist a finite part δ and $e, x \in \mathbb{N}$ such that

$$A = \{s : (\exists \tau \supseteq \delta)(\tau(x) \simeq s \,\&\, \tau \Vdash_\alpha F_e(x))\}.$$

Clearly if A is forcing α-definable on $\mathfrak{A}$, then $A \leq_e \mathcal{P}_\alpha$. The vice versa is not always true. As we shall see later the forcing α-definable sets coincide with the sets which are relatively α-intrinsic with respect to sequence $\{B_\gamma\}_{\gamma \leq \zeta}$.

PROPOSITION 4.2. *Let α be a recursive ordinal and let $A \subseteq \mathbb{N}$ be not forcing α-definable on $\mathfrak{A}$. Set $\xi = \max(\alpha + 1, \zeta)$. There exists an enumeration f of $\mathfrak{A}$ satisfying the following conditions:*

1. $f \leq_e A^+ \oplus \mathcal{P}_\xi$.
2. *If $\gamma \leq \xi$, then* $\mathcal{P}_\gamma^f \leq_e f \oplus \mathcal{P}_\gamma$.
3. $f^{-1}(A) \not\leq_e \mathcal{P}_\alpha^f$.

PROOF. We shall construct the enumeration f by steps. At each step q we shall define a finite part δ_q so that $\delta_q \subseteq \delta_{q+1}$ and take $f = \bigcup_q \delta_q$. We shall consider three kinds of steps. On steps $q = 3r$ we shall ensure that the mapping f is total and surjective. On steps $q = 3r + 1$ we shall ensure that f is ξ-generic and on steps $q = 3r + 2$ we shall ensure that f satisfies (3).

Let $\gamma_0, \gamma_1, \ldots$ be a recursive enumeration of all ordinals less than ξ. For every natural number n set $Y_n = X_{(n)_1}^{\gamma_{(n)_0}}$. Notice that the sequence $\{Y_n\}$ is T-reducible to $\mathcal{P}_\xi$.

Let δ_0 be the empty finite part and suppose that δ_q is defined.

a) Case $q = 3r$. Let x_0 be the least natural number which does not belong to $\mathrm{dom}(\delta_q)$ and let s_0 be the least natural number which does not belong to the range of δ_q. Set $\delta_{q+1}(x_0) \simeq s_0$ and $\delta_{q+1}(x) \simeq \delta_q(x)$ for $x \neq x_0$.

b) Case $q = 3r + 1$. Consider the set Y_r. Check whether there exists an element ρ of Y_r such that $\delta_q \subseteq \rho$. If the answer is positive, then let δ_{q+1} be the least extension of δ_q belonging to Y_r. If the answer is negative then let $\delta_{q+1} = \delta_q$.

c) Case $q = 3r + 2$. Let x_q be the least natural number which does not belong to $\mathrm{dom}(\delta_q)$. Consider the set

$$C_r = \{s : (\exists \tau \supseteq \delta_q)(\tau(x_q) \simeq s \,\&\, \tau \Vdash_\alpha F_r(x_q))\}.$$

Clearly C_r is forcing α-definable on $\mathfrak{A}$ and hence $C_r \neq A$. Notice that $C_r \leq_e \mathcal{P}_\alpha$ uniformly in r and δ_q. Therefore, since $\alpha < \xi$, the set C_r is recursive in $\mathcal{P}_\xi$ uniformly in r and δ_q. Let s_0 be the least natural number such that

$$s_0 \in C_r \,\&\, s_0 \notin A \lor s_0 \notin C_r \,\&\, s_0 \in A.$$

Suppose that $s_0 \in C_r$. Then there exists a τ such that

(2) $$\delta_q \subseteq \tau \ \& \ \tau(x_q) \simeq s_0 \ \& \ \tau \Vdash_\alpha F_r(x_q).$$

Let δ_{q+1} be the least τ satisfying (2).

If $s_0 \notin C_r$, then set $\delta_{q+1}(x_q) \simeq s_0$ and $\delta_{q+1}(x) \simeq \delta_q(x)$ for $x \neq x_q$. Notice that in this case we have that $\delta_{q+1} \Vdash_\alpha \neg F_r(x_q)$.

From the construction above it follows immediately that $f = \bigcup_q \delta_q$ is e-reducible to $A^+ \oplus \mathcal{P}_\xi$ and hence it satisfies (1).

Let $\gamma \leq \xi$. Then there exists an e such that $\mathcal{P}_\gamma^f = \{x : f \models_\gamma F_e(x)\}$. Since f is ξ-generic, we can rewrite the last equality as $\mathcal{P}_\gamma^f = \{x : (\exists \tau \subseteq f)(\tau \Vdash_\gamma F_e(x))\}$ Therefore $\mathcal{P}_\gamma^f \leq_e f \oplus \mathcal{P}_\gamma$.

It remains to show that $f^{-1}(A) \not\leq_e \mathcal{P}_\alpha^f$. Towards a contradiction assume that $f^{-1}(A) \leq_e \mathcal{P}_\alpha^f$. Then there exists an r such that

$$A = \{f(x) : f \models_\alpha F_r(x)\}.$$

Consider the step $q = 3r + 2$. By the construction we have that

$$\delta_{q+1}(x_q) \notin A \ \& \ \delta_{q+1} \Vdash_\alpha F_r(x_q) \vee \delta_{q+1}(x_q) \in A \ \& \ \delta_{q+1} \Vdash_\alpha \neg F_r(x_q).$$

Hence by the genericity of f

$$f(x_q) \notin A \ \& \ f \models_\alpha F_r(x_q) \vee f(x_q) \in A \ \& \ f \models_\alpha \neg F_r(x_q).$$

A contradiction. ⊣

The following theorem is an abstract version of Theorem 2.6.

THEOREM 4.3. *Let α be a recursive ordinal and let $A \subseteq \mathbb{N}$ be not forcing α-definable on $\mathfrak{A}$. Set $\xi = \max(\alpha + 1, \zeta)$ and let Q be a total set such that $A^+ \oplus \mathcal{P}_\xi \leq_e Q$. Then there exists an enumeration f of $\mathfrak{A}$ satisfying the following conditions:*

1. *$f \leq_e Q$.*
2. *The enumeration degree of $f^{-1}(\mathfrak{A})$ is total, i.e. it contains a total set.*
3. *For all $\gamma \leq \zeta$, $f^{-1}(B_\gamma) \leq_e (f^{-1}(\mathfrak{A}))^{(\gamma)}$ uniformly in γ.*
4. *$f^{-1}(A) \not\leq_e (f^{-1}(\mathfrak{A}))^{(\alpha)}$.*
5. *$(f^{-1}(\mathfrak{A}))^{(\xi)} \equiv_e Q$.*

PROOF. According Proposition 4.2 there exists an enumeration g of $\mathfrak{A}$ such that $g \leq_e Q$, $\mathcal{P}_\xi^g \leq_e Q$ and $g^{-1}(A) \not\leq_e \mathcal{P}_\alpha^g$. Since $A^+ \leq_e Q$, we have also that $(g^{-1}(A))^+ \leq_e Q$.

From Theorem 2.6 it follows that there exists a total set F such that the following assertions are true:

(i) $g^{-1}(\mathfrak{A}) \leq_e F$.
(ii) For all $\gamma \leq \zeta$, $g^{-1}(B_\gamma) \leq_e F^{(\gamma)}$ uniformly in γ.
(iii) $g^{-1}(A) \not\leq_e F^{(\alpha)}$.
(iv) $F^{(\xi)} \equiv_e Q$.

We shall construct the enumeration f so that $f^{-1}(\mathfrak{A}) \equiv_e F$. Let s and t be two distinct elements of $\mathbb{N}$. Fix also two numbers x_s and x_t such that $g(x_s) \simeq s$ and $g(x_t) \simeq t$.

For $x \in \mathbb{N}$ set

$$f(x) \simeq \begin{cases} g(x/2) & \text{if } x \text{ is even,} \\ s & \text{if } x = 2z + 1 \text{ and } z \in F, \\ t & \text{if } x = 2z + 1 \text{ and } z \notin F. \end{cases}$$

Since "$=$" and "$\neq$" are among the underlined predicates of $\mathfrak{A}$, we have that $F \leq_e f^{-1}(\mathfrak{A})$. To prove that $f^{-1}(\mathfrak{A}) \leq_e F$ consider the predicate R_i of $\mathfrak{A}$. Let $x_1, \ldots, x_{r_i}$ be arbitrary natural numbers. Define the natural numbers $y_1, \ldots, y_{r_i}$ by means of the following recursive in F procedure. Let $1 \leq j \leq r_i$. If x_j is even then let $y_j = x_j/2$. If $x_j = 2z + 1$ and $z \in F$, then let $y_j = x_s$. If $x_j = 2z + 1$ and $z \notin F$, then let $y_j = x_t$. Clearly

$$\langle x_1, \ldots, x_{r_i} \rangle \in f^{-1}(R_i) \iff \langle y_1, \ldots, y_{r_i} \rangle \in g^{-1}(R_i).$$

Since $g^{-1}(\mathfrak{A}) \leq F$, from the last equivalence it follows that $f^{-1}(R_i) \leq_e F$. So we obtain that $f^{-1}(\mathfrak{A}) \leq_e F$.

To prove (2) it is sufficient to show that if $\gamma \leq \zeta$, then $f^{-1}(B_\gamma) \leq_e F^{(\gamma)}$ uniformly in γ. Denote by E_f the set $f^{-1}(=)$. Clearly for all $\gamma \leq \zeta$ we have that E_f and $g^{-1}(B_\gamma)$ are e-reducible to $F^{(\gamma)}$ uniformly in γ. Let us fix a $\gamma \leq \zeta$. From the definition of f it follows that

$$f^{-1}(B_\gamma) = \{x : (\exists y \in g^{-1}(B_\gamma))(\langle x, 2y \rangle \in E_f)\}.$$

Therefore $f^{-1}(B_\gamma) \leq_e F^{(\gamma)}$ uniformly in γ.

It remains to see that $f^{-1}(A) \not\leq_e F^{(\alpha)}$. Assume that $f^{-1}(A) \leq_e F^{(\alpha)}$. Clearly

$$g^{-1}(A) = \{x : 2x \in f^{-1}(A)\}.$$

Then $g^{-1}(A) \leq_e f^{-1}(A) \leq_e F^{(\alpha)}$. A contradiction. $\dashv$

DEFINITION 4.4. Let Q be a total subset of $\mathbb{N}$ and $\xi < \omega_1^{CK}$. An enumeration f of $\mathfrak{A}$ is ξ, Q-*acceptable* (with respect to the sequence $\{B_\gamma\}_{\gamma \leq \zeta}$) if f satisfies the following conditions:

(i) The enumeration degree of $f^{-1}(\mathfrak{A})$ is total.
(ii) $(\forall \gamma \leq \zeta)(f^{-1}(B_\gamma) \leq_e (f^{-1}(\mathfrak{A}))^{(\gamma)})$ uniformly in γ.
(iii) $(f^{-1}(\mathfrak{A}))^{(\xi)} \equiv_e Q$.

THEOREM 4.5. *Given a total set Q such that $\mathcal{P}_\xi \leq_e Q$, one can construct a ξ, Q-acceptable enumeration $f \leq_e Q$.*

PROOF. Repeat the proof of the previous theorem without bothering about the set A. $\dashv$

THEOREM 4.6. *Let $\alpha < \omega_1^{CK}$ and let $A \subseteq \mathbb{N}$. Let $\xi = \max(\alpha+1, \zeta)$. Suppose that $Q \geq_e \mathcal{P}_\xi$, Q is a total set and for all ξ, Q-acceptable enumerations f of $\mathfrak{A}$ we have that $f^{-1}(A) \leq_e (f^{-1}(\mathfrak{A}))^{(\alpha)}$. Then A is forcing α-definable on $\mathfrak{A}$.*

PROOF. First we shall show that $A^+ \leq_e Q$. By the previous theorem there exists an enumeration g of $\mathfrak{A}$ such that $g \leq_e Q$ and g is ξ, Q-acceptable. Then $g^{-1}(A) \leq_e (g^{-1}(\mathfrak{A}))^{(\alpha)}$. By the monotonicity of the enumeration jump we can conclude that

$$\left(g^{-1}(A)\right)' \leq_e \left(g^{-1}(\mathfrak{A})\right)^{(\alpha+1)} \leq_e \left(g^{-1}(\mathfrak{A})\right)^{(\xi)} \leq_e Q.$$

Since $(g^{-1}(A))^+ \leq_e (g^{-1}(A))'$, we get that $(g^{-1}(A))^+ \leq_e Q$. Therefore both A and $\mathbb{N} \setminus A$ are enumeration reducible to Q and hence $A^+ \leq_e Q$.

Assume that A is not forcing α-definable on $\mathfrak{A}$. Applying Theorem 4.3 we obtain an ξ, Q-acceptable enumeration f such that $f^{-1}(A) \not\leq_e (f^{-1}(\mathfrak{A}))^{(\alpha)}$. A contradiction. ⊣

§5. **Normal form of the forcing definable sets.** In this section we shall show that the forcing definable sets on the structure $\mathfrak{A}$ coincide with the sets which are definable on $\mathfrak{A}$ by means of a certain kind of *positive* recursive Σ_α^0 formulae. This formulae can be considered as a modification of the formulae introduced in [1], which is appropriate for their use on abstract structures.

Let $\mathcal{L} = \{T_1, \ldots, T_k\}$ be the first order language corresponding to the structure $\mathfrak{A}$. So every T_i is an r_i-ary predicate symbol. Let $\{P_\gamma\}_{\gamma \leq \zeta}$ be a recursive sequence of unary predicate symbols intended to represent the sets B_y. We shall suppose also fixed a sequence $\mathbb{X}_0, \ldots, \mathbb{X}_n, \ldots$ of variables. The variables will be denoted by the letters X, Y, W possibly indexed.

Next we define for $\alpha < \omega_1^{CK}$ the Σ_α^+ formulae. The definition is by transfinite recursion on α and goes along with the definition of indices (codes) for every formula. We shall leave to the reader the explicit definition of the indices of our formulae which can be done in a natural way.

DEFINITION 5.1.

(i) Let $\alpha = 0$. The elementary Σ_α^+ formulae are formulae in prenex normal form with a finite number of existential quantifiers and a matrix which is a finite conjunction of atomic predicates built up from the variables and the predicate symbols $T_1, \ldots, T_k$ and P_0.

(ii) Let $\alpha = \beta + 1$ and $\alpha \leq \zeta$. An elementary Σ_α^+ formula is in the form

$$\exists Y_1 \ldots \exists Y_m M(X_1, \ldots, X_l, Y_1, \ldots, Y_m),$$

where M is a finite conjunction of atoms of the form $P_\alpha(X_i)$ or $P_\alpha(Y_j)$, Σ_β^+ formulae and negations of Σ_β^+ formulae with free variables among $X_1, \ldots, X_l, Y_1, \ldots, Y_m$.

(iii) Let $\alpha = \beta + 1$ and $\alpha > \zeta$. An elementary Σ_α^+ formula is in the form

$$\exists Y_1 \ldots \exists Y_m M(X_1, \ldots, X_l, Y_1, \ldots, Y_m),$$

where M is a finite conjunction of atoms of Σ_β^+ formulae and negations of Σ_β^+ formulae with free variables among $X_1, \ldots, X_l, Y_1, \ldots, Y_m$.

(iv) Let $\alpha = \lim \alpha(p)$ be a limit ordinal and $\alpha \leq \zeta$. The elementary Σ_α^+ formula are in the form

$$\exists Y_1 \ldots \exists Y_m M(X_1, \ldots, X_l, Y_1, \ldots, Y_m),$$

where M is a finite conjunction of atoms of the form $P_\alpha(X_i)$ or $P_\alpha(Y_j)$ and $\Sigma_{\alpha(p)}^+$ formulae with free variables among $X_1, \ldots, X_l, Y_1, \ldots, Y_m$.

(v) Let $\alpha = \lim \alpha(p)$ be a limit ordinal and $\alpha > \zeta$. The elementary Σ_α^+ formula are in the form

$$\exists Y_1 \ldots \exists Y_m M(X_1, \ldots, X_l, Y_1, \ldots, Y_m),$$

where M is a finite conjunction of $\Sigma_{\alpha(p)}^+$ formulae with free variables among $X_1, \ldots, X_l, Y_1, \ldots, Y_m$.

(vi) A Σ_α^+ formula with free variables among $X_1, \ldots, X_l$ is an r.e. infinitary disjunction of elementary Σ_α^+ formulae with free variables among $X_1, \ldots, X_l$.

Notice that the Σ_α^+ formulae are effectively closed under existential quantification and infinitary r.e. disjunctions.

Let Φ be a Σ_α^+ formula with free variables among $W_1, \ldots, W_n$ and let $t_1, \ldots, t_n$ be elements of $\mathbb{N}$. Then by $\mathfrak{A} \models \Phi(W_1/t_1, \ldots, W_n/t_n)$ we shall denote that Φ is true on $\mathfrak{A}$ under the variable assignment v such that $v(W_1) = t_1, \ldots, v(W_n) = t_n$.

DEFINITION 5.2. Let $A \subseteq \mathbb{N}$ and let $\alpha < \zeta$. The set A is *formally α-definable* on $\mathfrak{A}$ with respect to the sequence $\{B_\gamma\}_{\gamma < \zeta}$ if there exists a Σ_α^+ formula Φ with free variables among $W_1, \ldots, W_r, X$ and elements $t_1, \ldots, t_r$ of $\mathbb{N}$ such that for every element s of $\mathbb{N}$ the following equivalence holds:

$$s \in A \iff \mathfrak{A} \models \Phi(W_1/t_1, \ldots, W_r/t_r, X/s).$$

We shall show that every forcing α-definable set is formally α-definable.

Let var be an effective mapping of the natural numbers onto the variables. Given a natural number x, by X we shall denote the variable $var(x)$.

Let $y_1 < y_2 < \cdots < y_k$ be the elements of a finite set D, let Q be one of the quantifiers $\exists$ or $\forall$ an let Φ be an arbitrary formula. Then by $Q(y : y \in D)\Phi$ we shall denote the formula $QY_1 \cdots QY_k \Phi$.

LEMMA 5.3. *Let $D = \{w_1, \ldots, w_r\}$ be a finite and not empty set of natural numbers and x, e be elements of $\mathbb{N}$. Let $\alpha < \omega_1^{CK}$. There exists an uniform recursive way to construct a Σ_α^+ formula $\Phi_{D,e,x}^\alpha$ with free variables among $W_1, \ldots, W_r$ such that for every finite part δ such that $\mathrm{dom}(\delta) = D$ the following equivalence is true:*

$$\mathfrak{A} \models \Phi_{D,e,x}^\alpha(W_1/\delta(w_1), \ldots, W_r/\delta(w_r)) \iff \delta \Vdash_\alpha F_e(x).$$

PROOF. We shall construct the formula $\Phi^\alpha_{D,e,x}$ by means of effective transfinite recursion on α following the definition of the forcing.

1) Let $\alpha = 0$. Let $V = \{v : \langle v, x \rangle \in W_e\}$. Consider an element v of V. For every $u \in D_v$ define the atom Π_u as follows

a) If $u = \langle 0, \langle i, x^u_1, \ldots, x^u_{r_i} \rangle \rangle$, where $1 \le i \le k$ and all $x^u_1, \ldots, x^u_{r_i}$ are elements of D, then let $\Pi_u = T_i(X^u_1, \ldots, X^u_{r_i})$.

b) If $u = \langle 2, x_u \rangle$ and $x_u \in D$, then let $\Pi_u = P_0(X_u)$.

c) Let $\Pi_u = W_1 \neq W_1$ in the other cases.

Set $\Pi_v = \bigwedge_{u \in D_v} \Pi_u$ and $\Phi^\alpha_{D,e,x} = \bigvee_{v \in V} \Pi_v$.

2) Let $\alpha = \beta + 1$ Let again $V = \{v : \langle v, x \rangle \in W_e\}$ and $v \in V$. For every $u \in D_v$ define the formula Π_u as follows:

a) If $u = \langle 0, e_u, x_u \rangle$, then let $\Pi_u = \Phi^\beta_{D,e_u,x_u}$.

b) If $u = \langle 1, e_u, x_u \rangle$, then let

$$\Pi_u = \neg \left[\bigvee_{D^* \supseteq D} (\exists y \in D^* \setminus D) \Phi^\beta_{D^*,e_u,x_u} \right].$$

c) If $\alpha \le \zeta$, $u = \langle 2, x_u \rangle$ and $x_u \in D$ then let $\Pi_u = P_\alpha(X_u)$.

d) Let $\Pi_u = \Phi^\beta_{\{0\},0,0} \wedge \neg \Phi^\beta_{\{0\},0,0}$ in the other cases.

Now let $\Pi_v = \bigwedge_{u \in D_v} \Pi_u$ and set $\Phi^\alpha_{D,e,x} = \bigvee_{v \in V} \Pi_v$.

3) Let $\alpha = \lim \alpha(p)$ be a limit ordinal. Let $V = \{v : \langle v, x \rangle \in W_e\}$. Consider a $v \in V$. For every element u of D_v we define the formula Π_u as follows:

a) If $u = \langle 0, p_u, e_u, x_u \rangle$, then let $\Pi_u = \Phi^{\alpha(p_u)}_{D,e_u,x_u}$.

b) If $\alpha \le \zeta$, $u = \langle 2, x_u \rangle$ and $x_u \in D$, then let $\Pi_u = P_\alpha(X_u)$.

c) Let $\Pi_u = \Phi^{\alpha(0)}_{\{0\},0,0} \wedge \neg \Phi^{\alpha(0)}_{\{0\},0,0}$ in the other cases.

Set $\Pi_v = \bigwedge_{u \in D_v} \Pi_u$ and $\Phi^\alpha_{D,e,x} = \bigvee_{v \in V} \Pi_v$. An easy transfinite induction on α shows that for every α the Σ^+_α formula $\Phi^\alpha_{D,e,x}$ satisfies the requirements of the lemma. ⊣

THEOREM 5.4. Let $\alpha < \omega^{CK}_1$ and let $A \subseteq \mathbb{N}$ be forcing α-definable on $\mathfrak{A}$. Then A is formally α-definable on $\mathfrak{A}$.

PROOF. Suppose that for all $s \in \mathbb{N}$ we have that

$$s \in A \iff (\exists \tau \supseteq \delta)(\tau(x) \simeq s \ \& \ \tau \Vdash_\alpha F_e(x)),$$

where δ is a finite part, e, x are fixed elements of $\mathbb{N}$. Let $D = \mathrm{dom}(\delta) = \{w_1, \ldots, w_r\}$ and let $\delta(w_i) = t_i$, $i = 1, \ldots, r$. Consider a finite set $D^* \supseteq D \cup \{x\}$. By the previous lemma

$$\mathfrak{A} \models \exists(y \in D^* \setminus (D \cup \{x\})) \Phi^\alpha_{D^*,e,x}(W_1/t_1, \ldots, W_r/t_r, X/s)$$

if and only if there exists a finite part τ such that $\mathrm{dom}(\tau) = D^*$, $\tau \supseteq \delta, \tau(x) \simeq s$ and $\tau \Vdash_\alpha F_e(x)$. Hence we have that for all $s \in \mathbb{N}$ the following equivalence is true:

$$s \in A \Longleftrightarrow \mathfrak{A} \models \bigvee_{D^* \supseteq D \cup \{x\}} \exists (y \in D^* \setminus (D \cup \{x\})) \Phi^{\alpha}_{D^*,e,x}(W_1/t_1, \ldots, W_r/t_r).$$

From here we can conclude that A is formally α-definable on $\mathfrak{A}$.   ⊣

THEOREM 5.5. *Let $A \subseteq \mathbb{N}$. Suppose that $\alpha < \omega_1^{CK}$ and $\xi = \max(\alpha + 1, \zeta)$. Let Q be a total set such that $\mathcal{P}_\xi \leq_e Q$. Then the following are equivalent:*

1. *A is relatively α-intrinsic with respect to the sequence $\{B_\gamma\}_{\gamma<\zeta}$.*
2. *For every ξ, Q-acceptable enumeration f of $\mathfrak{A}$, $f^{-1}(A) \leq_e (f^{-1}(\mathfrak{A}))^{(\alpha)}$.*
3. *A is forcing α-definable on $\mathfrak{A}$.*
4. *A is formally α-definable on $\mathfrak{A}$.*

PROOF. The implication $(1) \Rightarrow (2)$ is obvious. The implication $(2) \Rightarrow (3)$ follows from Theorem 4.6. The implication $(3) \Rightarrow (4)$ follows from the previous theorem. The last implication $(4) \Rightarrow (1)$ can be proved by transfinite induction on α.   ⊣

The characterization of the relatively α-intrinsic sets can be obtained from the theorem above by taking $\zeta = 0$ and $B_0 = \mathbb{N}$. Frome here one can easly derive the Normal Form of the relatively intrinsically Σ^0_α sets, obtained in [2] and [3]. Moreover we can get a slight improvement of the upper bound of the level of genericity compared to that obtained in [3]. Namely the following is true:

COROLLARY 5.6. *Suppose that $\mathfrak{A}$ is a structure with recursively enumerable underlined predicates and $\alpha < \omega_1^{CK}$. Let $A \subseteq \mathbb{N}$ and let for all enumerations f of $\mathfrak{A}$ such that $(f^{-1}(\mathfrak{A}))^{(\alpha+1)} \equiv_e \emptyset^{(\alpha+1)}$ we have that $f^{-1}(A)$ is enumeration reducible to $(f^{-1}(\mathfrak{A}))^{(\alpha)}$. Then A is relatively α-intrinsic on $\mathfrak{A}$.*

The last corollary generalizes the respective result [3, Corollary V.18], where the same upper bound is obtained under the additional condition that A is a $\Delta^0_{\alpha+1}$ set.

REFERENCES

[1] C. J. ASH, *Generalizations of enumeration reducibility using recursive infinitary propositional sentences*, **Annals of Pure and Applied Logic**, vol. 58 (1992), no. 3, pp. 173–184.

[2] C. J. ASH, J. F. KNIGHT, M. MANASSE, and T. SLAMAN, *Generic copies of countable structures*, **Annals of Pure and Applied Logic**, vol. 42 (1989), no. 3, pp. 195–205.

[3] J. CHISHOLM, *Effective model theory vs. recursive model theory*, **The Journal of Symbolic Logic**, vol. 55 (1990), no. 3, pp. 1168–1191.

[4] S. B. COOPER, *Partial degrees and the density problem. II. The enumeration degrees of the Σ_2 sets are dense*, **The Journal of Symbolic Logic**, vol. 49 (1984), no. 2, pp. 503–513.

[5] ———, *Enumeration reducibility, nondeterministic computations and relative computability of partial functions*, **Recursion Theory Week (Oberwolfach, 1989)** (K. Ambos-Spies, G. H. Müller, and G. E. Sacks, editors), Lecture Notes in Mathematics, vol. 1432, Springer, Berlin, 1990, pp. 57–110.

[6] K. McEvoy, *Jumps of quasi-minimal enumeration degrees*, **The Journal of Symbolic Logic**, vol. 50 (1985), no. 3, pp. 839–848.

[7] H. Rogers, Jr., **Theory of Recursive Functions and Effective Computability**, McGraw-Hill Book Company, New York, 1967.

[8] G. E. Sacks, **Higher Recursion Theory**, Springer-Verlag, Berlin, 1990.

[9] I. N. Soskov and V. Baleva, *Regular enumerations*, **The Journal of Symbolic Logic**, vol. 67 (2002), no. 4, pp. 1323–1343.

FACULTY OF MATHEMATICS AND COMPUTER SCIENCE
SOFIA UNIVERSITY
BLVD. "JAMES BOURCHIER" 5
1164 SOFIA, BULGARIA
E-mail: soskov@fmi.uni-sofia.bg
URL: www.fmi.uni-sofia.bg/fmi/logic/soskov
E-mail: vbaleva@fmi.uni-sofia.bg
URL: www.fmi.uni-sofia.bg/fmi/logic/vbaleva

MARTIN-LÖF RANDOM AND PA-COMPLETE SETS

FRANK STEPHAN

Abstract. A set A is Martin-Löf random iff the class $\{A\}$ does not have Σ_1^0-measure 0. A set A is PA-complete if one can compute relative to A a consistent and complete extension of Peano Arithmetic. It is shown that every Martin-Löf random set either permits to solve the halting problem K or is not PA-complete. This result implies a negative answer to the question of Ambos-Spies and Kučera whether there is a Martin-Löf random set not above K which is also PA-complete.

§1. Introduction. Gács [3] and Kučera [7, 8] showed that every set can be computed relative to a Martin-Löf random set. In particular, for every set B there is a Martin-Löf random set A such that $B \leq_T A \equiv_T B \oplus K$ where K is the halting problem. A can even be chosen such that the reduction from B to A is a weak truth-table reduction, Merkle and Mihailović [12] give a simplified proof for this fact.

A natural question is whether it is necessary to go up to the degree of $B \oplus K$ in order to find the random set A. Martin-Löf random sets can be found below every set which is PA-complete, so there are Martin-Löf random sets in low and in hyperimmune-free Turing degrees. A set A is called PA-complete if one can compute relative to A a complete and consistent extension of the set of first-order formulas provable in Peano Arithmetic. An easier and equivalent definition of being PA-complete is to say that given any partial-recursive and $\{0, 1\}$-valued function ψ, one can compute relative to A a total extension Ψ of ψ. One can of course choose Ψ such that also Ψ is $\{0, 1\}$-valued.

Extending all possible $\{0, 1\}$-valued partial-recursive functions ψ is as difficult as to compute a $\{0, 1\}$-valued DNR function. A diagonally nonrecursive (DNR) function f satisfies $f(x) \neq \varphi_x(x)$ whenever $\varphi_x(x)$ is defined. Kučera

Frank Stephan did most of this work while he held positions in Heidelberg and Sydney. In Heidelberg, F. Stephan was supported by the Deutsche Forschungsgemeinschaft (DFG), Heisenberg grant Ste 967/1–1. In Sydney, F. Stephan worked for the Kensington Research Laboratory of the National ICT Australia which is funded by the Australian Government's Department of Communications, Information Technology and the Arts and the Australian Research Council through Backing Australia's Ability and the ICT Centre of Excellence Program.

Logic Colloquium '02
Edited by Z. Chatzidakis, P. Koepke, and W. Pohlers
Lecture Notes in Logic, 27

[7] showed that one can compute relative to any Martin-Löf random set A a DNR function f but that f is not $\{0, 1\}$-valued: Taking a sufficiently large c, $f(x)$ is just the value of the string $1A(0)A(1)\ldots A(x+c)$ interpreted as a binary number. For all x where $\varphi_x(x)$ is defined, the Kolmogorov complexity of $f(x)$ is strictly larger than that of $\varphi_x(x)$ and it follows that $f(x) \neq \varphi_x(x)$.

So Martin-Löf random sets and PA-complete sets have in common that one can compute relative to them DNR functions. Therefore, it is a natural question whether their degrees coincide. Kučera [7] showed that this is not the case: while the measure of the class of Martin-Löf random sets is 1, the measure of the PA-complete sets is 0. Since the notion PA-complete is invariant with respect to Turing equivalence, there are Turing degrees containing Martin-Löf random sets but no PA-complete sets. It remains to ask whether there is at least an inclusion: does every Turing degree containing a PA-complete set also contain a Martin-Löf random set? Kučera [7, 8] answered this question also negatively and constructed several examples of PA-complete sets A such that no set below A is both, Martin-Löf random and PA-complete. Ambos-Spies and Kučera [1, Open Problem 3.5] asked whether there is an $A \not\geq_T K$ which fails to have this property. The negative answer to this question is the main result of the present work. So the PA-complete sets constructed by Kučera share the desired property with all PA-complete sets not above K.

From this result it also follows that the uniform constructions of Gács [3] and Kučera [7, 8] to reduce a given set B to a Martin-Löf random set A are optimal with respect to the Turing degree of A: For many sets B, in particular for the PA-complete sets B, one cannot avoid that the set A is above K.

§2. Only the random sets above K are PA-complete.

Before stating the main result, some explanations on the used notions of effective measure, randomness and PA-completeness are given.

MEASURES. The measure used is the standard measure for the infinite product $\{0, 1\}^\infty$ of $\{0, 1\}$ where the measure of $\{0, 1\}$ is 1 and of each $\{b\} \subseteq \{0, 1\}$ is $\frac{1}{2}$. In particular, the measure of the class of all sets B which extend a binary string σ is $2^{-|\sigma|}$ where $|\sigma|$ is the length of the string σ, that is, the cardinality of its domain. While in standard measure theory, every singleton $\{A\}$ has measure 0, this does no longer hold for effective versions like the Σ_1^0-measure. Most singletons $\{A\}$ are not effectively measurable. The measurable ones are contained in a sequence of classes which are effectively measurable and whose measure goes to 0. More formally, A is contained in a class of Σ_1^0-measure 0 iff there is a sequence $U_0, U_1, \ldots$ of subclasses of $\{0, 1\}^\infty$ such that

- every U_n has measure 2^{-n} or less;
- the U_n are uniformly Σ_1^0, that is, there is a recursively enumerable set R of pairs (σ, n) such that a set B is in U_n iff there is an m with $(B(0)B(1)\ldots B(m), n) \in R$;
- $A \in U_n$ for all n.

MARTIN-LÖF RANDOM SETS are those sets which are not contained in a class of Σ_1^0-measure 0. Equivalently, one can say that A is Martin-Löf random iff $\{A\}$ does not have Σ_1^0-measure 0.

PA-COMPLETE SETS are those sets which permit to compute a complete and consistent extension of Peano Arithmetic. Equivalently, one can say that a set A is PA-complete iff every $\{0, 1\}$-valued partial-recursive function ψ has a total $\{0, 1\}$-valued extension Ψ which is computable relative to A.

The interested reader might consult the books of Li and Vitányi [10], Odifreddi [13] and Soare [15] for a formal definition of the further concepts mentioned in this paper. Martin-Löf random sets are named after their inventor Martin-Löf [11].

The main result says that there are two types of Martin-Löf random sets: the first type are the computationally powerful sets which permit to solve the halting problem K; the second type of random sets are computationally weak in the sense that they are not PA-complete. Every set not belonging to one of these two types is not Martin-Löf random.

THEOREM. *Let A be Martin-Löf random. Then one can compute relative to A a complete and consistent extension of Peano-Arithmetic if and only if one can solve the halting problem relative to A.*

PROOF. The theorem is proven by considering any $A \not\geq_T K$ which is PA-complete and showing that such an A cannot be Martin-Löf random. As a first step, one constructs a partial-recursive $\{0, 1\}$-valued function ψ. Since A is PA-complete, A permits to compute a total $\{0, 1\}$-valued extension of ψ. In the second step, one uses some properties of ψ and the fact that $A \not\geq_T K$ for the construction of a Martin-Löf test which witnesses that A is not Martin-Löf random.

CONSTRUCTION OF ψ. The goal of the construction of ψ is that the class of oracles B such that φ_e^B is a total extension of ψ is small. More precisely, the measure of this class and also the measure of almost every approximation to it should be below 2^{-e-1}. Furthermore, it is sufficient to assign to every e an interval I_e such that the same holds for the class of oracles B for which φ_e^B is total on I_e.

Now the construction is given in detail. The partial function ψ is undefined on $\{0, 1, 2, 3\}$. On the intervals $I_e = \{2^{e+2}, 2^{e+2} + 1, \ldots, 2^{e+3} - 1\}$, one defines ψ in stages as below, the construction of ψ on I_e does not interact with the construction on any other $I_{e'}$, $e' \neq e$. The above mentioned approximations are obtained by considering in stage s only those numbers where ψ has already been defined in previous stages and only those oracles, which are computed on I_e with use at most s. Here the use of φ_e^B at input x is defined as follows: $\text{use}(e, x, B)$ is the supremum of all t such that the computation $\varphi_e^B(x)$ either needs at least time t or queries B at t; note that $\text{use}(e, x, B) = \infty$ if the computation $\varphi_e^B(x)$ does not halt. Before starting the construction, let $a_0 = \min(I_e)$ and ψ be undefined everywhere.

STAGE s of the construction of ψ on I_e. Let $P_{e,s,b}$ be the Δ_1^0-class of all oracles B such that, for all $x \in I_e$,

- $\varphi_e^B(x)$ halts and $\text{use}(e, x, B) \leq s$;
- if $x < a_s$ then $\varphi_e^B(x) = \psi(x)$;
- if $x = a_s$ then $\varphi_e^B(x) = b$.

Compute the measures $d_{e,s,0}$ of $P_{e,s,0}$ and $d_{e,s,1}$ of $P_{e,s,1}$. If $d_{e,s,0}+d_{e,s,1} > 2^{-e-1}$

- then choose $b \in \{0,1\}$ such that $d_{e,s,b} \leq d_{e,s,1-b}$, let $\psi(a_s) = b$ and update $a_{s+1} = a_s + 1$
- else let ψ unchanged and let $a_{s+1} = a_s$.

This completes stage s.

PROPERTIES OF ψ. It is easy to see that ψ is partial-recursive. Furthermore, whenever in stage s a new value for ψ on I_e is defined, the measure of the class of oracles B for which φ_e^B is consistent with ψ before but not after stage s is at least 2^{-e-2}. One defines in at most $2^{e+2} - 2$ many stages s a new value $\psi(a_s)$ because after $2^{e+2} - 2$ times defining a new value, the measure of the class of the oracles B for which φ_e^B is $\{0,1\}$-valued and consistent with ψ is at most 2^{-e-1}. It follows that $a_s \in I_e$ for all stages s. So the procedure to define ψ on I_e terminates eventually without having used up the entries on I_e completely. For all stages s where no new value $\psi(a_s)$ is defined it holds that $d_{e,s,0} + d_{e,s,1} \leq 2^{-e-1}$.

A IS NOT MARTIN-LÖF RANDOM. Since A is PA-complete, there is an A-recursive $\{0,1\}$-valued total function Ψ which extends ψ. By the Padding Lemma [13, Proposition II.1.6], there is a recursive ascending one-one sequence $e_0, e_1, \ldots$ of programs which compute Ψ relative to A: $\Psi = \varphi_{e_k}^A$ for all k. Furthermore, let $b_0, b_1, \ldots$ be a one-one enumeration of K.

Now one defines for every s the following numbers $e(s)$ and $r(s)$ in dependence of s: $e(s) = e_{b_s}$ and $r(s)$ is the first stage $t > s$ where $P_{e(s),t,0} \cup P_{e(s),t,1}$ has at most the measure $2^{-e(s)-1}$. Note that the membership of B in $P_{e(s),r(s),0} \cup P_{e(s),r(s),1}$ depends only on the values of B at arguments up to $r(s)$, thus this class is a Σ_1^0-class. Since the construction is effective in s, one can define the sequence

$$U_n = \bigcup_{s \text{ where } e(s) \geq n} P_{e(s),r(s),0} \cup P_{e(s),r(s),1}$$

and has that these classes U_n are uniformly Σ_1^0-classes. The measure of U_n is bounded by the sum over all $2^{-e(s)-1}$ where $e(s) \geq n$. Since the mappings $s \to b_s$ and $k \to e_k$ are one-one, so is the mapping $s \to e(s) = e_{b_s}$. Thus the infinite sum $2^{-n-1} + 2^{-n-2} + \cdots = 2^{-n}$ is an upper bound for the measure of U_n. It follows that the sequence of the U_n is a Martin-Löf test.

For given k, let $f(k) = \max\{\text{use}(e_k, x, A) : x \in I_{e_k}\}$. The function f can be computed relative to A. Since $K \not\leq_T A$, there are infinitely many $k \in K$ such that $s > f(k)$ for the s with $k = b_s$. It follows for these k, s and all $x \in I_{e_k}$ that $\varphi^A_{e_k}(x)$ halts and use$(e_k, x, A) \leq r(s)$. So $A \in P_{e(s).r(s),0} \cup P_{e(s).r(s),1}$. Since A is in infinitely many classes $P_{e(s).r(s),0} \cup P_{e(s).r(s),1}$, A is also in all classes U_n. It follows that A is not Martin-Löf random. $\dashv$

EXAMPLES. A natural example for a Martin-Löf random set of the first type is Chaitin's Ω. Ω is in the Turing-degree of K. One among several definitions for Ω is that there is a universal prefix-free Turing machine M such that Ω is the set of positions n where the n-th binary digit of the halting probability is 1. Formally, the halting probability of M is the measure of the Π^0_1-class of those sets B such that there is an n for which $M(B(0)B(1) \cdots B(n))$ halts. Martin-Löf random sets are immune and thus not recursively enumerable, but Ω is as close to being recursively enumerable as possible: The left cut $\{\sigma \in \{0, 1\}^* : \sigma$ is lexicographically before $\Omega(0)\Omega(1)\ldots\}$ is a recursively enumerable set of finite strings. Kučera and Slaman [9] showed that all Martin-Löf random sets with recursively enumerable left cut are Ω-like, that is, every such set can be constructed in the same way as Chaitin constructed Ω.

One can relativize the definition of Ω to oracles, so Ω^K is the probability that a fixed universal prefix-free machine with access to the oracle K holds. This relativized construction gives a set which is Martin-Löf random relative to K and satisfies $(\Omega^K)' \equiv_T K \oplus \Omega^K$ by a result of Kautz [6]. This equivalence gives immediately $\Omega^K \not\geq_T K$. So, Ω^K is an example for a Martin-Löf random set of the second type.

§3. Applications. Arslanov's Completeness Criterion [2] says that every recursively enumerable set which permits to compute a fixed-point-free function is already above K. A fixed-point-free function f has the property that $W_x \neq W_{f(x)}$ for all x. Jockusch, Lerman, Soare and Solovay [4] showed that Arslanov's Completeness Criterion also holds with DNR functions in place of fixed-point-free functions. They furthermore generalized the result: For every sequence $A_1, A_2, \ldots, A_n$ of sets such that $A_1 <_T A_2 <_T \cdots <_T A_n, A_n \not\geq_T K$, A_1 is recursively enumerable and each A_{m+1} is recursively enumerable relative to A_m for $m = 1, 2, \ldots, n - 1$, one cannot compute a DNR function relative to A_n. The corresponding result does not hold for Martin-Löf random sets because each of them can compute a DNR function [7]. But the main result of the present work gives a natural variant of Arslanov's completeness criterion: A Martin-Löf random set A is above K iff one can compute a $\{0, 1\}$-valued DNR function relative to A.

Call a set B to be DNR iff its characteristic function is an (automatically $\{0, 1\}$-valued) DNR function. The degrees of DNR sets are closed upward: Let $\{a_0, a_1, \ldots\}$ be a recursive set such that $\varphi_x(x) = 2$ for all $x \in \{a_0, a_1, \ldots\}$ and let $B \leq_T A$ be a DNR set. Then one can construct a further DNR set C by $C(x) = B(x)$ for $x \notin \{a_0, a_1, \ldots\}$ and $C(a_k) = A(k)$ for all k. The set C

has the same Turing degree as A. So, the Turing degrees containing a DNR set coincide with those of PA-complete sets. In particular, the Turing degrees containing both, a DNR set and a Martin-Löf random set, coincide with the upper cone of Turing degrees above the one of K. This answers negatively Open Problem 3.5 of Ambos-Spies and Kučera [1]. Furthermore, as Ambos-Spies and Kučera already noted, their Open Problem 3.6 is also solved by the negative answer to Open Problem 3.5: Whenever a Turing degree contains a DNR set and a Martin-Löf random set, then every Turing degree above this one contains such sets, too.

Kučera [8] constructed a PA-complete set $A <_T K$ such that the Turing degree of A does not contain a Martin-Löf random set. Since there is a Martin-Löf random set below A, one can conclude that the Turing degrees of Martin-Löf random sets are not closed upward [1]. This result is even strengthened.

PROPOSITION. *For every set $A \not\geq_T K$ there is a set $B \geq_T A$ such that the Turing degree of B does not contain a Martin-Löf random set.*

PROOF. The sets above K are exactly the sets which permit to compute a function majorizing the modulus c_K of convergence of any fixed given enumeration of K. Since $A \not\geq_T K$, no such majorizing function is A-recursive. By the Hyperimmune-Free Basis Theorem of Jockusch and Soare [5], there is a PA-complete set $B \geq_T A$ which is hyperimmune-free relative to A. In particular, every total B-recursive function is majorized by a total A-recursive function. Therefore no total B-recursive function majorizes c_K. Thus $B \not\geq_T K$. It follows that the Turing degree of B does not contain a Martin-Löf random set.   ⊣

Scott and Tennenbaum [14] showed that the Turing degree of a PA-complete set cannot be minimal. Since the Turing degree of K is not minimal, one only has to consider sets $A \not\geq_T K$. The traditional way to prove this result is to take a Martin-Löf random set B below A and then to consider the set $C = \{x : 2x \in B\}$. For these sets A, B, C one has $\emptyset <_T C <_T B \leq_T A$. On the one hand, the main result of the present work permits to conclude that already B satisfies $\emptyset <_T B <_T A$; one does not need the set C. On the other hand, the proof that $C <_T B$ is less involved than the proof that $B <_T A$. So it is a matter of taste which proof one prefers.

Gács [3] and Kučera [7, 8] showed that one can find for any set B a Martin-Löf random set A such that $B \leq_T A \equiv_T B \oplus K$. In this result, K cannot be replaced by any set $C \not\geq_T K$: Given such a set $C \not\geq_T K$, there is a set B which is PA-complete, above C and not above K. Then all Martin-Löf random sets A with $A \geq_T B$ satisfy $A \geq_T K$. In particular, such a set A is not in the Turing degree of $B \oplus C$.

Acknowledgment. The author would like to thank Bill Gasarch, André Nies and Sebastiaan A. Terwijn for interesting comments.

REFERENCES

[1] KLAUS AMBOS-SPIES and ANTONÍN KUČERA, *Randomness in computability theory*, **Computability Theory and its Applications. Current Trends and Open Problems. Proceedings of a 1999 AMS-IMS-SIAM Joint Summer Research Conference, Boulder, Colorado, USA, June 13–17, 1999** (Peter A. Cholak, Steffen Lempp, Manuel Lerman, and Richard A. Shore, editors), Contemporary Mathematics, vol. 257, American Mathematical Society, Providence, 2000, pp. 1–14.

[2] MARAT M. ARSLANOV, *Some generalizations of a fixed-point theorem*, **Soviet Mathematics**, (1981), no. 25, pp. 1–10, translated from **Izvestiya Vysshikh Uchebnykh Zavedenij, Matematika**, (1981), no. 228, pp. 9–16.

[3] PÉTER GÁCS, *Every sequence is reducible to a random one*, **Information and Control**, vol. 70 (1986), no. 2-3, pp. 186–192.

[4] CARL G. JOCKUSCH, JR., MANUEL LERMAN, ROBERT I. SOARE, and ROBERT M. SOLOVAY, *Recursively enumerable sets modulo iterated jumps and extensions of Arslanov's completeness criterion*, **The Journal of Symbolic Logic**, vol. 54 (1989), no. 4, pp. 1288–1323.

[5] CARL G. JOCKUSCH, JR. and ROBERT I. SOARE, Π^0_1 *classes and degrees of theories*, **Transactions of the American Mathematical Society**, vol. 173 (1972), pp. 33–56.

[6] STEVEN M. KAUTZ, *Degrees of Random Sets*, Ph.D. thesis, Cornell University, 1991.

[7] ANTONÍN KUČERA, *Measure, Π^0_1-classes and complete extensions of* PA, **Recursion Theory Week, Proceedings of a Conference Held in Oberwolfach, West Germany, April 15–21, 1984** (Heinz-Dieter Ebbinghaus, Gert H. Müller, and Gerald E. Sacks, editors), Lecture Notes in Mathematics, vol. 1141, Springer, Berlin, 1985, pp. 245–259.

[8] ———, *On the use of diagonally nonrecursive functions*, **Logic Colloquium 1987. Proceedings of the Colloquium Held in Granada, Spain, July 20–25, 1987** (Heinz-Dieter Ebbinghaus, José Fernandez-Prida, Manuel Garrido, Daniel Lascar, and Mario Rodríguez Artalejo, editors), Studies in Logic and the Foundations of Mathematics, vol. 129, North-Holland, Amsterdam, 1989, pp. 219–239.

[9] ANTONÍN KUČERA and THEODORE A. SLAMAN, *Randomness and recursive enumerability*, **SIAM Journal on Computing**, vol. 31 (2001), no. 1, pp. 199–211.

[10] MING LI and PAUL VITÁNYI, **An Introduction to Kolmogorov Complexity and its Applications**, 2 ed., Springer, New York, 1997.

[11] PER MARTIN-LÖF, *The definition of random sequences*, **Information and Computation**, vol. 9 (1966), pp. 602–619.

[12] WOLFGANG MERKLE and NENAD MIHAILOVIĆ, *On the construction of effective random sets*, **Mathematical Foundations of Computer Science 2002, 27th International Symposium, MFCS 2002, Warsaw, Poland, August 26–30, 2002, Proceedings** (Krzysztof Diks and Wojciech Rytter, editors), Lecture Notes in Computer Science, vol. 2420, Springer, Berlin, 2002, pp. 568–580.

[13] PIERGIORGIO ODIFREDDI, **Classical Recursion Theory**, North-Holland, Amsterdam, 1989.

[14] DANA SCOTT and STANLEY TENNENBAUM, *On the degrees of complete extensions of arithmetic*, **Notices of the American Mathematical Society**, vol. 7 (1960), pp. 242–243.

[15] ROBERT I. SOARE, **Recursively Enumerable Sets and Degrees**, Springer, Berlin, 1987.

SCHOOL OF COMPUTING AND DEPARTMENT OF MATHEMATICS
NATIONAL UNIVERSITY OF SINGAPORE
3 SCIENCE DRIVE 2, SINGAPORE 117543
REPUBLIC OF SINGAPORE
E-mail: fstephan@comp.nus.edu.sg

LEARNING AND COMPUTING IN THE LIMIT

SEBASTIAAN A. TERWIJN

Abstract. We explore two analogies between computability theory and a basic model of learning, namely Osherson and Weinstein's model theoretic learning paradigm. First, we build up the theory of model theoretic learning in a way analogous to the way computability theory is built up. We then discuss Δ_2-definability of predicates on classes and prove a limit lemma for continuous functionals.

§1. Introduction. The notion of limit computability crops up in a natural way in the study of the arithmetical hierarchy and the notion of relative computability, and has been extensively studied over the last decades, see e.g. the monographs [9, 16]. The idea of learning as a limit process is also central to a large number of models of learning, in particular those introduced by Gold [3], whose paradigm became known under the phrase "learning in the limit". Osherson and Weinstein [12] introduced a model of learning in the limit of first order sentences over models of a first order theory. Below we describe Osherson and Weinstein's model, and we state two results linking Δ_2-definability to limit computability. Before we do so we discuss some related work.

The relation between computation and definability by logical formulas is one of the cornerstones of computability theory, cf. Odifreddi [9], Rogers [14]. Interestingly, prior to his fundamental contribution to the theory of induction [3], Gold already wrote about limit computability [2], apparently unaware of earlier work of Shoenfield [15]. The book by Kelly [6] and the paper Gasarch et al. [1] contain topological characterizations of classes of functions that have a classifier, which is a method for deciding whether a function is in the given class or not. Although technically different, some of these results are similar in form to results such as Theorem 1.2 and Proposition 2.6 below. The role of limit computability in mathematics in general is investigated in the project by Hayashi [4].

A Δ_2-formula is a formula that is equivalent (over some theory and in a language given by the context) to both a Σ_2-formula (i.e. a formula in existential-universal form) and a Π_2-formula (i.e. a formula in universal-existential form).

The author was supported by the Austrian Research Fund FWF (Lise Meitner grant M699-N05).

Logic Colloquium '02
Edited by Z. Chatzidakis, P. Koepke, and W. Pohlers
Lecture Notes in Logic, 27
© 2006, Association for Symbolic Logic

The following lemma is a basic result from computability theory. Let ω be the set of natural numbers. A set $A \subseteq \omega$ (which we identify with its characteristic function) is *limit computable* if there is a computable function f such that for all n, $A(n) = \lim_{s \to \infty} f(n, s)$.

LEMMA 1.1 (Shoenfields Limit Lemma [15]). *A set $A \subseteq \omega$ is definable by an arithmetical Δ_2-formula if and only if A is limit computable.*

The Δ_2-definable arithmetical sets can alternatively be characterized as the class of sets computable relative to the halting set K (Post).

The second result of the type above we discuss is the result of Osherson, Stob, and Weinstein [11] proving that a sentence is learnable over the models of a theory T if and only if the sentence is equivalent to a Δ_2-sentence over T. (See Theorem 1.2 below.) We first describe Osherson and Weinstein's model theoretic learning paradigm [12]. For a survey of this and related models of learning see e.g. [13, 10]. An interesting link between this paradigm and the theory of belief revision was provided in Martin and Osherson [7, 8].

For a first order theory T, $\mathrm{mod}(T)$ denotes the set of *countable* models of T. Since only countable models are considered, we may assume they all have the set of natural numbers ω as their universe. Fix a language $\mathcal{L}$ of finite signature that includes identity, and a countable set $\{v_i : i \in \omega\}$ of variables. A convenient but inessential assumption we make is that $\mathcal{L}$ has no function symbols. The *basic formulas* of $\mathcal{L}$ are the atomic formulas and the negations thereof. An *assignment* is a function $d : \{v_i : i \in \omega\} \to \omega$ that is onto, so that every element in the model is assigned at least one variable. Given a model $M \in \mathrm{mod}(T)$ and an assignment d, the *environment* for M and d is the ω-sequence with e whose i-th element is $e_i = \{\beta \text{ basic} : \mathrm{var}(\beta) \subseteq \{v_0, \ldots, v_{i-1}\} \wedge M \models \beta(d \restriction i)\}$. That is, the environment e lists all basic facts of that are true in M, in the order determined by the assignment d. The environment determined by M and d is denoted by $[M, d]$. If $\sigma \sqsubseteq d$ is a finite initial segment of d, the finite initial segment of $[M, d]$ induced by it is denoted by $[M, \sigma]$. Clearly, if e is an environment for both M and N, then M and N are isomorphic. The set of finite initial segments of environments, i.e. segments of the form $e \restriction n$, is denoted by SEQ.

A *learner* is defined to be any function from SEQ to the set of $\mathcal{L}$-sentences. So in this setting a learner is conceived of as something that conjectures an $\mathcal{L}$-sentence after having seen a finite basic part of a model of a given theory.

Given a sentence φ and a theory T, we say that a learner Φ *learns* φ on $\mathrm{mod}(T)$ if for every $\tau \in \mathrm{SEQ}$, $\Phi(\tau) \in \{\varphi, \neg\varphi\}$, and for every $M \in \mathrm{mod}(T)$ and every environment e for M

- $\theta = \lim_{n \to \infty} \Phi(e \restriction n)$ exists,
- $M \models \theta$.

That is, Φ learns φ on $\mathrm{mod}(T)$ if for any model M of T, Φ can determine in the limit the truth value of φ in M, given any enumeration of the basic truths of M.[1] A sentence φ is *learnable* over $\mathrm{mod}(T)$ if some learner Φ learns φ over $\mathrm{mod}(T)$.

THEOREM 1.2 (Osherson, Stob, and Weinstein [11]). *For any theory T, a sentence φ is learnable over $\mathrm{mod}(T)$ if and only if it is equivalent over T to both a Σ_2-sentence and a Π_2-sentence.*

Theorem 1.2 has as a consequence that there exists a computable universal learner, that is able to learn any first-order sentence over a set of models $\mathrm{mod}(T)$, given T as an oracle. Universal learners also exist for the basic setting of Golds model, but they are in general not computable: Stephan and Terwijn [18] proved that a Turing degree contains a universal text learner if and only if the degree is greater than or equal to $0''$.

In Section 2 below we also use the notion of oracle (or relative) computability to introduce some new learning-theoretic notions that parallel basic notions in computability theory. In particular we define analogues for (relative) computability and computable enumerability in the model theoretic setting. We also point out where the analogy breaks down. In Section 3 we consider a higher order analogy between the two theories. We prove a limit lemma for functionals and discuss definability on classes.

§2. A first analogy.

We start by defining some learning theoretic concepts that parallel basic notions in computability theory. In learning theory, a change in hypothesis of the learner is called a *mind change*. When counting mind changes, it is customary to allow an initial empty hypothesis "?", such that the change from "?" to the formulation of a first hypothesis does not count as a mind change.

DEFINITION 2.1. We have the following notions of learnability of sentences:

one-shot learnable $\equiv$ learnable with 0 mind changes, starting with '?' Example: any basic formula, any sentence in T.

Σ_1-*learnable* $\equiv$ learnable with at most 1 mind change, starting with 'φ is false'.

Π_1-*learnable* $\equiv$ learnable with at most 1 mind change, starting with 'φ is true'.

one-shot learnable relative to a set of sentences S $\equiv$ one-shot learnable when validity of every $\varphi \in S$ in M is known.

The last notion serves as an analogue of the notion of relative computability. A special case is when S consists of all $\forall$-sentences, giving the analogous notion of learnability relative to the halting set K, which we can also think of as a

---

[1]This definition of learning the truth of a given sentence φ has obvious variations that can be defined by using sets different from $\{\varphi, \neg\varphi\}$, see e.g., [10].

∀-oracle.[2] The analogy between the several notions is summarized in the following table:

Computability theory	Model theoretic learning
$A \subseteq \omega$	model M
$n \in A$	$M \models \varphi$
approximating function f	learner Φ
computable	one-shot learnable
limit computable	learnable in the limit
computably enumerable, Σ_1^0-definable	Σ_1-learnable See also Proposition 2.3
co-c.e., Π_1^0-definable	Π_1-learnable See also Proposition 2.3
A-computable	one-shot φ-learnable
Limit Lemma: K-computable = limit computable	Proposition 2.4: one-shot learnable relative ∀-oracle = learnable in the limit
Posts theorem: computable $= \Sigma_1^0 \cap \Pi_1^0$	Proposition 2.5
K-computable $= \Sigma_2^0 \cap \Pi_2^0$	Proposition 2.6
limit computable $= \Sigma_2^0 \cap \Pi_2^0$	Theorem 1.2

Below we will use the following notion, which is a specification of the notion of confirmability in [11] to Σ_1-sentences:

DEFINITION 2.2. A sentence φ is Σ_1-*confirmable* in T if for every model M in $\mathrm{mod}(T \cup \{\varphi\})$ there is a Σ_1-sentence ψ_M such that

- $M \models \psi_M$,
- $T \cup \{\psi_M\} \models \varphi$.

PROPOSITION 2.3. *A sentence φ is Σ_1-learnable if and only if it is equivalent (over the background theory T) to a Σ_1-sentence. A sentence φ is Π_1-learnable if and only if it is equivalent (over T) to a Π_1-sentence.*

PROOF. We only prove the difficult direction of the first part. The second part of the proposition can be proved in a similar way. Let Φ be a learner

---

that Σ_1-learns φ. The proof that φ is equivalent to a Σ_1-sentence follows the scheme of the proof of Theorem 1.2, with some simplifications. First we prove that φ is Σ_1-confirmable in T. So let $M \in \mathrm{mod}(T \cup \{\varphi\})$. Let σ be a locking sequence for Φ and M, i.e. a sequence σ such that for all extensions $\tau \sqsupseteq \sigma$, $\Phi([M, \tau]) = \Phi([M, \sigma]) = \varphi$. Such a sequence is easily seen to exist. Let χ be the conjunction of all formulas in $[M, \sigma]$, and let ψ_M be the Σ_1-sentence obtained by quantifying out the free variables in χ with an existential quantifier. Clearly $M \models \psi_M$, and if $N \models T \cup \{\psi_M\}$ then there is a sequence σ' such that $[N, \sigma'] = [M, \sigma]$, so that $\Phi([N, \sigma']) = \Phi([M, \sigma]) = \varphi$, and hence $N \models \varphi$ since Φ Σ_1-learns φ on $\mathrm{mod}(T)$.

Now to conclude the proof, we use a compactness argument to show that φ is Σ_1-learnable if and only if φ is equivalent over T to a Σ_1-sentence. (cf. [11, p. 669]). The 'if' part is immediate. For the 'only if' part, consider the set

$$\Sigma = \{\psi_M : M \in \mathrm{mod}\left(T \cup \{\varphi\}\right)\}.$$

We claim that φ is equivalent to a finite disjunction Δ of sentences in Σ. By definition of ψ_M, every such finite disjunction entails φ. Suppose conversely that for every finite $\Delta \subseteq \Sigma$ the theory $T \cup \{\varphi\} \cup \{\neg\theta : \theta \in \Delta\}$ is consistent. Then by the compactness and Löwenheim-Skolem theorems there is a countable model N of $T \cup \{\varphi\} \cup \{\neg\theta : \theta \in \Sigma\}$, which contradicts $N \models \psi_N$. ⊣

PROPOSITION 2.4. *A sentence φ is one-shot learnable with an oracle for $\forall$-sentences if and only if it is learnable in the limit.*

PROOF. This follows from Theorem 1.2. The proof is similar to that of Proposition 2.3. ⊣

PROPOSITION 2.5. *For any sentence φ the following are equivalent:*

(i) *φ is one-shot learnable,*
(ii) *φ is both Σ_1-learnable and Π_1-learnable*
(iii) *φ is both equivalent (over the background theory T) to a Σ_1-sentence and a Π_1-sentence.*

PROOF. (i)$\Rightarrow$(ii) is immediate. (ii)$\Leftrightarrow$(iii) follows from Proposition 2.3. For (ii)$\Rightarrow$(i), when φ is both Σ_1 and Π_1-learnable then it is one-shot learnable in the following way: Start with hypothesis '?' and wait until the Σ_1 and the Π_1-algorithm have the same output. ⊣

In order to "relativize" Proposition 2.5 to a $\forall$-oracle, it will be useful to have the following notion of relativized structure. Given the first order language $\mathcal{L}$ and an $\mathcal{L}$-structure M, we want to define a new language $\mathcal{L}^\forall$ and structure $M^\forall$ such that the $\forall$-formulas from the first setting become atomic in the second. This is easily achieved as follows. For every n-ary predicate $R(x_1, \ldots, x_n)$ in

$\mathcal{L}$ and every partition of the variables $x_1, \ldots, x_n$ into $y_1, \ldots, y_k$ and $z_1, \ldots, z_l$ ($l > 0$), we have a new l-ary predicate $R^{\forall y_1, \ldots, y_k}(z_1, \ldots, z_l)$ in $\mathcal{L}^\forall$, with the intended meaning

$$R^{\forall y_1, \ldots, y_k}(z_1, \ldots, z_l) \iff \forall y_1, \ldots, y_k R(x_1, \ldots, x_n).$$

(R is assumed to be at least binary here. There is no loss in not considering unary predicates.) The structure $M^\forall$ is the same as M, except that there are now *more* atomic formulas since the language has changed. Since environments enumerate in the limit all basic facts, an environment for $M^\forall$ acts as an oracle for all $\forall$-formulas for M.

PROPOSITION 2.6. *A sentence φ is one-shot learnable with an oracle for $\forall$-sentences if and only if it is both equivalent (over T) to a Σ_2-sentence and a Π_2-sentence.*

PROOF. We obtain this by "relativizing" Proposition 2.5 to a $\forall$-oracle:

φ is one-shot learnable with an $\forall$-oracle $\qquad\qquad \iff$
φ is one-shot learnable over $M^\forall$ $\qquad\qquad\qquad\quad \iff$
φ equivalent to both a Σ_1 and a Π_1-sentence in $\mathcal{L}^\forall$ $\quad \iff$
φ equivalent to both a Σ_2 and a Π_2-sentence in $\mathcal{L}$

The second equivalence holds by Proposition 2.5.   ⊣

Say that φ is Σ_2-*learnable* if it is Σ_1-learnable with a $\forall$-oracle, and that φ is Π_2-*learnable* if it is Π_1-learnable with a $\forall$-oracle. From Propositions 2.4, 2.5, and 2.6 we see that φ is learnable if and only if it is both Σ_2-learnable and Π_2-*learnable*.

EXAMPLE 2.7. Let T be the theory of linear orders R. Then $\varphi = \exists x \forall y R x y$ is not learnable. It is not even Π_2-learnable. However, φ is Σ_2-learnable.

EXAMPLE 2.8. Let $\mathcal{L} = \langle \leq, R, S \rangle$ be a signature with three binary predicates, and let T be the theory saying that $\leq$ is a linear order. Let $\varphi = \exists x \forall z \exists y \geq z R x y$ be the sentence saying that there is an infinite R-section, and let T contain the axiom $\neg\varphi \leftrightarrow \exists x \forall z \exists y \geq z S x y$, so that every model of T either has an infinite R-section or an infinite S-section. Then φ is not learnable over $\mathrm{mod}(T)$, but since φ is Δ_3 over T, it is learnable with an $\forall$-oracle.

EXAMPLE 2.9. One can ask for which theories T the above analogy is closest. Suppose that the language $\mathcal{L}$ contains the language of arithmetic, and that T is a recursive axiomatization of basic computability theory. Then the questions about one-shot learnability, Σ_1-learnability, etc. become identical to the questions of computability theory. So in this sense model theoretic learning is more general than computability theory. One can prove e.g. the following. In analogy to the existence of Turing incomparable c.e. sets (Friedberg-Muchnik) there exist Σ_1-learnable formula's φ and ψ (i.e. sets of Σ_1-sentences parameterized by one formula) such that neither φ is one-shot learnable relative to ψ, nor the other way round.

Despite the analogy between the two theories, there are also differences. One is in the analogy between the Limit Lemma and Proposition 2.4. In the first case we have a set A given by a computable approximation: $A(x) = \lim_s f(x, s)$. Seen as a learning problem, the task is to "learn" (compute) in the limit the value $A(n)$ from computable "data" only. Because the "learner" f is uniformly computable, it occurs in the language $\mathcal{L}$ and hence it can be quantified over, which gives an easy proof that A is K-computable. This possibility does not exist in the learning model, which makes the proof of Proposition 2.4 more difficult.

In the next section we look at a closer, more faithful analogy than the one given in this section, by jumping one level higher at the computability theory side, namely from sets of natural numbers to subsets of 2^ω.

§3. A second analogy. In this section we work with the first-order language of arithmetic with one extra unary predicate A, to be interpreted by a subset of ω. We denote this language by $\mathcal{L}$. A formula $\varphi \in \mathcal{L}$ need not be arithmetical if A is interpreted by a nonarithmetical set. If the interpretation of A is fixed, $\mathcal{L}$ is the language of all formulas that are arithmetical in A. We also write $A \models \varphi$ if $\varphi(A)$ is true. (Note that it would be more correct to write $\langle \omega, A \rangle \models \varphi$, since the universe will always be interpreted by ω, and only the interpretation of A varies.)

In the following, a functional will be a mapping $\Phi : 2^\omega \times \omega \to \omega$. The notion of continuity of functionals is related to the notion of relative computability in a natural way: A mapping Φ from 2^ω to 2^ω or ω is continuous if and only if if there exists an oracle X such that Φ is computable relative to X. (The set X is obtained by a suitable coding of the modulus of continuity.)

DEFINITION 3.1. For a continuous functional Φ we write $\lim \Phi = \varphi(\mathcal{A})$ if for every $A \in \mathcal{A}$, $\lim_{s \to \infty} \Phi(A, s)$ exists and

$$\lim_{s \to \infty} \Phi(A, s) = 1 \iff \varphi(A),$$
$$\lim_{s \to \infty} \Phi(A, s) = 0 \iff \neg\varphi(A).$$

We mostly write $\Phi^A(s)$ instead of $\Phi(A, s)$, to stress that we think of A as an oracle.

We now have the following analogy at the level of classes of sets.[3]

---

[3]It was pointed out to us that the left hand side of this table is essentially the paradigm of classification from [1, 17]. It is possible to explicitly translate all these paradigms into each other, using a sufficiently expressive language.

Computability theory	Model theoretic learning
a class $\mathcal{A} \subseteq 2^\omega$	$\mathrm{mod}(T)$, the countable models of a first-order theory T
$A \in \mathcal{A}$	$M \in \mathrm{mod}(T)$
oracle for A	environment $e =$ oracle for basic truths in M
$\varphi \in \mathcal{L}$, defines the class $\varphi(\mathcal{A}) = \{A \in \mathcal{A} : A \models \varphi\}$	φ sentence in the language of M
continuous functional	learner
Φ such that $\lim \Phi = \varphi(\mathcal{A})$ "Φ learns φ on $\mathcal{A}$"	Φ learns $\varphi \equiv$ $\forall M \forall e$ for M (Φ learns φ from e)

We have the following limit lemma for functionals:

PROPOSITION 3.2. *For every $\mathcal{A} \subseteq 2^\omega$ and every formula $\varphi \in \mathcal{L}$ the following are equivalent*:

(i) *There exists a partial recursive functional Φ such that $\lim \Phi = \varphi(\mathcal{A})$*,
(ii) *There exists a total recursive functional Φ such that $\lim \Phi = \varphi(\mathcal{A})$*,
(iii) *There exists a Δ_2-formula $\psi \in \mathcal{L}$ such that $\{A \in \mathcal{A} : \varphi(A)\} = \{A \in \mathcal{A} : \psi(A)\}$*.

PROOF. (i)$\Rightarrow$(ii) We can easily turn a partial recursive functional Φ into a total one $\hat{\Phi}$ by a looking back technique:

$$\hat{\Phi}^A(s) = \begin{cases} \Phi^A(t) & \text{for the largest } t \text{ such that } \Phi^A(t) \text{ converges in } s \text{ steps,} \\ 0 & \text{if such } t \text{ does not exist.} \end{cases}$$

Then $\hat{\Phi}$ is total and $\lim_s \hat{\Phi}^A(s)$ equals $\lim_s \Phi^A(s)$ whenever this last limit exists. In particular the limits are equal for every $A \in \mathcal{A}$.

(ii)$\Rightarrow$(iii) Suppose that $\lim \Phi = \varphi(A)$. Then for every A in $\mathcal{A}$,

$$\varphi(A) \iff \lim_{s \to \infty} \Phi^A(s) = 1$$
$$\iff \exists t \forall s \geq t \Phi^A(s) = 1$$
$$\iff \forall t \exists s \geq t \Phi^A(s) = 1.$$

The last equivalence holds because $\lim_s \Phi^A(s)$ exists for every $A \in \mathcal{A}$. Since Φ is total recursive we see that φ is equivalent on $\mathcal{A}$ to both a Σ_2 and a Π_2-formula.

(iii)$\Rightarrow$(i) Suppose that ψ is a formula such that for every $A \in \mathcal{A}$

$$\psi(A) \iff \exists u \forall v S^A(u, v)$$
$$\neg\psi(A) \iff \exists x \forall y R^A(u, v),$$

where S and R are computable predicates. Say that $\psi(A)$ is *falsified* up to n at stage s if $\forall u \leq n \exists v \leq s \neg S^A(u, v)$, and likewise for $\neg \psi(A)$ (with S replaced by R). Define the partial recursive functional Φ as follows: To define $\Phi^A(n)$, search for a stage s such that either $\psi(A)$ or $\neg \psi(A)$ are falsified up to n at stage s. If $\psi(A)$ is falsified up to n let $\Phi^A(n) = 0$, and $\Phi^A(n) = 1$ otherwise. It is easy to see that Φ is total on $\mathcal{A}$. ⊣

The proof of Proposition 3.2 relativizes to an arbitrary oracle X: there is a partial X-recursive functional Φ such that $\lim \Phi = \varphi(A)$ if and only if here exists a Δ_2^X-formula ψ (i.e. a Δ_2-formula in the language $\mathcal{L}$ with an extra predicate for X) such that $\{A \in \mathcal{A} : \varphi(A)\} = \{A \in \mathcal{A} : \psi(A)\}$. Osherson et al. [11] showed that for special classes $\mathcal{A}$, namely for $\mathcal{A}$ the set of models of a first order theory, it is possible to obtain part (iii) from Proposition 3.2 even if the learner Φ from part (i) is not recursive. We formulate a recursion theoretic version of this result in the next theorem, using the Stone topology (see Keisler [5, p. 59]). The basic closed sets of this topology on 2^ω are the sets

$$\{A \in 2^\omega : A \models \varphi\}$$

where φ is an $\mathcal{L}$-sentence.[4]

THEOREM 3.3. *Let $\mathcal{A} \subseteq 2^\omega$ be compact in the Stone topology on 2^ω, and let $\varphi \in \mathcal{L}$. Then the following are equivalent:*

(i) *There exists a partial continuous functional Φ such that $\lim \Phi = \varphi(\mathcal{A})$,*
(ii) *There exists a Δ_2-formula $\psi \in \mathcal{L}$ such that $\{A \in \mathcal{A} : \varphi(A)\} = \{A \in \mathcal{A} : \psi(A)\}$.*

PROOF. The implication (ii)⇒(i) holds for every $\mathcal{A} \subseteq 2^\omega$ by relativizing the proof of (iii)⇒(i) in Proposition 3.2. The proof of (i)⇒(ii) is along the same lines as the proof of Theorem 1.2. The compactness of $\mathcal{A}$ is exactly what is needed (twice). ⊣

We end by showing that the condition of compactness in Theorem 3.3 in general cannot be deleted. We first prove a simple lemma.

LEMMA 3.4. *Let φ be the $\mathcal{L}$-sentence $\forall x \exists y (y \geq x \wedge y \in A)$ expressing that A is infinite. Then φ is not equivalent to a Σ_2-sentence, and for every finite initial segment σ we have*

(1) $$(\forall \psi \in \Sigma_2)(\exists A \sqsupset \sigma)[\varphi(A) \not\leftrightarrow \psi(A)].$$

PROOF. This is a straightforward diagonalization argument. It can also be shown that φ is complete for the Π_2 sentences of $\mathcal{L}$ in a natural sense,[5] but the property (1) will be sufficient here. Fix a Σ_2-sentence ψ, say $\psi = \exists x \forall y (R^A(x, y))$ for some recursive predicate R that uses A as an oracle. If

---

[4]This is called the Stone topology because it is obtained by considering the Stone space of the Lindenbaum algebra of the language $\mathcal{L}$.

[5]Namely, for every Π_2-sentence ψ in $\mathcal{L}$ there is a recursive functional Φ (Φ can even be chosen to be uniform in a code of ψ) such that for every set A, $\psi(A) \Leftrightarrow \varphi(\Phi(A))$.

there is a finite set $A \sqsupseteq \sigma$ such that $\psi(A)$ holds then we are done. Otherwise, we have

(2) $$(\forall A \sqsupseteq \sigma)\left[A \text{ finite} \longrightarrow \forall x \exists y \left(\neg R^A(x, y)\right)\right].$$

We use the property (2) to build an infinite set $C \sqsupseteq \sigma$ with $\neg \psi(C)$. We build $C = \bigcup_s \sigma_s$ using a finite extension construction such that

(3) $$(\forall s)\left(\exists y \le |\sigma_{s+1}|\right)\left[\neg R^{\sigma_{s+1}}(s, y)\right].$$

Stage $s = 0$: Set $\sigma_0 = \sigma$. Stage $s + 1$: Denote by $\sigma_s \widehat{\ } \emptyset$ the infinite sequence obtained by extending σ_s with infinitely many 0's. Since this is a finite set, by (2) we have $\exists y (\neg R^{\sigma_s \widehat{\ } \emptyset}(s, y))$. Let y be the smallest such y, and let u be the use of the computation of $\neg R^{\sigma_s \widehat{\ } \emptyset}(s, y)$, i.e. u is the largest number used in that computation. Define $\sigma_{s+1} = \sigma_s \widehat{\ } 0^u \widehat{\ } 1$, that is the sequence obtained by concatenating u 0's to σ_s followed by a 1. Then we also have $\neg R^{\sigma_{s+1} \widehat{\ } \emptyset}(s, y)$ since the 1 that is added in σ_{s+1} is above the use u.

Now $C = \bigcup_s \sigma_s$ is infinite because every σ_s adds a new 1, and by (3) it satisfies $\neg \psi(C)$.   ⊣

PROPOSITION 3.5. *There exist $A \subseteq 2^\omega$, a continuous functional Φ, and sentence φ such that* (i) *from Theorem 3.3 holds but* (ii) *does not.*

PROOF. Note that by Proposition 3.2 Φ will have to be nonrecursive. Let φ be the sentence $\forall x \exists y (y \ge x \wedge y \in A)$. By the property (1) we can choose for every $\psi \in \Sigma_2$ an A_ψ such that $\varphi(A_\psi) \not\leftrightarrow \psi(A_\psi)$. Now if we pick the A_ψ in some canonical way, and we define

$$\mathcal{A} = \left\{ A_\psi : \psi \in \Sigma_2^0 \right\},$$

then φ will be learnable over $\mathcal{A}$ (i.e. $\lim \Phi = \varphi(\mathcal{A})$ for some continuous functional Φ), *provided* that we can read off ψ from A_ψ. To obtain this, we make $\mathcal{A}$ *self-describing*: Under a suitable coding of all formulas we want that if n is the least number in A_ψ, then n is a code of ψ.

So, given ψ with code $n = \ulcorner \psi \urcorner$, we define A_ψ as follows: See if there is a finite set $A \sqsupseteq 0^{(n-1)} \widehat{\ } 1$ such that $\psi(A)$. If so, let A_ψ be such an A. If not, pick for A_ψ some infinite $A \sqsupseteq 0^{(n-1)} \widehat{\ } 1$ with $\neg \psi(A)$. Such A exists by (1).

Since the collection $\mathcal{A}$ of all A_ψ is self-describing, and given ψ we can decide (noneffectively) which decision has been made in the definition of A_ψ, and hence whether A_ψ is infinite or not, there is a learner Φ with $\lim \Phi = \varphi(\mathcal{A})$. Furthermore, since $\varphi(A_\psi)$ if and only if $\neg \psi(A_\psi)$, item (ii) from Theorem 3.3 does not hold.   ⊣

Acknowledgement. We thank Frank Stephan and the referee for many comments and pointers to the literature.

REFERENCES

[1] W. GASARCH, M. PLESZKOCH, F. STEPHAN, and M. VELAUTHAPILLAI, *Classification using information*, **Annals of Mathematics and Artificial Intelligence**, vol. 23 (1998), pp. 147–168.

[2] E. M. GOLD, *Limiting recursion*, **The Journal of Symbolic Logic**, vol. 30 (1965), pp. 28–48.

[3] ———, *Language identification in the limit*, **Information and Control**, vol. 10 (1967), pp. 447–474.

[4] S. HAYASHI, *Limit computable mathematics*, www.shayashi.jp/PALCM/index-eng.html.

[5] H. J. KEISLER, *Fundamentals of model theory*, **Handbook of Mathematical Logic** (J. Barwise, editor), North-Holland, 1978, pp. 47–103.

[6] K. KELLY, **The Logic of Reliable Inquiry**, Oxford University Press, 1995.

[7] E. MARTIN and D. OSHERSON, *Scientific discovery based on belief revision*, **The Journal of Symbolic Logic**, vol. 62(4) (1997), pp. 1352–1370.

[8] ———, **Elements of Scientific Inquiry**, MIT Press, 1998.

[9] P. G. ODIFREDDI, **Classical Recursion Theory**, Studies in logic and the foundations of mathematics Vol. 125 and Vol. 143, North-Holland, 1989, 1999.

[10] D. OSHERSON, D. DE JONGH, E. MARTIN, and S. WEINSTEIN, *Formal learning theory*, **Handbook of Logic and Language** (J. van Benthem and A. ter Meulen, editors), Elsevier, 1997, pp. 737–775.

[11] D. OSHERSON, M. STOB, and S. WEINSTEIN, *A universal inductive inference machine*, **The Journal of Symbolic Logic**, vol. 56(2) (1991), pp. 661–672.

[12] D. OSHERSON and S. WEINSTEIN, *Identification in the limit of first order structures*, **Journal of Philosophical Logic**, vol. 15 (1986), pp. 55–81.

[13] ———, *Paradigms of truth detection*, **Journal of Philosophical Logic**, vol. 18 (1989), pp. 1–42.

[14] H. ROGERS, **Theory of Recursive Functions and Effective Computability**, McGraw-Hill, 1967.

[15] J. R. SHOENFIELD, *On degrees of unsolvability*, **Annals of Mathematics**, vol. 69 (1959), pp. 644–653.

[16] R. I. SOARE, **Recursively Enumerable Sets and Degrees**, Springer, 1987.

[17] F. C. STEPHAN, *One-sided versus two-sided classification*, **Archive for Mathematical Logic**, vol. 40 (2001), pp. 489–513.

[18] F. C. STEPHAN and S. A. TERWIJN, *The complexity of universal text-learners*, **Information and Computation**, vol. 154(2) (1999), pp. 149–166.

TECHNICAL UNIVERSITY OF VIENNA
INSTITUTE OF DISCRETE MATHEMATICS AND GEOMETRY
WIEDNER HAUPTSTRASSE 8–10 / E104
A–1040 VIENNA, AUSTRIA
E-mail: terwijn@logic.at
URL: www.logic.at/people/terwijn/

Lecture Notes in Logic

General Remarks

This series is intended to serve researchers, teachers, and students in the field of symbolic logic, broadly interpreted. The aim of the series is to bring publications to the logic community with the least possible delay and to provide rapid dissemination of the latest research. Scientific quality is the overriding criterion by which submissions are evaluated.

Books in the Lecture Notes in Logic series are printed by photo-offset from master copy prepared using LaTeX and the ASL style files. For this purpose the Association for Symbolic Logic provides technical instructions to authors. Careful preparation of manuscripts will help keep production time short, reduce costs, and ensure quality of appearance of the finished book. Authors receive 50 free copies of their book. No royalty is paid on LNL volumes.

Commitment to publish may be made by letter of intent rather than by signing a formal contract, at the discretion of the ASL Publisher. The Association for Symbolic Logic secures the copyright for each volume.

The editors prefer email contact and encourage electronic submissions.

Editorial Board

Editorial Policy

1. Submissions are invited in the following categories:
i) Research monographs   iii) Reports of meetings
ii) Lecture and seminar notes   iv) Texts which are out of print
Those considering a project which might be suitable for the series are strongly advised to contact the publisher or the series editors at an early stage.

2. Categories i) and ii). These categories will be emphasized by Lecture Notes in Logic and are normally reserved for works written by one or two authors. The goal is to report new developments quickly, informally, and in a way that will make them accessible to non-specialists. Books in these categories should include
– at least 100 pages of text;
– a table of contents and a subject index;
– an informative introduction, perhaps with some historical remarks, which should be accessible to readers unfamiliar with the topic treated;
In the evaluation of submissions, timeliness of the work is an important criterion. Texts should be well-rounded and reasonably self-contained. In most cases the work will contain results of others as well as those of the authors. In each case, the author(s) should provide sufficient motivation, examples, and applications. Ph.D. theses will be suitable for this series only when they are of exceptional interest and of high expository quality.

Proposals in these categories should be submitted (preferably in duplicate) to one of the series editors, and will be refereed. A provisional judgment on the acceptability of a project can be based on partial information about the work: a first draft, or a detailed outline describing the contents of each chapter, the estimated length, a bibliography, and one or two sample chapters. A final decision whether to accept will rest on an evaluation of the completed work.

3. Category iii). Reports of meetings will be considered for publication provided that they are of lasting interest. In exceptional cases, other multi-authored volumes may be considered in this category. One or more expert participant(s) will act as the scientific editor(s) of the volume. They select the papers which are suitable for inclusion and have them individually refereed as for a journal. Organizers should contact the Managing Editor of Lecture Notes in Logic in the early planning stages.

4. Category iv). This category provides an avenue to provide out-of-print books that are still in demand to a new generation of logicians.

5. Format. Works in English are preferred. After the manuscript is accepted in its final form, an electronic copy in LATEX format will be appreciated and will advance considerably the publication date of the book. Authors are strongly urged to seek typesetting instructions from the Association for Symbolic Logic at an early stage of manuscript preparation.

LECTURE NOTES IN LOGIC

From 1993 to 1999 this series was published under an agreement between the Association for Symbolic Logic and Springer-Verlag. Since 1999 the ASL is Publisher and A K Peters, Ltd. is Co-publisher. The ASL is committed to keeping all books in the series in print.

Current information may be found at http://www.aslonline.org, the ASL Web site. Editorial and submission policies and the list of Editors may also be found above.

Previously published books in the *Lecture Notes in Logic* are:

1. *Recursion Theory.* J. R. Shoenfield. (1993, reprinted 2001; 84 pp.)

2. *Logic Colloquium '90; Proceedings of the Annual European Summer Meeting of the Association for Symbolic Logic, held in Helsinki, Finland, July 15–22, 1990.* Eds. J. Oikkonen and J. Väänänen. (1993, reprinted 2001; 305 pp.)

3. *Fine Structure and Iteration Trees.* W. Mitchell and J. Steel. (1994; 130 pp.)

4. *Descriptive Set Theory and Forcing: How to Prove Theorems about Borel Sets the Hard Way.* A. W. Miller. (1995; 130 pp.)

5. *Model Theory of Fields.* D. Marker, M. Messmer, and A. Pillay. (First edition, 1996, 154 pp. Second edition, 2006, 155 pp.)

6. *Gödel '96; Logical Foundations of Mathematics, Computer Science and Physics; Kurt Gödel's Legacy. Brno, Czech Republic, August 1996, Proceedings.* Ed. P. Hajek. (1996, reprinted 2001; 322 pp.)

7. *A General Algebraic Semantics for Sentential Objects.* J. M. Font and R. Jansana. (1996; 135 pp.)

8. *The Core Model Iterability Problem.* J. Steel. (1997; 112 pp.)

9. *Bounded Variable Logics and Counting.* M. Otto. (1997; 183 pp.)

10. *Aspects of Incompleteness.* P. Lindstrom. (First edition, 1997. Second edition, 2003, 163 pp.)

11. *Logic Colloquium '95; Proceedings of the Annual European Summer Meeting of the Association for Symbolic Logic, held in Haifa, Israel, August 9–18, 1995.* Eds. J. A. Makowsky and E. V. Ravve. (1998; 364 pp.)

12. *Logic Colloquium '96; Proceedings of the Colloquium held in San Sebastian, Spain, July 9–15, 1996.* Eds. J. M. Larrazabal, D. Lascar, and G. Mints. (1998; 268 pp.)

13. *Logic Colloquium '98; Proceedings of the Annual European Summer Meeting of the Association for Symbolic Logic, held in Prague, Czech Republic, August 9–15, 1998.* Eds. S. R. Buss, P. Hájek, and P. Pudlák. (2000; 541 pp.)

14. *Model Theory of Stochastic Processes.* S. Fajardo and H. J. Keisler. (2002; 136 pp.)

15. *Reflections on the Foundations of Mathematics; Essays in Honor of Solomon Feferman.* Eds. W. Seig, R. Sommer, and C. Talcott. (2002; 444 pp.)

16. *Inexhaustibility; A Non-exhaustive Treatment.* T. Franzén. (2004; 255 pp.)

17. *Logic Colloquium '99; Proceedings of the Annual European Summer Meeting of the Association for Symbolic Logic, held in Utrecht, Netherlands, August 1–6, 1999.* Eds. J. van Eijck, V. van Oostrom, and A. Visser. (2004; 208 pp.)

18. *The Notre Dame Lectures.* Ed. P. Cholak. (2005, 185 pp.)

19. *Logic Colloquium 2000; Proceedings of the Annual European Summer Meeting of the Association for Symbolic Logic, held in Paris, France, July 23–31, 2000.* Eds. R. Cori, A. Razborov, S. Todorčević, and C. Wood. (2005; 408 pp.)

20. *Logic Colloquium '01; Proceedings of the Annual European Summer Meeting of the Association for Symbolic Logic, held in Vienna, Austria, August 1–6, 2001.* Eds. M. Baaz, S. Friedman, and J. Krajíček. (2005, 486 pp.)

21. *Reverse Mathematics 2001.* Ed. S. Simpson. (2005, 401 pp.)

22. *Intensionality.* Ed. R. Kahle. (2005, 265 pp.)

23. *Logicism Renewed: Logical Foundations for Mathematics and Computer Science.* P. Gilmore. (2005, 230 pp.)

24. *Logic Colloquium '03: Proceedings of the Annual European Summer Meeting of the Association for Symbolic Logic, held in Helsinki, Finland, August 14–20, 2003.* Eds. V. Stoltenberg-Hansen and J. Väänänen. (2006; 407 pp.)

25. *Nonstandard Methods and Applications in Mathematics.* Eds. N.J. Cutland, M. Di Nasso, and D. Ross. (2006; 248 pp.)

26. *Logic in Tehran: Proceedings of the Workshop and Conference on Logic, Algebra, and Arithmetic, held October 18–22, 2003.* Eds. A. Enayat, I. Kalantari, M. Moniri. (2006; 341 pp.)

27. *Logic Colloquium '02: Proceedings of the Annual European Summer Meeting of the Association for Symbolic Logic and the Colloquium Logicum, held in Münster, Germany, August 3–11, 2002.* Eds. Z. Chatzidakis, P. Koepke, and W. Pohlers. (2006; 359 pp.)

Printed and bound by CPI Group (UK) Ltd, Croydon, CR0 4YY

23/10/2024

01777672-0007